NIELS BOHR AND THE QUANTUM ATOM

The Bohr Model of Atomic Structure 1913–1925

AF734959

HELGE KRAGH

Aarhus University, Denmark

OXFORD
UNIVERSITY PRESS

OXFORD
UNIVERSITY PRESS

Great Clarendon Street, Oxford, OX2 6DP,
United Kingdom

Oxford University Press is a department of the University of Oxford.
It furthers the University's objective of excellence in research, scholarship,
and education by publishing worldwide. Oxford is a registered trade mark of
Oxford University Press in the UK and in certain other countries

© Helge S. Kragh 2012

The moral rights of the author have been asserted

First Edition published in 2012

All rights reserved. No part of this publication may be reproduced, stored in a retrieval system, or transmitted, in any form or by any means, without the prior permission in writing of Oxford University Press, or as expressly permitted by law, by licence or under terms agreed with the appropriate reprographics rights organization. Enquiries concerning reproduction outside the scope of the above should be sent to the Rights Department, Oxford University Press, at the address above

You must not circulate this work in any other form
and you must impose this same condition on any acquirer

Published in the United States of America by Oxford University Press
198 Madison Avenue, New York, NY 10016, United States of America

British Library Cataloguing in Publication Data
Data available

Library of Congress Control Number: 2012932673

ISBN 978–0–19–965498–7

CONTENT

PREFACE

Modern physics is essentially based on a remarkable series of discoveries and theories dating from the two decades between 1895 and 1915. Integrating Rutherford's nuclear model with elements of Planck's quantum theory, Niels Bohr's theory of the atom paved the way for the later quantum mechanics, but the atomic theory itself had but a short lifetime. This makes it only more attractive from a historical point of view, because one is able to follow the theory through its entire life, from its birth in 1913 through its adolescence around 1918 to its declining years in 1924–1925. The present work is largely limited to this period. It is a fairly detailed and complete account of how a most important scientific theory came into being, how it was further developed, and how it eventually was abandoned to be replaced by an even better theory that in important ways built on it. Of course, this better theory is still with us.

The book is written in connection with the 2013 centennial of Bohr's theory, which does not, however, imply that it is of a celebratory nature. Although there are good reasons to celebrate Bohr and his theory, these reasons are not what motivated the book. It is primarily a detailed history of Bohr's atomic theory and only secondarily about the physicist who created the theory. In other words, it is not biographical, although it can perhaps be characterized as a biography of a scientific theory. Because this theory was the brainchild of and so intimately connected with the physicist Niels Bohr, he cannot avoid appearing prominently throughout the book. If there is another main figure in the book, it will be Arnold Sommerfeld, the German physicist who did so much to promote and develop Bohr's ideas.

The historical study of Bohr's theory of atomic structure is far from new. In fact, some of the finest historical studies, due in particular to Léon Rosenfeld, John Heilbron, Thomas Kuhn, Paul Forman, and Ulrich Hoyer, were published less than a decade after Bohr's death in 1962. Since then, historians of science have continued to investigate the old quantum theory of atoms, molecules, and radiation, but at a slower pace. Of course, the present book draws on the extensive secondary literature, including some papers I wrote on the subject many years ago. *Niels Bohr and the Quantum Atom* aspires to give a comprehensive picture of the theory in all its major facets, although with a focus on those parts of the theory which were due to Bohr or in which he was significantly involved. It is impossible to write such a history without taking into regard the complex technical and conceptual matters that were in many ways the very heart of the theory. On the other hand, I do not deal with technicalities for the sake of technicalities. A substantial part of the book will be accessible to readers with no background in physics.

While technical in places, the book offers a broader picture of Bohr's theory than the one which can be found elsewhere in the scholarly literature. Thus, it deals extensively with the reception of the theory (Chapter 3) and also with the criticism launched against it by physicists of a more conservative inclination (Section 4.5). To get a full impression of a new scientific theory, one needs to look not only at its followers but also at its critics. Bohr's theory scored its most important victories in atomic structure and related spectroscopic experiments, which were also the fields that led to its decline in about 1924. However, as originally conceived it was not limited to the domain of atomic architecture but was ambitiously announced as a theory of the constitution of matter, whether atomic or molecular. The chemical aspects of Bohr's theory are rarely given much attention in the writings of either physicists or historians of science, which in my view tends to distort the overall picture and impact of the theory. Molecular spectroscopy was almost as important in the development as atomic spectroscopy, and the views of the chemists are as enlightening as those of the physicists. Major parts of the Chapters 6 and 7 deal with the chemical aspects and the ways in which the chemical community responded to the theory.

The book is chronologically organized, describing how Bohr and his colleagues, especially in Germany, thought about and developed the surprising new quantum theory of atoms. It starts however with a chapter outlining earlier atomic theories that to some extent, if mostly indirectly, provided the background for Bohr's breakthrough in 1913. Of these earlier theories, the nuclear atom of Rutherford was of direct and crucial significance to Bohr, while J. J. Thomson's mechanical model of the atom and later J. Nicholson's atomic model inspired him in both a positive and negative sense. Bohr's theory was highly original, but of course it was not created out of nothing.

The picture of the young Bohr that emerges in this book is one of an extremely hard-working, competitive, ambitious, and determined physicist. In his later years Bohr became a quantum sage, the oracle from Blegdamsvej in Copenhagen, and is often portrayed as more a philosopher than a physicist. Although Bohr in a sense always thought 'philosophically' about foundational problems in physics, in his younger days he was unquestionably and solely a physicist. He was a visionary physicist, but his visions were restricted to the realm of physics. Although a theorist, he was almost obsessively occupied with experimental tests, which to him counted more than (but were not, of course, contrary to) philosophical reflections. His famous correspondence principle, which is detailed in Chapter 5, could not be tested experimentally, yet its consequences could and were tested. The correspondence principle is one of the most fascinating parts of Bohr's theory, not least because it was so closely connected to his personal way of thinking. It was a prime example of what Einstein called Bohr's 'musicality' and 'unique instinct and tact'.

From a philosophical point of view, the Bohr atom of the old quantum theory is far less interesting than the mature quantum mechanics that emerged in 1925 and since then has served as a stable foundation of physics. Indeed, during the lifetime of the Bohr atom it was ignored by professional philosophers, many of whom may have been unaware of it. All the same, later philosophers have taken a keen interest in Bohr's theory and analyzed it from different points of view. In an appendix of the book (Chapter 9) I offer an overview of some of these considerations, which include how Bohr's theory appears in the theories of scientific development argued by T. S. Kuhn, I. Lakatos and other philosophers. I am well aware that this topic is covered only incompletely and provisionally.

An important source for any serious study of Bohr's physics is the many letters that are preserved at the Niels Bohr Archive in Copenhagen, appropriately located in the historic institution that was established for Bohr in 1921. Some of these letters, if far from all of them, are reproduced in the invaluable 12-volume series *Niels Bohr's Collected Works* (1972–2007), the first volumes of which I have used extensively. I am grateful to Finn Aaserud, director of the Niels Bohr Archive, for granting me access to the rich materials included in the archive. Much of this material, if not all of it, is included in the Archive for History of Quantum Physics to which I have had electronic access through the Max Planck Institute for History of Science in Berlin.

Helge Kragh
Aarhus University, Denmark

1

Atomic Theories Before 1913

The first really successful theory of atomic structure was proposed by Niels Bohr in an epoch-defining paper in *Philosophical Magazine* of July 1913. It was this theory that established atomic theory as a fundamental and progressive field of physics intimately connected with spectroscopy and the new and still mysterious quantum theory. But although rational atomic theory, in the sense of a scientific theory dealing with the internal structure of the atom, dates from the beginning of the twentieth century, ideas of complex atoms and their structure can be found much earlier.[1] Many of the theories of this earlier period were speculative suggestions with little or no foundation in experiment. Some of them were of a philosophical rather than scientific nature. They were all short-lived, but of course some of the theories lived longer and were developed to a higher level than others. The vortex theory of the second half of the nineteenth century and J. J. Thomson's electron theory were among the more successful of the pre-Bohr atomic theories. The Bohr atom, revolutionary as it turned out to be, was part of a long tradition in atom-building and to some extent influenced by earlier conceptions of atomic architecture.

It is important to realize that until about World War I atomic theory was not only a small field of physics, it was also not highly regarded in the physics community. It is telling that at the first (and only) International Congress of Physics, held in Paris in 1900 in connection with the World Exhibition, only one of the 92 invited papers, namely the one of J. J. Thomson, dealt explicitly with atomic structure. The models proposed in the early period were rarely meant to be realistic pictures of the atom, but merely illustrations of mechanisms that might help in understanding some aspect or other of physical phenomena. Referring to the period around 1910, it has been said that 'for the average physicist of the time, speculations about atomic structure were something like speculations about life on Mars—very interesting for those who liked this kind of thing, but without much hope of support from convincing scientific evidence and without much bearing on scientific thought and development'.[2] A decade later the situation was quite different, the change in attitude being to a large extent a result of Bohr's quantum theory of atomic structure.

Not only were internally structured atoms not universally accepted in the first decade of the twentieth century, there was in the period still scattered opposition to the very existence of atoms and molecules, ranging from uncompromising anti-atomism to cautious scepticism. Some positivistic minded physicists and chemists, including Pierre Duhem in France, Ernst Mach in Austria, and Georg Helm and Wilhelm Ostwald in Germany, rejected a realist interpretation of atomic theory. The unobserved atoms might be heuristically useful concepts, but they regarded them as 'metaphysical' rather than real constituents of matter. Although opposition was declining, in 1913 Jean Perrin in France still felt it necessary to emphasize that atoms do exist and that the opponents of the atomic theory, 'which until recently were numerous, have been convinced and have abandoned one after the other the sceptical position that was for a long time legitimate and no doubt fruitful'.[3] This was true in the case of Ostwald, who in 1908 admitted the existence of atoms, but not in the cases of Duhem and Mach: they both died in 1916, convinced that atoms were nothing but constructs of the human mind.

In 1913—the year that Bohr published his atomic theory—Duhem criticized the confidence with which the 'school of neo-atomists' spoke of their hypotheses of the inner structure of matter. 'We do not share this confidence', he wrote. 'We are not able to recognize in these hypotheses a clairvoyant vision of what there is beyond sensible things; we regard them only as *models*.'[4] Bohr, too, regarded his theory of the atom as a model. The crucial difference was it was not *only* a model, but a model *of* something that really existed. As to Duhem, he never mentioned Bohr's atomic theory, of which he may have been unaware.

1.1 PRE-ELECTRON ATOMIC SPECULATIONS

The Daltonian atom of the early nineteenth century was a primitive elementary body with no internal constitution, the atoms of the different elements having nothing structural in common. What distinguished a hydrogen atom from an atom of lead was solely its atomic weight, a measurable property. Although this was the kind of atom that appealed to most chemists, from an early date there were suggestions that the atoms themselves were somehow complex bodies. The English physician and chemist William Prout argued in 1815–1816 that the atomic weights indicated a common composition of the elements, namely that all atoms were made up of unit particles, which he identified with hydrogen atoms. Prout's hypothesis was taken up and modified in various ways by several chemists, first and most effectively by the Scotsman Thomas Thomson who promoted it in a work of 1825 ambitiously entitled *An Attempt to Establish the First Principles of Chemistry by Experiments*. However, the hypothesis remained controversial throughout the century and was rejected by leading chemists from Berzelius to Mendeleev.[5] Not only was it speculative, but increasingly accurate determinations of the atomic weights contradicted the original form of the hypothesis. Thus the atomic weight of chlorine turned out to be close to 35.5 (in terms of the weight of hydrogen), a value which evidently posed problems for believers in Prout's hypothesis.

All the same, the general idea of a material unity in nature—that all matter ultimately consists of structures made up of a primitive particle or *protyle*—remained popular. The hypothesis was widely conceived as too attractive to be wrong. The introduction of spectroscopy and the periodic system of the elements inspired further interest in neo-Proutean hypotheses, which more often than not were seen in connection with the evolutionary worldview that was so popular during the Victorian era.

In some cases atomic speculations related to the Proutean tradition had a distinct air of Pythagoreanism, as in the works of the Danish–American chemist and polymath Gustavus Hinrichs, one of several precursors of the periodic system. Hinrichs, who combined his atomic theory with numerological considerations of spectroscopic and astronomical data, was convinced that all chemical elements were composed of a single substance.[6] For this basic element he proposed the name *pantogen*, assuming it to have an atomic weight of either 0.5 or 0.25. While Hinrich's works were not well known, and 'pantogen' never caught on, somewhat similar ideas were suggested by several leading scientists in England in particular. The astronomer Joseph Norman Lockyer and the chemists William Crookes and Thomas Carnelley were among the most articulate and visionary advocates of evolutionary neo-Prouteanism. However, although this tradition in speculative atomic theory significantly influenced the first electron models of the atom, it did not include definite models of the composition of atoms. Moreover, neo-Prouteanism and related ideas were mainly of interest to chemists, astronomers and amateur scientists, whereas most physicists chose to ignore them. For this reason they need not be further examined.

Models of the internal architecture of atoms were proposed many years before the discovery of the electron. Some of them were based on hypothetical electrical particles, while other models assumed neutral but equally hypothetical subatomic constituents. These early atomic models had in common that they were speculative, and some of them very much so, and also that they made very little impact on mainstream science. In some cases they were merely casual speculations of a philosophical nature. For example, this was the case with the ideas of the Danish physicist and engineer Ludvig August Colding, who is better known for his contributions to 'the imperishability of forces' or what became known as the law of energy conservation. In an unpublished note of 1854, he pictured atoms, or what he called 'molecules', as analogous to planetary systems. 'Many facts seem to me to indicate that every molecule constitutes an infinitely small planetary system, be it with or without a central body', he wrote. 'Each of these small planets has a characteristic rotation about its axis, and this rotation determines both the electric tension and magnetic polarity of the particle.'[7]

By the mid-nineteenth century the ether was generally assumed to play an important role in microphysics, whether based on atoms or not. While the ether was usually considered a homogeneous imponderable medium, there was no scarcity of ideas assuming other forms—for example a corpuscular ether. To get an impression of mechanical ether atoms in the speculative tradition, consider the ideas of two scientists from German-speaking Europe. In 1857, Ferdinand Redtenbacher, an Austrian-born director of the Polytechnic College in Karlsruhe, announced an atomic theory based on what he called 'dynamids'.[8] According to his model, matter consisted of ponderable atomic particles surrounded by shells of imponderable ether. The material particles were kept together by a hypothetical mechanical force analogous to Newtonian gravitation, while the ether particles were assumed to be mutually repulsive and attracted by the massive atomic core. It was such a system of a massive core and minute ether particles, arranged in shells, that he called a dynamid. Redtenbacher related his dynamid theory to contemporary ideas of heat, gases, elasticity, and optics, and discussed on this basis various expressions for the dispersion of light. It was a speculative atomic theory, but one with physical applications.

About thirty years later the respected Swiss botanist Carl Wilhelm von Nägeli proposed a detailed atomic theory that had some qualitative features in common with the older one of Redtenbacher, in particular that it was based on a corpuscular ether governed by mechanical forces of both a repulsive and attractive nature.[9] Nägeli pictured the atom as a tightly packed system of billions of tiny ether particles ('amers'), some of which were ponderable and would therefore tend to agglomerate into an atomic core. The ponderable ether core was surrounded with an ether atmosphere of density decreasing with the distance, a *Schwerätherhülle*. According to Nägeli, his etherial atomic model offered an explanation of several chemical problems, including the nature of affinity and the combination of atoms into molecules. He also thought that it was suggestive with regard to physiology and biology in general.

From about 1850 views of ether and matter became increasingly based on electrical rather than mechanical theory. One of the first suggestions of electrical atoms was made by Richard Laming, an English physician and amateur physicist, who between 1828 and 1851 postulated the existence of subatomic, unit-charged particles. According to Laming, the atom was composed of a material core surrounded by an 'electrosphere' of concentric shells of electrical particles of both charges.[10] This kind of corpuscular electrical theory was unusual in England but fairly popular among those German physicists in favour of electrical actions propagating instantaneously over a distance. In 1846 a fundamental force law of this kind was proposed by Wilhelm Weber, who at the time served as professor of physics in Leipzig. Weber conceived his force law as the core of a unified theory of the future that might possibly lead to an explanation of all of nature. By the 1860s he had developed an electrical theory according to which the neutral ether consisted of positive and negative particles orbiting around each other. Moreover, he extended his picture of the ether to an analogous picture of the chemical atoms.[11]

In his later work, some of it unpublished and of a fragmentary nature only, Weber considered the ponderable atom to be structured like a planetary system, with a large number of tiny electrically charged particles revolving around a heavy massive part. The system was kept together by electrical forces satisfying his force law. In a paper of 1871 he explained:

> Let *e* be the positively charged electrical particle, and let the negative particle, carrying an opposite charge of equal amount, be denoted –*e*. Let only the latter be associated with the massive atom, whose mass is so large that the mass of the positive particle may be considered negligible. The particle –*e* may then be considered as being at rest, while just the particle *e* moves around the particle –*e*.[12]

Weber came to the conclusion that the chemical elements were composed of an equal number of positive and negative particles revolving around each other and possibly also performing vibrations. In this way he thought that the mass might be explained in terms of electricity and that an understanding of the periodic system was within reach. Moreover, he speculated that the chemical elements, if they were composites of electrical particles, might possibly be decomposed into lighter elements. The dream of the alchemists received justification from electrodynamics! According to Maxwell's electromagnetic field theory a circulating electrical particle would lose energy and hence cause the atom to become unstable and eventually collapse, but this problem (which later appeared prominently in atomic theory) did not appear in Weber's alternative theory.

Independent of Weber, Robert Grassmann, a brother and collaborator of the mathematician Hermann Grassmann, developed somewhat similar ideas of ether atoms consisting of electrical doublets. He considered chemical atoms to be composed of a positive particle surrounded by a spherical shell of polarized ether doublets. Although Robert Grassmann's ideas received little attention, they were critically reviewed by the physicist and pioneering psychologist Gustav Fechner, a close friend of Weber and himself an advocate of atomism.[13] Fechner had for a long time been interested in atomic theory, both in its scientific and philosophical aspects. As early as 1828 he suggested a dynamical model of the atom in close analogy to the solar system and governed by Newton's law of gravitation. The atoms, he said in this early work, 'simulate in small dimensions the situations of the astronomical objects in large dimensions, being animated in any case by the same forces'.[14]

The theories of Weber, Grassmann, and other German scientists were based on hypothetical electrical particles. When the electron was turned into a real particle at the end of the century, physicists were generally puzzled that it existed in a negative form only. The neutrality of the ether seemed to require complete charge symmetry and yet the positive electron was conspicuously missing. Apparently without knowing of the earlier work of Weber and Grassmann, the British–Australian physicist William Sutherland suggested in 1899 that the ether consisted of doublets of negative and positive electrons held tightly together, particles for which he coined the name 'neutron'. As he wrote in a paper two years later, 'In the free æther the positive and negative electron revolving... round their centre of inertia form what I have proposed to call the neutron, the electric doublet, which gives the æther its chief electric and magnetic properties'.[15] By that time atomic models were no longer based on purely hypothetical entities. It was now generally agreed that atoms contained a large number of electrons, all of them carrying the same negative charge, but there was no agreement as to the number or arrangement of the electrons.

1.2 FROM VORTEX ATOM TO ELECTRON ATOM

The atomic model developed by the famous Cavendish physicist Joseph John Thomson in the early years of the twentieth century can with some justification be called the first modern model of the atom. Contrary to earlier models it was based on an experimentally known entity, the electron, and for this and other reasons it could be subjected to experimental tests. While Thomson's electron dates from his celebrated investigation of cathode rays in 1897, his electron atom did not simply grow out of these experiments. There were other and even more important roots, for Thomson had for several years been convinced that the atom was a complex body made up of a single primordial particle or substance. He was in important respects a loyal follower of Prout, a heritage he proudly admitted. Moreover, as a theoretical entity the electron antedates the 1897 experiments, which explains why we can find ideas of electrons and electron atoms (as well as the name 'electron') in the literature even before that year.

The leading electron theorist Joseph Larmor argued in 1894 that electrons—which he pictured as 'singular points in the ether'—were the primordial units of all matter. The following year he went a step further, now suggesting 'a molecule [atom] to be made up of, or to involve, a steady configuration of revolving electrons'.[16] His picture was not unlike the one that Weber had earlier proposed on a non-Maxwellian and more speculative basis. In 1894 Larmor did not make it clear whether or not he conceived of the electron as a subatomic particle, and it is quite possible that at the time he did not. However, two years later he did.

Thomson's unitary idea of matter consisting of subatomic electrical charges owed a debt to his general predisposition toward neo-Prouteanism and, in particular, to his earlier work on the vortex theory of atoms. According to this theory, first proposed by William Thomson (Lord Kelvin) in 1867, atoms might be conceived of as vortical modes of motion in a perfect, all-pervading fluid.[17] The fluid was generally taken to be identical to the ether. For about two decades the ambitious and mathematically complex vortex theory attracted much interest among mathematically inclined British physicists, including Peter G. Tait, Augustus Love, William Hicks, Micaih Hill, and J. J. Thomson. It was applied to a variety of physical and chemical problems, such as line spectra, affinity, chemical combination, the behaviour of gases, and even gravitation. Although not convinced of its truth, Maxwell was full of praise of the vortex theory because of its methodological virtues and ontological parsimony. In a deservedly famous article on 'Atom' for the 1875 edition of *Encyclopaedia Britannica*, he singled out Kelvin's vortex model as by far the most attractive picture of atomic constitution. 'The greatest recommendation of this theory, from a philosophical point of view', he said, 'is that its success in explaining phenomena does not depend on the ingenuity with which its contrivers "save appearances", by introducing first one hypothetical force and then another'.[18]

Among those who found the vortex atom attractive was also the mathematician and statistician Karl Pearson, who however preferred an alternative version of the ultimate atom. In 1885 he proposed than an atom might be a differentiated spherical part of the ether, or perhaps a vacuum within the ether, pulsating with a natural frequency.[19] He found the conception of spherical atoms to be promising with respect to the understanding of a wide range of phenomena, including chemical affinity and spectral lines. Six years later he modified his ideas into a theory of 'ether squirts', point atoms from which ether continuously flowed into space.[20] In addition to the ether squirts, acting as points of positive matter, he postulated the existence of negative matter in the form of sinks that absorbed ether. Although Pearson developed his ambitious atomic ether theories in considerable mathematical detail and attempted to link them to experimental knowledge, compared to the vortex atom they attracted very little interest.

In a prize essay of 1882 young J. J. Thomson examined theoretically the question of the stability of vortices arranged at equal intervals around the circumference of a circle.[21] Using standard perturbation theory he found after lengthy calculations that the configurations with $n = 2, 3, 4, 5$, and 6 vortices would be dynamically stable, but that seven vortices on the same ring could not form a stable system. For larger n he relied on an analogy with experiments with floating magnetized needles, which the American physicist Alfred Mayer, at the Stevens Institute of Technology in Hoboken, New Jersey, had made in 1878 (Figure 1.1). That Mayer's experiments could be taken to illustrate the periodic system of the elements had first been pointed out by Kelvin. Although Thomson, like most other physicists, abandoned the vortex atom programme about 1890, the idea continued to guide him and appeal to him. Thus, in a paper of 1890 he linked the periodic system with the vortex atomic model and pointed out the suggestive similarity between an arrangement of vortices and the regularity found among the chemical elements.

The vortex atom approach greatly influenced Thomson's thinking about the complexity of the elements. In 1897 he no longer thought of the vortex atom as a realistic model, yet his new primordial particle, the electron, had more than a little similarity with the vortices of the old theory. In his seminal paper of October 1897, Thomson suggested that that the atom consisted of a large number of electrons (which he insisted on calling 'corpuscles'), possibly held together by an unspecified central force. In this first version of the Thomson model, the atom was pictured as just an aggregation of electrons and so, assuming Coulomb forces between the electrons, there was no attractive force to keep the atom from exploding. Thomson had to invent the needed attractive force, and this he did. 'If we regard the chemical atom as an aggregation of a number of primordial atoms [electrons]', he wrote, 'the problem of finding the configurations of stable equilibrium for a number of equal particles acting on each other according to some law of force... is of great interest in

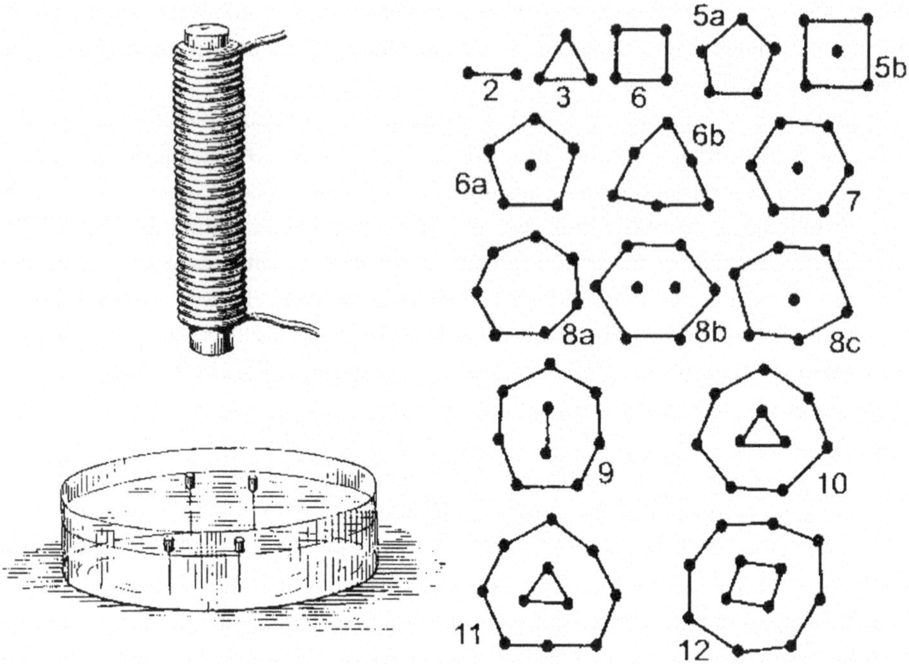

Fig. 1.1. Mayer's experiments with magnetized needles. To the left, J. J. Thomson's illustration in Thomson 1907, p. 111. To the right, some of Mayer's magnet configurations. Some of the configurations are unstable. For example, five needles may arrange themselves in a square with a central needle but a slight mechanical disturbance will make the system turn into the stable pentagon configuration.

connection with the relation between the properties of an element and its atomic weight'.[22]

In 1897 Thomson only knew the charge-to-mass ratio e/m of the cathode ray electrons and therefore had to assume that the particles were subatomic, with a charge numerically equal to that of the hydrogen ion. Two years later the assumption was confirmed when he and his research students at the Cavendish Laboratory succeeded in determining the charge of the electron, which led to a mass for the particle of the order of one-thousandth of a hydrogen atom. The same year, in an address to the British Association for the Advancement of Science, Thomson expounded his atomic model in a fuller and more confident way. What became sometimes known (rather inappropriately) as the 'plum pudding' model, he explained as follows:

> I regard the atom as containing a large number of smaller bodies which I shall call corpuscles; these corpuscles are equal to each other; the mass of a corpuscle is the mass of the negative ion in a gas at low pressure, i.e. about 3×10^{-26} of a gramme. In the normal atom, this assemblage of corpuscles forms a system which is electrically neutral. . . . [T]he negative effect is balanced by something which causes the space through which the corpuscles are spread to act as if it had a charge of positive electricity equal in amount to the sum of the negative charges on the corpuscles.[23]

It was this picture that Thomson developed into a quantitative and sophisticated atomic model over the next few years (Section 1.4). In his book *Electricity and Matter*, based on the Silliman lectures he gave at Yale University in May 1903, he provided a full if mostly qualitative account of the theory.

Thomson's model was the most important of the electron atomic models of the early twentieth century, but it was not the only one. Shortly after Thomson's announcement of the cathode-ray electron in the spring of 1897, Kelvin suggested his 'Aepinus atom', named after Franz Aepinus, a eighteenth-century natural philosopher from Germany who had pioneered a one-fluid electrical theory. Kelvin pictured the atom as a number of 'electrions' embedded in a globe of positive electricity, a picture which had much in common with Thomson's but nonetheless differed from it. For example, Kelvin's hypothetical electrions did not have the same mass and charge as the electrons, and they were thought to be subject to an ad hoc force law more complicated than the ordinary Coulomb force. In work between 1902 and 1907 Kelvin applied the Aepinus model in an attempt to explain or illustrate the puzzling phenomenon of radioactivity, which according to him was probably triggered by etherial waves or some other external agency.[24] Very few physicists found the Aepinus model to be of any value. In spite of being the most elaborate attempt of the period to explain radioactivity in intra-atomic terms, and in spite of being proposed by a physicist of the highest possible distinction, the ideas of the aging Kelvin had almost no impact on the further development of atomic theory. British physicists referred respectfully to it, but without taking it seriously or developing it.

In his book *Electrons* of 1906, the physicist Oliver Lodge, then at Birmingham University, surveyed the various candidates of atomic structure at the time. Apart from the ideas of the Thomson–Kelvin type he mentioned the possibility that the atom 'consists of a kind of interlocked admixture of positive and negative electricity, indivisible and inseparable into units'.[25] This may have been a reference to the picture of the atom favoured by Philipp Lenard, at the time professor of physics at the University of Kiel and the recipient of the 1905 Nobel Prize for his work on cathode rays. Based on his studies of the absorption of cathode rays in gases, Lenard suggested in 1903 that the interior of the atom was mostly empty space.[26] To explain the experimental results he assumed the atom to be composed of impenetrable 'dynamids', a dynamid being a kind of tightly bound neutral doublet consisting of a negative and a positive electron. As mentioned, the idea of intra-atomic dynamids, or at least the name, had been introduced by Redtenbacher half a century earlier, but Lenard did not refer to his predecessor. The constituent dynamids were much smaller than the atom. He estimated the radius of a dynamid to be at most 3×10^{-12} cm, implying that the atom was nearly empty: 'The space occupied by a cubic metre of solid platinum is empty, in the same sense that celestial space traversed by light is empty, except for the proper volume of the dynamids, which cannot in all exceed a cubic millimetre.'[27] In this respect, if in no other, he anticipated the later Bohr–Rutherford atom. Indeed, this is what a few patriotic German physicists claimed after

World War I, suggesting that Rutherford's atom was merely a particular version of Lenard's.[28]

Lenard found that the number of dynamids in an atom was proportional to the atomic weight, but did not offer a value for the factor of proportionality. Moreover, he assumed the dynamids to be in rapid rotation, which he thought might cast light on the nature of radioactivity. Although he was little concerned with spectroscopic evidence, he outlined a mechanism according to which the atom would emit characteristic spectral lines when free electrons returned to equilibrium in the dynamic atomic structure. Lenard's atomic hypothesis of 1903 was qualitative and rather vague. Not only did it not address spectroscopic issues, it also did not connect with issues of chemistry. For example, he did not give the number of dynamid units in either hydrogen or other elements, nor did he suggest how his atoms might combine into molecules. For these and other reasons Lenard's work exerted but little influence on the further development of atomic structure and almost none on the British atom builders. His model of the atom was no more successful than Kelvin's Aepinus atom in attracting interest from other physicists.

Yet another conception of the atom, for a brief while popular among some physicists, was to view it as, in the words of a French amateur physicist, 'a miniature solar system composed of particles revolving round one another without touching and incessantly pursuing their eternal course under the influence of the forces which direct them'.[29] In this case the positive electricity was assumed to be located in the hypothetical positive electrons that could still be considered plausible particles, if undetected ones, in the early years of the new century. Primarily with the aim of explaining the mechanism of line spectra, 24-year-old James Jeans proposed in 1901 that the atom consisted of a large number of positive and negative electrons, supposedly differing only in the sign of their charge.[30] Jeans speculated that the electrons formed shells of alternating charges in the atoms, with the outermost layer consisting purely of negative electrons. To secure dynamical equilibrium he furthermore suggested modifying Coulomb's law at very small distances, which he did by introducing in an ad hoc manner a repulsive force independent of the sign of the electronic charges. Although Jeans's atom was ephemeral, it was the first attempt to interpret spectral laws on the basis of a definite model of atomic structure.

An atomic model similar to the one of Jeans was argued by Lodge, who thought that 'The whole of the atom may be built up of positive and negative electrons interleaved together, and nothing else'.[31] Electrons in a state of violent motion would imply a loss of radiation energy, causing the atom to decay in a kind of atomic earthquake. To Lodge's mind this did not speak to the disadvantage of the model, for in that way he could offer a qualitative explanation of radioactivity of a kind similar to the one proposed by Kelvin. Of course, on this picture one would expect all elements to be radioactive, but this was just what many physicists believed at the time. In the first decade of the twentieth century it was often assumed that radioactivity was a common property of atoms, only exhibited more strongly in heavy elements like radium, thorium, and uranium. Models of the type suggested by Jeans and Lodge were short-lived. Their explanatory force was limited, they made use of ad hoc assumptions, and they presupposed the existence of positive electrons for which there was no experimental evidence. (There were, however, a few claims of evidence[32].) Compared to Thomson's model, the Jeans–Lodge atom had little to offer.

Atomic models such as those mentioned, and most of those to be mentioned, were very much a British speciality. According to the Victorian tradition, models served heuristic purposes rather than representing some reality of nature. They were first of all mental illustrations, formulated mathematically and based on the established laws of mechanics and electrodynamics, if often supplemented with hypothetical forces. A model were not to be taken literally, but were seen as a method or picture that offered some insight into the inner workings of nature. Speaking of the vortex model of atoms, Larmor said in an address of 1900 to the British Association for the Advancement of Science:

> The value of such a picture may be held to lie, not in any supposition that this is the mechanism of the actual world laid bare, but in the vivid illustration it affords of the fundamental postulate of physical science, that mechanical phenomena are not parts of a scheme too involved for us to explore, but rather present themselves in definite and consistent correlations, which we are able to disentangle and apprehend with continually increasing precision.[33]

This was a philosophy that governed British physics not only in the era of the vortex model, but also in the first two decades of the twentieth century. We shall meet it in later chapters.

Philosophical Magazine, a commercially published journal founded in 1798, emerged as the premier journal for atom-builders in the British tradition. Models of a similar kind were rare among French and German physicists, who generally favoured a more phenomenalist approach and looked upon dynamical models with some distrust. They might share the Britons' enthusiasm over the new physics based on electrons and ether, but typically without engaging in model making of the elaborate kind favoured by British physicists. For example, the atomic models of Lenard and Stark completely lacked the mathematical framework that was such a characteristic feature of British models of the Kelvin–Thomson tradition.

In 1901 Walter Kaufmann, a physicist at the University of Bonn, gave an address to the Association of German Scientists and Physicians (Versammlung deutscher Naturforscher und Ärtzte) in which he surveyed the state and promises of electron physics. Much in the spirit of Thomson and Lodge he concluded that 'the electrons would be the long-sought-for "primordial atoms" whose different groupings would form the chemical elements, and the old alchemists' dream of the transformation of the elements would be brought a good deal nearer realisation'.[34] He added that a mathematical treatment of the stability of the intra-atomic electrons might even lead to an explanation of the periodic system. Yet Kaufmann refrained from advocating a particular model of the atom corresponding to an arrangement of the electrons, and he did not engage in the mathematical work to find the electron configurations.

Augusto Righi, a prominent professor of physics at the University of Bologna, may serve as another example of how physicists on the Continent conceived the structure of the atom. Electrons, he said in a popular book of 1909, are 'the elements of construction in the architecture of the atom', which at the time was completely uncontroversial. More specifically, 'It may be admitted that a material atom is nothing but a system consisting of a certain number of positive and an equal number of negative electrons, and that the latter, or at least some of them, move about the remaining portion like satellites'.[35] Characteristically, he did not go further than such generalities and felt no desire to construct definite models out of the positive and negative electrons.

1.3 THE RISE AND FALL OF THE THOMSON MODEL

In work performed between 1903 and 1904, Thomson transformed his crude picture of the atom into a quantitative and sophisticated atomic model.[36] From a physical point of view, the model consisted of a sphere of atomic dimension and uniformly filled with a positive fluid of uniform charge density; within the sphere a large number of point-like negative electrons moved in rings around the centre. Unlike the electrons, the positive sphere was hypothetical, assumed to be frictionless and without mass. Chemically and spectroscopically inert, its only function was to provide an elastic force upon the electrons and thus keep the atom together. According to Thomson and most contemporary physicists, even the lightest atoms were highly complicated structures, the simplest one (hydrogen) being a congeries of about $n = 1000$ electrons.

The models that Thomson examined mathematically were mostly planar, but this was merely a simplifying assumption. He was well aware that to obtain more realistic models he would have to consider three-dimensional structures, as he did for a small number of electrons, from $n = 1$–8 (Figure 1.2). In any case, the function of his model was basically heuristic, to help physicists visualizing physical phenomena and thereby suggest new experimental and theoretical ideas. 'My object', he said, 'has been to show that stable arrangements of corpuscles would have many properties in common with real atoms, and I have attempted to illustrate the properties by considering a special case chosen solely on the ground of simplicity'.[37]

It was an important feature of the 1904 model that the electrons were arranged on rings with circular motion providing the necessary centrifugal force. However, according to Maxwellian electrodynamics accelerating electrons will emit radiation energy and so, it would seem, the atom would eventually collapse. This problem had been considered by Larmor in an important paper of 1897 in which he calculated the power radiated by an accelerated charge to be

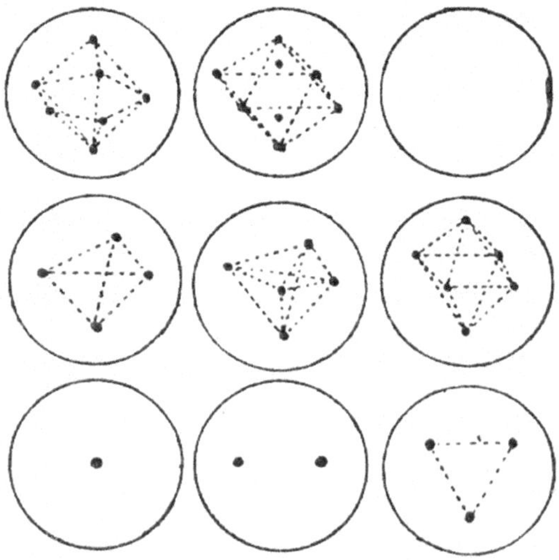

Fig. 1.2. Thomson atoms in three dimensions, as pictured by J. J. Thomson in a lecture to the Royal Institution in 1905.

$$\frac{dW}{dt} = \frac{2}{3}\frac{e^2\gamma^2}{c^3},$$

where c denotes the velocity of light in vacuum and γ is the acceleration. In the case of a single electron of charge e revolving with speed v on a circle of radius a, meaning that $\gamma = v^2/a$, the energy loss would be

$$\frac{dW}{dt} = \frac{2}{3}\frac{e^2v^4}{c^3a^2}.$$

Larmor realized that if atoms were composed of orbiting electrons, as he believed they were, they might quickly lose their orbital energy through radiation.[38] However, he did not consider this an insurmountable problem, among other reasons because he could prove that for extra electrons the effect will not be additive. On the contrary, for a large number of orbiting electrons the amount of energy radiated could be made arbitrarily small. The same kind of argument was used by Thomson with respect to his model atom. Considering the case of n electrons on the same ring, he proved that the radiation drain catastrophe reduced drastically with n. For example, taking the radiation from a single revolving electron as unity, he found that the radiation from a circle with six electrons moving with a speed of $v = 0.001c$ would be only 10^{-16}. For the general case of n equidistantly placed electrons on a single ring he derived the expression

$$\left(\frac{dE}{dt}\right)_n = \frac{2ce^2\beta^{2n+2}}{a^2}\frac{(n+1)n^{2n+3}}{(2n+1)!},$$

where the relative velocity $\beta = v/c$ was assumed to be small.

Of course, electrodynamic stability was not enough; the model atom also had to be mechanically stable. Let us follow a few of the steps in Thomson's long and complicated paper of 1904. The model atom has n electrons arranged at equal angular intervals around a circle of radius a, the ring being placed concentrically in the sphere of positive electricity of radius b. The atom is assumed to be neutral; that is, the positive charge is ne. The ring might be at rest or rotate uniformly with angular velocity $\omega = v/a$. For such a system it is readily shown that the equilibrium condition is

$$\left(\frac{a}{b}\right)^3 = \frac{S_n}{4n} + \frac{m\omega^2 a^3}{ne^2}, \quad \text{where} \quad S_n = \sum_{j=1}^{n-1} \frac{1}{\sin(j\pi/n)}.$$

To determine the stability of the equilibrium configurations Thomson followed a method that closely resembled the one he had applied for vortex atoms more than twenty years earlier and which had its origin in celestial mechanics. The method was to calculate the vibrational frequencies q of the electrons when subjected to a small perturbation. If all the q values turn out to be real, the perturbed electrons will perform small oscillations about the equilibrium positions of the form exp(iqt), implying that the configuration is stable. On the other hand, if any of the qs contain an imaginary part the disturbance of the perturbed electron will increase exponentially and the equilibrium will be unstable. To carry out this programme Thomson derived general expressions for the frequencies of n electrons arranged equidistantly on a ring. There are $3n$ possible frequencies, $2n$ arising from vibrations in the plane and n from vibrations perpendicular to it. Thomson found that the rotation of the ring stabilized the equilibrium system against perturbations at right angles to the plane.

As an illustration, consider the two cases $n = 2$ and $n = 6$. For $n = 2$ Thomson's formulae resulted in four planar frequencies,

$$q = \sqrt{\frac{6e^2}{mb^3} + \omega^2}, \quad q = \sqrt{\frac{2e^2}{mb^3}} \pm \omega, \quad q = 0,$$

and two transversal frequencies,

$$q = \sqrt{\frac{2e^2}{mb^3}}, \quad q = \omega.$$

Since all the frequencies are real, the two-electron system will be stable. In the case of $n = 6$ one of the frequencies turns out to be imaginary and for this reason six electrons distributed uniformly along a ring will not in general form a mechanically stable system.

As general results of his lengthy calculations Thomson pointed out firstly that the ring structure can be stabilized by internal electrons and secondly that the stability may depend on a critical angular velocity. For example, for $n = 6$ the system will stabilize if ω becomes greater than about $3e^2/mb^3$. Thomson described the general picture of his planar model atom as follows:

> We have thus in the first place a sphere of uniform positive electrification, and inside this sphere a number of corpuscles arranged in a series of parallel rings, the number of corpuscles in a ring varying from ring to ring: each corpuscle is travelling at high speed round the circumference of the ring in which it is situated, and the rings are so arranged that those which contain a large number of corpuscles are near the surface of the sphere, while those in which there is a smaller number of corpuscles, are more in the inside.[39]

For atoms with a large number of electrons, Thomson devised a graphical approximation method by means of which he could find the stable configurations. He assumed that the number of rings was a minimum, so that the outer rings were filled up with as many electrons as possible before new electrons were added to the internal structure. In this way he was led to ring structures such as those shown in Table 1.1. As Thomson pointed out, similarly to what he had done in his earlier work on the vortex atom, the electron configurations provided a striking analogy to the periodic system. If physical and chemical properties of the elements were associated with certain structures of electrons, one would expect that elements with, for example, 39, 58, and 80 electrons belonged to the same chemical group: 'Such a series would belong to elements which are in the same group according to the periodic law, i.e., these elements form a series which, if arranged according to Mendeléef's table would all be in the same vertical column.'[40] Contrary to later ideas of atomic chemistry, Thomson associated valency and other chemical properties with internal electron structures and not with the electrons in the outermost ring.

Thomson's model was considered attractive not only because it promised a reduction of all matter to electrons, but also because it was able to explain, if only in a vague and qualitative manner, a wide range of physical and chemical phenomena. The most important of these phenomena were radioactivity, photoelectricity, emission and

Table 1.1 Select electron configurations in Thomson atoms from $n = 9$ to $n = 96$. The symbol n (x_1, x_2, x_3, ...) means a total of n electrons with x_1 electrons in the innermost ring, x_2 electrons in the second ring, x_3 electrons in the third ring, etc.

9 (1,8)	10 (2,8)	
21 (1,8,12)	22 (2,8,12)	23 (2,8,13)
37 (1,8,12,16)	38 (2,8,12,16)	39 (2,8,13,16)
56 (1,8,12,16,19)	57 (2,8,12,16,19)	58 (2,8,13,16,19)
78 (1,8,12,16,19,22)	79 (2,8,12,16,19,22)	80 (2,8,13,16,19,22)
94 (1,5,11,15,17,21,24)	95 (1,5,11,15,18,21,24)	96 (1,6,11,15,18,21,24)

dispersion of light, the normal Zeeman effect, and the periodic system of the elements. In addition, Thomson could explain experiments on the scattering of beta particles on matter by assuming that the basic mechanism in the scattering process was a collision between a beta electron and a bound atomic electron.[41] The observed deflection arose by multiple scattering; that is, the collective result of numerous individual electron–electron scatterings. Although Thomson's theory of beta scattering did not rely critically on his atomic model—the positive sphere of electricity played almost no role—it was consistent with it. Apart from indicating a possible explanation of the periodic system, Thomson's ideas of atomic structure also included a theory of valency in rough agreement with the one proposed by the German chemist Richard Abegg. In general his ideas attracted favourable attention among chemists.[42] In spite of the considerable explanatory force of Thomson's theory, its explanations had more of a character of analogies than of deductions. Its explanatory breadth was not accompanied by a proper predictive power.

Physicists in the early years of the nineteenth century did not recognize radioactivity as a spontaneous and inherently probabilistic phenomenon, as a later generation would do. Like Kelvin's Aepinus atom and Lodge's atomic model, Thomson's model built on electrical particles, and it had the advantage that it provided a qualitative explanation of radioactivity in terms of a rearrangement of the atomic electrons. In a stable atom the rings of electrons would rotate with a high velocity, but because of the small radiation drain the velocity would slowly diminish and eventually become subcritical. As Thomson explained:

> When, after a long interval, the velocity reaches the critical velocity, there will be what is equivalent to an explosion of the corpuscles, the corpuscles will move far away from their original positions, their potential energy will decrease, while their kinetic energy will increase. The kinetic energy gained in this way might be sufficient to carry the system out of the atom, and we should have, as in the case of radium, a part of the atom shot off. In consequence of the very slow dissipation of energy by radiation the life of the atom would be very long.[43]

This sketch remained the essential explanation of radioactivity within the framework of the Thomson atom and was for a while accepted by Rutherford, among others. However, the explanation was evidently a sketch only. For one thing, it failed to account for the exponential decay law; for another it offered no explanation of why radioactivity was a phenomenon limited to elements heavier than lead (if indeed it was). In addition, it referred loosely to parts of the atom being shot off, without making it explicit what these parts were. By 1908 alpha particles were identified with doubly charged helium ions, which according to Thomson were atomic systems including spheres of positive electricity. How were these systems expelled from the mother atom? In spite of these and other problems, the radiation-drain hypothesis remained popular for several years.

From the very beginning the Thomson model was plagued by conceptual as well as experimental problems. The imponderable sphere of positive electricity was a mathematical artifice, a ghost-like entity whose only function was to keep the electrons together. In a letter to Lodge of April 1904 Thomson admitted that 'I have . . . always tried to keep the physical conception of the positive electrification in the background'.[44] In the same letter he expressed the hope of explaining the positive electricity as an epiphenomenon due to the negative electrons:

> When one considers that all the positive electricity does, on the corpuscular theory, is to provide an attractive force to keep the corpuscles together, while all the observable properties of the atom are determined by the corpuscles, one feels, I think, that the positive electrification will ultimately prove superfluous and it will be possible to get the effects we now attribute to it, from some property of the corpuscles.

Alas, Thomson did not succeed in either explaining or explaining away the positive electricity. On the contrary, his further research showed that the number of electrons was much smaller than originally assumed and that the positive electricity therefore could not be weightless or nearly so; it had to be substantial. But what was it?

In an important paper of 1906 Thomson analyzed the experimental data on the scattering of various kinds of radiation (light, X-rays, and beta rays) in relation to his theory of the atom with the aim of estimating the number of electrons in real atoms.[45] The chief result of his analysis was surprising as well as disturbing, namely that the number of electrons n must be of the order of the atomic weight. With this result he decimated the electronic population of atoms by a factor of about one thousand! The consequences were discomfiting and nearly disastrous to the original Thomson model because they demonstrated that the positive electricity made up by far the majority of the mass of the atom. According to Lodge, Thomson's paper was 'the most serious blow yet dealt at the electric theory of matter'.[46] The small number of electrons reopened the radiation problem: physicists could no longer count on the reduction of the radiation drain caused by many electrons, at least not in the case of hydrogen and the other lightest elements. If the radiation-drain mechanism of radioactivity were maintained, it would seem to imply that the light elements such as hydrogen, helium, and lithium should be particularly radioactive, in stark contrast to experimental knowledge.

There was another reason why the conclusion of 1906 undermined (or ought to have undermined) the credibility of Thomson's 'plum pudding model'. With the assumption of thousands of electrons in even the lighter atoms, there was no way in which a reasonably exact match could be established between the models atoms and those really existing. But if there were only, say, four electrons in a helium atom it meant that the model of a helium atom could be confronted with the chemical and physical properties of helium. In the case of the lightest elements it could no longer be argued that the number of electrons was too large or that three-dimensional calculations were not technically feasible. Thomson evaded the problem by ignoring it. This kind of exact comparison between a particular model and a particular atom was not part of his style of physics. As Bohr later said, 'Things needed not to be very exact for Thomson, and if it resembled a little, it was so'.[47]

By the early twentieth century it was desirable for a candidate theory of atomic structure to account for line spectra and their regularities, but in fact none of the models available at the time were able to do so. By and large, spectra were outside the scope of atomic theory and Thomson's model was no exception. Thomson did not even attempt to calculate the frequencies of spectral lines from the vibrations of the atomic electrons. The difficulty, which was not particular to the Thomson atom, was highlighted by several British physicists, including Jeans and Lord Rayleigh. In a paper of 1897 Rayleigh had analyzed the problem in general terms, concluding that vibrating systems of electrical charges would almost always result in formulae involving the square of the vibration frequencies. The problem was that the empirical formulae of Rydberg, Balmer, Ritz, and other spectroscopists were simple expressions in the first power of the frequency. Reconsidering the problem within the framework of an idealized Thomson model, in 1906 Rayleigh saw no way to escape the conclusion. As if in a state of desperation, he suggested that 'the frequencies observed in the spectrum may not be frequencies of disturbance or of oscillations in the ordinary sense at all, but rather form an essential part of the original constitution of the atom as determined by conditions of stability'.[48] With hindsight, Rayleigh happened to anticipate one of the key features of Bohr's later atomic theory.

With the recognition that atoms contain only relatively few electrons it became problematic to attribute the origin of the often very large number of spectral lines to the simultaneous vibrations of a single atomic source. Within the framework of the classical Thomson atom, where the positive sphere was inert, there were too few vibrational modes to account for the numerous lines in, for example, the spectrum of iron. According to many physicists about 1910, the origin of each line was most likely to be found in the vibrations of different sources that independently emitted light waves at different times. Moreover, it became common to think about spectral lines as being associated with ionization, namely, as caused by electrons recombining with the positive parts of the atoms. 'When an atom of an element is giving out its spectrum', wrote Thomson, 'it is surrounded by a swarm of corpuscles; and combinations... might be expected to be formed.' The lines would be due 'not to the vibrations of corpuscles inside the atom, but of corpuscles vibrating in the field of force outside the atom'.[49]

In a book of 1907, *The Corpuscular Theory of Matter*, Thomson gave a comprehensive exposition of his theory of atomic structure, now revised in accordance with the finding that atoms contained only a small number of electrons. He concluded that the number of optically active electrons, as manifested in refraction, dispersion, and spectra, was equal to the number of chemically active or valence electrons. We have to suppose, he wrote, that

> there are in the atom some corpuscles equal in number to the valency which are especially easily moved. To represent the mobility of these corpuscles let us suppose that they are placed in a shell of positive electricity of small density around the much denser core which

> contains the rest of the corpuscles and the equivalent quantity of positive electricity. Thus we may picture the atom as having a crowded centre, surrounded by a rarefied atmosphere through which a few corspuscles are scattered, the positive electricity in the atmosphere being equivalent to the negative charge on the corpuscles scattered through it.[50]

This picture of the atom differed in significant respects from the one advocated a few years earlier. First, the positive charge was no longer weightless, but made up the major part of the mass of the atom. Second, the atom was conceived as an atomic core surrounded by an atmosphere or shell of low charge density. And third, the few chemically and optically active electrons moved on the surface of the atom.

In about 1910 Thomson had quietly abandoned his original atomic model and begun to focus his research on the positive electricity in the form of positive rays or what continental physicists preferred to call canal rays. At the 1909 meeting of the British Association he no longer defended the plum pudding model but vaguely suggested that the atom contained negative as well as positive elementary charges, both kinds being ponderable. Shortly later Thomson's picture of the atom faced a new and grave difficulty, namely its inability to explain the scattering experiments with alpha rays performed in Manchester by Hans Geiger and Ernest Marsden under Rutherford's supervision (see Section 1.5). Although these experiments were highly important, the demise of the Thomson atom was not simply caused by them. The refutation of the classical Thomson process was a gradual process, during which anomalies and conceptual problems accumulated until most physicists, including Thomson himself, realized that it could not be developed into a satisfactory state. Although Thomson never officially buried his model of 1904, he changed it so many times and so drastically that it became a new model.[51]

This new model (of which more will be said in Section 3.5) pictured the atom as a conglomerate of positive and negative particles held together by forces known and unknown. In contrast to what he had done in such detail in 1904, in his new model Thomson made no attempt to calculate the configurations of the particles. The model was designed in such a way that it could reproduce some of the new quantum phenomena, such as Einstein's law of photoelectricity, but this was only possible by adopting suitable hypotheses of an ad hoc nature. In his presentation at the 1913 Solvay conference Thomson provided an explanatory sketch of valency, and he argued that his theory led to electron configurations for the simpler atoms that corresponded to the known periodicity of the elements. According to Thomson, many chemical properties could be understood in his model as being due to a dipole–dipole interaction caused by the mobility of the atomic electrons. He proposed that the total number of electrons in the light elements was of the same order as the atomic weight A. Among his examples were H = 1, He = 2, Li = 5, C = 6, O = 8, F = 11, Na = 13 and Cl = 19.[52]

Thomson's conclusion that $n \sim A$ was not accepted by all atom-builders and it did not imply the immediate death of his model. In 1911, the year that Rutherford proposed his nuclear atom, Harold Wilson at McGill University, Montreal, argued from the

Thomson model that the electrons were distributed in concentric layers within the positive charge. He concluded that $n \cong 8A$ was good approximation.[53] Hydrogen would have eight electrons, and for the alkali metals his best offer was Na = 142, K = 320, Rb = 600 and Cs = 1020. Wilson's defense of the Thomson model was one of the last of its kind. By the early 1910s the classical Thomson atom was no longer the subject of research or taken seriously as a realistic picture of the atom, and yet it continued to live on for several years. For example, it speaks to the appeal of the model that Owen Richardson, in the 1916 edition of his textbook *The Electron Theory of Matter*, covered it in great detail, in fact in greater detail than the Bohr model.[54] Readers of Richardson's book would not have suspected that Thomson's theory of the atom already belonged to the past.

From about 1904 to 1910 the Thomson atom was generally accepted as the best candidate of an atomic theory. This was particularly the case in England, but the model also attracted some attention on the Continent, where several physicists expressed interest. In a lecture in Göttingen of 1909, Max Born dealt with Thomson's atomic model, which he found fascinating because of its 'remarkable agreement' with the periodic system and other phenomena of nature. What appealed to him was the spirit of the model, not its details. According to Born: 'Thomson's atomic model . . . is like a piano excerpt from the great symphonies of luminating atoms. Although it may seem in many ways to be crude and imprecise, yet it gives us a starting point for understanding this mighty music.'[55] Given that Thomson's model was unable to account for the spectra it is ironic that Born spoke of it as a symphony of luminating atoms.

In his 1906 lectures at Columbia University, published three years later as *The Theory of Electrons*, Hendrik A. Lorentz considered a generalized version of the Thomson atom.[56] Rather than adopting Thomson's uniformly charged positive sphere, he assumed that the charge density varied in some unknown manner with the distance from the centre. As a special case Lorentz analyzed in detail a system of four electrons situated at the corners of a regular tetrahedron whose centre coincided with the centre of the positive sphere. Spatial models of the Thomson atom were investigated some years later by Arthur Erich Haas in Austria and Ludwig Föppl in Germany.[57] Their laborious calculations did not result in new physical insights and were primarily mathematically motivated. Neither Haas nor Föppl was concerned with comparing their three-dimensional equilibrium structures with the physical and chemical properties of real elements. It is telling that Föppl's extensive work, which was part of his dissertation at the University of Göttingen, was done at the suggestion of the mathematician David Hilbert and published in a mathematics journal.

Whereas Haas's work on the equilibrium configurations according to 'Thomson's spirited theory' was of no importance, a paper that he published two years earlier merits attention because it was the first attempt to apply quantum theory to the structure of atoms.[58] Haas made use of Thomson's picture of the hydrogen atom, or what he took to be Thomson's picture, but assumed that the single electron would revolve along the surface of the positive sphere. It should be noted that in 1911 it was not generally accepted that the hydrogen atom contains only one electron. Thomson, for one, did not say so and neither did Rutherford in his paper on the nuclear atom of 1911. Because Haas's electron moved on the surface of the positive sphere, it was subjected to an electric force that might just as well come from a positive charge concentrated in the centre. From this point of view his model can be seen as equivalent to the slightly later nuclear atom. Haas assumed the potential energy of the electron to be given by $e^2/b = h\nu$, where ν denotes the frequency of revolution. From this follows an expression for Planck's constant h, namely

$$h = 2\pi e\sqrt{mb}.$$

Guided by Balmer's expression for the hydrogen spectrum, and making use of some rather arbitrary assumptions, Haas also obtained the formula

$$R = 16\pi^2 \frac{me^4}{ch^3},$$

where R is Rydberg's spectroscopic constant. The latter expression happens to be of the right order of magnitude and is, in fact, exactly eight times larger than the value Bohr derived in 1913. Haas also considered the possibility that the mass of the electron was of electromagnetic origin, which was a natural assumption at the time. In this case he derived for Planck's constant

$$h = \frac{2\pi e^2}{c}\sqrt{\frac{2b}{3r}},$$

where r is the classical radius of the electron, $r = e^2/mc^2 \cong 3 \times 10^{-15}$ m.

Haas's theory attracted the critical attention of leading physicists, including H. A. Lorentz, Max Planck, Arnold Sommerfeld, and Paul Langevin. It was discussed at the 1911 Solvay congress, where Sommerfeld objected that Planck's constant of action should not be derived from atomic quantities. Rather than basing h on a special model of the atom, he preferred a general and model-independent theory of the constant. Sommerfeld's view corresponded to the one that Bohr adopted in his atomic theory of 1913: h is an irreducible constant that can be used to explain atomic constants while the opposite approach of Haas and Thomson is illegitimate. Haas's model was also considered by his compatriot Arthur Schidlof, who modified it by assuming that, in the case of many-electron atoms, a part of the negative electricity was located at the centre of the positive sphere.[59] On this basis he found a more general expression of h in terms of atomic quantities and one which he thought conformed better to Thomson's atomic theory.

Johannes Stark, the professor of physics at the Technische Hochschule in Aachen, was one of the few physicists who supported an atomistic conception of light similar to the light quantum hypothesis proposed by Einstein in 1905.[60] Like other physicists at the time he was interested in the problem of the distribution of positive and negative electricity in atoms, but it was not in this context that he found the energy quanta of Planck and Einstein useful. From experiments on radioactivity and light emitted by positive rays he suggested in 1910 that atoms consisted of electrons and a massive 'atomic ion' corresponding to the chemical atom. The positive charge of the atomic ion was not distributed uniformly over the atom's surface but concentrated in quanta at certain points on it. For the positive quantum Stark introduced the term 'archion' and on this basis he developed a purely qualitative theory of the constitution of atoms and molecules.[61] The chemical atom, he claimed, consists solely of positive archions, which are characteristic of the element, and negative electrons that may be dissociated from the atom. Stark developed his theory of an *Atomdynamik* (atomic dynamics) based on electrons and archions into an elaborate system, but it failed to convince other physicists. After about 1914 the archion fell into oblivion.

In spite of the failure of Stark's system of atoms and molecules it deserves mention in the present context because it involved quantum theory related to spectra. Discussing the emission of absorption bands, he made use of Planck's constant by assuming that the difference in potential energy of two states of the valence electron was given by multiples of $h\nu$, where ν is the frequency of the emitted light. He argued that all lines in a band spectrum must satisfy $\lambda \geq hc/\epsilon$, where ϵ is a binding energy characteristic for the valence electron. This may look a little like Bohr's formula of 1913, except that Stark was not concerned with the line spectra of atoms but only with the band spectra of molecules. Nonetheless, in his Nobel lecture of 1919, Stark maintained that his assumption involving Planck's constant 'forms the starting point of Bohr's theory of the emission of serial lines'.[62]

1.4 PLANETARY ATOMS

What the American historian of ideas John Theodore Merz in 1896 called the 'astronomical view of molecular phenomena' has a long tradition in the history of science and ideas.[63] Postulating that the microcosm is structured in analogy with the macrocosm, and that the two realms of nature are governed by the same laws, the view reappeared in a new version when scientists in the nineteenth century began to speculate about the internal composition of atoms. As mentioned in Section 1.1, explicit analogies between atoms and the planetary system appeared decades before the discovery of the electron. Sometimes little more than metaphors, planetary and similar micro–macro analogies continued to be popular in the first decades of the twentieth century, if perhaps more in the popular than the scientific literature. In his survey of 1906, Lodge included a picture of the atom as a kind of solar system, with the electrons revolving 'like asteroids' around a solar concentration of positive electricity.[64] In a textbook published two years later, Sophus M. Jørgensen, a prominent Danish professor of chemistry, said of the atom that it 'is, in fact, now considered to be a nucleus of positive electricity, around which negative electrons rotate with immense velocities in definite paths, like the planets in the solar system'.[65]

The first scientist to propose a planetary atomic model based on electrons may have been the French physical chemist Jean Perrin, a physics Nobel laureate of 1926 for his work on Brownian motion and related phenomena. In a popular lecture of 1901 he suggested the following picture of the atom:

> Each atom will be constituted, on the one hand, by one or several masses very strongly charged with positive electricity, in the manner of positive suns whose charge will be very superior to that of a corpuscle, and, on the other hand, by a multitude of corpuscles, in the manner of small negative planets, ... [with] the total negative charge exactly equivalent to the total positive charge, in such a way that the atom in electrically neutral.[66]

On the basis of this picture Perrin indicated that it might have spectroscopic applications and also that it might suggest an explanation of the enigmatic radioactivity: the outermost electrons—the 'Neptunes of the system' as he called them—would tend to leave the electric attraction from the atomic sun. However, his model was nothing but a rough sketch and probably not meant to be more than that. Thus he did not attempt to calculate the configurations of the planetary electrons and showed no interest in the stability of their orbits. Nor did he consider a possible mechanism for the emission of light.

Of much greater interest is the model proposed by the Japanese physicist Hantaro Nagaoka, although his was a 'Saturnian' rather than a planetary model of the atom.[67] Nagaoka had done post-doctoral studies at the universities of Berlin and Munich and there had become acquainted with Maxwell's 1856 essay on the mechanical stability of Saturn's system of rings. In this work, for which he was awarded the Adams Prize, Maxwell had concluded that Saturn's central body, surrounded by a rotating ring with a large number of separate satellite particles, would remain stable if the angular velocity of the ring was sufficiently high. In a paper of 1904 in *Philosophical Magazine*, Nagaoka, acknowledging his indebtedness to Maxwell's analysis, replaced Saturn's central body with a tiny positive charge and the multitude of satellites with electrons. The atomic system, he wrote,

> ... consists of a large number of particles of equal mass arranged in a circle at equal angular interval and repelling each other with forces inversely proportional to the square of distance; at the centre of the circle, place a particle of large mass attracting the other particles according to the same law of force. If these repelling particles be revolving with nearly the same velocity about the attracting centre, the system will generally remain stable, for small disturbances, provided the attracting force be sufficiently great.[68]

The Japanese physicist calculated that the central particle had to have a charge of at least 10 000 times the numerical charge of the electron. Nagaoka's elaborate calculations were primarily aimed at explaining the frequencies of band spectra (for which he found 'a close resemblance'), but he also thought that his model was suggestive with regard to radioactivity, resonance, luminescence and 'chemical affinity and valency, electrolysis, and many other subjects connected with atoms and molecules'. Although he derived theoretical spectral formulae from his model, he did not compare them with measured spectra. Like Thomson and other model builders of the period, Nagaoka safeguarded his conclusions by adding that 'the actual arrangement in a chemical atom may present complexities which are far beyond the reach of mathematical treatment'.

Published in the leading journal of atomic theory, Nagaoka's atom was well known and attracted some interest. For example, it was positively evaluated by Henri Poincaré in his book *La Valeur de la Science* of 1908, where the French mathematician called the model 'a very interesting attempt, but not wholly satisfactory'.[69] However, Nagaoka's calculations were severely criticized by George A. Schott, a physicist at the University College of Wales, who argued that the assumptions of Nagaoka were inconsistent and that the model could not possibly lead to the claimed agreement with experiments.[70] Schott proved in a general way that a system like Nagaoka's would be unable to generate the number of waves observed in either discrete spectra or band spectra. In reply to Schott's critique Nagaoka argued that it rested on a misunderstanding of his 'ideal atom', but the rejoinder had no effect. The hypothesis of the atom consisting of a 'central body charged with positive electricity, while the satellites are all negatively electrified' was also criticized by Thomson, although without mentioning Nagaoka by name.[71] By 1909 the Saturnian atom had disappeared from the scene of physics, abandoned even by Nagaoka himself, who recognized that the large number of electrons required by his theory disagreed with the empirically based conclusions of Thomson and other physicists. A few years later the Saturnian atom would reappear, if in an entirely different dressing, with the Rutherford–Bohr nuclear theory.

A new atomic theory, which had some features in common with Nagaoka's, was proposed in 1911 by John William Nicholson, at the time a lecturer at the Cavendish Laboratory.[72] (The following year he was appointed professor of mathematics at King's College, London.) His paper on the constitution of the atom appeared some months after Rutherford had introduced the idea of an atomic nucleus, but although Nicholson was aware of Rutherford's work he was not inspired by it and did not believe the two versions of the nuclear model to have much in common. Nicholson's atom was much closer to Thomson's model, both in spirit and computational details, and has been aptly described as a Thomson atom 'with the dimensions of the positive sphere shrunk from atomic size to one much smaller than the radius of the electron'.[73] Incidentally, whereas Rutherford did not refer to the central charge as a 'nucleus', Nicholson did, although he did not invent the name.[74]

Nicholson's model was different from earlier conceptions of the atom, not only because it offered definite constitutions of the chemical elements but also because it relied on astronomical evidence rather than laboratory evidence. The ambitious aim of Nicholson's theory was to derive all the atomic weights of the elements from combinations of certain proto-atoms, which he supposed existed in free form in the stellar realm only. He considered the massive positive nucleus to be purely of electromagnetic origin, hence much smaller than the electron, and located at the centre of the atom with rings of electrons revolving around it.

In agreement with the earlier evolutionary views of his compatriots Crookes and Lockyer, Nicholson was convinced that terrestrial matter had evolved from simpler forms that still existed in the stars and nebulae and could be studied by means of the spectroscope. To understand the architecture of atoms, the physicist would have to look to the heavenly regions. The atomic model proposed by Nicholson was mainly concerned with the four primary elements that were supposed to exist in the nebulae and the Sun's corona. Having concluded that a ring atom with only one electron could not exist, his simplest atom ('coronium') consisted of a single ring of two electrons rotating about a nucleus of charge $+2e$. His four primary elements, their proposed symbols, nuclear charges and atomic weights were as shown in Table 1.2:

Although Nicholson's three-electron 'hydrogen' was closely related to the chemical element hydrogen, he did not conceive the two atoms as being identical. Coronium was the simplest possible atomic system and ordinary hydrogen was seen as a kind of polymer of the primordial 'hydrogen'. Somewhat confusingly he chose the symbol H for this form of hydrogen. Whereas H, Nu, and Pf were thought of as constituents of the chemical elements, this was not the case with coronium, which was assumed to reside in the solar corona only. For this reason he did not assign it a chemical symbol. In his paper of 1911 Nicholson derived the weights and compositions of all of the elements (!), simple examples being He = NuPf, Li = 3Nu2H and Be = 3Pf2H.

Nicholson's theory was not only about the constitution of elements, but also, and in his later publications increasingly so, about the spectral lines emitted by the primitive atoms and caused by their vibrating ring electrons. His general method was to calculate the frequencies of the vibrations in coronium, nebulium, and protofluorine atoms and then to compare the results to unassigned lines occurring in nebular and coronal spectra. In this way he was able to account for most of these lines and also to predict new lines in both types of spectrum. For example, in the case of nebulium he predicted the existence of a line of wavelength 4353 Å, which shortly afterwards was found in a nebular spectrum.[75] He similarly predicted a line of wavelength 6374.8 Å due to protofluorine, which agreed nicely with the later discovery of $\lambda = 6374.6$ in the spectrum of the solar corona. Naturally he took these confirmations as support for his theory.

Table 1.2 Nicholson's four primary elements

Element	Symbol	Nuclear charge	Atomic weight
Coronium	–	$2e$	0.51282
'Hydrogen'	H	$3e$	1.008
Nebulium	Nu	$4e$	1.6281
Protofluorine	Pf	$5e$	2.3615

In 1914 Nicholson extended his list of primary elements, now including 'protohydrogen' with a single electron revolving around a positive unit charge and 'archonium' with a nucleus of charge $+6e$ and an atomic weight of about 2.9. Ordinary hydrogen was assumed to be an evolution product of protohydrogen, but with a nucleus of a more complicated structure, consisting of x 'positive electrons' and x–1 negative electrons. In general, Nicholson stressed the difference between his celestial primary elements and those found on Earth, a difference he in part ascribed to the complex nature of the atomic nucleus. 'The nuclei in terrestrial atoms are not simple', he said in 1914, 'they consist of a complicated system of positive and negative charges closely packed together, and not a mere positive charge'.[76] Nicholson saw his belief in what he called 'sub-elements' vindicated by astrospectroscopic measurements. For example, he found the mass of archonium to be 2.945 and calculated that the doubly charged positive archonium ion should radiate with a wavelength of 3729 Å. When the French physicist Charles Fabry and his colleagues Henri Bourget and Henri Buisson detected a double line of wavelength 3726–3729 Å in the Orion nebula and attributed it to an unknown gas with an atomic weight of approximately 3, Nicholson concluded that the existence of archonium had now been confirmed.[77]

In his attempts to explain the line spectra and determine the dimensions of the primary atoms, Nicholson was led to introduce Planck's constant into his theory. Up to this time the quantum of action had always been associated with energy, in the form $E = h\nu$, but now Nicholson extended its physical meaning. In 1912 he derived a relation between the potential energy of p rotating electrons and the frequency of rotation ν, namely

$$p\frac{mr^2 4\pi^2 \nu^2}{\nu} = nh,$$

where n is a whole number and r the radius of the electron ring. Interpreting the relation as a quantization of the angular momentum L, he argued that in the case of a five-electron ring (protofluorine) the angular momentum could be written as $L = 25h/2\pi$; for the rings of three and four electrons (hydrogen and nebulium) he similarly found $L = 18h/2\pi$ and $L = 22h/2\pi$. Inductive reasoning led him to the generalization that the angular momentum of simple atoms could only change by integral multiples of the quantity $h/2\pi$, that is,

$$L = n\frac{h}{2\pi}, n = 1, 2, 3, \ldots$$

Nicholson offered the following picture of the radiating atom:

> The quantum theory has apparently not been put forward as an explanation of 'series' spectra, consisting of a large number of related lines given by comparatively simple atoms. Yet... we are led to suppose that lines of a series may not emanate from the same atom, but from atoms whose internal angular momenta have, by radiation or otherwise, run down by

> various discrete amounts from some standard value. For example, on this view there are various kinds of hydrogen atoms, identical in chemical properties and even in weight, but different in their internal motions.[78]

No wonder that Bohr, when he came across Nicholson's atomic theory, found it to be interesting as well as disturbingly similar to his own ideas. Although today forgotten or only recalled by historians of physics and astronomy, in the period 1913–1916 Nicholson's atom was a rival to Bohr's, and Nicholson the chief critic of Bohr's ideas of the quantum atom.[79]

1.5 RUTHERFORD'S NUCLEAR ATOM

Ernest Rutherford's scientific reputation rested on his pioneering contributions to the study of radioactivity, a field of research in which he immersed himself fully. When he was awarded the Nobel Prize in 1908, it was in chemistry and for his work on radioactive decay. He was not particularly interested in atomic models except that until about 1908 he was generally in favour of a model of the kind proposed by Thomson, which he found useful in explaining certain qualitative features of radioactive phenomena. In broad conformity with Thomson's model, he conceived the atoms of even the lightest elements as conglomerates of thousands of electrons. In *Radioactive Transformations*, a book based on the Silliman Lectures of 1905, he included a section on atomic constitution in which he, in a general and guarded way, endorsed a Thomson-like picture of the atom. As Rutherford phrased it, 'The mobile electrons constitute, so to speak, the bricks of the atomic structure, while the positive electricity acts as the necessary mortar to bind them together'.[80] Although he found this 'a somewhat arbitrary arrangement', at the time he could see no better alternative.

Only in 1910 did Rutherford turn seriously to atomic theory, primarily as a result of his deep interest in the behaviour and nature of alpha particles.[81] In 1908, in joint work with his assistant Thomas Royd, he had definitively shown the alpha particle to be identical with a doubly charged helium ion, although at the time the composition of the ion was unknown. In the same year Hans Geiger, a German physicist working with him in Manchester, reported preliminary results of the scattering of alpha particles on metal foils. Geiger noticed an appreciable scattering and the following year he investigated the matter more fully in collaboration with 21-year-old Ernest Marsden. The two physicists found that heavier metals were far more effective as reflectors than light ones and that a thin platinum foil reflected (that is, scattered by an angle $\phi > 90°$) one of every 8,000 of the alpha particles striking it.

The experiments induced Rutherford to investigate the scattering of alpha particles and to compare the results with Thomson's 1910 theory of the scattering of beta particles. This theory, according to which the beta electrons were multiply scattered through small angles, seemed to agree nicely with experiments made by James Arnold Crowther, a young Cambridge physicist. According to Crowther, Thomson's theory was brilliantly confirmed and his experiments implied that the number of electrons in an atom was about three times the atomic weight. For example, Crowther found that aluminium, with atomic weight 27, must have 85 electrons. Suspecting that Crowther's interpretation was biased in favour of the Thomson theory, Rutherford set about to find a unified theory that could account for the scattering of both beta and alpha particles.

According to Thomson, the alpha particle was of atomic dimensions and his best estimate was that it contained 10–12 electrons. Rutherford, on the other hand, came to the conclusion that the alpha particle must be considered a point particle, like the electron. Because the alpha particle was a helium atom deprived of its two electrons, this view implied, in effect, a nuclear model of the helium atom. Rutherford reached this important conclusion before he developed his scattering theory based on the idea of point-like alpha particles. By late 1910 he was focusing on a new picture of atomic structure consistent with the scattering experiments (Figure 1.3). In a letter to the American radiochemist Bertram Boltwood of 14 December 1910 he wrote: 'I think I can devise an atom much superior to J. J.'s, for the explanation of and stoppage of α and β particles, and at the same time I think it will fit in extraordinarily well with the experimental numbers.'[82] Rutherford presented his new and superior atomic model, primarily based on the scattering experiments by Geiger and Marsden, in a landmark paper in the *Philosophical Magazine* of May 1911.

In this paper Rutherford concluded that in order to produce the observed deflections of $\phi > 90^\circ$, scattering had to take place in a single encounter between the alpha particle and a highly charged and concentrated mass. He therefore suggested that the atom contained at its centre a massive charge *Ne* surrounded by a cloud of the opposite charge. Since the results of his calculations were independent of the sign of the charge, the nucleus could just as well be a concentration of electrons embedded in a positive fluid, not unlike an extreme case of the Thomson atom: 'Consider an atom which contains a charge $\pm$ *Ne* at its centre surrounded by a sphere of electrification containing a charge $\mp$ *Ne* supposed uniformly distributed throughout a sphere of Radius *R*. . . . For convenience, the sign [of the central charge] will be assumed to be positive.' Rutherford admitted that the experimental evidence did not rule out the possibility that 'a small fraction of the positive charge may be carried by satellites extending some distance from the centre'.[83]

Based on this picture of the atom, Rutherford derived a formula that expressed the number of charged (alpha or beta) particles y scattered a certain angle ϕ at a distance from the scattering material. The formula related $y = y(\phi)$ to the mass and velocity of the incident particles, the number of atoms in a unit volume of the scatterer, and the nuclear charge N of the scatterer. As Rutherford demonstrated, in the case of alpha particles in particular his formula agreed very well with the experimental data obtained in Manchester. Although the scattering of alpha particles was the most important

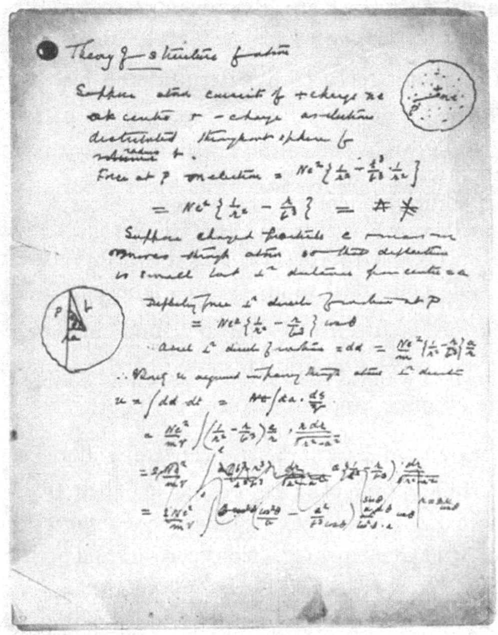

Fig. 1.3. Rutherford's early calculations of the scattering of alpha particles, and a sketch of the nuclear atom as he saw it in early 1911 (Heilbron 1968). He pictured a heavy atom as consisting of a central positive charge surrounded by an atmosphere of negative electrons, but without caring about the configurations of the electrons.

evidence for the nuclear atom, he also found good agreement in the case of beta scattering. Taken together the data indicated 'that the value of this central charge for different atoms is approximately proportional to their atomic weights, at any rate for atoms heavier than aluminium'. According to Rutherford's analysis, the gold atom had a charge of $N \cong 100$, which agreed with what he suspected was a general approximation, namely $A/2 < N < A$.

The nuclear atom introduced by Rutherford in the late spring of 1911 did not make a splash in the world of physics. Surprisingly from a later perspective, the model was met with indifference and scarcely considered to be a new theory of the constitution of the atom. It was not mentioned in the proceedings of the first Solvay congress, taking place in the autumn of 1911 with Rutherford as a participant, nor did it receive much attention in the physics journals. The second edition of the Cavendish physicist Norman Campbell's *Modern Electrical Theory*, with a preface dated March 1913, included a chapter on the structure of the atom in which the theories of Thomson and Stark were singled out, but with no mention of Rutherford's nuclear atom. Campbell was not impressed by the state of research in atomic structure: 'It cannot be pretended that we have at present any but a most rudimentary theory of the atom; none of the suggestions which have been made as to its structure lead in any case to the possibility of calculating in detail one property of an atom from a knowledge of the others; no quantitative relations can be deduced.'[84]

Rutherford himself does not seem to have considered his discovery as the epoch-making event that it turned out to be. For example, in his massive 1913 textbook on radioactivity, titled *Radioactive Substances and their Radiations*, with the preface being dated October 1912, there were only two references to the nuclear atom and its implications. He now declared the nucleus to be positively charged, surrounded by electrons 'which may be supposed to be distributed throughout a spherical volume or in concentric rings in one plane'. The nucleus was extremely small, but not point-like. On the contrary, Rutherford pictured it as a complex body held together by what would become known as nuclear forces, the first example of strong interactions. He wrote as follows:

> Practically the whole charge and mass of the atom are concentrated at the centre, and are probably confined within a sphere of radius not greater than 10^{-12} cm. No doubt the positively charged centre of the atom is a complicated system in movement, consisting in part of charged helium and hydrogen atoms [alpha particles and protons]. It would appear as if the positively charged atoms of matter attract one another at very small distances for otherwise it is difficult to see how the component parts at the centre are held together.[85]

It is customary to speak of Rutherford's atomic model, but in 1911 there was not really such a model, at least not in the sense of 'atomic model' that was ordinarily adopted at the time. This observation goes a long way in explaining the initial lack of interest in the nuclear atom.

Rutherford presented his theory primarily as a scattering theory and realized that, considered as a theory of atomic structure, it was most incomplete. First and foremost, it could offer no suggestion of how the electrons were arranged, the very issue that was central to atomic models. 'The question of the stability of the atom proposed need not be considered', he wrote, 'for this will obviously depend upon the minute structure of the atom, and on the motion of the constituent charged parts'.[86] His nuclear atom was impotent when it came to chemical questions such as valency and the periodic system, and it fared no better when it came to physical questions such as spectral regularities and the dispersion of light. An atomic theory anno 1911 would be considered really convincing only if it included the system of electrons. After all, it was agreed that this part of the atom was responsible for the majority of the atomic phenomena that could be tested experimentally.

The status of Rutherford's theory improved in the spring of 1913, when Geiger and Marsden published new data on the scattering of alpha particles that were in excellent agreement with the scattering formula. 'We have completely verified the theory given by Prof. Rutherford', they concluded. The experimental results afforded 'strong evidence of the correctness of the underlying assumptions that an atom contains a strong charge of the centre of dimensions, small compared with the diameter of the atom'.[87] The new experiments sharpened the relationship between the nuclear charge and atomic weight, which Rutherford now took to be $N \cong A/2$. The work of Geiger and Marsden confirmed Rutherford's atomic model considered as a scattering theory, but not as a theory of atomic constitution. The results of the two Manchester physicists were as irrelevant for the electronic configurations as Rutherford's atom was silent about them.

Nor did the scattering data help to solve the instability problem of the nuclear atom, a serious problem that Rutherford was well aware of but decided to ignore. In July 1913 Lodge reminded him of it: 'The astronomical view of the structure of the atom, with a minute, positive, massive nucleus and a neutralising number of electrons subject to the inverse square law is in many ways attractive. But there are difficulties about it, chief among which is a question of stability . . . I suppose you have your own ways of getting over these difficulties.'[88] Actually Rutherford did not have ways to get over the difficulties, but fortunately Niels Bohr did. In the same month that Lodge penned his letter, there appeared in *Philosophical Magazine* a paper in which the Rutherford nuclear model was extended into a quantum Bohr–Rutherford atomic model. The paper provided a solution to the instability problem, although a highly unconventional solution of a kind that Lodge and most other physicists found hard to accept.

1.6 ISOTOPY AND ATOMIC NUMBER

The existence of isotopes—species of the same chemical element with different atomic weights—was not foreseen by any of the theories of atomic structure, whether Thomson's, Nagaoka's, Rutherford's, or Nicholson's. The idea was however anticipated by Crookes as early as 1886, when he suggested to the British Association that 'when we say that the atomic weight of, for instance, calcium is 40, we really express the fact that, while the majority of the calcium atoms have an actual weight of 40, there are not a few which are represented by 39 or 41, a less number by 38 and 42, and so on'.[89] On the same occasion he speculated that the hypothetical helium, supposed to be of atomic weight 0.5, might be the prime matter. Crookes's spirited idea was one more attempt to save Prout's attractive hypothesis from ugly experimental facts. It led him to interesting speculations about 'meta-elements', but the idea turned out to be unviable.

It was primarily the perplexing study of radioactive decay series some two decades later that first indicated the possibility of atoms with different properties belonging to the same element.[90] Some of the substances found in the radioactive series and identified by their radioactive properties turned out to have a strong chemical resemblance to other elements; in fact they were inseparable from them and yet they were not identical. In desperation, some scientists grouped several radio-elements (say, radium emanation, actinium emanation, and thorium emanation) into the same place in the periodic system; others chose to extend the periodic system to accommodate the new radio-elements. By 1910, Frederick Soddy, Rutherford's former collaborator, concluded that radium, mesothorium 1, and thorium X, although of different atomic weights and radioactive properties, were not merely chemically similar, but chemically identical. Soddy did not believe that the hypothesis of different species of the same element was restricted to the radioactive elements in the upper part of the periodic system. When he coined the word 'isotope' in late 1913, he related it to Rutherford's nuclear atom:

> The same algebraic sum of the positive and negative charges in the nucleus, when the arithmetical sum is different, gives what I call 'isotopes' or 'isotopic elements', because they occupy the same place in the periodic table. They are chemically identical, and save only as regards the relatively few physical properties which depend upon atomic mass directly, physically identical also.[91]

The Polish physical chemist Kasimir Fajans, who had spent a year 1910–1911 working with Rutherford in Manchester, had in an earlier paper suggested essentially the same hypothesis.[92] He called a group of chemically identical elements a 'pleiade', but, while 'isotope' caught on, 'pleiade' did not. In fact, Fajans may have been alone in using the term. Contrary to Soddy, at the time he did not accept Rutherford's nuclear atom, but argued that alpha particles were expelled from the outer layer of the atom. He found Nicholson's atomic theory to be 'extremely tempting' and superior to Rutherford's.

Radioactive decay was not the only phenomenon that pointed towards isotopy; so did the positive rays investigated by J. J. Thomson. Francis Aston, who served as Thomson's assistant in parts of his research programme, analyzed positive rays of neon, known to have the atomic weight 20.2. Surprisingly, Aston's experiments revealed not only rays corresponding to atomic weight 20 but also weaker rays corresponding to atomic weight 22. In the absence of a proper explanation, he suggested he had found what he called 'meta-neon', possibly a new inert gas.[93] He made the announcement at the 1913 meeting of the British Association in Birmingham, the same meeting where Bohr's atomic theory was first introduced and discussed by leading physicists. After a brief period of confusion, Aston and Soddy realized that what Aston had discovered was a heavy isotope of neon. Thomson at first thought that the recorded species of atomic weight 22 might be the neon hydride compound NeH_2 and only reluctantly agreed that neon was probably a mixture of chemically inseparable species with different atomic weights. Understandably, he did not agree with Soddy's interpretation of isotopy in terms of the nuclear model of the atom.

Before 1913 the order of the elements in the periodic system was universally taken to be given by the atomic weight. Although this caused some anomalies, such as that related to the 'reversed' atomic weights of tellurium (Te = 127.6) and iodine (I = 126.9), the convention or dogma of atomic weight being the defining property of a chemical element was rarely questioned. Rutherford's argument that the number of positive charges in the nucleus varied with the atomic weight as $N \cong A/2$ did not question the standard view, it merely sharpened it and connected it to the atomic nucleus.

According to Charles Galton Darwin, who at the time was a lecturer at Manchester University, the 1913 scattering experiments of Geiger and Marsden convinced Rutherford and his group that the nuclear charge was the defining quantity of a chemical element.[94] The idea certainly was in the air, but it took until November 1913 before it was explicitly formulated, and then from the unlikely source of a Dutch amateur physicist. Trained as a lawyer, Antonius van den Broek had since 1907 published articles on radioactivity and the periodic system. Guided by neo-Proutean speculations he thought that consecutive elements in the periodic system differed by two units in weight, an idea which he combined with the relation $N \cong A/2$ made plausible by the Manchester experiments. In a short communication to *Nature* dated November 10 he disconnected the ordinal number from the atomic weight and instead identified it with the nuclear charge N (or Z, as it subsequently became symbolized). This hypothesis, he said, 'holds good for Mendeleev's table but the nuclear charge is not equal to half the atomic weight'.[95]

Van den Broek's suggestion was quickly adopted by Soddy, Bohr, and Rutherford and his group. As Rutherford pointed out, the idea of an 'atomic number'—a word that he possibly coined—had already been used by Bohr in his atomic theory and it received convincing confirmation in the new X-ray spectroscopic experiments of Henry Moseley.[96] In an address of 1934 celebrating the centenary of Mendeleev's birth, Rutherford credited Bohr for first having recognized the significance of an ordinal number for the chemical elements: 'The idea that the nuclear charge of an element might be given by its ordinal or atomic number was first suggested and used by Bohr in developing his theory of spectra. By a strange oversight, Bohr himself gave the credit of this suggestion to van den Broek, who later discussed the applicability of this conception to the elements in general.'[97]

What was often referred to as 'van den Broek's hypothesis' remained controversial for several years, one of the reasons being that it restricted—disciplined—the number of chemical elements. Some chemists and physicists, Nicholson among them, denied that the atomic number defined the place of an element in the periodic system and thus limited the number of elements in a period. The recognized Swedish spectroscopist Johannes Robert ('Janne') Rydberg admitted the notion of an ordinal number different from the atomic weight, but argued that it was two units greater than the one adopted by van den Broek. According to Rydberg, in the first period there were four elements rather than just hydrogen and helium, lithium should be element number 5, beryllium element number 6, and so forth (Figure 1.4). In his periodic system of 1913, he included the hypothetical coronium and nebulium among the light elements.[98]

As pointed out by Soddy and others, the new notion of atomic number fitted very well with the notion of isotopy and both fitted with Rutherford's nuclear model of the atom. The atomic number also implied a new meaning of the term 'element' that in some respects differed rather drastically from the one traditionally accepted by the

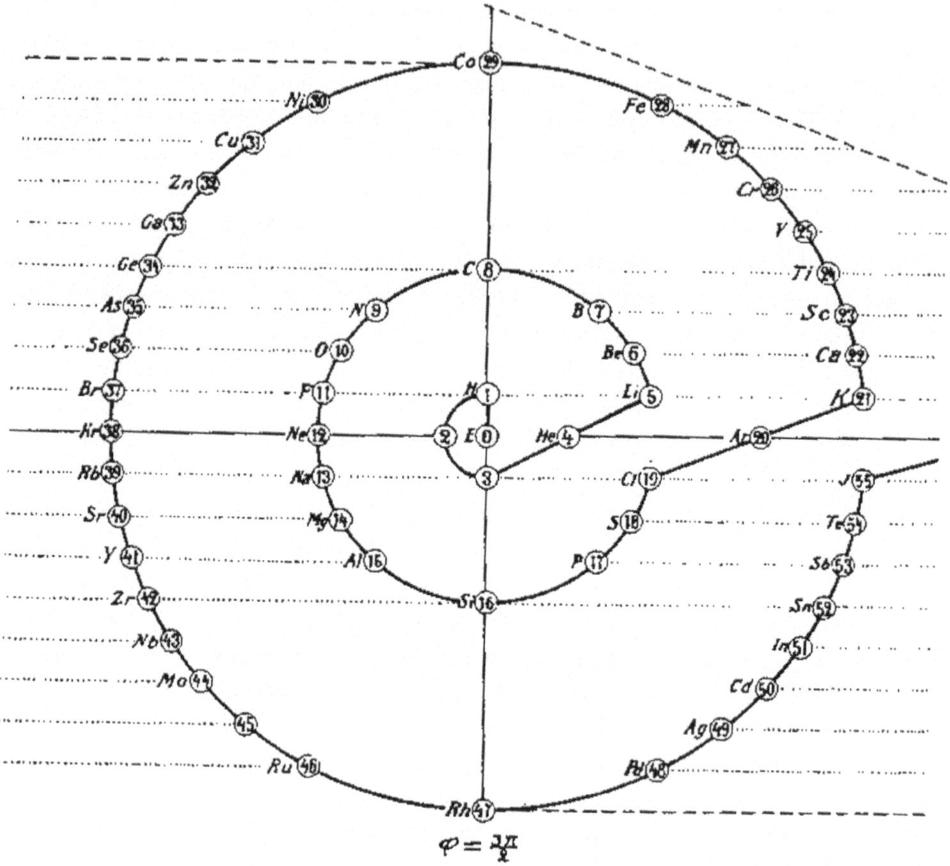

Fig. 1.4. Part of Rydberg's periodic system of 1913, based upon an ordinal number different from the one adopted by van den Broek, Rutherford, and Bohr. Notice the two elements between hydrogen and helium, and also that Rydberg considered the electron (with ordinal number 0) a kind of chemical element.

chemists. For this reason isotopy and the atomic number became controversial in some chemical circles. Only after World War I was the atomic number officially adopted as the defining quantity of a chemical element. In 1921 the German Atomic Weight Commission decided to base its new table on the atomic number, and two years later the International Committee on Chemical Elements followed the same policy.[99]

When Bohr developed the final version of his atomic theory in the spring of 1913, he was aware of most of the work done by Thomson, Nagaoka, Nicholson, Rutherford, Soddy, and van den Broek. Although his theory was highly original, some of this work influenced his thinking and the way he formulated his theory.

Notes for Chapter 1

1. This chapter is based on Kragh 2010a.

2. Andrade 1958, p. 442.

3. Perrin 1913, p. 295. On Perrin's very influential book and the late anti-atomism in France and elsewhere, see Nye 1972.

4. Duhem 1913, p. 71.

5. On Prout's hypothesis and its role in nineteenth-century conceptions of atoms and elements, see Farrar 1965, Kragh 1982, and Brock 1985.

6. Hinrich's ideas are described in Zapffe 1969 and van Spronsen 1969.

7. Note of 2 March 1854, reproduced in Dahl 1972, p. 177. The term 'molecule' should not be understood in its modern, chemical sense. Throughout the nineteenth century the term, especially as used by physicists, often meant just a very small unit particle, what others would call an atom.

8. Redtenbacher 1857. See also Rosenberger 1965, vol. 3, pp. 554–5, a reprint of a work first published between 1886 and 1890.

9. Nägeli 1884, pp. 681–820. See also Kragh 1989a. Rosenberger 1965 includes accounts of the atomic theories of Fechner, Weber, Grassmann, Nägeli, and other nineteenth-century scientists.

10. Laming 1845. For background on Laming and his ideas, see Farrar 1969.

11. Descriptions of Weber's electro-atomic research programme and its connection to the works of other German physicists can be found in Wise 1981 and Schönbeck 1982.

12. Weber 1871, p. 44, as quoted in Mehra and Rechenberg 1982a, p. 169.

13. Fechner 1864. Grassmann 1862. On the electro-atomic ideas of the Grassmann brothers, see Kuntze 1909.

14. Fechner 1828, p. 275, as quoted in Mehra and Rechenberg 1982a, p. 169.

15. Sutherland 1899. Sutherland 1901, p. 272. The eminent physical chemist Walther Nernst reinvented the ether-neutron or possibly took it over from Sutherland (Nernst 1907, p. 392). On the problem of the positive electron until the discovery of the positron in 1932, see Kragh 1989b.

16. Larmor 1895, p. 741. Parts of this and the following section rely on material in Kragh 1997a and Kragh 2003a.

17. A full historical account of the vortex atom theory can be found in Kragh 2002a, which includes references to the primary literature.

18. Maxwell 1965, Part II, pp. 444–84, on p. 471.

19. Pearson 1885. See also Porter 2004, pp. 179–92.

20. Pearson 1891.

21. Thomson 1883.

22. Thomson 1897, p. 313. See also Davis and Falconer 1997. The history of the electron, and the several changes in the meaning of 'electron', is discussed in Arabatzis 2006.

23. Thomson 1899, p. 565.

24. For example (William) Thomson 1904.

25. Lodge 1906, p. 149.

26. Lenard 1903. For background on Lenard's atomic hypothesis, which was closely connected to his work on cathode rays and the photoelectric effect, see Wheaton 1978.

27. Lenard 1903, p. 739.

28. For example Gehrcke 1923, who spoke of the 'Lenard–Rutherford atomic model' and the 'Lenard–Rutherford–Bohr conception of atomic structure'.

29. LeBon 1905, p. 247, who concluded that all material atoms were slowly degrading into the ether, 'the final nirvana to which all things return after a more or less ephemeral existence' (p. 315).

30. Jeans 1901.

31. Lodge 1903 and Lodge 1906, p. 148.

32. The French physicist Jean Becquerel, a son of Henri Becquerel of radioactivity fame, claimed to have obtained experimental evidence for positive electrons, but his claim was generally disbelieved (Kragh 1989b).

33. Larmor 1900, p. 625. On the British model-making tradition, see Heilbron 1977b.

34. Kaufmann 1901, p. 15, with English translation in *The Electrician* **48** (1901), 95–7.

35. Righi 1909, pp. 151–2. Like several other physicists at the time, Righi argued that the new view of the composite atom implied the possibility of transmutation of elements, indeed that all elements were probably radioactive.

36. Thomson 1904a and Thomson 1904b. For historical studies of the Thomson atom, see Heilbron 1977a and Kragh 1997a.

37. Thomson 1904b, p. 119.

38. On Larmor and his formula, see Warwick 1993.

39. Thomson 1904a, p. 254.

40. Thomson 1907, p. 114.

41. Thomson 1906 and Thomson 1910, with historical analysis in Heilbron 1968.

42. See Stranges 1982 and, for the role of chemical considerations in Thomsons research programme, Sinclair 1987.

43. Thomson 1904c, p. 265. On the problem of subatomic explanations of radioactivity, see Kragh 1997b.

44. Quoted in Davis and Falconer 1997, p. 195.

45. Thomson 1906, which according to Heilbron (1968, p. 269) is 'one of the most important papers on atomic structure ever written'.

46. Lodge 1906, p. 194.

47. Interview of Niels Bohr by Thomas S. Kuhn, Léon Rosenfeld, Erik Rüdinger, and Aage Petersen on 31 October to 14 November 1962, Niels Bohr Library & Archives, American Institute of Physics, College Park, MD USA, www.aip.org/history/ohilist/4517_1.html.

48. Rayleigh 1906, p. 123. The model Rayleigh used as an illustration was a kind of Thomson atom, but with an infinite number of 'electrons' smeared out over the entire positive sphere. Other pre-1913 attempts of making sense of spectroscopic regularities in terms of atomic structure are discussed in Maier 1964 and in Carazza and Robotti 2002.

49. Thomson 1906, p. 774.

50. Thomson 1907, p. 166.

51. Thomson 1913a and Thomson 1921.

52. Thomson 1921, p. 13.

53. Wilson 1911.

54. Richardson 1916 (second edition of Richardson 1914), pp. 559–79. See also Section 3.1.

55. Born 1909, p. 1031.

56. Lorentz 1952, pp. 120–3, 294–300.

57. Haas 1912. Föppl 1912.

58. Haas 1910a and Haas 1910b, reprinted in Hermann 1965a, pp. 27–60.

59. Schidlof 1911.

60. On Stark's support of Einstein's light quantum hypothesis and the views of the two physicists on this matter, see Mehra and Rechenberg 1982a, pp. 99–105.

61. Stark 1910, especially pp. 67–95. Whereas physicists ignored Stark's ideas of atomic and molecular constitution, for a while they attracted considerable interest among chemists. See Stranges 1982, pp. 192–200. For contemporary reviews, highlighting the chemical applications of Stark's theory, see Campbell 1913, pp. 341–8, and Hahn and Holmes 1915.

62. Stark 1911, p. 108. Stark's Nobel lecture can be found on http://nobelprize.org. Fujisaki 1982 argues that Bohr was influenced by Stark's theory (see also Section 2.3).

63. Merz 1965, vol. 1, p. 354.

64. Lodge 1906, p. 150, who did not endorse the picture.

65. Jørgensen 1908, p. 26. In what superficially looks like an anticipation of the Bohr–Rutherford model, Jørgensen probably had Nagaoka's model in mind. He seems to have thought, if so mistakenly, that this kind of atomic model was generally accepted.

66. Perrin 1901, p. 460. See also Nye 1972, pp. 83–5.

67. On Nagaoka and his atomic model, see Yagi 1964 and Conn and Turner 1965, pp. 111–19.

68. Nagaoka 1904, p. 445.

69. Poincaré 1958, p. 109.

70. Schott 1904. Schott 1907.

71. Thomson 1905, p. 142.

72. Nicholson 1911a. On Nicholson's life and career, see Wilson 1956.

73. McCormmach 1966, p. 165, which gives details of Nicholson's atomic model and its relation to Bohr's theory of the atom.

74. Jørgensen 1908, quoted above, may have been the first to call the positive centre a nucleus.

75. Nicholson 1911b and Nicholson 1912a.

76. Nicholson 1914a, p. 487. For a detailed account of the later version of Nicholson's atomic theory, see Wilson 1916, pp. 15–28.

77. Bourget et al. 1914. Nicholson 1914b.

78. Nicholson 1912b, p. 730.

79. On Nicholson's sustained critique of Bohr's theory, see Section 3.6 and Kragh 2011a.

80. Rutherford 1906, p. 266.

81. For details on Rutherford's road to the nuclear atom, see Heilbron 1968. See also Andrade 1958.

82. Quoted in Badash 1969, p. 235.

83. Rutherford 1911, p. 670 and p. 687.

84. Campbell 1913, p. 349. The first edition of the book, issued in 1907, contained a detailed review of Thomson's atomic theory, without mentioning alternatives. According to Campbell this was justified, 'for there is no rival theory of importance in the field' (p. 232).

85. Rutherford 1913a, p. 620.

86. Rutherford 1911, p. 671.

87. Geiger and Marsden 1913, p. 606, and Conn and Turner 1965, pp. 150–63.

88. Letter from Lodge to Rutherford, quoted in Wilson 1983, p. 322.

89. Crookes 1886, p. 569.

90. The history of isotopy in the period 1910–1915 is covered in Brock 1985, pp. 196–216, and Bruzzaniti and Robotti 1989. On the casual connection to occultism, see Hughes 2003.

91. Soddy 1913, p. 400. Reprinted together with other historical papers on radiochemistry and isotopes in Romer 1970 (pp. 251–2). The 'positive and negative charges in the nucleus' is a reference to protons and electrons.

92. Fajans 1913.

93. Aston 1913. Due to World War I, Aston was only able to continue his work on isotopes, leading to the invention of the mass spectrograph, in 1919. Brock 1985. pp. 205–15.

94. Darwin 1955.

95. Van den Broek 1913, p. 373. For details on van den Broek and his work, see Snelders 1974 and Scerri 2007, pp. 165–9.

96. Rutherford 1913b.

97. Rutherford 1934, p. 636. Bohr referred to van den Broek in his second communication on atomic structure, stating the hypothesis in a form that referred only to the number of electrons in a neutral atom and not to the atomic nucleus (Bohr 1913c, p. 477). See also Bohr 1932, where he describes the recognition of the atomic number and the existence of isotopes in the years 1912–1913.

98. Rydberg 1913. Pauli 1994 (originally published 1955).

99. On the controversial reconceptualisation of chemical elements in the period, see Kragh 2000.

2

On the Constitution of Atoms and Molecules

Atomic theory experienced a dramatic shift with the publication in the summer of 1913 of Niels Bohr's model of the atom. Although in some respects with roots in the Thomson tradition and 'based on the methods of mid-Victorian Cambridge physics',[1] it also marked a break with this tradition. The model initiated an entirely new and immensely fruitful development in the study of atomic, subatomic, and molecular phenomena. First of all, it made crucial use of the quantum theory of Planck and Einstein, making it clear that an understanding of atomic structure had to build on principles of a non-classical nature. How these principles related to the existing quantum theory remained unclear, but with his treatise of 1913 Bohr made a most important contribution to the process that eventually led to quantum mechanics.

Bohr's atomic theory is often conceived in a rather narrow way, essentially referring to his model of hydrogenic atoms governed by the quantum postulates of stationary states and the mechanism for emission and absorption of light by transitions from one quantum state to another. This is the historical legacy of the theory, which proved to be viable and of central importance in the construction of our current picture of the atom built on quantum mechanics. A century later, Bohr's visualizable planetary atom is still a favoured model, which for pedagogical reasons appears in physics and chemistry textbooks at high school level. However, it is important to recall that Bohr's three papers, comprising a total of 71 pages in *Philosophical Magazine*, carried the common title 'On the Constitution of Atoms and Molecules', thus indicating that the work was meant to be more than just a new physical theory of the structure of the hydrogen atom. Molecules—and thereby chemistry—were an important part of Bohr's very ambitious theory of matter, which also included aspects of radioactivity, X-rays, optics, magnetism, and more. The sequence of papers covered a broad spectrum of subjects, and only some of them successfully as seen from a later perspective.

In this chapter we shall follow the scientific path of the young Danish physicist, from his early studies at the University of Copenhagen to the publication of his so-called trilogy. Bohr did not originally have an interest in atomic structure, a subject that was foreign to him until he became a research assistant under Rutherford in the early spring of 1912. On the other hand, his research into the electron theory of metals made him well prepared to look at atoms through different eyes to his contemporaries. Rutherford's role in the creative phase of Bohr's theory is well documented and was of the utmost importance to the ambitious Dane, who in the summer of 1912 had ready a framework for his new conception of atomic structure: what is known as the Manchester memorandum. However, it took the best part of a year before he was able to hand Rutherford the complete manuscripts of his three papers. What these papers were about will be explained in the four last sections of this chapter, which direct particular attention to those aspects that related to spectroscopy and atomic chemistry.

2.1 THE TRAINING OF A YOUNG PHYSICIST

Niels Henrik David Bohr was born in Copenhagen in 1885.[2] His mother, Ellen Adler, belonged to a wealthy and influential Jewish banking family, while his father, Christian Bohr, was a promising physiologist who had done postdoctoral studies in Leipzig under the anti-vitalistic German physiologist Karl Ludwig. On the basis of his innovative studies on the physiology of respiration, in 1890 he was appointed professor at Copenhagen University. Internationally recognized for his work on the exchange of oxygen and carbon dioxide between living organisms and the environment, Christian Bohr was twice nominated for the Nobel Prize in physiology or medicine. However, in neither of the cases did he receive the prestigious prize. Young Niels Bohr was raised in an intellectually stimulating environment, which not only included his father and his circle of friends from the university and the Royal Danish Academy of Sciences and Letters, but also his younger brother Harald, with whom he had and continued to have a very close relationship. While Niels pursued a career in physics, the equally bright Harald chose pure mathematics as his calling and soon became an internationally known mathematician with strong connections to David Hilbert and his famous school of mathematics in Göttingen.

In 1903 Niels Bohr enrolled at Copenhagen University to study physics as his major subject and mathematics, astronomy, and chemistry as his minor subjects. At that time Denmark could boast only a single university professor in physics. Christian Christiansen, a colleague and friend of Christian Bohr, enjoyed an international reputation both as a textbook writer and researcher. His *Elemente der Theoretischen Physik*, which appeared in 1894, was a translation of a Danish original from 1887–1889, which marked the beginning of theoretical physics in Danish science. Written in the tradition of Kirchhoff and Helmholtz, the book was remarkable for its inclusion of the most recent international research. It appeared in a revised edition as late as 1921, although by that time it was outdated.[3] As a textbook for Danish students Christiansen wrote *Lærebog i Fysik* (*Textbook in Physics*), the third edition of which formed the basis of the course he taught young Bohr and the few other physics students in Copenhagen. His most significant contribution to physics was an experimental investigation of anomalous dispersion, a subject that came to play an important role in the electron theories of the late 1890s and later also in the new quantum theory. Another of his important experimental studies, dating from 1884, resulted in the insight that cavities with small holes provide a method to produce nearly blackbody-distributed radiation. Christiansen published these works in internationally recognized journals, such as the *Annalen der Physik und Chemie*, for which reason they attracted the attention of leading physicists including Helmholtz and Boltzmann.

Niels Bohr's view of physics was to some extent influenced by his teacher Christiansen, for whom he had great respect. As to his more general outlook and interests, he received inspiration from other sources as well. Together with his brother Harald, from whom he was inseparable, he was for several years active in a discussion group called 'Ekliptika', a circle of twelve young Copenhagen students founded in 1905. The group met regularly to read and debate a wide range of issues, many of them of a general philosophical nature. The organizer of the Ekliptika group was Edgar Rubin, who was later to become a professor of psychology. Several of the members of Ekliptika, including Niels Bohr, were influenced by the philosophy professor Harald Høffding, who at the time had won international recognition and was Denmark's best-known philosopher. Much of what Bohr knew of philosophy came to him from Høddfing, with whom he was also acquainted through his father. A great deal has been written about the controversial Bohr–Høffding relationship, especially in connection with the later complementarity principle and sometimes also the correspondence principle. The only relevance to the present context, limited as it is to early atomic theory, is the suggestion that Bohr's idea of acausal and discontinuous quantum jumps was somehow indebted to the thoughts of the famous Danish nineteenth-century existentialist philosopher Søren Kierkegaard. The influence, if there was one, was supposed to have been mediated through Høffding.[4] Suffice to say that this is a speculative hypothesis that lacks documentation as well as, in my view, plausibility. Whatever the influence of Høffding (and Kierkegaard, or other philosophers) on Bohr's later thinking, there is no reason to believe that his atomic theory of 1913 was even remotely inspired by what he happened to know of philosophy prior to that year.

While still a student in Copenhagen, Bohr published his first scientific paper, which was based on a prize problem on methods of determining the surface tension of liquids that the Danish Academy of Sciences had proposed. In early 1907 Bohr won the Academy's gold medal for his predominantly experimental research, and the following year he prepared a more elaborate version of his work for publication. His prize essay resulted in two papers, both published in the *Philosophical Transactions* of the Royal Society.[5] The first and longer of them, which appeared in early 1909, was mainly experimental, while another paper of 1910 was purely theoretical. Both of these early papers involved a heavy dose of mathematical analysis. The same year that Bohr's first paper appeared in print he completed his master's degree, which included the writing of a paper on a problem set by the professor, Christiansen. The problem that Christiansen chose for his student dealt with the application of electron theory to explain the physical properties of metals. With a satisfactory essay and equally satisfactory oral examinations, Bohr earned his master's degree at the end of 1909. By that time he had become seriously interested in the electron theory of metals, as shown by a letter to his brother in the spring of that year: 'At the moment I am wildly enthusiastic about Lorentz's (Leiden) electron theory'.[6] This was a subject he wanted to examine in depth in a forthcoming doctoral dissertation, which at the time was the necessary entrance ticket for admission to the academic world.

The electron theory of metallic conduction was established on a new basis with an ambitious theory proposed by the Leipzig physicist Paul Drude in 1900.[7] Drude assumed the existence of electric carriers of both charges in metals, some of them free and others bound; he pictured the first as constituting a kind of gas, which could be analyzed by applying the kinetic theory of gases developed by Maxwell and Boltzmann. In this way he was able to derive a theoretical expression for the Wiedemann–Franz law established on an empirical basis in 1853. According to this law the ratio of the heat conductivity K and the electrical conductivity σ was the same for all metals at the same temperature, that is, $K/\sigma \cong$ const. In 1872 the Danish physicist Ludvig Lorenz generalized the law to $K/\sigma \cong \text{const} \times T$, where T is the absolute temperature. Drude found theoretically the expression

$$\frac{K}{\sigma} = \frac{4}{3}\left(\frac{3R}{2e}\right)^2 T,$$

where R denotes the gas constant and e the elementary charge. He thus derived from theory the Wiedemann–Franz–Lorenz law and explained the empirical constant appearing in it.

While Drude assumed 'electrons' of both charges, within the next five years Hendrik Antoon Lorentz developed a more sophisticated theory on the assumption that the negative electrons were the only carriers of electric conduction. He further assumed that the positive particles (ions) remained fixed in the metal and that ion–electron collisions could be treated as collisions between elastic hard spheres. Although Lorentz's theory was mathematically attractive and more complete than Drude's, its empirical success was limited. For example, Lorentz derived the Wiedemann–Franz–Lorenz law, but unfortunately with a numerical factor (8/9 rather than 4/3) that made the agreement with measurements less precise. There were other empirical and conceptual problems with Lorentz's theory, such as there were with the alternative theories proposed by J. J. Thomson, William Sutherland, James Jeans, Richard Gans, and others. One of the most serious problems was its inability to account for the specific heats and their variation with temperature, a phenomenon which had been explained by Einstein in 1906 in terms of quantized thermal vibrations of the atoms. In Einstein's theory the electrons did not contribute to the specific heat.

Another of the problems related to the new and poorly understood quantum theory, which at the time essentially meant Planck's radiation law for heat radiation. Planck's experimentally confirmed law ought to be derivable from electron theory, or at least be compatible with it, but apparently it was not. To make things worse, in a lecture of 1908 Lorentz proved rigorously that existing electron theory must lead to the classical Rayleigh–Jeans law, which reproduces the observed blackbody spectrum only for long wavelengths.[8] By 1910 the electron theory of metals was a focal point for theoretical physicists—promising and at the same time intriguingly problematic. Bohr's choice of topic for his dissertation was just right.

On 13 May 1911 Bohr successfully defended his doctoral thesis, a thorough and densely written work that impressed his opponents and Denmark's small physics community. In accordance with the university statutes it was written in Danish and for this reason it was not well known, or known at all, by physicists outside Scandinavia. (Only in 1921 did it become possible to write dissertations in German, French, or English, and even then only by permission.) It had deserved better, for at the time it was the most complete and critical analysis of the electron theory of metals in the physics literature.

What makes Bohr's dissertation of more than passing interest is its comprehensiveness, its sharp critical perspective, and his acute awareness that important parts of classical physics were irreparably in trouble. It is also worth pointing out that it was a highly mathematical work and thus contradicts, like other of his early publications, the claim that Bohr was a poor mathematician.[9] Moreover, the dissertation is of interest because it provides information about what Bohr knew at the time and how he assessed contemporary physicists and problems in physics. For example, he was familiar with Thomson's *Corpuscular Theory of Matter* and thus also with the Thomson atomic model, although at the time there is no indication that he showed interest in the internal constitution of atoms. Thomson's book included two substantial chapters on the theory of metallic conduction, followed by a chapter on 'the arrangement of corpuscles in the atom' in which Thomson gave a thorough if largely qualitative account of his atomic theory. One can assume that Bohr read the former chapters with greater interest than the latter. Another physicist to whom Bohr referred in his dissertation was Einstein, whose papers of 1909 and 1910 signified that 'it seems impossible to explain the law of heat radiation if one insists upon the fundamental assumptions underlying the electromagnetic theory'. Bohr continued:

> This is presumably due to the circumstance that the electromagnetic theory is not in accordance with the real conditions and can only give correct results when applied to a large number of electrons (as are present in ordinary bodies) or to determine the average motion of a single electron over comparatively long intervals of time (such as in the calculation of the motion of cathode rays) but cannot be used to examine the motion of a single electron within short intervals of time.[10]

The problem with the electromagnetic theory was that it did not apply to the interaction between electrons or interactions between electrons and the radiation field. On the other hand, Bohr never doubted that the theory was valid for the radiation field itself.

The aim of Bohr's thesis was, as he said, 'to carry out the calculations for the various phenomena that are explained by the presence of free electrons in metals in as great generality as possible, while retaining the fundamental points of view underlying the theory of Lorentz'. While he admitted Lorentz's theory to be 'mathematically very perfect', he also noted that 'the physical assumptions on which it is based can hardly be expected to be valid, even approximately, for actual metals'.[11] His extension of the existing theory consisted in part in assuming that the force between ions and electrons varied with the distance according to r^{-n}, where n is arbitrary. With this generalization he was able to improve on some of Lorentz's results. More importantly, he realized that several phenomena were outside the reach of the electron theory, irrespective of its degree of sophistication and level of generality. There was something seriously wrong—something fundamental. In the introduction he wrote that there were

> . . . many properties of bodies impossible to explain if one assumes that the forces which act within the individual molecules . . . are of such a kind [mechanical]. Apart from several generally known examples of this, for instance calculations of the heat capacity of bodies and of the law of heat radiation for short wavelengths, in what follows we shall encounter another example of it, namely in our discussion of the magnetic properties of bodies.[12]

Bohr discussed at length the problems of magnetism, namely that neither free nor bound electrons contributed to the magnetic properties of matter, at least not to diamagnetism. This he considered a crucial failure, and he returned to it in the very last sentence of the thesis: 'It does not seem possible, at the present stage of development of the electron theory, to explain the magnetic properties of bodies from this theory'.[13] Bohr realized that the classical laws of physics were inadequate when applied to rapidly moving electrons, but he did not yet relate the difficulties to the structure of atoms, a topic he does not seem to have been interested in at the time.

Bohr had effectively discovered that according to classical statistical mechanics the net magnetization of a collection of electrons in equilibrium will vanish. It follows that classical physics cannot account for magnetism, whatever its form. The importance of this insight was not noticed at the time, although to Bohr it indicated that magnetism was a quantum phenomenon. Only in 1921 did the Dutch physicist Hendrika J. van Leeuwen independently rediscover what in modern physics literature is referred to as the Bohr–van Leeuwen theorem.[14]

For Bohr, now a doctor of philosophy ('dr. phil.'), it was natural to continue his studies into the electron theory abroad, and his chosen destination was Cambridge, where Thomson resided. Although critical of Thomson's electron theory of metallic conduction, he had the greatest admiration for the famous English physicist who, so he said in his dissertation, 'has contributed so much to the *experimental* foundation of electron theory'.[15] Financed by a stipend from the Carlsberg Foundation, in September

1911 Bohr arrived in Cambridge, eager to work with Thomson and secure his career in theoretical physics.

2.2 FROM CAMBRIDGE TO MANCHESTER

Bohr's stay in Cambridge did not live up to his high expectations. In an interview shortly before his death in 1962, he recalled the period as 'absolutely useless'.[16] This must be considered an exaggeration, but on the other hand there is no doubt that after a month or two his initial enthusiasm had evaporated and been replaced by a growing disappointment. J. J. Thomson, his hero, showed little interest in the ideas of the young and awkward Dane. Thomson's primary research field was no longer the electron theory of metals but exciting experiments with positive rays, which were of an entirely different nature and far from Bohr's interests. The busy director of the Cavendish Laboratory did not particularly welcome the fact that his new research student criticized his work on conduction in metals. In a long and informative letter of 1 December 1911 to Carl Wilhelm Oseen, a young Swedish physics professor with whom Bohr had close relations, Bohr pointed out a serious error in Thomson's theory: 'I mentioned it to Thomson three weeks ago. He said he would think about it, but I have not heard from him since'.[17] The Bohr brothers had first met Oseen in the summer of 1911 at a congress of Scandinavian mathematicians in Copenhagen. The Swedish physicist, who had read Niels Bohr's dissertation with interest (as a Swede he had no difficulties with the Danish language), was among the first to recognize the genius of his colleague in Copenhagen.

Nor was Thomson seriously interested in the complex arguments of Bohr's dissertation: 'Thomson has so little time; I gave him the [translation of the] dissertation when I came, but he hasn't read it yet. I have only talked with him a few moments about certain points, and I don't yet know whether he will agree with me or not'.[18] The fruitless and time-consuming efforts to get the dissertation published in English was a constant source of worry for him, both in Cambridge and during his subsequent stay in Manchester. He attempted to get it published first by the Royal Society and then by the Cambridge Philosophical Society, but none of these plans worked out.[19] In spite of his lack of fruitful interaction with Thomson, he continued to value him. Thus after having visited Rutherford he wrote in a letter to his fiancée, Margrethe Nørlund, that Rutherford was not quite of the same caliber as the Cavendish physicist: 'J. J. Thomson is a tremendously great man, and I have learnt such an enormous amount from his lectures; I like him so much . . .'[20]

Although Bohr's stay in Cambridge was far from what he had looked forward to, it was not quite as 'absolutely useless' as he recalled late in life. For one thing, he attended lectures by Thomson, Larmor, and Jeans, although at the time he did not speak with either Larmor or Jeans. More importantly, he established contacts with several physicists of the younger generation who shared some of his own research interests. One of them was Australian-born William Lawrence Bragg, who was Bohr's junior by five years and in 1915 would become the youngest Nobel laureate ever. At the time a student in Cambridge, Bragg followed some of the same courses as Bohr, including Jeans' course on statistical mechanics and radiation theory. He happened to meet Bohr after one of Jeans' lectures, a meeting he reported to his father, William Henry Bragg: 'I got an awful lot from a Dane who had seen me asking Jeans questions, and after the lecture came up to me and talked over the whole thing. He was awfully sound on it, and most interesting, his name was Böhr, or something that sounds like it'.[21] In later recollections Bragg wrote: 'J. J. Thomson gave us stimulating fireworks. I also got very excited over some lectures of Jeans, because they opened up a new world of statistical mechanics. After them a strange young man used to draw me aside and explain at enormous length just where Jeans was wrong. This was Bohr!'[22]

During his stay in Cambridge Bohr also met with Nicholson and the Birmingham physicist Samuel McLaren, both of whom had recently published papers on the electron theory of metals. Whereas Bohr praised the methods used by McLaren as 'extremely interesting', he bluntly dismissed Nicholson's paper as 'perfectly crazy'. About his conversation with Nicholson, he wrote to Oseen: 'He was extremely kind, but with him I shall hardly be able to agree about very much'.[23] At that time Nicholson had just published his atomic theory, but apparently the subject did not enter their conversation. Bohr only became aware of Nicholson's planetary theory of atoms about a year later, as we shall see in Section 2.3.

Bohr's letter to Oseen ended with a remark of considerable interest. 'At the moment I am very enthusiastic about the quantum theory (I mean its experimental aspects)', he wrote; 'but I am not yet sure if this is not because of lack of knowledge; the same I can say, but only to a still higher degree, about my relation to the magneton theory'. The 'magneton' Bohr referred to was a unit magnetic moment M_0 that the French physicist Pierre Weiss, director of the physics laboratory at the Zurich Polytechnic, had recently introduced as a way to explain his measurements of the magnetic properties of various salts.[24] Weiss's hypothesis of the magneton as a physical reality, and not merely a number, aroused immediate interest. It was discussed during the Solvay congress from 30 October to 3 November 1911, where Paul Langevin, interpreting it in terms of revolving electrons and linking it to Planck's constant, derived an expression of M_0 in rough agreement with Weiss's experimental data. The Weiss–Langevin magneton was the quantity

$$M_0 = \frac{e}{12mc}\frac{h}{2\pi}.$$

Bohr, who was aware of the Solvay discussions, may have been thinking along similar lines.[25] For an orbiting electron the magnetic moment is given by $M = (e/2mc)L$, where L is the angular momentum and e the charge of the electron. Bohr hinted at a connection between the magneton and the quantum of action, possibly by applying

Planck's relation to the kinetic energy E_k of the orbiting electron; that is, $E_k = h\nu$. Now the kinetic energy and angular momentum of a circulating electron are related as $E_k = \pi\nu L$, where ν is the orbital frequency. It then follows that the magneton is

$$M_0 = \frac{e}{2mc}L = \frac{e}{mc}\frac{h}{2\pi},$$

or twelve times the Weiss–Langevin magneton. Bohr may have seen the discreteness of M and L as a possible solution to the dilemma of the existing electron theory of metals, namely its inability to account for magnetism. What is presently known as the Bohr magneton, a quantity first defined by 20-year-old Wolfgang Pauli in 1920, is one half of the magneton considered by Bohr in 1912.[26]

In March 1912 Bohr terminated his stay in Cambridge and moved to Manchester to do research under Rutherford, whom he had first seen at a dinner in the Cavendish Laboratory in October 1911. The following month, while spending a weekend in Manchester with the physiologist James Lorrain Smith, a friend of Bohr's father, he got a chance to meet Rutherford personally.[27] On this occasion the two physicists talked about the recent Solvay conference, but not about Rutherford's nuclear atom. The transfer to Manchester was not motivated by a desire to work on atomic structure, radioactivity, or any of the other topics that Rutherford and his group cultivated so successfully. In the spring of 1912 the main focus of Bohr's mind was still on the electron theory of metals, not on the structure of the nuclear atom. On the other hand, in Manchester he also worked experimentally on radioactivity, a subject he soon came to see as closely related to the structure of the atom. In this line of work he received much help and inspiration from George von Hevesy, a Hungarian physical chemist who worked in Rutherford's group and with whom Bohr established a lifelong friendship.

'I remember, as if it were yesterday', Bohr said in his Faraday Lecture of 1930, 'the enthusiasm with which the new prospects for the whole of physical and chemical science, opened by the discovery of the atomic nucleus, were discussed in the spring of 1912 among the pupils of Rutherford'.[28] More clearly than other scientists at the time, Bohr realized that while radioactive disintegration was rooted in the nucleus, the ordinary physical and chemical properties of matter had to be explained by the surrounding system of electrons. The general view in Manchester was that beta particles had their origin in the extra-nuclear cluster of electrons,[29] but this was not a view shared by Bohr. He believed that they came from the nucleus, as he argued in public in his trilogy of 1913 (see Section 2.7). Moreover, he came to the conclusion that while in alpha decay a radioactive element descends in the periodic system by two units, in beta decay it raises by one unit (in terms of the atomic number, $\Delta Z = -2$ and $\Delta Z = 1$). At about the same time similar ideas were suggested by Soddy, Fajans, and a few other scientists, but without associating them with Rutherford's nuclear model as Bohr did.

On the other hand, Hevesy, who in the autumn of 1912 investigated the chemical properties of elements involved in radioactive transformations, was acquainted with and sympathetic to the ideas of Rutherford and Bohr. He was consequently inclined to interpret his results in accordance with these ideas. In January 1913 he wrote to Bohr regarding a new paper he had written on the valencies of radioactive elements: '*Where* do these [alpha and beta] particles come from, was the question you were so much occupied with.... My results seem to support the view that they stem from the interior of the atoms'.[30] He further commented in a somewhat surprising way on Bohr's new version of the hydrogen atom. Hevesy could understand how an atom with many electrons arranged in rings could be in an equilibrium state, but 'How can a positive atomic nucleus and *one* rotating electron be in equilibrium? I cannot imagine at all what would prevent the electron from falling into the centre of the atom?' In his letter of reply Bohr expressed his agreement with Hevesy's suggestion that radioactivity was purely a nuclear phenomenon:

> The beautiful results were exactly what I had expected according to the point of view I hold on the constitution of the atoms... By the phenomena of radioactivity we observe an explosion of the nuclei... the chemical and physical properties of the new elements formed will only depend of the charge of the new nuclei; which latter again will depend on the charge of the rays expelled. The latter relation is just the one you have found in your experiments; and your results were therefore what I had expected and hoped.[31]

In a letter to Fajans from the spring of 1913, Rutherford informed him about the ongoing work of the Danish physicist: 'Bohr of Copenhagen has been working at the general theory of atoms built on my model and appears to have made definite progress.... He is a very capable fellow, and there will soon appear a very interesting paper by him in the Phil. Mag'.[32] In fact, it was only after having studied Bohr's very interesting paper that Fajans became convinced of the truth of the nuclear atom. In a letter to Rutherford of 13 December 1913 he said: 'I have followed Bohr's papers with extraordinary interest, and now I no longer doubt the complete correctness of your atomic theory. The reservations I expressed in my last letter have been entirely removed by Bohr's work'.[33]

It is difficult to know precisely what Bohr's thoughts were on these matters since he only published them over a year later, when he included them in the second part of his great trilogy. Concerning the relationship between radioactivity and the periodic system, Hevesy said in a letter of October 1913: 'Though [Alexander] Russell was already interested in the problem and I started the valency experiments when Bohr came to Manchester, no doubt he encouraged us both very much and if we trace the origin of the above ideas to their origin, we will find them in Bohr's mind, as pointed out to me by himself in his usual modest way'.[34] In fact, Bohr wanted to publish his innovative ideas in 1912, but was dissuaded by Rutherford who 'warned with characteristic caution against overstressing the bearing of the atomic model and extrapolating from comparatively meagre experimental data'.[35]

What really changed Bohr's mind in the direction of atomic structure was not so much radioactivity as it was absorption theory. The change was to a large extent caused by a paper on the absorption of alpha particles written by Charles Galton Darwin, a 25-year-old theorist working with Rutherford (and a grandson of *the* Charles Darwin, of evolution fame). The aim of Darwin's paper was to examine theoretically the energy loss of alpha particles as they were absorbed in or scattered on thin metal foils and air. Naturally for a Manchester physicist (but not for other physicists) he adopted Rutherford's nuclear model, which 'supposes the atom to consist of a cluster of electrons held by an unknown field of forces round a central charge'.[36] As to the cluster of electrons he made parallel calculations assuming that they were either distributed homogeneously in a spherical atomic volume or over the surface of the sphere. Darwin's analysis rested on a couple of further assumptions, in particular that the atomic electrons could to a good approximation be considered to be free. Contrary to scattering, which was thought to be due to the nucleus, the free electrons were chiefly responsible for the absorption.

The result of Darwin's analysis was a formula for the loss of energy that included two atomic parameters: the atomic radius and the number of electrons outside the nucleus. Comparing the formula with experimental data he obtained a rough agreement, but not one that was quantitatively convincing. For example, he found atomic radii that decreased with atomic weight and differed substantially from the commonly accepted values derived from the kinetic theory of gases. Assuming the hydrogen atom to contain only one electron, his formula resulted in an atomic radius of about 5×10^{-8} cm, an order of magnitude too large.

In a letter to Harald of 12 June Bohr reported that Darwin's theory was 'very unsatisfactory in the basic conception', although not in its reliance on the nuclear atom to which he referred for the first time: 'In the last few years he [Rutherford] has worked out a theory of atomic structure which seems to have a much more solid basis than anything that we had formerly'.[37] Only a week later, in another letter to his brother, Bohr explained that he was now working on atomic structure and what he had found so far was 'perhaps a little bit of reality'.[38] Sensing that his ideas were important and that other physicists might be on the same track, he was 'eager to finish it in a hurry'. However, his critical improvement of Darwin's absorption theory proved more difficult than expected and only reached completion after he had left Manchester and returned to Copenhagen. The paper, communicated to *Philosophical Magazine* by Rutherford, was dated August 1912 but it took until January 1913 before it appeared in print.

Bohr objected to Darwin's theory that the loss of velocity of an alpha particle would depend on the motion of the electrons during the collision and, for this reason, the electrons could not be considered as free. Instead he treated the electrons as bound to the nucleus by an elastic force, hence oscillating with characteristic frequencies. As he saw it, it should be possible to get 'information about the internal structure of atoms' from the experimentally determined velocity loss of alpha particles, namely information in the form of frequencies. Importantly, he related the frequency of the oscillators to Planck's quantum hypothesis:

> According to Planck's theory of radiation we further have that the smallest quantity of energy which can be radiated out from an atomic vibrator is equal to $\nu \cdot k$, where ν is the number of vibrations per second and $k = 6.55 \cdot 10^{-27}$. This quantity must be expected to be equal to, or at least of the same order of magnitude as, the kinetic energy of an electron just sufficient to excite the radiation.[39]

The quantity k is obviously Planck's constant, which was usually denoted h. From an empirical point of view Bohr's theory was an improvement over Darwin's, although its agreement with experiments was far from perfect. Relying on data for dispersion and refraction in gases obtained by Clive and Maude Cuthbertson at University College, London, he concluded 'with great certainty' that a hydrogen atom contained only one electron and helium two electrons. This he interpreted as support of Rutherford's nuclear atom:

> The value $r = 2$ for the number of electrons in a helium atom, indicated by experiments on dispersion and on absorption of α-rays, is what we, adopting Rutherford's theory of atoms, necessarily must conclude from the behaviour of α-rays, according to which helium atoms formed from α-particles will contain 2 electrons outside the central nucleus.[40]

Bohr was aware that this number of electrons in helium did not agree with the value inferred from Drude's dispersion theory of 1904, from which he calculated $r = 2.3$. However, as he wrote in a letter to Clive Cuthbertson: 'As to the question of the interpre[ta]tion of the result according to Drude's theory, I am of the opinion... that the number of electrons per atom calculated from the theory need not to be whole numbers, as the theory in question uses assumptions as to the motions of the electrons which hardly are fulfilled in the actual atoms'.[41] By that time Bohr had reached the conclusion that atomic electrons did not perform mechanical oscillations, as assumed in Drude's theory, but that dispersion was a quantum phenomenon to be explained in terms of transitions between stationary energy states.

Apart from hydrogen and helium, for the lighter atoms Bohr obtained reasonably good values for the number of electrons, such as 14 for aluminium and 18 for molecular oxygen, to be compared with $A/2 = 13.5$ and $A = 16$. For the heavier elements the agreement was less satisfactory: tin = 38, gold = 61, and lead = 65, as compared to $A/2 = 59.3$, 98.5, and 103.5, respectively.

2.3 THE MANCHESTER MEMORANDUM

In late June and early July 1912, shortly before returning to Copenhagen, Bohr prepared a draft document in which he summarized in a rather sketchy fashion his most recent ideas about atomic structure. He may have sent it to Rutherford on 6 July, for a short letter of that date says that 'I send the remarks concerning the structure and stability of molecules for which you kindly asked'.[42] The handwritten notes, consisting of six sheets of paper and often referred to as the Manchester memorandum, were evidently written in haste. They carried the title 'On the Constitution of Atoms and Molecules', the very same title that he chose for the trilogy of articles that appeared in *Philosophical Magazine* a year later.[43] In this important document the structure and stability of atomic systems were indeed the key issues.

Bohr started by briefly outlining Rutherford's atomic model, or rather his own version of it—a positive nucleus surrounded by one or more rings of electrons. (Rutherford had not suggested a ring model.) He then noticed that in such an atom mechanical stability requires the electrons to be rotating around the nucleus. 'By an analysis analogous to the one used by Sir J. J. Thomson', he said, 'it can very simply be shown that a ring [such] as the one in question possesses no stability in the ordinary mechanical sense . . . and the question of stability may [must] therefore be treated from a quite different point of view'.[44] It should be emphasized that the 'question of stability' dealt with by Bohr in the memorandum was restricted to mechanical stability. He did not mention the radiation instability caused by the accelerating electrons, although he was undoubtedly aware of it. Radiative instability would soon become an important issue, but by the summer of 1912 this was not yet the case (Figure 2.1).

Not only did Bohr find that the Rutherford ring atom was mechanically unstable, he also noticed that it had no characteristic radius to define the size of the atom. 'There is nothing to determine a quantity of dimension, a "length"', and nothing 'to allow from mechanical considerations to discriminate between the different radii and times of vibration [rotation].' Considering a ring determined only by the charge of the nucleus and the number of electrons, he wrote that it 'can rotate with an infinitely great number of different times of rotation, according to the assumed different radius of the ring; and there seems to be nothing . . . to allow from mechanical considerations to discriminate between the different radii and times of vibration'. This was an important point, for all experience indicates that the size of real atoms does not vary arbitrarily, and nor does the frequency of the radiation they emit. Without providing calculations, Bohr discussed how many electrons a single ring could contain before its became unstable. It is 'immediately seen', he said, that the number is seven, for this is the largest number for which the total energy of an electron (the sum of the kinetic energy and the potential energy) remains negative and thus allows it to be in a bound state with the nucleus.

> This, together with the fact that inner rings of electrons in Prof. Rutherford's atom-model will have only very little influence (and always to the worse) on the stability of outer rings seems to be a very strong indication of a possible explanation of the periodic law of the chemical properties of the elements (the chemical properties are assumed to depend on the stability of the outermost ring, the 'valence electrons') by help of the atom-model in question.

That is, Bohr argued that because the addition of electrons within a ring would reduce its stability, the building up of the electron system would at some stages occur with the formation of a new external ring. This idea connected the outermost electrons with the valence of the elements in an attractive way, in contrast to the case in Thomson's

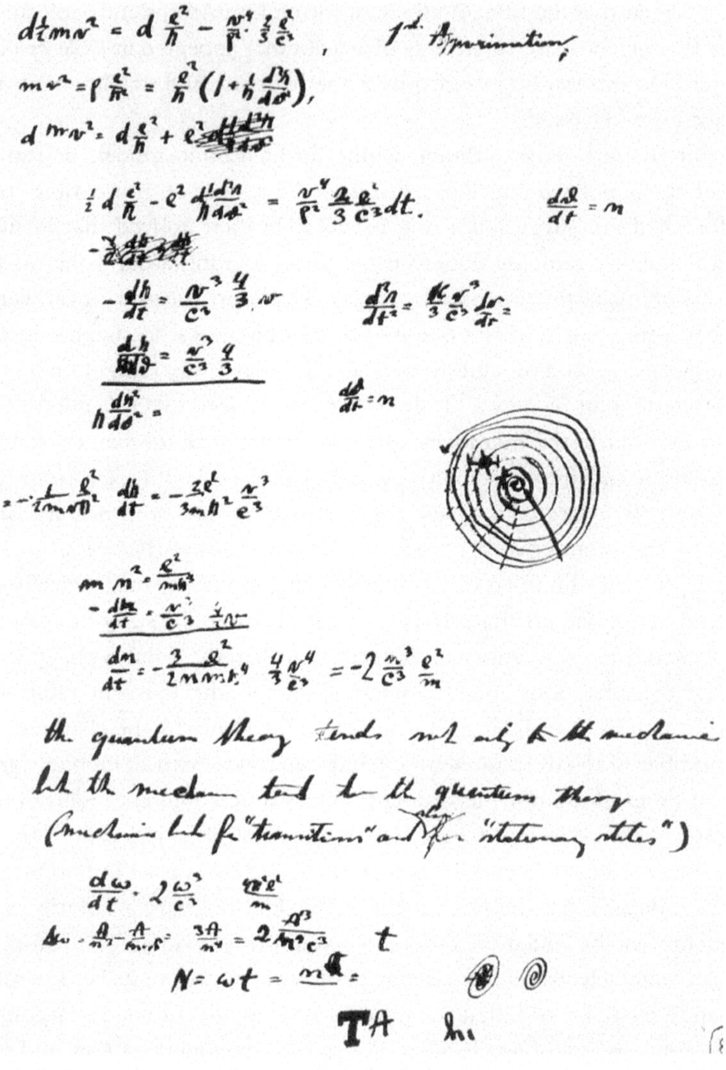

Fig. 2.1. Bohr's draft calculations from 1913 of the radiation emitted by a rotating electron, causing it to spiral towards the nucleus. The sentence reads: 'The quantum theory tends not only to the mechanics but the mechanics tend to the quantum theory (mechanics both for "transitions" and for "stationary states")'.

model of the atom. Recall that in the Thomson atom valency and other chemical properties were assumed to depend on the number of electrons in the internal rings. Bohr realized the advantage of his scheme over that of Thomson, as he explained in a footnote: 'The difference in this respect between the atom-model considered and J. J. Thomson's atom-model is very striking, and seems to make it impossible to give a satisfactory explanation of the periodic law from the last mentioned atom-model'.

With regard to the problem of fixing a definite size of the atom, Bohr felt it necessary to introduce a new hypothesis:

> This hypothesis is: that there, for any stable ring (any ring occurring in the natural atoms), will be a definite ratio between the kinetic energy of any electron in the ring and the time [frequency] of rotation. This hypothesis, for which there will be given no attempt of a mechanical foundation (as it seems hopeless), is chosen as the only one which seems to offer a possibility of an explanation of the whole group of experimental results which gather about and seem to confirm conceptions of the mechanism of the radiation as the ones proposed by Planck and Einstein.

In agreement with what he had already stated in his dissertation of 1911, he added that classical mechanics was known to fail in the domain of atoms: 'It seems to be rigorously proved that the mechanics is not able to explain the experimental facts in problems dealing with single atoms'. Bohr explained his special hypothesis in connection with his discussion of the hydrogen molecule, stating it in the Planck-like form

$$E = K\nu,$$

where E denotes the kinetic energy of an electron moving in a ring with radius r, given by $E = \frac{1}{2}mv^2 = \frac{1}{2}m(2\pi\nu r)^2$. K is a new constant and ν denotes the electron's frequency of revolution. Bohr's relation $E = K\nu$ is of course very similar to Planck's formula, but nowhere in the memorandum did Bohr indicate the value of K; nor is there any reason to suspect that he knew it, except that he assumed $K \approx h$. However, it is quite likely that he adopted the relationship $K \cong 0.6h$ and that he based it on empirical data.[45] Only in his paper of 1913 did he provide theoretical arguments that $K = \frac{1}{2}h$.

For a ring in a state of equilibrium the centrifugal and attractive forces must be the same, meaning that

$$m(2\pi\nu)^2 r = X\frac{e^2}{r^2},$$

where $X = 1$ if there is only a single electron and $X > 1$ for more electrons. For example, Bohr calculated $X = 1.75$ for the helium atom and $X = 1.049$ for the hydrogen molecule. The equation does not determine the radius or the frequency, but only the product $r^3\nu^2$ of the two quantities. Introducing the $E = K\nu$ hypothesis yields expressions for r and ν that depend only on the constant K, the computable factor X, and the atomic constants m and e. It also leads to an expression for the work required to remove an electron from the ring, namely

$$W = X^2\frac{me^4\pi^2}{2K^2}.$$

Bohr did not write down this expression, but he stated that for one mole of matter 'the energy of a system containing a ring of n electrons acted on by a central force e^2/r^2X is equal to $-nX^2A$, where A is approximately equal to 1.3×10^{-11} erg and $NA = 1.9 \times 10^5$ cal'. The quantity N denotes Avogadro's number, close to 6×10^{23}, and 1 erg = 10^{-7} J = 0.24×10^{-7} cal. The formula enabled him to calculate the heat released when two moles of hydrogen atoms combine into one mole of diatomic molecules according to $H + H \rightarrow H_2$. The value he obtained, 3.8×10^4 cal, was of the right order of magnitude. Using the same kind of reasoning, he calculated that the corresponding process for helium, $He + He \rightarrow He_2$, would be endothermic by about 3×10^5 cal, implying that diatomic helium molecules cannot exist in nature. This agreed with experimental knowledge, as no such molecules had ever been detected. Bohr further proposed electronic structures—'figurations' as he called them—for the molecules O_2 and O_3, and also for the compounds H_2O, CH_4, and C_2H_2, but in these cases only in qualitative versions (see also Sections 2.4 and 2.6).

The stability and heat of formation of simple compounds such as H_2 constituted one class of phenomena that Bohr thought provided support for his ideas of atomic structure. Ever careful to connect his theoretical ideas with experimental data he cited three other relevant relations:

1. the periodic variation of the atomic volume, meaning the ratio between atomic weight and density, that the German chemist Lothar Meyer had first demonstrated in 1870—this variation could be qualitatively understood on the basis of Bohr's assumption that external electron rings were successively added to the atom as it grew heavier;
2. 'Whiddington's law', due to the Cambridge physicist Richard Whiddington, according to which the minimum velocity of electrons capable of exciting characteristic X-rays was proportional to the element's atomic weight;
3. 'Bragg's law', named after W. Henry Bragg, that the stopping power of metal foils for alpha rays was approximately proportional to the square root of the atomic weight.

Bohr did not account for these laws in the memorandum, but he would return to them in his published paper of the following year. Parts II and III of his trilogy were to a large extent elaborations of issues dealt with in the 1912 memorandum to Rutherford.

In general the Manchester memorandum was full of ideas, which were, however, sketches and proposals rather than explicitly worked out. The only exception was the case of the hydrogen molecule, which at the time could not be compared with experiments on the heat of formation. The main message of the memorandum was that the stability of the atom required the introduction of quantum considerations, as preliminarily given by the $E = K\nu$ hypothesis.

Soon after having returned to Copenhagen Bohr became the teaching assistant to Martin Knudsen, the new professor of physics who had been appointed after Christiansen's retirement. Knudsen was internationally known for his experimental investigations of gases at very low pressure and one of the chosen few who had been invited to the 1911 Solvay meeting where the quantum theory had been the main issue. However, he had no interest in the new quantum theory or, for that matter, in other areas of theoretical physics that occupied the mind of Bohr, some fourteen years younger.[46] Busy with his duties as lecturer and teaching assistant, Bohr had little time to develop the ideas that had germinated during his fruitful stay in Manchester. In a letter to Rutherford of 4 November 1912 he expressed the hope of finishing his 'paper on the atoms' in a few weeks, but this turned out to be much too optimistic. Among the reasons for the delay were not only his teaching obligations, but also that he kept changing the plan and content of his paper as he came across new experiments and theories of relevance to it.

One of the problems concerned the number of electrons in the simplest atoms and their relation to dispersion data, about which he wrote to Rutherford on 6 July 1912: 'I saw some difficulties in the explanation of the experimental results concerning the dispersion, which I had not seen before'.[47] The difficulties were not easily solved, and Bohr returned to them four months later:

> I have made some progres[s] with regard to the question of the dispersion. The number of electrons in a Hydrogen- and a Helium-atom calculated from the dispersion seems thus to work out nearer to respectively 1 and 2, if the forces acting on the electrons are assumed to vary inversely as the square of the distance, than if they, as in Drude's theory, are assumed to be of the elastic type. I have however just in this calculation met with some serious trouble arising from the instability of the system in question which has not allowed the execution of the calculation in the extension desirable.[48]

Bohr found that the use of a Coulomb force instead of an elastic force was insufficient to give the right number of electrons in both hydrogen and helium, and it took him some time to obtain a reasonably satisfactory solution. 'My suggestion is now', he reported to Rutherford in a letter of June 1913, 'that the frequency of vibration perpendicular to the plane of the ring (which frequency according to the theory can be calculated on the ordinary mechanics) in a hydrogen-molecule is of the same order of magnitude as the frequencies in the plane of the ring, but in a helium-atom much greater and therefore of no sensible influence on the dispersion'.[49]

Importantly, at the end of 1912 Bohr became aware of Nicholson's atomic theory, realizing its disturbing similarity to his own view of the constitution of the atom. At first he thought that the two theories could not both be right, presumably because the radii and frequencies deduced by Nicholson were so very different from those he found on his own model. However, he soon reached a different conclusion. As he wrote in a Christmas postcard to his brother Harald, Nicholson's theory was not incompatible with his own but rather stood in a complementary relationship to it. 'For, the latter's [Bohr's] calculations should be valid for the final or classical state of the atoms, while Nicholson seems to be concerned with the atoms while they radiate, i.e., while the electrons are about to lose their energy, before they have occupied their final positions.'[50] This is also what he said in a letter to Rutherford about a month later, where he pointed out the similarities and dissimilarities between the two conceptions of the nuclear atom:

> In his calculations, Nicholson deals, as I, with systems of the same constitution as your atom-model; and in determining the dimensions and the energy of the systems he, as I, seeks a basis in the relation between the energy and the frequency suggested by Planck's theory of radiation. The state of the systems considered in my calculations are however—between states in conformity with the relation in question—characterized as the one in which the systems possess the smallest possible amount of energy, i.e. the one by the formation of which the greatest possible amount of energy is radiated away. It seems therefore to me to be a reasonable hypothesis, to assume that the state of the systems considered in my calculations is to be identified with that of the atoms in their permanent (natural) state . . . According to the hypothesis in question the states of the systems considered by Nicholson are, contrary, of a less stable character; they are states passed during the formation of the atoms, and are states in which the energy corresponding to the lines in the spectrum characteristic for the element in question is radiated out.[51]

Bohr emphasized the difference between Nicholson's theory, which was essentially concerned with the spectra and formation of atoms, and his own theory dealing with the permanent state of atoms: 'I do not at all deal with the question of calculation of the frequencies corresponding to the lines in the visible spectrum'. Considering the nature and content of Bohr's theory as it appeared in print half a year later, this is a most remarkable statement. But notice that Bohr does not speak of spectral lines in general, only of those in the visible part of the spectrum.[52]

Only in mid-February 1913 did Bohr realize that his ideas of atomic constitution could be fruitfully extended to a theory of light emission. A letter to Hevesy dated 7 February reveals no sign of change but merely summarizes Bohr's ideas as stated in the Manchester memorandum. Although in this letter Bohr relates the energy of the radiation emitted during the binding of an electron and 'the frequency of rotation of the electron considered in its final orbit' in the form $E = h\nu$, this does not amount to a quantum theory of light emission. For Bohr adds that the constant in front of the frequency 'is not exactly equal to Planck's constant, but differ[s] from it by a numerical factor'.[53] Thus in the early days of February 1913 Bohr's picture of the radiation mechanism was still essentially classical and in line with the commonly accepted view. This view was that the spectral lines are produced by electrons vibrating or rotating with the same frequencies that appear in the spectrum. The mechanical frequencies of atomic oscillators—at the time generally taken to be electrons—correspond to the optical frequencies. Bohr's view was also fairly orthodox in another respect, since he thought that atoms must be ionized or dissociated before they would emit radiation.

'As soon as I saw Balmer's formula, the whole thing was immediately clear to me'.[54] This was what Bohr told Léon Rosenfeld, his later assistant and close collaborator, and what he said at several other occasions. The spectral laws of Balmer and Rydberg were well known and frequently discussed during the first decade of the century. They not only appeared in scientific papers but also in textbooks, one of them being the textbook by Christiansen that Bohr used while a student (Figure 2.2).[55] Here the formula of the Swiss schoolteacher Johann J. Balmer, dating from 1884, was stated in the standard form

$$\lambda = A\frac{n^2}{n^2 - 4},$$

with $A = 3647$ Å and $n = 3, 4, \ldots$ It is therefore remarkable that Bohr insisted that he did not know about the formula before he was made aware of it by his colleague Hans Marius Hansen in February 1913. In an interview shortly before his death, he recalled:

> I didn't know anything about the spectral formulae. Then I looked it up... I discovered it, you see. Other people knew about it, but I discovered it for myself. And I found then that there was this very simple thing about the hydrogen spectrum. I was just reading the book of Stark, and at that moment I felt now we will just see how the spectrum comes.[56]

Bohr might not have 'known' Balmer's formula in the sense that he took notice of it as relevant for his own work, but it is most unlikely that he was literally unaware of it. Although he had undoubtedly seen the formula, he might have forgotten about it. At any rate, as far as we can tell it was during a conversation with Hansen that he realized its significance. According to one of the versions of Balmer's formula, the frequencies of the visible lines in the hydrogen spectrum can be written as

$$\nu = cR\left(\frac{1}{2^2} - \frac{1}{n^2}\right), \; n = 3, 4, \ldots$$

where R is the empirical Rydberg's constant. Around 1910 it was known that Balmer's formula could be extended to other series in the hydrogen spectrum, and some spectroscopists followed the Swiss physicist Walther Ritz in suspecting a generalized formula of the form $\nu = cR(1/m^2 - 1/n^2)$, $n > m$.[57] Balmer's formula was an eye-opener to Bohr, who immediately realized that it could be interpreted as the difference between two energy terms. Not only did Balmer's formula indicate as much in the case of hydrogen, but Rydberg had also shown that the frequencies of various spectral series could be expressed as the difference between two spectral terms, namely in the form

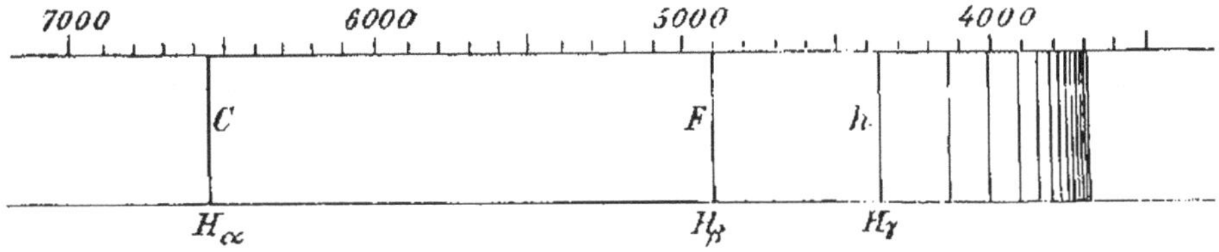

Fig. 2.2. The Balmer spectrum as reproduced in the 1910 edition of C. Christiansen's textbook in physics, which Bohr had studied. The same illustration was used in Kramers and Holst 1923, p. 58. *Source*: Christiansen 1910, p. 354.

$$\nu = \frac{cR}{(1+p_1)^2} - \frac{cR}{(m+p_2)^2},$$

where $m = 2, 3, 4, \ldots$ and p_1 and p_2 are constants that characterize the series and the element in question. Formulae of this kind were discussed in Stark's *Atomdynamik*, where Bohr may have met them. We do not know precisely how he reasoned,[58] but as early as 6 March he sent a letter to Rutherford that included a revised first part of his paper. In contrast to the earlier drafts and letters, the new version was 'mainly dealing with the problem of emission of line spectra', as he informed Rutherford. Bohr continued:

> I have tried to show that it from such a point of view seems possible to give a simple interpretation of the law of the spectrum of hydrogen, and that the calculation affords a close quantitative agreement with experiments. (I have given reasons which show, that if the foundation of the theory is sound, we may assume that
>
> $$\frac{2\pi^2 me^4}{h^3} = 3.290 \times 10^{15}$$
>
> Putting your value $e = 4.65 \times 10^{-10}$, I get $h = 6.26 \times 10^{-27}$. Putting Millikan's value $e = 4.87 \times 10^{-10}$, I get $h = 6.76 \times 10^{-27}$. Unfortunately, however, Planck's constant is hardly known with any great accuracy.) The second chapter deals with the atoms, the third with molecules, and the last chapters with magnetism and some general considerations. I hope you will find that I have taken a reasonable point of view as to the delicate question of the simultaneous use of the old mechanics and of the new assumptions introduced by Planck's theory of radiation.[59]

With the new insight into the origin of line spectra, Bohr's theory differed even more markedly from Nicholson's than earlier. However, he suggested that perhaps the celestial spectra investigated by Nicholson were not the result of true emission of light but rather due to scattering of radiation. In that case, 'Nicholson's theory would fit exceedingly well in with the considerations of my paper', he reported to Rutherford.[60] The final version of the paper, only completed after Bohr had gone to Manchester to discuss it with Rutherford, was submitted to *Philosophical Magazine* on 5 April. This was the first part of the sequel of three papers that is collectively known as Bohr's trilogy.

2.4 QUANTUM JUMPS

Bohr's trilogy did not start with the two later postulates that were to become so famous, which he only introduced after he had offered the first of his derivations of the energy of an atomic state.[61] Instead he prefaced the first of the papers by emphasizing that all his considerations rested on Rutherford's nuclear atom, according to which 'the atoms consist of a positively charged nucleus surrounded by a system of electrons kept together by attractive forces from the nucleus'. As mentioned above, this view of the atom was at the time only accepted by a minority of physicists outside Manchester. In contrast to the Thomson atom, the electrons in Rutherford's atom were not in mechanically stable states. This problem Bohr associated with another one, namely that whereas Thomson's model included a natural length as given by the radius of the positive sphere, Rutherford's did not, because it is not possible to determine an atomic length solely by the charges and masses of electrons and nuclei.[62] As Bohr saw it, this necessitated the introduction of 'a quantity foreign to the classical electrodynamics, i.e. Planck's constant, or as it often is called the elementary quantum of action'. He reasoned that, 'By the introduction of this quantity the question of the stable configuration of the electrons in the atoms is essentially changed, as this constant is of such dimensions and magnitude that it, together with the mass and charge of the particles, can determine a length of the order of magnitude required'.[63]

To establish the mechanism of how an electron became bound to a nucleus, Bohr considered the simplest atom of the Rutherford type, an electron (charge $-e$) revolving with orbital frequency ω around a nucleus of charge E in a stationary orbit with a major axis of $2a$. Assuming that the electron, in spite of its accelerated motion, radiated no energy, he found from classical mechanics the frequency and major axis as given by the total energy W of the atom; that is, the energy necessary to remove the electron to a distance infinitely far from the nucleus. The expressions

$$\omega = \frac{\sqrt{2}}{\pi}\frac{W^{3/2}}{eE\sqrt{m}} \quad \text{and} \quad 2a = \frac{eE}{W} \tag{1}$$

show that all possible values for ω and a can be obtained by varying W. However, the distinct spectral lines indicate that a and ω assume definite values, implying that the energy of atoms must be restricted to definite values as well. Bohr next noted that according to Maxwellian electrodynamics the electron will radiate energy in agreement with Larmor's formula, with the result that W increases and a diminishes until the electron coalesces with the nucleus. The energy dissipated in this way, he noticed, 'will be enormously great compared with that radiated out by ordinary molecular processes'.[64] This was a problem that had played no role in either the Manchester memorandum or other of his earlier sketches, but now it became of pivotal significance.

As a remedy against the potential disaster of atomic radiative instability he appealed to Planck's quantum hypothesis in the version Planck had developed during the years 1911–1912, a version known as Planck's 'second theory'. Assuming—but not justifying—that during the binding of an electron a monochromatic radiation of frequency $\nu = \omega/2$ is emitted (ω is the orbital frequency in 'the final orbit'), he reasoned that 'the amount of energy emitted by the process considered is equal to $\tau h\nu$, where h is Planck's constant and τ an entire number' (p. 5). This he formulated as an energy quantization for stationary orbits,

$$W = \tau h \frac{\omega}{2} \tag{2}$$

From equation (1) he then obtained the values of W, ω, and a in terms of atomic constants:

$$W = \frac{2\pi^2 me^2 E^2}{\tau^2 h^2}, \qquad \omega = \frac{4\pi^2 me^2 E^2}{\tau^3 h^3}, \qquad a = \frac{\tau^2 h^2}{4\pi^2 meE} \tag{3}$$

That is, according to Bohr even the simplest atom should be capable of existing in an infinite number of distinct states, characterized by varying sizes and amounts of energy (Figure 2.3). Atoms existing in states given by high quantum numbers (values of τ) might even be fairly large, say with radii about 0.01 mm. This was a picture of the atom quite different from what physicists had been used to. The large hydrogen atoms described by Bohr's formula were later called 'Rydberg atoms'. Isolated Rydberg atoms were first observed in 1965, when scientists at the National Radio Astronomy Observatory in the USA detected radiation from hydrogen atoms in interstellar space undergoing transitions between levels near $\tau = 100$.

We are, Bohr said, 'led to assume that these configurations will correspond to states of the system in which there is no radiation of energy; states which consequently will be stationary as long as the system is not disturbed from outside'. In the case of the ground

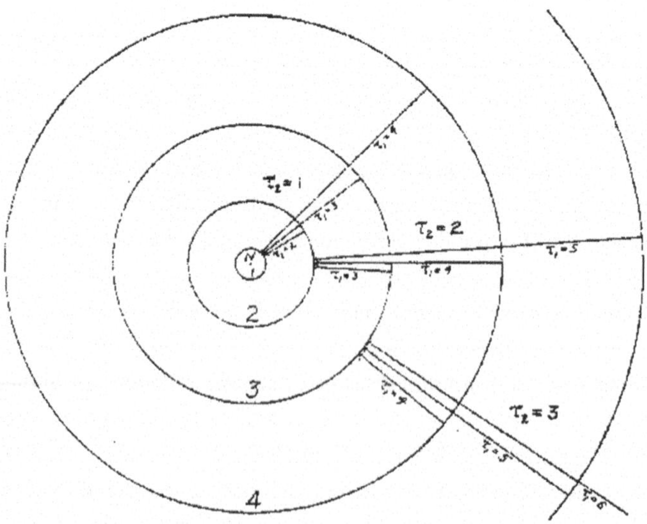

Fig. 2.3. Bohr's atomic model of hydrogen, possibly the first picture of its kind, as depicted in a paper of early 1915.

Source: Harkins and Wilson 1915, p. 1406.

state $\tau = 1$ he found the numerical values $a = 0.55 \times 10^{-8}$ cm, $\omega = 6.2 \times 10^{15}$ s^{-1}, and $W = 13$ eV. The values of a and ω were of the same order of magnitude as the linear dimension of atoms and the optical frequencies, respectively. More interestingly, W corresponded to the ionization energy of hydrogen, a quantity which at the time was only roughly known from experiments. As Bohr pointed out in the second part of the trilogy, Thomson had recently measured $W \cong 11$ eV in experiments with positive rays, a value that a few years later turned out to be inferior to the one predicted by Bohr's atomic theory.[65]

Before proceeding to the hydrogen spectrum Bohr dealt at some length with Nicholson's atomic theory, which he praised for its precise predictions of the spectral lines supposed to originate in the hypothetical nebular and coronal elements. However, he also raised serious objections against the theory, in particular that it rested on the classical assumption that the radiated frequency was the same as a mechanical frequency in the atomic system:

> As a relation from Planck's theory is used, we might expect that the radiation is sent out in quanta; but systems like those considered, in which the frequency is a function of the energy, cannot emit a finite amount of a homogeneous [monochromatic] radiation; for, as soon as the emission of radiation is started, the energy and also the frequency of the system is altered.[66]

Later in the paper Bohr, adopting a reconciliatory strategy, suggested that the difficulties of Nicholson's theory 'may be only formal'. Much as he had done in his earlier letters to Rutherford, he now proposed that the spectra studied by Nicholson were not due to emission but to scattering of radiation. In that case, 'we immediately understand the entirely different form for the laws connecting the lines discussed by Nicholson and those connecting the ordinary line-spectra considered in this paper'.[67]

In Bohr's original interpretation of equation (2) it meant that the atom emitted τ quanta of the same frequency $h\omega/2$, which would, however, imply a monochromatic radiation in contradiction of experience. Having critically examined Nicholson's theory he now realized that the radiation process must yield a *single* quantum corresponding to the reinterpretation

$$W = \tau\left(h\frac{\omega}{2}\right) = h\left(\tau\frac{\omega}{2}\right).$$

Thus, during the binding of a free electron to orbit number τ only one energy quantum of frequency $\tau\omega/2$ is emitted. Bohr now formulated what he called the two 'principal assumptions' upon which his own theory rested, namely:

> 1. that the dynamical equilibrium of the systems in the stationary states can be discussed by help of the ordinary mechanics, while the passing of the systems between different stationary states cannot be treated on that basis;

2. that the latter process is followed by the emission of a *homogeneous* radiation, for which the relation between the frequency and the amount of energy emitted is the one given by Planck's theory.[68]

As to the second assumption or postulate, Bohr remarked that it is 'in obvious contrast to the ordinary ideas of electrodynamics, but appears to be necessary to account for experimental facts'. The derivation of Balmer's formula was now an easy matter, as he just had to write down the energy difference between two stationary states and equal it to a quantum of energy,

$$W(\tau_2) - W(\tau_1) = h\nu. \tag{4}$$

For the hydrogen atom, where $E = e$, the result is

$$\nu = \frac{2\pi^2 me^4}{h^3}\left(\frac{1}{\tau_2^2} - \frac{1}{\tau_1^2}\right). \tag{5}$$

Two consequences followed immediately. Not only does the formula reproduce the ordinary Balmer series for $\tau_2 = 2$ and $\tau_1 = 3, 4, \ldots$ and the Paschen series in the ultraviolet for $\tau_2 = 3$ and $\tau_1 = 4, 5, \ldots$, it also predicts new series in the extreme ultraviolet ($\tau_2 = 1$) and infrared ($\tau_2 > 3$) regions 'which are not observed, but the existence of which may be expected'. In addition, what had hitherto been an empirical quantity, the Rydberg constant, was now derived theoretically as $cR = 2\pi^2 me^4/h^3 = 3.290 \times 10^{15}\ \mathrm{s}^{-1}$, the same value Bohr had reported to Rutherford in his letter of 6 March. The numerical value of Rydberg's constant was $R = 109\,740\ \mathrm{cm}^{-1}$, in excellent agreement with the measured value $R = 109\,675\ \mathrm{cm}^{-1}$. (The two numbers are identical to the accuracy within which the constants m, e, and h were known at the time.) As we shall see, many physicists were impressed by Bohr's derivation of Rydberg's constant, which according to Bertrand Russell 'was perhaps the most sensational evidence in favour of Bohr's theory when it was first published'.[69]

Bohr also came up with a clever explanation of why only 12 of the Balmer lines had been found in experiments with vacuum tubes, while astronomers had observed as many as 33 lines. The proposed explanation, based on the increased size of the atoms when in highly excited states, as given by equation (3), led him to conclude that spectral lines arising from hydrogen atoms in high quantum states required a large space of very small density. In the rarefied atmosphere of the stars there must exist hydrogen monster-atoms about thousand times as large as those of Earth ($2a = 1.3 \times 10^{-5}$ cm, corresponding to $\tau = 33$), a result of the large volume and low density. He therefore predicted that the large number of Balmer lines found in the emission spectra of celestial objects would never be observed in the laboratory, although it might be possible to detect the absorption lines.

Realizing that the relation $W = \frac{1}{2}\tau h\omega$ was conspicuously ad hoc, Bohr suggested a different and, he thought, more satisfactory way to derive it. This derivation was the first germ of what would later become the correspondence principle. Instead of equation (2) he assumed the more general relation

$$W = f(\tau)h\omega$$

and argued that, to agree with the form of the Balmer expression, the function $f(\tau)$ had to be proportional to the integer τ; that is $f(\tau) = b\tau$. He then considered a transition from the state $\tau = N$ to the state $\tau = N - 1$. The frequency of the radiation emitted becomes

$$\nu = \frac{\pi^2 me^2 E^2}{2b^2 h^3} \frac{2N-1}{N^2(N-1)^2},$$

while the frequency of revolution of the electron in the stationary state $\tau = N$ is

$$\omega_N = \frac{\pi^2 me^2 E^2}{2b^3 h^3 N^3},$$

and similarly for $\tau = N - 1$. The ratio between the mechanical frequencies of the two states is given by

$$\frac{\omega_N}{\omega_{N-1}} = \frac{(N-1)^3}{N^3},$$

which tends towards unity when $N >> 1$. 'If N is great the ratio between the frequency before and after the emission will be very near equal to 1; and according to the ordinary electrodynamics we should therefore expect that the ratio between the frequency of radiation and the frequency of revolution also is very nearly equal to 1' (p. 13). This gives the result that $b = \frac{1}{2}$, justifying the factor of $\frac{1}{2}$ introduced in the original derivation. For a passage from $\tau = N$ to $\tau = N - n$, where n is small compared to N, the result is $\nu = n\omega$. Foreshadowing the correspondence principle, he commented (p. 14):

> The possibility of an emission of a radiation of such a frequency may also be interpreted from analogy with the ordinary electrodynamics, as an electron rotating round a nucleus in an elliptical orbit will emit a radiation which according to Fourier's theorem can be resolved into homogeneous components, the frequencies of which are $n\omega$, if ω is the frequency of revolution of the electron.

The new interpretation of $W = \tau h\omega/2$ was 'not that the different stationary states correspond to an emission of different numbers of energy-quanta, but that the frequency of the energy emitted during the passing of the system from a state in which no energy is yet radiated out to one of the different stationary states, is equal to different multiples of $\omega/2$, where ω is the frequency of revolution of the electron in the state considered'. The earlier interpretation was now dismissed as only preliminary (which makes one wonder why he included it in the first place).

Bohr justified his result in yet another way, namely by quantizing the angular momentum L of the revolving electron. As mentioned in Section 1.4, Nicholson had previously restricted changes in the angular momentum to units of $h/2\pi$. Moreover, in a paper dealing with the quantum theory of diatomic gases that was published shortly before Bohr's, Paul Ehrenfest in Leiden had independently assumed the angular momentum of a rotating molecule to be quantized according to

$$\tfrac{1}{2}J(2\pi\omega)^2 = \tau\frac{h\omega}{2},$$

where J denotes the moment of inertia. However, he did not formulate the quantization as a general principle.[70] Noting that 'there obviously can be no question of a mechanical foundation of the calculations given in this paper', Bohr nonetheless interpreted his calculations 'by help of symbols taken from the ordinary mechanics'. For the case of an electron moving in a circular orbit he found

$$L = ma^2(2\pi\omega) = \tau\frac{h}{2\pi} \qquad (6)$$

In Bohr's words: 'The angular momentum of the electron round the nucleus in a stationary state of the system is equal to an entire multiple of a universal value, independent of the charge on the nucleus'. Since $\pi L = W/\omega$ for a circular orbit, the quantization of L is equivalent to the result of equation (2). While the quantization of the angular momentum played only a secondary role in the first part of the trilogy, in the second part it had become the 'main hypothesis' of the theory.[71]

In a section on absorption of radiation Bohr argued that his theory was able to account not only for the emission of light but also for absorption phenomena. These he regarded as the reverse of emission; that is, he attributed them to a transition from a lower-energy state to a higher one, whether bound or free. His general hypothesis was that 'a system of electrons will absorb a radiation of a frequency different from the frequency of vibration of the electrons calculated in the ordinary way'. Considering an electron in a stationary state to receive an energy quantum $h\nu$—as 'in experiments on ionization by ultra-violet light and by Röntgen rays'—he argued that in these cases it would enter an unbound state. If W denotes the negative of the binding energy, the ejected electron will acquire a kinetic energy $E_{\mathrm{kin}} = h\nu - W$, which is just Einstein's formula of 1905 for the photoelectric effect. Bohr further argued that experiments made by the American physicist Robert Williams Wood at Johns Hopkins University on absorption of light by sodium vapour could be qualitatively explained along similar lines. According to Bohr, Wood's results could be interpreted as a transition of the valence electron in the sodium atom from a bound to a free state.[72]

In the later literature Bohr's famous frequency relation $W_1 - W_2 = h\nu$ was sometimes seen as the reverse of Einstein's hypothesis of energy generation by absorption of light, and for this reason was referred to as the 'Einstein–Bohr condition'. For example, Sommerfeld called Bohr's condition 'a counterpart to Einstein's photoelectric equation'.[73] However, although there is an obvious formal similarity between the two equations, from a conceptual and historical perspective they are quite different. For one thing, Einstein's $h\nu$ was a light quantum, a parcel of energy, whereas Bohr's was the energy of a monochromatic electromagnetic wave. Moreover, Einstein's light quantum hypothesis was independent of any atomic model, while Bohr's frequency condition was an integral part of his theory of atomic structure, dependent as it was on the notion of stationary states.

While Bohr's theory of the hydrogen atom agreed impressively with spectroscopic measurements, the agreement did not amount to a confirmation of his prediction of the many discrete energy levels in the atom. In 1913 the experimental data necessary for a direct confirmation did not exist, but they were obtained in later experiments measuring the resonance potentials required to raise the electron from the ground state to higher states. Table 2.1 shows data from experiments on the radiation from atomic hydrogen made in 1923 by P. S. Olmstead and Karl T. Compton at Princeton University. As the two American physicists pointed out, 'the agreement with . . . the Bohr theory is exact'.[74] By that time this was a conclusion that surprised no one.

The hydrogen atom and other one-electron atomic systems constituted the core of Bohr's theory as he presented it in the first part of his trilogy. However, he also discussed systems in which more electrons were arranged in a ring, as he did in greater detail in the second part.[75] This he did by following the stability calculations of Thomson and Nicholson, but with the important difference that he did not base his

Table 2.1 Comparison of energy levels (in eV) in atomic hydrogen between measured values and values predicted by Bohr's theory. The last row corresponds to ionization.

Quantum transitions	Observed voltage	Calculated voltage
1–2	10.15	10.154
1–3	12.05	12.034
1–4	12.70	12.692
1–5	13.00	12.997
1–6	13.17	13.162
1–7	13.27	13.262
....	....	...
1-∞	13.54	13.539

Source: Olmstead and Compton 1923.

conclusions solely on mechanical considerations. His hypothesis was that 'the stability of a ring of electrons rotating round a nucleus is secured through the above condition of the universal constancy of the angular momentum, together with the further condition that the configuration of the particles is the one by the formation of which the greatest amount of energy is emitted' (p. 23). In a brief section on the spectra of elements with several electrons he pointed out that his theory provided an immediate explanation of the so-called combination principle formulated in 1908 by Ritz.[76] According to the Swiss physicist, the wave numbers (or frequencies) of the spectral lines of an element can in all cases be written as the difference between two spectral terms, each of which depends on an integer. Let n_1 and n_2 be two integers, and F_r and F_s two functions of these numbers. Then:

$$\frac{1}{\lambda} = \frac{\nu}{c} = F_r(n_1) - F_s(n_2).$$

By focusing attention not on the spectral lines, but on the terms, Ritz's combination principle greatly helped spectroscopists in organizing the spectra, but until Bohr's theory it remained phenomenological, without a basis in physical theory.

In the second part of the trilogy Bohr offered a more elaborate analysis of the distribution of electrons in coplanar rings round a nucleus. In the case of a single ring his calculations showed that the number of electrons increased only slowly with the nuclear charge. A ring of n electrons 'cannot rotate in a single ring round a nucleus of charge ne unless $n < 8$', he concluded, just as he had done in the Manchester memorandum a year earlier. To accommodate eight electrons in the same ring a nuclear charge of $Z \geq 10$ was required, implying a positive ion instead of a neutral atom. Bohr furthermore considered the tendency of separate rings of electrons to coalesce into a single ring. He argued that such coalescence will only occur when the rings contain the same number of electrons, and 'accordingly the numbers of electrons on inner rings will only be 2, 4, 8, ...'.[77] As we shall see in Section 2.6, these and similar results provided the seeds for a possible explanation of the periodic system on the basis of Bohr's new theory.

Bohr was not very happy about either of the derivations of the Balmer formula given in the trilogy as outlined above. In a lecture before the Danish Physical Society on 20 December 1913 he reconsidered his model of the hydrogen atom and its connection to Planck's quantum theory.[78] By that time he thought it was misleading to base the atomic model on Planck's second theory and preferred to use only the two postulates. As he pointed out, while in Planck's theory the frequency of an oscillator was independent of the amount of energy contained in the system of oscillators, in his own theory the frequency of revolution depended on the energy according to the Keplerian relationship $\omega^2 \sim W^3$. Moreover, while Planck assumed that an oscillator could emit several quanta of the same frequency at once, in Bohr's theory only a single quantum of energy was emitted during a transition.

While in the first part of the trilogy Bohr had emphasized that 'The foundation of the hypothesis [quantization of angular momentum] has been sought entirely in its relation with Planck's theory of radiation' (p. 25), he now endeavoured to loosen this relation. Although he kept to the concept of energy quanta as given by $E = h\nu$, he wanted to do without the Planckian notion of atomic oscillators. Already in the conclusion of the third part of the trilogy he had expressed his misgivings about 'The assumption of such vibrators . . . [which] involves the assumption of quasi-electric forces and is inconsistent with Rutherford's theory, according to which all the forces between the particles of an atomic system vary inversely as the square of the distance apart'.[79]

'We can not expect', he said in Copenhagen, 'that all cases of disagreement between the theoretical conceptions hitherto employed and experiment will be removed by the use of Planck's assumption regarding the quantum of the energy momentarily present in an oscillating system'.[80] As far as Planck's theory was concerned it only appeared indirectly in the fundamental assumption of quantum jumps, which can be written

$$\nu = \frac{1}{h}(W_1 - W_2). \tag{7}$$

Bohr further stressed that there were no observational reasons to keep to Planck's picture of oscillating charges. He appealed to empiricism: 'We stand here almost entirely on virgin ground, and upon introducing new assumptions we need only to take care not to get into contradiction with experiment'. What does 'direct observation' tell us about the mechanism causing the blackbody spectrum so precisely expressed by Planck's formula? According to Bohr, nothing:

> No one has ever seen a Planck's resonator, nor indeed even measured its frequency of oscillation; we can observe only the period of oscillation of the radiation which is emitted. It is therefore very convenient that it is possible to show that to obtain the laws of temperature radiation it is not necessary to make any assumptions about the systems which emit the radiation except that the amount of energy emitted each time shall be equal to $h\nu$, where h is Planck's constant and ν is the frequency of the radiation.[81]

In his lecture of December 1913 Bohr did not attempt to derive Balmer's formula, but only the coefficient appearing in it; that is, the Rydberg constant. This he did by comparing equation (7) with Balmer's formula, which suggests $W = Rhc/n^2$ for the nth stationary state. Adopting a correspondence approach similar to the one used in the trilogy, he expressed the frequency of the radiation arising from a transition from state $(n + 1)$ to state n as

$$\nu = Rc\left(\frac{1}{n^2} - \frac{1}{(n+1)^2}\right).$$

For large values of n this gives approximately $\nu = 2Rc/n^3$. According to the correspondence view this radiation frequency is equal to the frequency of revolution. Inserting $W = Rhc/n^2$ in equation (1) then gives

$$\omega^2 = \frac{2R^3h^3c^3}{\pi^2 e^4 mn^6} = \frac{4R^2c^2}{n^6}.$$

From this expression it follows that $R = 2\pi^2 me^4/h^3c$, the same expression that he had found in the trilogy.

Bohr's new atomic theory was a somewhat strange mixture of classical and quantum ideas based on two postulates that were only justified by the empirically successful use he made of them. Bohr himself, much aware of the arbitrary and unsatisfactory features of the theory, emphasized its 'preliminary and hypothetical character'.[82] In his Copenhagen address he was even more direct: 'The fact that the deficiencies of the atomic model we are considering stand out so plainly is . . . perhaps no serious drawback; even though the defects of other atomic models are much better concealed they must nevertheless be present and will be just as serious'. Bohr furthermore admitted that what he was proposing was not really an explanation, in the causal sense of the term, of how light is emitted from atoms. 'I am by no means trying to give what might ordinarily be described as an explanation', he said; 'nothing has been said here about how and why the radiation is emitted'.[83] This was a point to which he would return at later occasions. Thus, in 1922 he emphasized that his atomic theory 'does not attempt an "explanation" in the usual sense of this word, but only the establishment of a connexion between facts which in the present state of science are unexplained'.[84]

In Chapter 3 we shall see that many of Bohr's contemporaries found the theory to be puzzling for this reason; but we shall also see that they were nonetheless impressed by its explanatory and predictive power over a broad domain and in spectroscopy in particular.

2.5 SPECTROSCOPIC PUZZLES

According to Robert Millikan, the eminent American experimentalist and a Nobel laureate of 1923 for his measurements of the photoelectric effect and the electron's charge, the period 1912–1914 was 'comparable in importance with the period of the laws of Galilean–Newtonian mechanics some three centuries earlier'. Looking back on the development to which he had himself contributed so importantly, Millikan highlighted Bohr's theory as a major reason for the revolutionary change in the period: 'For the immense field of spectroscopy was essentially an unexplored dark continent prior to the advent of Bohr's theory. Bohr's equation has been the gateway through which hundreds of explorers have since passed into that continent until it has now become amazingly well mapped'.[85]

Indeed, Bohr's atomic theory scored its first and most spectacular successes in the area of optical spectroscopy, more precisely with respect to the spectra of hydrogen and other one-electron atomic systems. As mentioned above, in the trilogy he had predicted—or, as he phrased it, 'expected'—the existence of hydrogen lines arising from electron transitions to the states $\tau = 1$ and $\tau > 3$. In the early 1920s physicists at the Johns Hopkins University detected lines in the infrared region that agreed precisely with Bohr's predictions: In 1922 Frederick Brackett found the first two lines belonging to $\tau = 4$, and also three new members of the Paschen series. Two years later August Pfund found a single line representing the $\tau = 5$ series.[86] While these discoveries were unsurprising—given that Bohr's theory was solidly established at the time—the discovery of the ultraviolet series belonged to a different category.

In 1914 the experienced spectroscopist Theodore Lyman at the Jefferson Laboratory of Harvard University reported the observation of two new hydrogen lines in the ultraviolet region, but without mentioning Bohr or his theory.[87] Indeed, he may well have been unaware of the theory. Even after Bohr had become aware of the lines and referred to 'the series in the ultra-violet recently discovered by Lyman' as further confirmation of his theory,[88] Lyman refrained from considering the theoretical relevance of his discovery. While he did not refer to atomic theory in 1914, in papers of 1915 and 1916 he briefly mentioned that the spectra of hydrogen and helium 'have recently come into prominence through the theoretical researches of Bohr, Nicholson, and others'.[89] In his paper in the *Astrophysical Journal* he called the relations between the spectra of the two elements 'a fascinating subject for speculation' and said: 'In connection with Bohr's speculations it is important to observe that λ 1217, which forms the first member of the Ritz [Lyman] series, occupies exactly the same position when obtained from helium as when it is produced in hydrogen'.[90] In none of his papers did he credit Bohr's formula and he ignored his atomic theory. The American spectroscopist obviously preferred to deal with experimental facts rather than 'speculations'.

In a paper of 1896, also in the *Astrophysical Journal*, the American astronomer Edward Pickering at the Harvard College Observatory reported spectral lines (including $\lambda = 5411$, 4541, and 4200) from the star ξ Puppis. These followed a Balmer-like expression. He and Rydberg consequently attributed them to a new hydrogen series.[91] (Initially he thought that the lines might be due to a new element.) The series discovered by Pickering converged to the same limit as the Balmer series, which was one reason why Rydberg identified it as the diffuse series of hydrogen, supplementing the sharp Balmer series. What was generally thought to be new hydrogen lines were studied in the laboratory by Alfred Fowler, an astrophysicist at the Imperial College, London, and a former assistant of Norman Lockyer. In 1912 he succeeded in reproducing the line $\lambda = 4686$ Å in a discharge tube with a mixture of hydrogen and helium.[92] This line also appeared prominently in ξ Puppis and in the spectra from many nebulae.

Although Fowler was unable to find the 4686 line and other of Pickering's lines in tubes filled with hydrogen only, he persuaded himself (and others) that the new series was indeed due to hydrogen and not to helium. It seemed natural to assume that the 4686 line was the first of the lines in hydrogen's principal series. The wavelengths of the Pickering–Fowler series extended the Balmer–Ritz formula in the sense that they followed an expression of the form

$$\frac{1}{\lambda} = R\left(\frac{1}{(n_2/2)^2} - \frac{1}{(n_1/2)^2}\right). \qquad (8)$$

For example, the 4686 line corresponded to $n_1 = 4$ and $n_2 = 3$. Obviously, this was an expression for the hydrogen spectrum that flatly disagreed with Bohr's interpretation, where the denominators necessarily have to be integers. Half-integral quantum numbers would eventually turn up in quantum theory, and then cause problems of their own, but within the context of Bohr's early one-quantum atomic model they were strictly inadmissible.

As early as 6 March 1913, in his letter to Rutherford, Bohr argued that the lines found by Fowler were really due to helium and that the presence of hydrogen was only of indirect significance. 'Dr Bjerrum suggested to me that if my point of view was right the lines might also appear in a tube filled with a mixture of helium and chlorine (oxygen, or other electronegative substances); indeed it was suggested, that the lines might be still stronger in this case'.[93] Since the laboratories in Copenhagen were not equipped for doing such an experiment, Bohr asked Rutherford if it could possibly be performed in Manchester. Two weeks later Rutherford responded:

> Your ideas as to the mode of origin of spectrum and hydrogen are very ingenious and seem to work out well; but the mixture of Planck's idea with the old mechanics make [sic] it very difficult to form a physical idea of what is the basis of it. There appears to me one grave difficulty in your hypothesis, which I have no doubt you fully realise, namely, how does an electron decide what frequency it is going to vibrate at when it passes from one stationary state to the other? It seems to me that you would have to assume that the electron knows beforehand where it is going to stop. . . . I was much interested in your speculations in regard to Fowler's spectrum. I mentioned the matter to Evans here, who told me that he was much interested in it, and I think it quite possible that he may try some experiments on the matter when he comes back next term.[94]

In the first part of the trilogy Bohr explained that 'we can account naturally for these [Pickering–Fowler] lines if we ascribe them to helium' (p. 10). He simply took $E = 2e$ in equation (3) and could then rewrite the expression (8) as

$$\frac{1}{\lambda} = 4R\left(\frac{1}{n_2^2} - \frac{1}{n_1^2}\right) \quad \text{with} \quad R = \frac{2\pi^2 me^4}{h^3 c}.$$

In this way Bohr accommodated the new spectral series, interpreting it as due to the formation of He^+ ions. He further suggested reasons why the spectrum was not observed in ordinary helium tubes. Bohr did not simply ascribe the new lines to quantum transitions in the He^+ ions. In accordance with ideas common at the time he thought that the atoms first had to lose both of their electrons; the lines would then be emitted as an electron was captured to form He^+ in one or other stationary state.

In agreement with what Rutherford had told Bohr in his letter of 20 March, the question of the origin of the 4686 line was investigated by his assistant Evan Jenkin Evans, who succeeded in detecting the line in helium discharge tubes with no trace of hydrogen. 'For some time I have been investigating the origin of the 4686 line', Evans said in a preliminary report, stating that 'the experiments already carried out support Bohr's theory'.[95] Unconvinced that Bohr had really explained the 4686 line and other lines in the supposed hydrogen series, in the autumn of 1913 Fowler pointed out that the theoretical values, although very close to those observed, did not quite agree with them. Bohr's response to the challenge from London was to modify his analysis by taking into account the finite mass of the nucleus, which he did by replacing m with the reduced mass as given by

$$\mu = \frac{mM}{m+M} = \frac{m}{1+m/M},$$

where M is the mass of the nucleus. In this way the Rydberg constant will depend slightly on the mass of the element and, as Bohr showed, the discrepancies disappear.[96] The ratio of the Rydberg constants for a very heavy one-electron atom and the hydrogen atom becomes

$$\frac{R_\infty}{R_H} = 1 + \frac{m}{M_H}.$$

Fowler now conceded that Bohr's atomic theory gave a correct explanation. As he wrote to Bohr, 'Your letter published in last week's "Nature" struck me as a valuable addition to your Phil. Mag. paper of July'.[97] Although this was not the last word in the case of Fowler's lines, the dispute did much to promote Bohr's theory and make it known among British physicists at an early date. It took a little longer in Germany, but there, too, the resolution of the 4686 puzzle came to be seen as convincing evidence for Bohr's theory. 'Only now do I see that Bohr's theory is exactly right', Friedrich Paschen wrote to Sommerfeld. 'There is no doubt that Bohr's final formula is as accurate as the measurements can be made'.[98] Using a special kind of helium tube, Paschen succeeded in photographing the ionised helium components and showed that they agreed perfectly with the ones calculated on Bohr's theory (Table 2.2).

In his letter to *Nature* in which he introduced the reduced mass correction, Bohr pointed out that 'according to the theory helium must be expected to emit a series of lines closely, but not exactly, coinciding with the lines of the ordinary hydrogen spectrum'. The lines would correspond to transitions from $n_1 = 6, 8, 10, \ldots$ to $n_2 = 4$. Experiments

Table 2.2 Bohr's helium lines. The second column gives the wavelengths of the Pickering series as calculated by Bohr, and the third column the experimental values published by Paschen in 1916. The fourth column contains stellar lines found by Harry Plaskett in 1921. The table is adapted from Plaskett 1922.

Balmer hydrogen lines	Helium lines, Bohr theory	Helium lines, laboratory	Helium lines, stars
6562.79	6560.15	6560.13	6560.04
	5411.53	5411.55	5411.62
4861.33	4859.35	4859.34	4859.09
	4541.62	4541.61	4541.67
4340.47	4338.70	4338.70	4338.79
	4199.86	4199.86	4200.06
4101.74	4100.00	4100.05	4100.26

with pure helium conducted by Evans in late 1914 confirmed the series predicted by Bohr, which was one more spectroscopic triumph for the new atomic theory.[99]

While Fowler fully recognized Bohr's explanation of the 4686 line as due to 'proto-helium', he remained ambiguous with respect to the theory on which the explanation was based. As he wrote to Bohr:

> Evans's detection of the components near the Balmer lines seems to me to complete the evidence for the helium origin of the 4686 and Pickering lines. But, as I said in the Bakerian lecture (p 258), the assignment of the lines to proto-helium is 'independent of Bohr's theory'. . . . Your theory has the great merit of accounting for the lines in positions slightly different from those calculated by Rydberg, and of predicting Evans's lines. But perhaps we may find that some other theory will do the same thing. Meanwhile I am a warm supporter of your theory.[100]

Fowler's recognition that the 4684 line was indeed due to ionized helium, as predicted by Bohr, did not quite close the case. In some quarters doubts remained as to whether this was the true explanation. In a paper from March 1915 Thomas Ralph Merton, a spectroscopist at Imperial College, London, observed that Bohr's theory 'has given rise to a considerable amount of theoretical discussion'.[101] Spectroscopic experiments based on a new interference method suggested to him that the evidence provided by Evans was inconclusive and that the mass of the atom from which the 4686 line originated was much smaller than that of the helium atom. He found that it was only about one-tenth of the mass of a hydrogen atom and thus 'due to systems of subatomic mass'. What these unlikely systems might be, Merton wisely refrained from saying. Nor did he spell out the theoretical significance of his conclusion, although it obviously contradicted Bohr's explanation as well as other parts of established atomic physics.

Just as Bohr had reinterpreted the Pickering–Fowler lines in agreement with his theory, so he did with another series of spectral lines that Nicholson in early 1913 had suggested belonged to the hydrogen spectrum and had interpreted according to his own atomic theory.[102] The British astrophysicist had found that the frequencies of some hitherto unidentified lines in the spectra of a class of stars known as Wolf–Rayet stars could be expressed in a manner similar to the Balmer and Pickering–Fowler lines, namely as

$$\nu = Rc\left(\frac{1}{4} - \frac{1}{\left(n \pm \frac{1}{3}\right)^2}\right)$$

However, as Bohr saw it, the lines were more likely to be due to the doubly charged lithium ion Li^{2+}, just as the Pickering–Fowler lines were due to the He^+ ion. He simply rewrote Nicholson's expression as

$$\nu = Rc\left(\frac{1}{(n_2/3)^2} - \frac{1}{(n_1/3)^2}\right) = 9Rc\left(\frac{1}{n_2^2} - \frac{1}{n_1^2}\right)$$

where the factor 9 comes from the square of the charge of the lithium nucleus ($E = 3e$). Nicholson did not accept Bohr's reinterpretation, based as it was on a model of the lithium atom with two electrons in the inner ring and one electron in the outer ring. According to Nicholson, this model was dynamically impossible (see further discussion in Section 3.6). The cases of He^+ and Li^{2+} were different in the sense that whereas Bohr's prediction of the He^+ lines was confirmed experimentally, this was not the case with his reinterpretation of Nicholson's lines. There were no experimental data on the doubly charged lithium ion.

Bohr was well aware that if the atomic electrons revolved with velocities comparable to the velocity of light it might be necessary to replace the constant mass m with the varying relativistic mass $m(v)$ as given by Einstein's formula

$$m(v) = \frac{m}{\sqrt{1 - v^2/c^2}}$$

However, in the trilogy he saw no reason to introduce the small relativistic modification.[103] He only this this in a letter to Fowler of 15 April 1914 in which he reported the more general but still approximate formula for the spectral lines emitted by an electron bound to a nucleus of mass M and charge Ne.[104] The modified formula he offered was this:

$$\nu = \frac{2\pi^2 Mme^4N^2}{(M+m)h^3}\left(\frac{1}{n_1^2} - \frac{1}{n_2^2}\right)\left\{1 + \frac{\pi^2e^4N^2}{h^2c^2}\left(\frac{1}{n_1^2} - \frac{1}{n_2^2}\right)\right\}$$

Although the relativistic effect was very small ($\pi^2e^4/h^2c^2 \cong 1.3\times\ 10^{-5}$) it was not negligible, and it became important in connection with precision experiments made by William Edward Curtis, an astrophysicist and student of Fowler. According to Curtis's measurements of the Balmer spectrum the wavelengths of the lines deviated slightly from those predicted by the Balmer–Bohr formula.[105] Bohr hoped to explain the deviations by means of his relativistic modification, but had to admit that it only accounted for about one third of the deviations observed by Curtis. 'Might not the disagreement as to H_{α} in some way be connected with the doubling of the line?' he asked Fowler.[106] We shall consider the connection between Bohr's theory and the fine structure of the hydrogen spectrum in Section 4.4.

2.6 BOHR'S ATOMIC CHEMISTRY

Considerations and results of a chemical nature occupied an important place in Bohr's trilogy, which was after all a treatise on the constitution of atoms *and molecules*.[107] Whereas he dealt with the atomic structure of the chemical elements in Part I, in Part III he focused on the structure of simple molecules. The chemical aspects were in Bohr's mind from the very beginning of his project, undoubtedly in part inspired by his contact with Hevesy. As mentioned in Section 2.3, molecules figured prominently in the Manchester memorandum. In his letter to Hevesy of 7 February 1913, shortly before he became aware of the Balmer spectrum, Bohr said that he could 'explain not only the order of magnitude of the dimensions of the atoms, but also the way in which the atom-volumes vary with the valency of the element considered (i.e. with the number of electrons in the outermost ring)'. He continued:

> Besides a very suggestive indication of an understanding of the periodic system of the elements, the considerations in question leads [sic] to a theory of chemical-combinations, a theory which permits to follow the process of combining of atoms in detail, and applied on the most simple systems immediately gives the result, that 2 Hydrogen-atoms will combine into a molecule, while 2 Helium-atoms wont [sic]. . . . [B]y simple application of considerations as the above indicated, [I] have hope to obtain a knowledge of the structure of the systems of electrons surrounding the nuclei in atoms and molecules, and thereby hope of a detailed understanding of what we may call the 'chemical and physical' properties of matter.[108]

In dealing with atoms containing more than one electron Bohr realized that their constitution could not be fully determined by the conditions he had used in the case of hydrogen, that is, mechanical stability and the constancy of the angular momentum of the electrons. He therefore adopted an alternative approach, which can be reconstructed as consisting of four elements: (i) involved calculations of mechanical stability; (ii) applications of chemical knowledge, such as valency, reactivity, and ionization potentials; (iii) considerations related to spectroscopy; and (iv) simplicity assumptions, including that the electron rings are situated in the same plane through the nucleus even in the case of many electrons. In Bohr's words: 'On the general view of formation of atoms, however, and by making use of the knowledge of the properties of the corresponding elements, it will be attempted... to obtain indications of what configurations of the electrons may be expected to occur in atoms'.[109] There is little doubt that empirical knowledge of a chemical and physical nature played a greater role than calculations based on mechanical and quantum-theoretical principles.

The lightest atoms with atomic number between 2 and 4 were treated separately by Bohr and in general accordance with his treatment of the hydrogen atom. In the case of neutral helium he argued that stability required the two electrons to move in the same orbit, a configuration he referred to as 2(2) and for which he found the radius and binding energy to be $a = 0.571\ a_0$ and $W = 6.13\ W_0$ (a_0 and W_0 are the values for the hydrogen atom; $W_0 \cong -13$ eV). Since W for the configuration 2(3) turned out to be smaller than for 2(2), he concluded that a negative helium ion could not exist. For the energy of the helium ion He^+ he obtained $W = 4\ W_0$, meaning that both electrons were bound to the nucleus more firmly than in the case of hydrogen. The ionization energy for helium amounts to $(6.13 - 4)W_0 = 2.13\ W_0$ or about 27 eV, which Bohr noticed was of the same order of magnitude found experimentally by James Franck and Gustav Hertz in Berlin, namely 20.5 eV.[110] He also claimed to find support for his model in measurements of the ultraviolet absorption frequency in helium, where the Cuthbertsons had found a value of $5.9 \times 10\ s^{-1}$. Bohr's calculations based on the energy difference between He and He^+ in their ground states resulted in a frequency parallel to the orbital plane of $6.6 \times 10\ s^{-1}$, and from ordinary mechanics he obtained for the perpendicular frequency $20.3 \times 10\ s^{-1}$. He concluded that the frequency of dispersion 'may be regarded as corresponding to vibrations in the plane of the ring', whereas the perpendicular vibrations had only a negligible influence on dispersion.[111]

Bohr's calculations of helium and other atoms and molecules more complicated than hydrogen were crude and rested on an arbitrary mixture of quantum and classical considerations. In 1913 they did not result in convincing agreement with experiments. Later and much more sophisticated calculations did not materially improve the agreement. As we shall see, the normal helium atom never succumbed to Bohr's theory. His treatment of the lithium atom is another case that may be used to exemplify his eclectic and somewhat opportunistic approach. In this case he calculated the total binding energy for two configurations, one being a two-ring system and another in which the three electrons moved on the same ring. He obtained the values, in his notation,

$$\mathrm{W}[3(2,1)] = -16.02\,W_0 \text{ and } \mathrm{W}[3(3)] = -17.61\,W_0.$$

This means that the latter configuration is energetically favoured, but of course it contrasts with the chemical properties of lithium. For the Li^+ ion Bohr found $W = -15.13\ W_0$, which implied that it should be easier to ionize helium than lithium vapour! Characteristically, Bohr chose to ignore the result of his mechanical calculations and concluded that the (2,1) configuration was the correct one 'from a consideration of the chemical properties' of the element. His model of beryllium was no better, for here the theory showed that the configuration with four electrons revolving in a single ring was energetically favoured: he got $W = -37.04\ W_0$ for this configuration, and $W = -33.61\ W_0$ for the 4(2,2) structure.[112]

The same willingness to sacrifice mechanical calculations for chemical considerations appeared in Bohr's assignment of electrons in systems of rings. While he had found that the inner ring, to be mechanically stable, could accommodate no more than seven electrons, in the end he chose the number eight. The reason was obviously the known periodicity of the elements, with the first periods including eight elements. As to the number of electrons in the outer ring he did not even pretend to base it on calculations: 'The number of electrons in this ring is arbitrarily put equal to the normal valency of the corresponding element'.[113] This accounts for the change in the building-up scheme at nitrogen, which he assigned the configuration (4,3) rather than (2,5). He gave no reason for this change except that three outer electrons are necessary to account for nitrogen's tervalency. A similar case can be observed for phosphorus, to which he assigned the configuration (8,4,3) rather than (8,2,5).

Bohr also made no attempt to take into regard the distinction between the so-called principal and secondary valences (or normal valence and contravalence), concepts which the German physical chemist Richard Abegg had introduced in 1904. According to Abegg's 'rule of eight' the sum of the two valences was always equal to eight.[114] For example, chlorine has one negative normal valence (as in HCl) and seven positive contravalences (as in $NaClO_4$); similarly, nitrogen displays valencies from -3 (as in NH_3) to $+5$ (as in N_2O_5). Shortly after Bohr's paper had appeared, Hevesy asked him: 'How would you explain the difference between principal and secundary valencies. (Haupt und Nebenvalenzen.) Can the first be connected with the outermost ring, and the second one indirectly though with the other rings?'[115] His question remained unanswered.

For the lighter elements, up to manganese with 25 electrons, Bohr suggested configurations that he thought corresponded to the periodic system (Table 2.3). Although he did not assign electron arrangements to atoms heavier than chromium, based on the periodic system he suggested that 'elements of higher atomic weight [than about 60] contain a recurrent configuration of 18 electrons in the innermost rings'. Moreover, he argued that in some cases, such as the rare earth metals, the building up of electrons took place in an inner rather than the outer ring. In this way it would be possible to account for the striking chemical similarity of this group of elements. Finally he indicated an explanation of the 'observed increase of the electropositive character for an increase of atomic weight of the elements in every single group of the periodic system', say from beryllium to radium (p. 497). According to Bohr, this was a result of the increasingly weaker binding of the outer electrons as the number of rings increased.

The ideas of molecular structure included in the Manchester memorandum, and also in his letter to Hevesy of 7 February 1913, were only published in the third part of the

Table 2.3 Bohr's 1913c proposal of electron rings in chemical elements. In his table in *Philosophical Magazine* he did not assign chemical symbols to the structures.

H	1 (1)	F	9 (4, 4, 1)	Cl	7 (8, 4, 4, 1)
He	2 (2)	Ne	10 (8, 2)	Ar	18 (8, 8, 2)
Li	3 (2, 1)	Na	11 (8, 2, 1)	K	19 (8, 8, 2, 1)
Be	4 (2, 2)	Mg	12 (8, 2, 2)	Ca	20 (8, 8, 2, 2)
B	5 (2, 3)	Al	13 (8, 2, 3)	Sc	21 (8, 8, 2, 3)
C	6 (2, 4)	Si	14 (8, 2, 4)	Ti	22 (8, 8, 2, 4)
N	7 (4, 3)	P	15 (8, 4, 3)	V	23 (8, 8, 4, 3)
O	8 (4, 2, 2)	S	16 (8, 4, 2, 2)	Cr	24 (8, 8, 4, 2, 2)

trilogy. Bohr's general idea was to picture the covalent bond as a ring of two or more revolving electrons common to the two atoms forming a diatomic molecule:

> If we thus consider a neutral system containing two nuclei with great charges, it follows that in a stable configuration the greater part of the electrons must be arranged around each nucleus approximately as if the other nucleus were absent; and that only a few of the outer electrons will be arranged differently rotating in a ring round the line connecting the nuclei. The latter ring, which keeps the system together, represents the chemical 'bond'.[116]

For the simple symmetrical case of two nuclei of equal charges *Ne* and a ring of *n* electrons between them, he stated the equilibrium condition as

$$b = a\left[\left(\frac{4n}{N}\right)^{2/3} - 1\right]^{-1/2}, \tag{9}$$

where $2b$ is the distance between the nuclei and $2a$ the diameter of the ring. For systems containing several rings the effect of the inner rings would have to be taken into account, but Bohr chose to disregard this more general case with the excuse that it 'involves elaborate numerical calculations'.

The only molecule he discussed in detail was hydrogen, supposed to consist of two nuclei (protons) kept together by a ring of two electrons placed diametrically opposite. From equation (9) he got $b = a/\sqrt{3}$ and for the total energy of the molecule he calculated

$$-W = -\frac{e^2}{2a}F$$

with the numerical factor $F = 1.049$ (which he had given already in the Manchester memorandum). To determine the radius a, Bohr applied the quantum condition of the constancy of the angular momentum, $mva = h/2\pi$. In this way he arrived at the expression

$$W = \frac{4\pi^2 e^4 m}{h^2} F,$$

which he compared with the corresponding value W_0 for the hydrogen atom, as derived in the first part of the trilogy. The result was

$$(W - 2W_0) = 2W_0(F^2 - 1) = 0.20\ W_0.$$

Thus, 'it follows that the two hydrogen atoms combine into a molecule with emission of energy'.[117] With a value of Avogadro's number $N_A = 6.5 \times 10^{23}$ the theoretical result for the formation of one mole of H_2 came out as 60 kcal.

The American physical chemist Irving Langmuir, a researcher at the General Electric Research Laboratory and a future chemistry Nobel laureate, had recently made experiments from which he inferred the heat of formation of H_2 to be about 130 kcal mole^{-1}. Compared with Bohr's theoretical value this was of the right order of magnitude, but the numerical agreement was far from convincing.[118] However, a few months after the appearance of Bohr's paper Langmuir informed him that improved experiments had resulted in a value of 76 kcal mole^{-1}. Langmuir praised Bohr's theory of the hydrogen molecule as 'valuable and wonderfully suggestive' and found the disagreement between his new measurements and Bohr's calculations to be not intolerably high. (He did however find it 'very improbable' that the correct value could be as low as 60 kcal mole^{-1}.) In a letter to Langmuir Bohr described how the experimental results of 1912 had been 'the cause of much doubt for me as to the correctness of my assumptions on the constitution of the hydrogen molecule'.[119] At the same time he offered a recalculation using better values of the constants of nature. Bohr's revised result was $W = 63$ kcal mole^{-1}, only 17% away from the experimental value. Considering the tentative character of Bohr's molecular model this was a promising agreement, but the promise did not hold for long. More elaborate and accurate experiments made Langmuir conclude that 'it now becomes impossible to reconcile our experiments with the value $q = 63\,000$ [cal mole^{-1}] calculated according to the method of Bohr'.[120] As the final value for the heat of formation at constant volume he arrived at 84 kcal mole^{-1}.

Bohr also dealt with absorption lines in molecular hydrogen, assuming them to have their origin in the vibrations of the electron ring and the system of nuclei. In the first case he found an absorption frequency of 3.7×10^{15} s^{-1}, which he compared with the value 3.5×10^{15} s^{-1} calculated from dispersion experiments made by the Cuthbertson couple.[121] Further considering the molecule ions H_2^- and H_2^+, with three and one electrons rotating between the nuclei, respectively, he concluded that while the first one was mechanically stable the latter would be unstable against a displacement of the electron perpendicular to the plane of its orbit. Since the H_2^+ ion had been identified in Thomson's experiments with positive rays, this 'may therefore at first sight be considered as a serious difficulty for the present theory' (p. 867). However, Bohr suggested a solution, satisfactory to his own mind, that avoided the problem:

> A possible explanation, however, might be sought in the special conditions under which the systems are observed. We are probably dealing in such a case not with the formation of a stationary system by a regular interaction of systems containing single nuclei . . . but rather with a delay in the breaking up of a configuration brought about by the sudden removal of one of the electrons by impact of a single particle.

In another part of his paper he argued that 'in the breaking up of a hydrogen molecule by slowly increasing the distance apart of the nuclei, we obtain two *neutral* hydrogen atoms and not a positively and a negatively charged one' (p. 870).

In accordance with what he had said in the Manchester memorandum, in the trilogy Bohr argued that a helium molecule (He_2) would be unable to exist, and so would a chemical combination of helium and hydrogen. Although the hypothetical HeH molecule, bound together by a ring of three electrons, was energetically allowed, he gave reasons why 'the configuration in question cannot . . . be considered to represent a possible molecule of hydrogen and helium' (p. 867). Although Bohr focused on the simple hydrogen molecule, he saw no reason why it should not be possible to understand, at least in principle, other and more complicated molecules along similar lines of reasoning. Admitting that his conception of the covalent bond was premature and might have to be abandoned, he nonetheless proposed structures for the molecules HCl, H_2O, and CH_4 (Figure 2.4) For methane his structure was tetrahedral, with the C–H bonds represented by four rings each of which carried two electrons; the remaining two electrons were supposed to circle around the central carbon nucleus in an orbit of small radius. Referring to the possible structure of methane and similar compounds, he confessed that his theory was inadequate: 'The closer discussion of such questions, however, is far out of the range of the present theory' (p. 874).

In spite of its obvious inadequacies, Bohr's atomic chemistry of 1913 marked an important advance in chemical philosophy. It was the first theory of real atoms and

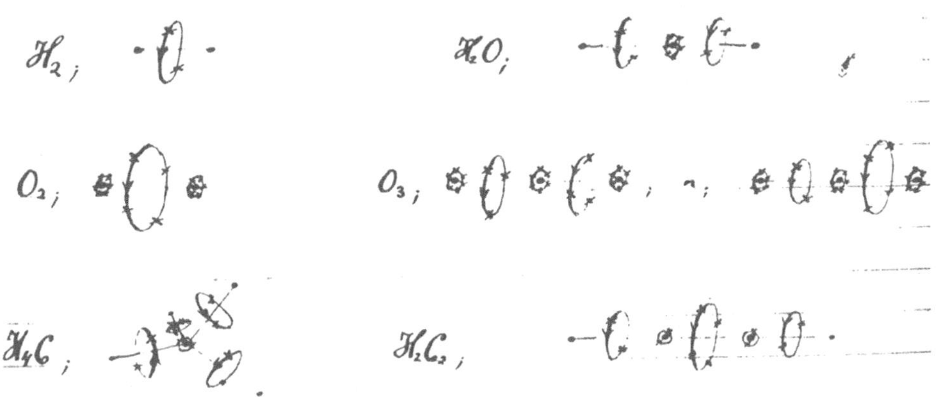

Fig. 2.4. Drawings from the Manchester memorandum showing how Bohr conceived the molecular structure of hydrogen, oxygen, ozone, water, methane, and ethyne.

Source: Hoyer 1974, p. 160.

molecules that offered an explanation of their structure in terms of configurations of electrons. The dynamical conception of the chemical bond was based on physical calculations which were, however, too complex to be of any use in molecules except the most simple ones. As it soon turned out, the covalent bond did not yield to Bohr's theory, which was, in this respect, a failure.[122] With the exception of a paper of 1919 on the possible existence of triatomic hydrogen (Section 3.2), Bohr did not follow up his ideas of molecular structure. Nor did these ideas catch the attention of contemporary chemists, who generally ignored them. This was not because they were unaware of them, for the three papers of the trilogy were extensively abstracted in the abstract journal of the London Chemical Society.[123] It was rather because they found the theory to be difficult and of little use to chemical problems. Among the very few chemists who responded to Bohr's atomic chemistry was Ernst Buchner at the Chemical Laboratory of the University of Amsterdam, who in 1915 suggested a modification of Bohr's electron configurations of simple molecules.[124] As Buchner pointed out, it was difficult to arrange the electrons in NH_3 in accordance with Bohr's principles and, at the same time, represent the radical NH_4 (or the corresponding ion NH_4^+) in agreement with chemical knowledge.

2.7 OTHER ASPECTS OF THE TRILOGY

'On the Constitution of Atoms and Molecules' was a sequel of papers rich in ideas and results. Although the atomic model and its use in explaining simple spectra and the electron configuration of elements had by far the greatest impact, the three papers included other innovative considerations that are worth mentioning.

In the second part of the trilogy Bohr briefly discussed the distinct X-ray lines that Charles Barkla and others had found were characteristic of the elements emitting the X-rays. Bohr suggested that the lines had their origin in a mechanism similar to the one producing the optical lines, only involving electron transitions to the inner rings of the atom. He ascribed the characteristic K-lines, known to have higher frequencies than lines belonging to other series, to electrons settling down in the innermost ring. Disregarding for reasons of simplicity the other atomic electrons, he considered an electron rotating round a nucleus of charge Ne. To remove the bound electron by the impact of a cathode-ray electron, the latter particle needs to have a minimum velocity equal to the velocity of the bound electron. According to Bohr's theory, this velocity is given by $2\pi e^2 N/h$. With the generally accepted relation $A = 2N$ between atomic weight and atomic number Bohr obtained

$$v_{\min} = \frac{\pi e^2}{h} A \cong 10^8 \mathrm{cms}^{-1},$$

in full agreement with the experimental result found by Whiddington. Bohr was well aware that his discussion of the atomic origin of the K-rays was only a 'first approximation', as he said to Rutherford. 'I have omitted any attempt to determine the numbers of electrons on the innermost rings, by comparing the characteristic Röntgen-radiation of the different types.'[125] It would soon turn out that Bohr's atom was even more useful in the study of characteristic X-rays than he imagined. But in the summer of 1913 he could not anticipate the results of Moseley's research programme on X-ray spectroscopy.

At the very end of the third part of the trilogy Bohr referred to the relationship between his atomic theory and Weiss's theory of magnetons. This was a subject which interested him greatly and which often turned up in his correspondence. 'I have thought somewhat about magnetism these days', he wrote to Harald in August 1913, 'and I think I have found out a little about it'. However, he decided not to include these thoughts in the paper: 'I had considered including some remarks about it in the proof to Part II; but I give it up after all . . . '.[126] Bohr's ideas about magnetism are known in part from a draft originally intended for Part II in which he considered the magnetic moment arising from an electron moving in a circular orbit. Using the quantum of the angular momentum he calculated the total magnetic moment μ for a mole of matter containing N_A particles:

$$\mu = N_A M_0 = \frac{N_A eh}{4\pi mc} = 5610.$$

This was almost exactly five times the value of the molar magneton as determined experimentally by Weiss. In his book *The Electron Theory of Matter*, Owen Richardson calculated a value that was six times as large as Weiss's, adding in a footnote: 'I have learned from a conversation (July 1913) with Dr Bohr, who had made similar calculations, that a more exact experimental value of the magneton makes this ratio exactly five'.[127] Bohr thought that the coincidence might be significant: 'This simple relation may be considered as promising for a theory of magnetism based on the assumption used in my former paper; a detailed discussion, however, involves problems of great difficulty as it is necessary to introduce new assumptions about the behaviour of bound electrons in a magnetic field'.[128] He left it at that.

Considering radioactivity to be an integral part of his theory, Bohr included a section on radioactive phenomena in Part II. Most of the section was concerned with beta radiation, which Bohr argued was of nuclear origin. As mentioned in Section 2.2, Bohr had reached this conclusion as early as about June 1912. It was an original view at a time when it was generally assumed by physicists accepting the nuclear model that beta particles came from the rings of electrons surrounding the nucleus. Thus, Rutherford believed that instabilities in the electron system caused by, for example, the emission of alpha particles gave rise to orbital electrons escaping as beta particles. As these free electrons passed the bound electrons in the rings they would lose some of their energy in the form of gamma radiation. He distinguished between 'the instability of the central nucleus and the instability of the electronic distribution. The former type of instability leads to the expulsion of an α-particle, the latter to the appearance of β and γ-rays'.[129] Bohr apparently accepted the gamma ray mechanism, but he stressed that 'On the present theory it seems also necessary that the nucleus is the seat of the expulsion of the high-speed β-particles'.[130]

For one thing, Bohr argued from the constancy of the angular momentum that when an alpha particle was expelled from the nucleus and traversed the electron system, the result would not disturb the electron rings to such an extent that they became unstable. As he expressed it to Rutherford: 'From the point of view of the theory, the argument of the invariance of the angular momentum of the electron during the expulsion of an α particle seems to me strongly in support of the hypothesis that the β-rays have their origin in the nucleus'.[131] To this argument he added another one, based on the new notion of isotopy, a word yet to be coined. It was known that some radioactive substances, apparently belonging to the same element, emitted beta rays with different velocities. If the substances were isotopes they would have the very same electron systems and only differ in their atomic weights, meaning their nuclei. The phenomenon 'shows that β-rays as well as the α-rays have their origin in the nucleus' (p. 501).

Bohr's conclusion soon became generally accepted. Rutherford expressed his agreement with his former research student in a letter to *Nature* dated 6 December 1913 in which he said that he had 'thought for some time' that beta rays had the same origin as alpha rays.[132] In his paper introducing isotopes, Soddy arrived at the same conclusion, but without mentioning Bohr. A somewhat similar view was expressed by Marie Curie at the second Solvay conference in 1913, in a discussion about Thomson's new theory of the atom.[133] Curie distinguished between 'essential' and 'peripheral' electrons, with the first residing in the 'nuclear part' of the atom. However, what might look like a reference to Rutherford's nuclear model was really a version of Thomson's. She did not refer to the views of either Rutherford or Bohr.

Notes for Chapter 2

1. Heilbron 1981, p. 230, according to whom 'Bohr's atomic theory belongs to the program of semiliteral model making initiated by J. J. Thomson and based on the methods of mid-Victorian Cambridge physics'.

2. Major biographies of Niels Bohr include Pais 1991, Moore 1966, and Röseberg 1985. There is also much information of a biographical kind in Rozental 1967 and in the 12 volumes of the *Niels Bohr Collected Works* published between 1972 and 2007.

3. Christiansen and Müller 1921, the content of which did not include either quantum or relativity physics. The book includes a valuable biographical essay by Eilhard Wiedemann on Christiansen's life and career (pp. xi–xxiv).

4. For opposite views on the Bohr–Høffding relationship, that is, whether Høffding significantly influenced Bohr's thinking about quantum physics or not, see Faye 1991 and Favrholdt 1992. The relevant literature can be found in these two works. In agreement with Faye, Wise

1990 argues for a close correlation between Høffding's philosophy and Bohr's ideas of atomic structure, if not necessarily a direct inspiration from the philosopher to the physicist. A stronger (but not more convincing) claim for Høffding's role in forming Bohr's ideas of atomic structure and the correspondence principle has recently been made by Angelino 2010. The suggestion that Kierkegaard's philosophy inspired some of Bohr's ideas appears in Jammer 1966, Holton 1970, and Feuer 1974.

5. The handwritten prize essay and the two papers are reproduced in Rud Nielsen 1972, pp. 21–92.

6. Niels Bohr to Harald Bohr, 26 April 1909, in Rud Nielsen 1972, p. 503.

7. For the history of the electron gas theory of metals, see Kaiser 2001 and the literature cited in this source. See also Hoddeson and Baym 1980.

8. See, for example, Kuhn 1978 and Seth 2004.

9. This claim has been made by Mara Beller, according to whom 'Bohr's mastery of mathematics was very limited' (Beller 1999, p. 259). She further argues that this supposed lack of mathematical skills was partly responsible for his complementarity interpretation of quantum mechanics.

10. Rud Nielsen 1972, p. 378, which contains a full English translation of the dissertation. Bohr first referred to Planck's radiation law and its incompatibility with electron theory in his master thesis of 1909. For an analysis of the dissertation and its role in the formation of Bohr's ideas of atomic structure, see Heilbron and Kuhn 1969. The papers by Einstein that Bohr referred to were on the theory of radiation and light quanta.

11. Rud Nielsen 1972, p. 299.

12. The translation on p. 300 in Rud Nielsen 1972 does not follow the Danish text very well. The quoted translation, which largely agrees with the one in Heilbron and Kuhn 1969, is my own.

13. Rud Nielsen 1972, p. 395.

14. Van Leeuwen 1921, who did not refer to Bohr's earlier work of which she most likely was unaware. The Bohr–van Leeuwen theorem has found use in several branches of modern physics, including plasma physics and electrical engineering.

15. Rud Nielsen 1972, p. 298. Emphasis added.

16. Pais 1991, p. 121.

17. Rud Nielsen 1972, p. 429. See also Niels Bohr to Harald Bohr, 23 October 1911, in Rud Nielsen 1972, pp. 527–32. On Bohr's acquaintance with Oseen, see Rosenfeld and Rüdinger 1967, p. 39. More details on Oseen and his promotion of quantum and relativity physics in Sweden are given in Grandin 1999.

18. Rud Nielsen 1972, p. 427. In the letter to Harald of 23 October: 'In fact, Thomson has so far not been as easy to deal with as I thought the first day.... He has not yet had time to read my paper, and I don't know if he will accept my criticism. He has only talked to me about it a few times for a couple of minutes'.

19. As late as 1920 Bohr attempted to get the dissertation published in the United States. For his various unsuccessful efforts, see Rud Nielsen 1972, pp. 103–19.

20. Quoted in Rosenfeld and Rüdinger 1967, p. 45, who provide no date.

21. Undated letter quoted in Caroe 1978, p. 70. W. Henry Bragg was at the time at Leeds University. In 1915 father and son shared the Nobel Prize for their pioneering studies of crystal structures by means of X-rays.

22. W. L. Bragg, Autobiographical Notes, quoted in Jenkin 2008, pp. 319–20. Jeans's lecture course was given in the Lent term 1912.

23. Bohr to Oseen, 1 December 1911, in Rud Nielsen 1972, p. 427.

24. For the history of the magneton, see Hoddeson et al. 1992, pp. 384–94 and Okada 2002. For Weiss's magneton, which turned out to be a mistake, see Quédec 1988.

25. This suggestion is due to Heilbron and Kuhn 1969, where further arguments and sources can be found.

26. See Okada 2002.

27. Wilson 1983, p. 326. Aaserud 2007, p. 108. See also the 1962 interview with Bohr (www.aip.org/history/ohilist/4517_1.html), where Lorrain Smith is transcribed as 'Lawrence Smith'.

28. Bohr 1932, p. 353.

29. In his book on radioactivity completed in the fall of 1912 (preface dated October 1912), Rutherford distinguished between two types of instability, the one rooted in the nucleus (alpha rays) and the other in the outer system of electrons (beta and gamma rays). When an alpha particle is expelled from the nucleus, 'Part of the surplus energy of a ring of electrons is released in the form of a high speed β particle and part in the form of γ rays'. Rutherford 1913a, p. 622. See further Section 2.7.

30. Hevesy to Bohr, 15 and 29 January 1913, in Hoyer 1981, p. 528. Hevesy 1913.

31. Bohr to Hevesy, 7 February 1913, in Hoyer 1981, p. 530.

32. Rutherford to Fajans, 19 April 1913, as quoted in Peierls 1988, p. 232.

33. Fajans to Rutherford, 13 December 1913, quoted in Jensen 2000, p. 34.

34. Hevesy to Rutherford, 14 October 1913, as reproduced in Eve 1939, p. 225.

35. Bohr 1961, p. 1085. Similarly in the 1962 interview by Kuhn et al. (www.aip.org/history/ohilist/4517_1.html) as quoted in Pais 1991, pp. 126–7.

36. Darwin 1912, p. 901.

37. Rud Nielsen 1972, p. 555. Although this is the first reference to Rutherford's atomic model in Bohr's correspondence, he most likely knew of the model even while at Cambridge.

38. Letter of 19 June 1912, ibid., p. 559.

39. Bohr 1913a, pp. 26–7. The paper is reprinted in Thorsen 1987, pp. 50–71.

40. Bohr 1913a, p. 25. Cuthbertson and Cuthbertson 1910.

41. Bohr to Cuthbertson, 4 April 1913, in Thorsen 1987, p. 666. The letter was a reply to Cuthbertson to Bohr, 23 March 1913.

42. Hoyer 1981, p. 577.

43. The Rutherford memorandum is reproduced in Rosenfeld 1963, pp. xxi–viii and Hoyer 1981, pp. 135–58. Rosenfeld assumes that a page is missing in the extant document preserved at the Niels Bohr Archive. A careful analysis of the memorandum is given in Heilbron and Kuhn 1969, who note some mistakes in Bohr's arguments. Some of my quotations from the memorandum are slightly modified, but only with regard to Bohr's frequent danicisms and spelling mistakes.

44. Hoyer 1981, p. 136. The following quotations are from the same source.

45. For the arguments in support of $K \cong 0.6h$, see Rosenfeld 1963, p. xxxi, and Heilbron and Kuhn 1969, pp. 250–1.

46. Pais 1991, p. 133. Rud Nielsen 1963, p. 22.

47. Bohr to Rutherford, 6 July 1912, in Hoyer 1981, p. 577.

48. Bohr to Rutherford, 4 November 1912, in Hoyer 1981, p. 577.

49. Bohr to Rutherford, 22 June 1913, in Hoyer 1981, p. 587. For Bohr's considerations, see Hoyer 1973, pp. 189–93.

50. Niels Bohr to Harald Bohr, 23 December 1912, in Rud Nielsen 1972, p. 563.

51. Bohr to Rutherford, 31 January 1913, in Hoyer 1981, p. 579.

52. Fujisaki 1982 argues that Bohr was greatly influenced by the theories of Drude and Stark concerning molecular spectra and that he considered the molecular hydrogen spectrum in the ultraviolet and infrared regions months before he turned to the line spectrum of atomic hydrogen. Fujisaki's arguments are circumstantial and have no convincing support in the sources.

53. Bohr to Hevesy, 7 February 1913, in Hoyer 1981, p. 529.

54. Rosenfeld 1963, p. xxxix.

55. Christiansen 1910, where Balmer's formula appears on pp. 353–4. Also the second edition of 1903 included the Balmer formula (p. 456). Incidentally, the fourth edition of the same textbook, from 1915 and revised by Martin Knudsen, mentioned Bohr's new theory in connection with Planck's introduction of the quantum of action: 'This approach has later been used by N. Bohr in a theory which, among other things, provide an explanation of the location of the lines in the Balmer series and other spectral series' (p. 456).

56. Interview of 7 November 1962 by Thomas Kuhn, Léon Rosenfeld, Erik Rüdinger and Aage Petersen (Niels Bohr Library & Archives, American Institute of Physics, www.aip.org/history/ohilist/4517_1.html). It is possible that the book where Bohr looked up Balmer's formula was Stark's *Atomdynamik*, which was in his reference library (Stark 1911). Stark's book introduced the Balmer spectrum and formula on pp. 43–5.

57. This formula was actually inferred by Balmer, but in an unpublished manuscript only. See Banet 1970 for an exposition of Balmer's route to his spectral formulae for hydrogen.

58. A plausible reconstruction of Bohr's analysis in February–March 1913 is offered in Rosenfeld 1963, pp. xl–xli.

59. Bohr to Rutherford, 6 March 1913, in Hoyer 1981, p. 581. I have changed Bohr's spelling of 'vieuw' to 'view'.

60. Bohr to Rutherford, 21 March 1913, in Hoyer 1981, p. 584.

61. There are several analyses of the first part of the trilogy. See in particular Hirosige and Nisio 1964, Jammer 1966, pp. 76–88, Heilbron and Kuhn 1969, pp. 267–83, Hoyer 1974, pp. 131–49, Petruccioli 1993, pp. 50–64, and Arabatzis 2006, pp. 124–44. A concise overview of the trilogy and Bohr's further development of his theory is given in El'yashevich 1986.

62. It was well known at the time that a natural length existed in the form of the radius of the electron, as given by the theory of electromagnetism. However, this length ($e^2/mc^2 \cong 10^{-13}$ cm) involved the speed of light in vacuum.

63. Bohr 1913b, p. 2. Bohr was aware of the discussions at the first Solvay congress, to the published report of which he referred (Langevin and de Broglie 1912). See also Mehra 1975. As mentioned, the problem of the natural length was also included in the earlier Manchester memorandum.

64. Bohr 1913b, p. 4. Bohr calculated the rate of change of radius and frequency of an electron circulating about a nucleus, but did not include the calculations in his published work. See Hoyer 1974, pp. 143–9 and Hoyer 1981, pp. 243–8.

65. Bohr 1913c, p. 488. Thomson 1912, p. 218. In 1914 the German physicist Heinrich Rau, at the University of Würzburg, reported experiments from which he inferred an ionization energy

for hydrogen of about 13 eV (Rau 1914). One year later Bohr narrowed down his theoretical value to 13.6 eV, which remains the accepted value to this day (Bohr 1915a).

66. Bohr 1913b, p. 7. Bohr also referred to Haas's attempt of 1910 'to explain the meaning and the value of Planck's constant on the basis of J. J. Thomson's atom-model'.

67. Bohr 1913b, p. 24. As we shall see in Section 3.6, Bohr's attempt of reconciliation was not accepted by Nicholson, who in 1913–1915 seriously criticized Bohr's theory.

68. Bohr 1913b, p. 7. The two assumptions are generally known as Bohr's postulates, but he did not refer to them as such until several years later (e.g. Bohr 1921a, p. 1).

69. Russell 1927, p. 175.

70. Ehrenfest 1913. While Bohr knew about Nicholson's work, he was not aware of Ehrenfest's in time to include it in his work.

71. Bohr 1913b, p. 15 and Bohr 1913c, p. 477.

72. Bohr 1913b, pp. 16–17. Wood 1911, p. 513.

73. Sommerfeld 1922a, p. 52. In his paper introducing the new quantum mechanics, Heisenberg (1925, p. 879) similarly referred to 'the Einstein–Bohr frequency condition'.

74. Olmstead and Compton 1923.

75. A detailed examination of Bohr's stability calculations is given in Hoyer 1973 and Hoyer 1974, pp. 100–13.

76. Ritz 1908. On Ritz's important works in theoretical spectroscopy, see Carazza and Robotti 2002.

77. Bohr 1913c, p. 482 and p. 495.

78. Bohr's lecture was first published in *Fysisk Tidsskrift* **12** (1914), 97–114. An English translation appeared in Bohr 1922a, pp. 1–19 and is reprinted in Hoyer 1981, pp. 283–301.

79. Bohr 1913d, p. 874. On the relation between Planck's second theory of quanta and Bohr's atomic theory, see Hirosige and Nisio 1964, who argued that Planck's theory was of great importance to Bohr's. On this matter Heilbron and Kuhn (1969, p. 272) disagreed, but Kuhn (1978, note 33 on p. 320) later admitted the importance of Planck's second theory. See also the discussion in Achinstein 1993.

80. Bohr 1922a, p. 10.

81. Ibid., p. 11. Bohr's statement can be seen as an early (if restricted) expression of the observability principle that became important in the final phase of the old quantum theory (Section 8.6).

82. Bohr 1913b, p. 19.

83. Bohr 1922a, p. 10 and pp. 12–13. According to Pais 1991 (p. 148), Bohr's derivation of the Rydberg constant represented 'a triumph over logic'. In his perceptive study of Bohr's theory, Petruccioli calls the theory 'a brilliant demonstration of how all the rules of logic and methodology might be violated in the interests of a cavalier type of theoretical approach' (Petruccioli 1993, p. 63). On the claimed inconsistency of Bohr's atomic theory, see the appendix of the present book.

84. Bohr 1922a, p. v, preface dated May 1922.

85. Millikan 1951, p. 110.

86. Brackett 1922. Pfund 1924. It took until 1953 before the series corresponding to $\tau = 6$ was detected.

87. Lyman 1914.

88. Bohr 1915a, p. 6.

89. Lyman 1915, p. 370 and Lyman 1916, p. 91. The second paper was an extended version of the first. Lyman was guided by a formula of Walther Ritz from 1908, not by Bohr's formula of 1913. For details on the works of Ritz and Lyman, see Konno 2002.

90. Lyman 1916, p. 100.

91. Pickering 1896. For a detailed account of the complex case of the Pickering–Fowler lines, see Robotti 1983. See also Maier 1964, pp. 476–84, and, for a valuable review from a more contemporary perspective, Plaskett 1922.

92. Fowler 1912.

93. Hoyer 1981, p. 582. The chemist Niels Bjerrum, who at the time was a lecturer at the University of Copenhagen and the following year was appointed professor at the Agricultural and Veterinary College, was a personal friend of Bohr. In 1911–1912 he did pioneering work in the new field of chemical physics which included the application of quantum theory to rotating and vibrating molecules (see Section 6.3). Bohr was acquainted with Bjerrum's works and referred to them in the third part of the trilogy, in connection with the hydrogen molecule. On his part, Bjerrum referred to Bohr's atomic theory in a paper of 1914 dealing with the infrared spectra of carbon dioxide and other simple gases. See Bjerrum 1914, p. 749, with English translation in Bjerrum 1949, pp. 41–55.

94. Rutherford to Bohr, 20 March 1913, in Hoyer 1981, p. 583. In Bohr's theory the electron did not 'vibrate', contrary to what Rutherford seems to have thought.

95. Evans 1913, dated 11 August. This may have been the first published reference to Bohr's atomic theory.

96. Fowler 1913. Bohr 1913e. In March 1914 Bohr pointed out that while $R_H = 109\ 675\ \text{cm}^{-1}$, the value for the heavier elements would be $R_\infty = 109\ 735\ \text{cm}^{-1}$ (Bohr 1914a, p. 512). The philosopher Imre Lakatos (1970, p. 149) referred to Bohr's response as a case of 'monster-adjustment', that is, turning a counterexample into an example.

97. Fowler to Bohr, 27 October 1913, in Hoyer 1981, p. 503. Fowler publicly admitted the agreement in a note accompanying Bohr's paper in *Nature*. In his Bakerian Lecture of 2 April 1914 he used Bohr's expression $R_\infty/R_H = 1 + m/M$ to suggest a proton–electron mass ratio of $M/m = 1836 \pm 12$ (Fowler 1914, p. 258).

98. Paschen to Sommerfeld, 24 February 1915, in Eckert and Märker 2000, p. 500. Paschen 1916.

99. Evans 1915.

100. Fowler to Bohr, 6 March 1915, in Hoyer 1981, p. 509.

101. Merton 1915, p. 383.

102. Nicholson 1913b. Bohr 1913c, p. 490. Ernest Wilson (1916, p. 25) called the new Balmer-like series 'one of the most remarkable and important points of Nicholson's work'.

103. Bohr 1913c, p. 480.

104. Bohr to Fowler, 15 April 1914, in Hoyer 1981, p. 504. Bohr included the formula in a paper of 1915 (Bohr 1915c, p. 334). On Bohr's relativistic calculations, see Hoyer 1974, pp. 241–7, and the sheet reproduced in Hoyer 1981, p. 382. This was the first scientific use of the theory of relativity made by a Danish physicist.

105. Curtis 1914, who referred to Bohr's theory and also to his still unpublished modification of the simple Balmer formula, which he knew from the correspondence between Bohr and Fowler.

106. Bohr to Fowler, 28 April 1914, in Hoyer 1981, p. 506.

107. On these aspects, see Kragh 1977 on which the present section draws. Bohr's atomic chemistry has generally been ignored by historians of chemistry (but see Scerri 2007, pp. 188–97, on Bohr's 1913 version of the periodic system). More will follow in Chapter 6.

108. Bohr to Hevesy, 7 February 1913, in Hoyer 1981, p. 530. Bohr explicitly excluded gravitation and radioactivity from his notion of 'chemical and physical properties of matter'.

109. Bohr 1913c, p. 486.

110. Franck and Hertz 1913.

111. Bohr 1913c, p. 489. For Bohr's use of dispersion data, see Konno 2000.

112. As Langmuir commented: 'Bohr's original theory would thus indicate that lithium and beryllium (and all subsequent elements) should be inert gases having even greater stability than helium' (Langmuir 1921c, p. 341).

113. Bohr 1913c, p. 496.

114. Abegg 1904. Abegg, a professor of physical chemistry at Breslau, was among the first to give an electron treatment of chemical affinity and valence. J. J. Thomson discussed Abegg's rule of eight and its relation to his own atomic theory in *The Corpuscular Theory of Matter* (Thomson 1907, pp. 118–19). He had to admit that his theory was unable to explain the rule. Since Bohr was familiar with Thomson's book he presumably was aware of Abegg's two valences and the relation between them. On Abegg's work and its significance for the early attempts to base the periodic system on arrangements of electrons, see Stranges 1982, pp. 46–8, 80–4.

115. Hevesy to Bohr, 6 August 1913, in Hoyer 1981, p. 532. The spelling is Hevesy's.

116. Bohr 1913d, p. 862.

117. Bohr 1913d, p. 863.

118. Langmuir 1912. Bohr referred to Langmuir's result, which he found difficult to reconcile with his own.

119. Langmuir 1914, p. 188. Langmuir to Bohr, 2 December 1913, and Bohr to Langmuir, 17 December 1913, in Hoyer 1981, pp. 539–40.

120. Langmuir 1915, p. 452.

121. Cuthbertson and Cuthbertson 1910.

122. For more on this failure, see Sections 3.9 and 6.1. Using modern ideas Bohr's theory of molecules can be extended to provide a good picture of the covalent bond and molecular behaviour (Svidzinsky et al. 2005). Interesting as this observation is, from a historical point of view it is of course irrelevant.

123. The first paper was abstracted by J. F. Spencer and the two others by H. M. Dawson (Spencer 1913; Dawson 1913).

124. Buchner 1915, who argued that his revised electron configurations might explain some of the chemical analogies known from inorganic chemistry, such as the analogy between the ammonium ion and the alkali ions. See also Section 6.1. Bohr contemplated a response to Buchner's paper, but his response remained a draft (Archive for History of Quantum Physics, Bohr Scientific Manuscripts).

125. Bohr to Rutherford, 10 June 1913, in Hoyer 1981, p. 586.

126. Niels Bohr to Harald Bohr, 3 August 1913, in Rud Nielsen 1972, p. 565.

127. Richardson 1916 (second edition of Richardson 1914), p. 395.

128. Manuscript on magnetism, reproduced in Hoyer 1982, pp. 254–6.

129. Rutherford 1912, p. 461 and similarly in Rutherford 1913a, p. 622.

130. Bohr 1913c, p. 500.

131. Bohr to Rutherford, 10 June 1913, in Hoyer 1981, p. 586.

132. Rutherford 1913b. However, Rutherford distinguished between primary and secondary beta electrons. He found it 'very likely that a considerable fraction of the β rays which are expelled from radioactive substances arise from the external electrons... arising from the primary expulsion of a β particle from the nucleus'. See also Jensen 2000, pp. 34–7.

133. Soddy 1913. For Curie, see her discussion contribution in Thomson 1921, p. 56.

3

Reception and Early Developments

Having completed his new theory of atomic constitution Bohr was eager to promote and develop it, which he did with great determination and considerable success over the following years. It was very much his theory, and it was left to him to make it known and accepted. Given the radical nature of the postulates on which the theory rested, Bohr could be satisfied with how it was received in the physics communities in England and Germany, the two leading countries for physics at the time. Of course its victory was not complete, for many physicists resisted it and many more were indifferent or just ignorant of it. This was the case in France, where the theory left no mark for several years. Léon Brillouin was one of the few French physicists interested in quantum and atomic theory, and even he took up the Bohr theory only after World War I. He recalled that 'I heard about quanta, photons and the like at Munich just before the war, and then during the war I lost contact completely with physical work. When I was demobilized in 1919 I took the book of Sommerfeld [*Atombau and Spektrallinien*] and read it like a novel'.[1] Bohr's theory was covered in a review article of 1918 written by Léon Bloch, but no French physicist contributed to its development until after 1920.[2]

In spite of the lack of appreciation in France and some other countries, by the end of 1915 the majority of physicists doing research in atomic physics and related areas recognized that Bohr's theory constituted an important advance that might well define the course of future research. Five years later it had achieved a nearly paradigmatic status.

Bohr's model of the atom was successful in both a social and a scientific sense. What swayed otherwise sceptical physicists to accept it was primarily its empirical successes, that is, its remarkable ability to account for or predict phenomena that other theories failed to explain. In his publications of 1913–1915, Bohr emphasized the explanatory power of his theory rather than its foundation in the admittedly strange quantum postulates. It was this ability to explain known facts and predict new ones that attracted most attention and forced physicists to take the theory seriously. The theory got its best possible start by explaining the Pickering–Fowler lines and defending its explanation of the hydrogen and helium spectra against the objections of sceptical spectroscopists. At about the same time Bohr's theory received unexpected support from Moseley's and later Kossel's analyses of the characteristic X-rays. Other discoveries that were announced only after the theory had appeared, most notably the Stark effect and the

Franck–Hertz effect, could be understood within the framework of the theory. This kind of fertility, robustness, and explanatory breadth impressed theorists and experimentalists alike. By the end of 1915 the successes weighed more heavily than the failures or inadequacies, including the theory's problematic conceptual framework.

If the empirical power of Bohr's theory made it attractive, its theoretical basis in the quantum postulates made it, to the minds of many physicists, unattractive. The opposition to the Bohr atom was in part empirically based and in part of a conceptual nature. Only a few physicists concluded that the empirical arguments against the theory were reason enough to dismiss it *in toto*. But some thought that it promised more than it could deliver and, in addition, that it was methodologically unacceptable (see also Section 4.5). To British physicists, in particular, it was a stumbling block that it did not provide a dynamical mechanism that allowed a physical picture or idea of what went on inside the atom to be formed. Some critics felt, not entirely without justification, that the theory's basic architecture was messy, a patchy combination of incomparable elements from quantum theory and classical electrodynamics. This was a feeling that would only grow stronger with time. As Henry Margenau later phrased it, 'Bohr's atom sat like a baroque tower upon the Gothic base of classical electrodynamics'.[3] Bohr himself was well aware of the apparent inconsistency, but he saw it as a temporary necessity and resource rather than a weakness. In the end, the inconsistency became the basis of quantum mechanics.

3.1 THE EARLIEST RESPONSES

A few reactions to Bohr's atomic theory appeared even before the first part of the trilogy had appeared in print. As mentioned in Section 2.5, Rutherford was puzzled at how an electron could 'decide' which quantum transition to perform. Did it have a choice? A free will? He instinctly sensed the element of acausality associated with Bohr's atom, a feature which would move to the forefront of discussion several years later. Rutherford's uneasiness was shared by Paul Ehrenfest, who in a letter to Lorentz of 25 August 1913 expressed his immediate reaction to Bohr's theory of the atom in this way: 'Bohr's work on the quantum theory of the Balmer formula (in the *Phil. Mag.*), has driven me to despair. If this is the way to reach the goal, I must give up doing physics'.[4] Ehrenfest was thoroughly familiar with quantum theory, but Bohr's way of applying quantum concepts to atomic structure puzzled him. It did not appeal to him at all, and it took several years until he came to accept Bohr's approach. As late as in the spring of 1916 he thought of the Bohr atomic model as 'completely monstrous'.[5] When he converted to the Bohr theory two years later, he became one of its strongest supporters, a true devotee.

Arnold Sommerfeld, professor of theoretical physics in Munich, was among the physicists to whom Bohr sent preprints of his paper in the July issue of *Philosophical*

Magazine. At the time he received the preprint, Sommerfeld had already read the paper, as recalled by Léon Brillouin:

> When Bohr's theory on the hydrogen atom was published in 1913, Sommerfeld immediately saw the importance of this new idea. I happened to be in his office when he opened the issue of Philosophical Magazine, which had just arrived; he glanced through it and told me, 'There is a most important paper here by N. Bohr, it will mark a date in theoretical physics'. And soon after, Sommerfeld started applying his own 'quantum of action method' to rebuild a consistent theory of Bohr's atom.[6]

However, there is little doubt that the recollection, told 36 years after the event, is to some extent a reconstruction. If Sommerfeld had really considered Bohr's paper to be of revolutionary importance, it is hard to understand why it took him so long to make the theory part of his own research work and the work of his assistants and students in Munich. We get a more authentic picture of Sommerfeld's first impression of Bohr's theory in a postcard he sent to Bohr on 4 September:

> I thank you very much for sending me your highly interesting work, which I have already studied in Phil. Mag. The problem of expressing the Rydberg–Ritz constant by Planck's *h* has for a long time been on my mind. Though for the present I am still rather sceptical about atomic models in general, calculating this constant is undoubtedly a great feat... From Mr. Rutherford, whom I hope to see in October, I may perhaps learn more details about your plans.[7]

Sommerfeld's scepticism did not evaporate instantly, and only at the end of 1914 did he seriously engage himself with the new atomic theory, which he soon extended in a most fruitful way.

The 1913 meeting of the British Association of the Advancement of Science, which took place in Birmingham 10–17 September, provided a golden opportunity for Bohr to get his new theory of atoms and molecules on the scientific agenda. Rutherford had suggested to the organizers that Bohr should be invited to take part in the discussion on radiation,[8] but due to his new position at the University of Copenhagen Bohr was uncertain about his possibility of going to Birmingham. However, realizing the importance of the meeting, at the last minute he decided to attend. He must have been pleased to listen to Oliver Lodge's presidential address, which included a reference to the 'very remarkable' agreement between the observed spectrum of hydrogen and the one calculated on the basis of Bohr's theory. 'Quantitative applications of Planck's theory, to elucidate the otherwise shaky stability of the astronomically constituted atom, have been made', Lodge said. 'One of the latest contributions to this subject is a paper by Dr Bohr in Philosophical Magazine for July this year'.[9] Lodge sensed that modern atomic and quantum theory presented a danger to the essential continuity in nature, and he did not like the challenge. As a believer in the continuous ether as the ultimate reality, he was bound to dislike the Bohr atom and its discontinuous quantum jumps (not to mention its disregard of the ether).

Although Bohr did not give a formal paper in Birmingham, he participated in some of the discussions and also gave a brief account of his theory on the 12th of September. According to the description in *Nature*:

> [Bohr's] scheme for the hydrogen atom assumes several stationary states for the atom, and the passage from one state to another involves the yielding of one quantum. Dr. Bohr also emphasised the difficulty of Lorentz's scheme for distinguishing between matter and the radiator . . . Prof. Lorentz intervened to ask how the Bohr atom was mechanically accounted for. Dr. Bohr acknowledged that this part of his theory was not complete, but the quantum theory being accepted, some sort of scheme of the kind suggested was necessary.[10]

Bohr's intervention took place in the discussion following Jeans' exposition of the problems of radiation theory, in which Jeans had given an account of Bohr's 'most ingenious and suggestive, and I think we must add convincing, explanation of the laws of spectral series'.[11] Jeans had only recently become sympathetic to quantum theory. Two years earlier, at the first Solvay congress, he had attempted to account for Planck's constant of action on classical grounds. Still at the Birmingham meeting he tentatively suggested a dynamical interpretation of the quantity, namely

$$h \cong 2\pi \frac{(4\pi e)^2}{c}$$

or, in terms of the later fine-structure constant α,

$$\frac{1}{\alpha} \cong 16\pi^2 \cong 158.$$

Jeans' interpretation of h went contrary to Bohr's thinking, for according to him Planck's constant was an irreducible constant of nature that could not be explained in terms of other constants. Although Jeans found Bohr's theory convincing, he was less happy with its foundation in the two quantum postulates. 'The only justification at present put forward for these assumptions is the very weighty one of success', he said.[12] He also noted that it had difficulties in explaining 'the Zeeman effect and interference'. Immediately after Jeans had given his address, a joyful Bohr wrote to his wife in Copenhagen:

> Jeans, who opened the discussion of the radiation problem, gave a very beautiful and kind presentation of my theory. I think that he is convinced that there is at least some reality behind my considerations. Just as little as in Cambridge, I have been able to talk much with him; he is very reticent, but I have come to like him a lot and think he is such a noble man. You should know how charming Lorentz is and how he during the discussion in the wisest and, at the same time, most amiable and certain way reproached J. J. Thomson and others

of the old school. I am going to meet him this afternoon at Sir Oliver Lodge and look very much forward to it.[13]

It was not only British physicists who became aware of Bohr's theory through Jeans' presentation. In a report on the Birmingham meeting in the *Physikalische Zeitschrift*, Paul Ewald included in full the part of Jeans' review in which he dealt with Bohr's theory.[14] He also gave a few more details on Bohr's critical remarks to Lorentz's notion of resonators and material particles. According to Ewald's report, Bohr argued that the relationship between the two concepts could be understood on the basis of his new model of the atom: 'The atom belongs to "matter" when the electron moves in a stationary orbit round the positive nucleus; the atom is a "resonator" at the time of transition from one orbit to another, that is, at the time it radiates'.[15]

Jeans spoke even more positively, and in greater detail, about Bohr's theory in his influential report on radiation and quantum theory that appeared the following year. The new quantum theory of atoms, as exposed in the 'very remarkable and intensely interesting Papers [sic] by Dr Bohr, of Copenhagen', appeared prominently in the report. As Jeans phrased it, Bohr's fundamental assumption 'is not inconsistent with the quantum-theory and is closely related to it'.[16] Although Jeans expressed some reservation with respect to the applicability of Bohr's theory to more complex atoms, he praised it for having opened a rich field by the use of quantum theory to problems of atomic structure. Moreover, he showed, more clearly and in greater detail than Bohr had done, that the photoelectric effect as interpreted by Einstein 'is now seen to be a necessary logical extension of Bohr's theory of absorption'.[17]

In another part of the Birmingham meeting, McLaren, by then a professor of mathematics at University College, London, referred briefly to Bohr in connection with the theory of magnetism.[18] Shortly afterwards he called attention to Bohr's use of Planck's constant as a measure of the angular momentum of revolving electrons, suggesting that Bohr's theory gave support to his own idea of an elementary magnetic quantity or magneton. According to McLaren, 'Bohr's postulate of a natural unit of angular momentum was very prominent' at the meeting of the British Association. He called it 'an idea of the first importance'.[19] Nicholson, on the other hand, was more guarded, if not yet negatively disposed towards Bohr's theory. Noting its recent success in explaining the Pickering–Fowler lines, he nonetheless opined—prophetically as it turned out—that 'The real test of his theory will lie in its capacity to account for the *usual* spectrum of helium'.[20] As to McLaren's theory of magnetons it was part of an ambitious electrodynamical theory of gravity that in scope and spirit was entirely different from Bohr's ideas. McLaren regarded the magneton as 'an inner limiting surface of the æther, formed like an anchor-ring', a notion which belonged to an altogether different framework of thought than the one of Bohr's theory.[21]

Unsurprisingly, Rutherford was much in favour of the new atomic theory, which complemented and justified his own earlier theory of the nuclear atom. But Rutherford's research interest was radioactivity and the atomic nucleus, not the electron system with which Bohr's theory was primarily concerned, and his early explicit support of Bohr's theory was consequently limited to a few general remarks. On the other hand, he left no doubt about his high opinion of the theory. Thus, in a paper with John M. Nuttall in the October issue of *Philosophical Magazine* he referred to the hydrogen and helium models 'assumed by Bohr in a recent interesting paper on the constitution of atoms, and [which have] been shown by him to yield very promising results'.[22] Although advocating Bohr's theory, Rutherford was reluctant to comment on the central parts of the theory dealing with spectroscopy and the quantum postulates. For example, in a paper of February 1914 he mostly dealt with Bohr's ideas of the nucleus and the nuclear origin of beta rays.[23] Only at the end of the paper did he refer to the quantum atomic theory. There can be no doubt, he concluded, 'that the theories of Bohr are of great interest and importance to all physicists as the first definite attempt to construct simple atoms and molecules and to explain their spectra'.[24] In another paper from the same time he expressed the feeling of a growing number of physicists:

> There no doubt will be much difference of opinion as to the validity of the assumptions made by Bohr in his theory of the constitution of atoms and molecules, but a very promising beginning has been made on the attack of this most fundamental of problems, which lies at the basis of Physics and Chemistry.[25]

The responses from German physicists were fewer and came later than those of their British colleagues. Shortly after the meeting of the British Association in Birmingham a corresponding meeting of the Association of German Scientists and Physicians (Gesellschaft deutscher Naturforscher und Ärzte) took place in Vienna. Speakers included Einstein, James Franck, Max von Laue, Johannes Stark, and Max Born. None of the published addresses referred to Bohr's theory, but there is little doubt that it was discussed informally.[26] Hevesy was present and in a letter written during the congress he told Bohr that he had just had a conversation with Einstein and told him that it was now certain that the Pickering–Fowler spectrum belonged to helium, in agreement with Bohr's theory:

> When he heard this he was extremely astonished and told me: 'Than the frequency of light does not depand at all on the frequency of the electron – (I understood him so??) And this is an *enormous achiewement*. The theory of Bohr must be then wright.' I can hardly tell you how pleased I have been and indeed hardly anything else could make me such a pleasure than this spontaneous judgement of Einstein.[27]

A few weeks later, Hevesy reported his conversation with Einstein in a letter to Rutherford, making essentially the same point: 'When I told him about the Fowler spectrum the big eyes of Einstein looked still bigger and he told me "Then it is one of the greatest discoveries"'.[28]

Perhaps—just perhaps—Einstein's positive reaction to Bohr's theory was influenced by his own earlier ideas of emission and absorption of light. In a letter to Lenard of November 1905 he mentioned the possibility that 'the emission or absorption of each individual spectral line is connected with a definite state of the emitting or absorbing centre (atom), which is characteristic for it'. Based on this letter, it has been speculated that at the time Einstein might have had the Bohrian idea of discrete energy states of the atom in which it does not radiate.[29] But this is admittedly a speculation.

That Bohr's theory was known in Germany is further indicated by a letter of October in which Geiger congratulated Bohr on his new work and asked for preprints for a colleague at the Physikalisch-Technische Reichsanstalt in Berlin.[30] Later the same month, from 27 to 31 October, the second Solvay conference on physics convened in Brussels. While the theme of the first conference had been radiation theory and quanta, the theme of the second was *La Structure de la Matière*, the structure of matter. Several of the talks dealt with atomic and molecular physics, but Bohr's new theory—which in a sense integrated the themes of the two conferences—was not mentioned in the published version of the talks and discussions.[31] Among the participants were Jeans, Sommerfeld, Einstein, Lorentz, Thomson, and Rutherford, all of whom were acquainted with Bohr's quantum atom or had at least heard of it. While Thomson in his address on the structure of atoms did refer to Bohr, it was not to his atomic theory but to an earlier work on the collision between charged particles and atomic electrons. Likewise, when Rutherford referred to the ideas of 'van den Broek and Bohr', it was to the new notion of atomic number.[32] Bohr's atomic theory did not make a splash in Brussels, not even a ripple.

3.2 A DIGRESSION ON TRIATOMIC HYDROGEN

Among the talks at the 1913 meeting of the British Association that Bohr attended was one by Thomson with the enigmatic title 'X_3 and the Evolution of Helium'. In his experiments with positive rays Thomson had detected atomic or molecular ions with a mass-to-charge ratio *m/e* three times that of the ordinary hydrogen ion H^+, and these he suggested were ions of a substance X_3 which was most likely triatomic hydrogen. He first stated the $X_3 = H_3$ hypothesis in 1911 and gave further arguments for it in his Bakerian Lecture of 1913 and in a monograph of the same year, *Rays of Positive Electricity*.[33] What Thomson had found was the H_3^+ ion, but he was convinced that it was merely the charged version of a stable molecule that was chemically inert or nearly so. Realizing that the existence of H_3 conflicted with ordinary ideas of valency (not to mention chemical experience), he suggested that these ideas would have to be modified in the light of modern theories of atomic structure: 'If... we regard an atom of hydrogen as consisting of a positive nucleus and one negative corpuscle, it will exert forces analogous to those excited by a magnet and I can see no reason why a group of three of these arranged so that their axes form a closed ring should not form a stable arrangement'.[34]

During the discussion in Birmingham following Thomson's talk Bohr suggested as an alternative the bold hypothesis that X_3 might possibly be a super-heavy isotope of hydrogen of atomic weight 3. That is, he effectively predicted what was later called tritium or hydrogen-3. Nearly fifty years later Bohr recalled the incident as follows: 'I just took up the question of whether in hydrogen one could have what you now call tritium. And then I saw that it was a way to show this by its diffusion in palladium. Hydrogen and tritium will behave similarly but the masses are so different that they will get separated out'.[35] A few weeks after the meeting of the British Association Hevesy wrote to Rutherford about Bohr's suggestion, which was not kindly received by the physicists and chemists gathered in Birmingham:

> In the following discussion Bohr – in his usual modest way – suggested the possibility of X_3 being an H atom with *one* central charge, but having a three times so heavy nucleus than Hydrogen.... The general appearance was that he [Thomson] told something highly ingenious and Bohr something very stupid. Just the contrary was the case. So I felt bound to stick up for Bohr and explained the meaning of Bohr's suggestion in more concrete terms, saying that Bohr's suggestion is that X_3 is possibly a chemically non-separable element from Hydrogen... Of course it is not very probable, but still a very interesting suggestion, which should not be quickly dismissed.[36]

Thomson was not the only prominent physicist in favour of triatomic hydrogen. Experiments made by Johannes Stark in Germany led to the same conclusion. Stark believed that the molecule and its corresponding ion agreed with his own unorthodox ideas of valence and molecular constitution, indeed provided strong support for them.[37] According to him, a triatomic ring of hydrogen atoms could be formed by three atoms and three electrons held in an equilibrium position by electric lines of force. The electrons shared by two hydrogen atoms would form a chemical bond between them. (Contrary to Lewis's later theory of shared electrons, Stark needed only one electron to form a bond.) Neither physicists nor chemists showed much interest in Stark's version of triatomic hydrogen.

As for Bohr, for a while he remained interested in the question of a form of hydrogen of atomic weight 3, whether a heavy isotope or a molecule. The mass correction he had introduced in connection with the hydrogen and helium spectra made it possible, at least in principle, to distinguish between different isotopes of the same element. Bohr did not mention this possibility in his trilogy, but he referred to it in unpublished notes from 1913 or 1914 concerning the triatomic hydrogen molecule proposed by Thomson. He wondered, as he had done at the Birmingham meeting, if the substance detected by Thomson could be a hydrogen isotope with atomic weight 3 rather than triatomic hydrogen. Bohr's theory offered, he said in the notes,

> ...a possibility to discriminate between the above eventualities, since according to this theory the spectrum of an element of atomic weight 3 and, as hydrogen, containing one electron should show a spectrum the lines of which are closely but not exactly coinciding

> with the lines of the hydrogen spectrum; the difference in wave length being of an order open for detection.[38]

Together with H. M. Hansen, Bohr even seems to have made experiments to settle the question. This was the first mention of what is known as the isotope effect or isotopic spectral shift, a method that many years later was applied to detect the heavy hydrogen isotope deuterium.[39] Bohr first mentioned the isotope effect in public at the 1915 meeting of the British Association: 'In the case of spectral vibrations, there occurs a small term depending on the mass of the central nucleus, and accordingly we ought to look out for a small but perceptible difference between the spectra of two isotopes'.[40] In notes prepared for the meeting he was more specific:

> We shall expect a correction of the wave-lengths represented by the factor $1 + m/M$, where m and M are the masses of the electron and the nucleus respectively. For elements of high atomic weight this correction is very small, e.g. the difference in the correction for isotopic lead of atomic weight 206 and 208.4 is only $3 \cdot 10^{-8}$. On the other hand, the difference to be expected for the two isotopic neons of atomic weight 20 and 22 is $2.5 \cdot 10^{-6}$ and might perhaps be detectable.[41]

He further predicted an isotopic shift in spectra of diatomic molecules consisting of atoms of different masses M_1 and M_2 (such as HCl), where the frequency of vibration in the infrared bands would be proportional to the quantity

$$\nu_{vib} = \sqrt{\frac{M_1 + M_2}{M_1 M_2}}.$$

Thus, in the case of HCl a small isotopic shift should occur, revealing the different masses of the two chlorine isotopes Cl-35 and Cl-37.

In 1919, in a little known paper published by the Nobel Institute in Stockholm, Bohr offered a theoretical examination of the H_3 molecule along the same lines he had applied to diatomic hydrogen six years earlier (Figure 3.1). He had considered the constitution of H_3 as early as the autumn of 1915, but at the time without publishing his considerations (see Section 6.1). It is not known what caused him to reconsider the hypothetical molecule, but it may have been experiments made by two scientists at Harvard University, the physicist William Duane and the chemist Gerald Wendt.[42] Examining the action of alpha rays on pure hydrogen, Duane and Wendt found a marked volume contraction, which they tentatively ascribed to the process $3\ H_2 \rightarrow 2\ H_3$. Bohr was aware of this work, which he cited in his article. His model of 1919 consisted of three electrons rotating in a common circular orbit of radius a, with one nucleus at the centre and the other two symmetrically displaced from it at a distance b:

> The model of a hydrogen molecule containing three atoms, which, although in chemical sense it would not show the same degree of stability as the ordinary diatomic molecule, should still possess the possibility of a permanent existence if undisturbed by external

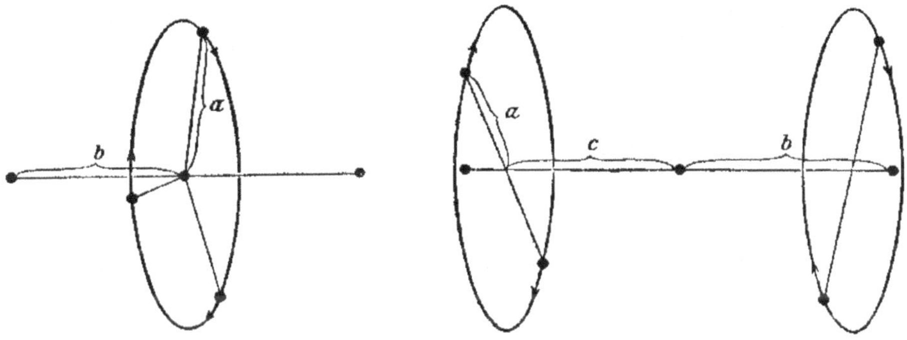

Fig. 3.1. Bohr's models of triatomic hydrogen (left) and its negative ion (right). In H_3, three electrons rotate in a common circular orbit of radius a, and the three nuclei are placed on the axis of the orbit, separated by the distance b. In H_3^-, $b = c$ and the four electrons circle between the two outer nuclei.

Source: Bohr 1919.

> agencies. This model may therefore possibly correspond to the molecule of a new modification of hydrogen, for the appearance of which under suitable conditions interesting evidence has been brought forward by Sir J. J. Thomson in his well known experiments on positive rays.[43]

In close conformity with his 1913 calculations of H_2, Bohr derived the condition for mechanical equilibrium of his H_3 model to be

$$\frac{e^2}{b^2} + \frac{e^2}{4b^2} = 3\frac{e^2 b}{(a^2 + b^2)^{3/2}}$$

from which it follows that $b = 1.123a$. For the orbit of the electrons and the total energy of the molecule in its ground state he found $a = 0.99\ a_0$ and $W = 3.06\ W_0$, where a_0 and W_0 are the corresponding values of the hydrogen atom.

Bohr further found that the process 3 $H_2 \rightarrow$ 2 H_3 was endothermic, but suggested that it might occur in a hydrogen gas ionized by, for example, the action of alpha rays. Apart from neutral H_2 molecules, such a gas would contain H^+ and H_2^-, leading to triatomic hydrogen by the exothermic process $H^+ + H_2^- \rightarrow H_3$. Moreover, he concluded that it would only take a small external action to break up the H_3 molecule into H and H_2. Bohr's theoretical investigation focused on H_3 rather than H_3^+, the reason being that for the latter system 'no configuration of mechanical equilibrium, in which the nuclei are at rest at finite distances from each other and the electrons move in circular orbits, can be formed'. If an electron were removed from H_3, the ion would split up as $H_3^+ \rightarrow H_2 + H^+$. Of course, his conclusion with respect to the positive molecule ran counter to Thomson's experimental claim. On the other hand, Bohr's calculations showed that H_3 could add another electron and form a H_3^- ion. For this ion he found $W = 3.43\ W_0$ and consequently predicted that it 'may exist permanently in the absence of external agencies'.

Although Bohr's model calculations made sense within the framework of his atomic theory, they concerned a hypothetical molecule and exerted very little influence on the further discussion of triatomic hydrogen and molecular constitution in general. While Thomson's discovery of H_3^+ was confirmed by other researchers, the neutral molecule remained controversial and became a hot topic among chemists and physicists in the 1920s. Wendt and his research student Robert Landauer were among those who thought the evidence in favour of 'active hydrogen' in the form of unstable H_3 to be convincing, not only for experimental reasons but also because the case was given theoretical weight by Bohr's calculations.[44] However, in spite of a good deal of experimental evidence for triatomic hydrogen, it was not accepted as a real molecule by the majority of chemists. New experiments and reconsiderations of the old evidence led by the mid 1930s to a consensus that H_3 did not, in fact, exist.

It remains to be said that today we know that neutral triatomic hydrogen does exist in an unstable form and that the positive ion is abundant in interstellar space. The spectra of both forms were identified in about 1980 and since then research on triatomic hydrogen has become a major interdisciplinary research field of great interest to astronomers, chemists, and physicists.

3.3 THE BRITISH SCENE

The main content of Bohr's sequence of papers in *Philosophical Magazine* was disseminated to wider circles of English-speaking physicists through *Science Abstracts*, the abstract journal issued by the Institution of Electrical Engineers. The first two papers were extensively abstracted by George De Tunzelmann, a London physicist and engineer, who summarized Bohr's theory as follows: 'The author's primary aim is to show that the introduction of Planck's constant, the elementary quantum of action, will serve, in Rutherford's model, to take the place of the radius of the positive sphere [in Thomson's model], and so make stability possible'. Curiously, the third paper received only a single line abstract, saying that it 'deals with systems containing several nuclei, on the same lines as in the earlier papers'.[45]

Owen Willans Richardson, a Cavendish physicist who in 1906 was appointed professor of physics at Princeton University, was a specialist in electron theory and the emission of electrons from hot bodies. (In 1929 he would receive the Nobel Prize for his work in this area, nominated by Bohr, among others.) He was acquainted with Bohr's theory of atomic structure not only from the papers in *Philosophical Magazine* but also from a conversation he had had with Bohr in Cambridge in July 1913, just at the time when the theory appeared.[46] In a book of 1914 on electron theory based on a series of lectures given in Princeton, *Electron Theory of Matter*, he included Bohr's new atomic theory, although in much less detail than he gave to the classical Thomson model. This was probably the earliest treatment of Bohr's theory in a regular textbook. The book was positively reviewed by Bohr, who used the opportunity to contrast the current development in atomic physics with the older but still surviving electromagnetic world view:

> In text-books only a few years old one finds great enthusiasm over what was called the future programme of the electromagnetic theory. It was believed that this theory constituted a final accomplishment of ordinary mechanics, and there appeared to be no limit to the application of the general principles of the theory... If at present we may speak of a programme for the future development, it would, perhaps, be to examine the constitution of the special atomic systems actually existing, and then, by means of the directly observable properties of matter, possibly to deduce the general principles. If so, the evolution would be exactly the reverse of that anticipated.[47]

Bohr valued the electromagnetic theory highly, but also recognized its limitations and deficiencies. Some years later, in a lecture on light and matter delivered to the Royal Danish Academy in Copenhagen, he said of the electromagnetic world view that it 'exhibits an inner harmony which is hardly attained by any other edifice within natural science'. All the same, this harmonious beauty did not guarantee truth, for the theory was detached from reality: 'While it has been possible to account for the general nature of a large number of physical phenomena, it has not been possible to explain in detail the observed properties of any of the elements that we learn about in chemical investigations'.[48]

Richardson was clearly impressed by the agreement of Bohr's theory with spectra, and noted that although Bohr's ideas 'frankly discards dynamical principles' they were nonetheless successful and promising. There is no doubt, he said, 'that this theory has been much more successful in accounting quantitatively for the numerical relationships between the frequencies of spectral lines than any other method of attack which has yet been tried'. Moreover:

> Although the assumptions conflict with dynamical ideas they are of a very simple and elementary character. The fact that they conflict with dynamics does not appear to be a valid objection to them, as there are a number of other phenomena, the temperature radiation for example, which show that dynamics is inadequate as a basis for complete explanation of atomic behaviour.[49]

In the second edition of 1916, which he prefaced on 11 January 1916, he dealt in much more detail with Bohr's theory, if still presenting it as merely an alternative to the Thomson model. He did not mention Sommerfeld's recent elaboration, which was probably unknown to him because of the war. Richardson rated the theory highly and dealt in considerable detail not only with the hydrogen atom, but also with many-electron atoms, the H_2 molecule, X-ray spectra etc. Yet he also covered Thomson's earlier theory, and that in even greater detail, carefully avoiding confronting the two theories. Given that Thomson and other physicists had abandoned the original Thomson model of 1904 several years ago, it is surprising that Richardson gave so much emphasis to it. (He only dealt briefly with Thomson's recent view of the structure of atoms and chemical combination.) Having presented the two theories, he left it to the reader to decide between the two alternatives. Although Richardson clearly valued Bohr's theory, apparently he did not fully realize its non-classical features and its disagreement with the classical electron theory on which most of the book was based.[50]

Richardson's *Electron Theory of Matter* was probably the first regular textbook that treated Bohr's theory, but it was not the first book that referred to it. In early 1914, George W. C. Kaye, a physicist at the National Physical Laboratory in Teddington, on the outskirts of London, and a former collaborator of J. J. Thomson, published a book on X-rays and their uses in which he included two references to Bohr's theory. Relegating Thomson's atomic theory to a footnote, he adopted the Bohr–Rutherford model according to which 'The outer electrons, by their number and arrangement, are responsible for the chemical and physical properties of the atom: the inner electrons have influence only on the phenomena of radioactivity'.[51] Kaye further mentioned Moseley's new and 'important deductions . . . bearing on Rutherford's and Bohr's theories of the structure of the atom'.[52]

Few British physicists realized how drastically Bohr's theory departed from conventional physics, for example that it denied the applicability of the principles of mechanics to systems of atomic dimensions. And many of those who did realize it, opposed the theory precisely for this reason. The Cavendish physicist Norman Robert Campbell recognized more clearly than most the radical nature of Bohr's atomic model. 'To attempt to explain Bohr's theory in terms of those principles [of classical physics] is useless', he pointed out in a review of January 1914.[53] Campbell praised the assumptions of the theory, which he saw as 'simple, plausible, and easily amenable to mathematical treatment; from them all the properties of any atomic system which does not contain more than one electron can be deduced uniquely'. As to more complex atomic systems Campbell admitted that the power of the theory was limited, but instead of regarding it a serious flaw he thought it was 'owing to the mathematical difficulties involved'.[54]

A more common response was to use the theory eclectically: to accept parts of it while ignoring or rejecting other parts. As an example, consider Herbert Stanley Allen, a physicist from King's College, University of London, who in a series of works in 1914–1915 investigated theoretically the effect of a magnetic force arising from the nucleus of an atom of the Bohr–Rutherford type. Allen apparently supported Bohr's atomic model: 'The success of Bohr's theory in explaining the ordinary Balmer's series in the spectrum of hydrogen, and especially in obtaining close agreement between the observed and the calculated values of Rydberg's constant, raises a strong presumption in its favour'.[55] However, the way he used parts of Bohr's theory to construct an atomic theory of his own was hardly in agreement with the views of Bohr. Not only did Allen assume that the magnetic field of the nucleus played an active part in the emission of radiation, he also suggested a picture of the atomic nucleus that differed from and was much more complex than the one proposed by Rutherford (and implicitly accepted by

Bohr). For example, in the case of heavy atoms Allen pictured the nucleus as a conglomerate of orbiting protons, alpha particles and beta particles that gave rise to an extra-nuclear magnetic field. He thought that the radius of the nucleus was much larger than the order of 10^{-15} m found by Rutherford—perhaps 1000 times as large.[56]

In his 1914 review of Bohr's theory, Campbell prophesied that 'theories of atomic structure will probably never be very interesting to chemists' because of the mathematical difficulties of calculations dealing with complex atoms and molecules.[57] This turned out to be seriously wrong, but at the time it was a reasonable assessment. Although chemists were generally reluctant to take up Bohr's ideas, many were aware of them, for instance from the detailed abstracts that appeared in the *Journal of the Chemical Society*.[58]

In a paper on radioactive elements from early 1914 Hevesy referred to Bohr's argument that the radii of isotopic ions are the same.[59] Another radiochemist, Frederic Soddy, introduced the Bohr–Rutherford atom to the chemists in his series of annual reports on radioactivity compiled at the request of the London Chemical Society. In the report for 1913, published in 1914, he adopted the new model, noting that the laws of electrodynamics did not apply to the interior of the atom. 'The model has been used with very considerable success, in conjunction with Planck's theory of quanta', he said, and it resulted in 'series relationships of the hydrogen and helium spectra, in striking accord with experimental determination'.[60] At the 1915 meeting of the British Association, taking place in Manchester under the shadow of the war, Bohr participated in a discussion on 'Radio-Active Elements and the Periodic Law' opened by Soddy. According to the summary account in *Nature*, at this occasion 'Dr N. Bohr pointed out that . . . properties depending on the outer rings of electrons would be the same for all isotopes'.[61]

Finally, Bohr's theory was of interest not only to physicists and chemists, but also, in its capacity of a theory of spectra, to some astronomers and astrophysicists. In a report on line spectra given to the Royal Astronomical Society in February 1914, the theory appeared prominently. As Fowler and Nicholson pointed out in their report, Bohr 'has given a remarkable theory of the hydrogen spectrum, which has led to a considerable amount of discussion'.[62] This was correct, but the quantum theory of the atom played only a very limited astronomical role at the time and was in fact unknown to the majority of astronomers.

Only in the early 1920s did the quantum atom move to the forefront of theoretical astrophysics. The first scientist to make innovative use of the Bohr atomic theory in an astronomical context was the Indian physicist Meghnad Saha, who in papers from 1920–1921 applied the 'modern theories of atomic structure and radiation' to the ionization of hydrogen and other elements in the solar atmosphere.[63] Then things went fast. Saha's pioneering work acted as a springboard for physicists and astronomers to develop quantitative models of stellar atmospheres on the basis of the ideas of Bohr, Sommerfeld, and other quantum theorists. By 1924, as a result principally of the work of Saha, E. Arthur Milne, and Ralph Fowler in Britain, John Eggert in Germany, and Henry Norris

Russell in the United States, the Bohr–Sommerfeld quantum atom had become an indispensable part of astrophysics.

3.4 MOSELEY AND THE CHARACTERISTIC X-LINES

Besides the successes from the spectra of hydrogen and helium, the strongest experimental support for Bohr's theory came from X-ray spectroscopy, a branch of science that did not yet exist when Bohr completed his trilogy. Although the support of X-ray spectroscopy was very important, initially it rested on a somewhat slender foundation.

The existence of monochromatic X-rays characteristic of the element emitting the rays had been known since 1906, when the phenomenon was discovered by Charles Glover Barkla, a physicist at the University of Liverpool.[64] Although Barkla could not determine the wavelengths of the characteristic rays he could study and classify them by means of their penetrating power. He soon found that there were two kinds of rays, which he named K and L radiation and where the first had a greater penetrating power than the latter. What was missing, among other things, was a method of determining the wavelength of the radiation, but such a method was provided after William Henry Bragg and his son William Lawrence Bragg in 1912 invented the X-ray spectrometer based on the reflection of X-rays on crystals.

In Manchester, Henry Gwyn Moseley, who was Bohr's junior by two years, set out to employ the method of the Braggs to measure and understand the wavelengths of the characteristic radiation. He had earlier collaborated with Darwin on X-ray diffraction, but from the summer of 1913 he pursued the new research programme alone.[65] Bohr knew Moseley, but it was only in July 1913 that he had a long discussion with him and told him about his new atomic theory. The two physicists evidently had shared interests, such as the periodic system and its relation to the atomic number. Moseley's research programme was to a large extent motivated by the possibility of confirming by means of X-ray spectroscopy van den Broek's hypothesis—or the van den Broek–Bohr hypothesis—of the atomic number. 'My work was undertaken for the express purpose of testing Broek's hypothesis, which Bohr has incorporated as a fundamental part of his theory of atomic structure', he wrote.[66]

Moseley constructed a new kind of X-ray tube where the targets could be easily interchanged and moved in position opposite to the cathode, to give out their characteristic rays. To determine the wavelengths he developed a photographic method. Having surmounted the inevitable experimental difficulties, in October 1913 he was ready to collect data, starting with the K lines from calcium to zinc. He found that there were two groups of lines, a stronger K_α and a weaker K_β. In a letter of 16 November he communicated his results—'extremely simple and largely what you would expect'—to Bohr and discussed their interpretation in terms of atomic structure.[67] For the K_α lines he found the relationship

$$\nu = \nu_0\left(\frac{1}{1^2} - \frac{1}{2^2}\right)K^2 \quad \text{with} \quad K = N - 1,$$

where ν_0 (= cR) is Rydberg's frequency constant and N (= Z) is the atomic number. Moseley thought that his data 'lend great weight to the general principles which you use, and I am delighted that this is so, as your theory is having a splendid effect on Physics [sic]'. Among the reasons for his optimism was not only the Balmer-like appearance of the K_α formula, but also that for two consecutive elements of atomic numbers N and $N + 1$ it followed from the formula that

$$\sqrt{\nu_{N+1}} - \sqrt{\nu_N} = \sqrt{\frac{3}{4}\nu_0} = const$$

This, he argued, 'proves your condition of constant angular momentum'. However, to explain the $(N - 1)$ factor he was led to the un-Bohrian view that 'the inner ring vibrates as a whole' and that it contains four electrons. He realized that it was not a very satisfactory picture, but at the time he could not think of a better alternative. Bohr did not accept the picture, which not only contradicted his own conception of the electron structure of the elements but also implied a change of the angular momentum of the inner electrons from $h/2\pi$ to h/π. 'Might it not rather be imagined that the radiation corresponds to a passing from a configuration in which the innermost ring contains n to one in which it contains $n - 1$ electrons?' he asked in reply, surprisingly adding that, 'For the moment I have stopped speculating on atoms'.[68]

A large part of Moseley's experimental work was not done in Manchester, but in Oxford, to where he had moved at the end of November 1913. He was not impressed by the attitude of the Oxford physicists towards atomic theory, as he reported in a letter to Rutherford early in the new year:

> Here there is no one interested in atom building. I should be glad to do something towards knocking on the head the very prevalent view that Bohr's work is all juggling with numbers until they can be got to fit. I myself feel convinced that what I have called the *h* hypothesis is true, that is to say one will be able to build atoms out of *e*, *m* and *h* and nothing else besides. Of the 3 varieties of this hypothesis now going Bohr's has far and away the most to recommend it, but very likely his special mechanism of angular momentum and so forth will be superseded.[69]

In the first sequel of Moseley's great work on 'The High-Frequency Spectra of the Elements', which appeared in December 1913 in Philosophical Magazine, he repeated the picture he had suggested to Bohr and generally emphasized that his analysis of the characteristic lines supported the Bohr–Rutherford model of the atom.[70] The second part appeared in the April 1914 issue and included a table of wavelengths of the lines in the L group from aluminium to gold. He provided a graphical representation of the K and L lines in what soon was to become known as a Moseley diagram (Figure 3.2). For the L_α lines he found a formula similar to that of the K_α lines:

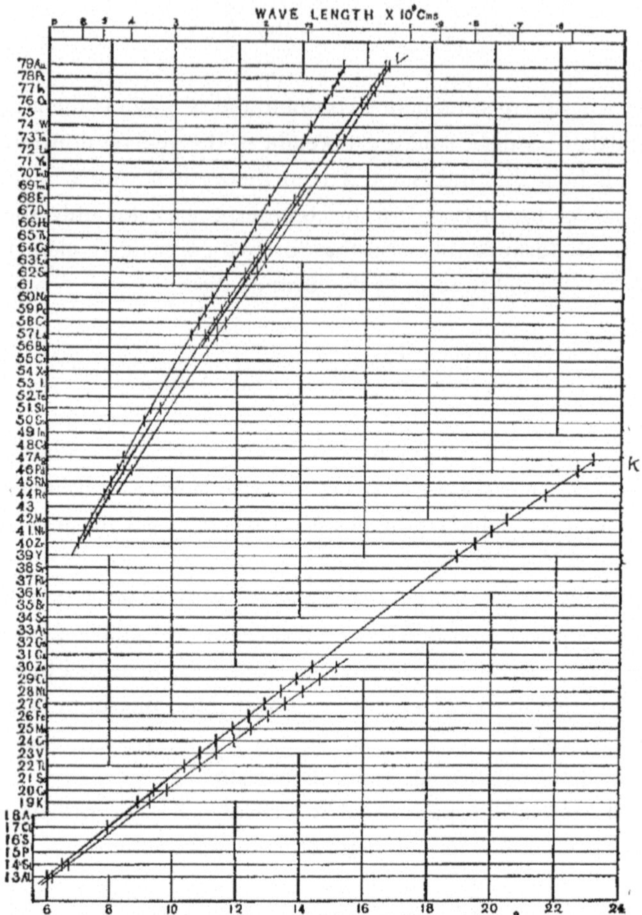

Fig. 3.2. The first Moseley diagram, showing the atomic numbers plotted against the square root of the X-ray frequencies for the K and L lines.

Source: Moseley 1914b, p. 709.

$$\nu = \nu_0 \left(\frac{1}{2^2} - \frac{1}{3^2} \right) K^2 \quad \text{with} \quad K \cong N - 7.4$$

In this work he did not address the question of the origin of the lines according to Bohr's theory, except that the L lines were supposed to come from a 'distinct vibrating system . . . situated the further from the nucleus'. Moseley consistently spoke of vibrating electrons rather than electrons jumping from one quantum state to another.

In spite of the shortcomings of the Bohr–Moseley theory, as it may be called, Moseley was convinced that his work provided solid support for Bohr's atomic theory. Critics of the Bohr atom outside the Manchester group such as Nicholson and Lindemann disagreed, if for different reasons. Frederick A. Lindemann, later Viscount Cherwell, argued that Moseley's data merely supported the hypothesis of an atomic number as suggested by van den Broek and Rutherford. 'The agreement of Bohr's constant with experimental data is not convincing to my mind in view of the large number of arbitrary assumptions in his derivation'.[71] By means of elaborate dimensional analyses he suggested that there were many ways in which results equivalent to Bohr's could be obtained, including some that avoided reference to quantum theory. Lindemann denied that experiments, whether in the X-ray or the optical region, provided unambiguous support for 'Dr Bohr's special assumptions'. These obviously did not appeal to his mind. Bohr immediately penned a brief reply in which he criticized the procedure adopted by Lindemann, and also Moseley responded, repeating that his experiments most certainly confirmed Bohr's theory.[72]

A more satisfactory explanation of the origin of the characteristic X-rays was only proposed in the autumn of 1914, and then by a physicist with no connection to Manchester.[73] Walther Kossel, a young Munich physicist trained by Lenard in Heidelberg, adopted Bohr's ring model of the atom but argued that the emission of X-rays was preceded by an ionization process that caused an inner ring to lose an electron. The resultant 'hole' would then be filled by an electron coming from an outer ring, with the energy difference between the rings appearing in the form of a distinct X-ray. According to Kossel, the K_α line arose from a transition from the L ring ($n = 2$) to the innermost K ring ($n = 1$), and K_β from a transition from the M ring ($n = 3$) to the K ring. Similarly, the L radiation was due to transitions $n > 2$ filling a vacancy in the L ring. It followed as an immediate consequence of Kossel's mechanism that, for example,

$$\nu(K_\beta) - \nu(K_\alpha) = \nu(L_\alpha)$$

and

$$\nu(K_\gamma) - \nu(K_\beta) = \nu(L_\beta) - \nu(L_\alpha),$$

in full agreement with experimental data. (The K_γ line was thought to be due to an electron jumping from $n = 4$ to $n = 1$.)

The mechanism proposed by Kossel was quickly adopted by Bohr, who in 1915 introduced it to physicists with no command of German or without access to the *Verhandlungen* of the German Physical Society.[74] Moseley may not have known about Kossel's work. He died in combat on 10 August 1915, hit by a Turkish bullet during the infamous Gallipoli campaign.

3.5 THE SILENCE OF THE OLD GUARD

At about the time Bohr's theory appeared in print, Thomson was still considered the recognized authority in atomic structure and his ideas were taken very seriously.[75] Resisting quantum theory as well as the nuclear model, he proposed in 1913 a new model of the atom that had only few similarities with the old plum-cake model dating back to 1904. He presented his new model to the British Association in September 1913 and in even greater detail to the Solvay congress the following month. According to the account in *Nature* of Thomson's reading of his paper Birmingham, it was 'a brilliant attempt to construct an atom which would account for some of the evidence for the quantum theory of energy . . . [and] it will be long before his illustration of the quantum theory by pin-pots is forgotten'. In fact, it did not take long.[76] A main feature of this second Thomson model was that the atom consisted of negative electrons bound together in stable equilibrium positions with positive particles in the form of hydrogen ions (protons) and alpha particles. The charged particles within the atom were assumed to be subject to two kinds of forces, a radial repulsive force varying inversely as the cube of the distance from the atomic centre and an inverse-square radial attractive force. Contrary to the ordinary Coulomb force, Thomson hypothesized that the attractive force was directive, namely confined to a number of radial tubes in the atom.

Making use of these and other assumptions Thomson succeeded, to his own satisfaction, to reproduce Einstein's equation for the photoelectric effect, including Planck's constant, which he characteristically expressed by atomic constants. Thomson found that

$$h = \pi\sqrt{Cme},$$

where C was a force constant of such a value that it secured the right value for h (e and m refer to the charge and mass of the electron). His model also provided an explanation of the production of X-rays and some of the data known from X-ray spectroscopy. Moreover, Thomson and others applied models of this kind to throw light on the nature of valency and other chemical phenomena, which for a period made the model popular among chemists.

In an influential theory proposed in 1914, Thomson introduced the later standard distinction between polar (ionic) and non-polar (covalent) chemical compounds, where the latter type consist of neutral atoms.[77] Although he did not base his theory on the Bohr–Rutherford model (to which he did not refer), his conception of atomic structure had qualitative features in common with it. Thus, he regarded the electrons as being arranged in concentric layers, where only the mobile electrons near the surface of the atom were responsible for the periodicity and optical spectra of the elements. Most chemical properties, he argued, could be understood from his model as being due to a dipole–dipole interactions caused by the mobility of the atomic electrons. While the inner layers were saturated with electrons, the outer one was not:

> There may, however, be a ring of corpuscles near the surface of the atom which are mobile and which have to be fixed if the atom is to be saturated. We suppose, moreover, that the number of corpuscles of this kind may be anything from 0 to 8, but that when the number

> reaches 8 the ring is so stable that the corpuscles are no longer mobile and the atom is so to speak self-saturated.[78]

Although Thomson's models of 1913–1914 were quite different from Bohr's, the two models addressed many of the same problems and were therefore, in a sense, rival conceptions of atomic structure. For example, Thomson found electron configurations for the simpler atoms that corresponded to the known periodicity of the elements, much as Bohr had done in the second part of his trilogy. From this point of view it may be considered surprising that Thomson simply chose to ignore Bohr's theory, which he did not mention in any of his works of 1913 or the following years. Increasingly isolated from mainstream physics, he consistently kept to his classical picture of the atom, modifying it from time to time in ways which were conspicuously ad hoc. Only in 1919 did he confront Bohr's atom, which at the time enjoyed universal acceptance among experts in atomic and quantum theory.

Thomson's late objections to the quantum atom were methodological rather than technical and presumably reflected his opinion when he first read Bohr's papers. Referring to Bohr's principle of discrete orbits or energy states characterized by quantum conditions, he said:

> This, however, is not the consequence of dynamical considerations; it is arithmetical rather than dynamical, and if it is true it must be the result of the action of forces whose existence has not been demonstrated. The investigation of such forces would be a problem of the highest interest and importance. By the use of this principle and a further one, that when an electron passes from one orbit to another it gives out radiation whose frequency is proportional to the difference of the energy of the electron in the two orbits, Mr. Bohr obtains an expression which gives with quite remarkable accuracy the frequencies of the lines in the four-line spectrum of hydrogen. It is, I think, however, not unfair to say that to many minds the arithmetical basis of the theory seems much more satisfactory than the physical.[79]

Thomson further objected, as others had done, that 'The vibrations which give rise to the spectrum do not on this theory correspond in frequency with any rotation or vibration in the atom when in the steady and normal state'. According to Thomson there was convincing experimental evidence, especially based on absorption spectra, that an electron in an unexcited state of the atom vibrated with the frequencies of its spectral lines. In short, as he saw it, Bohr's quantum atom was a mathematical construct with no basis in established physics. He kept to this view throughout his life, although eventually admitting that Bohr's theory had 'in some departments of spectroscopy changed chaos into order'.[80]

Bohr was not impressed by Thomson's new model of the atom, but he realized that it could be seen as an alternative to his own and therefore contemplated a response. A month after the meeting of the British Association, and after Thomson's paper had been published in the *Philosophical Magazine*, he wrote to Rutherford: 'As to the theory of the structure of atoms of Sir J. J. Thomson, I did not realise in Birmingham how similar many of his results are to those I had obtained', adding that 'this agreement has no foundation in the special atom-model used by Thomson but will follow from any theory which considers electrons and nuclei and makes use of Planck's relation $E = h\nu$'.[81] Bohr drafted a letter, apparently meant for *Nature*, in which he said as much, but did not send it.[82] In his letter to Rutherford of 16 October he elaborated his objections to Thomson's model as follows:

> Thus – quite apart from the fact that the assumption of repulsive forces varying inversely as the third power of the distance is in most striking disagreement with experiments on scattering of α–rays, – Thomson finds a value for the fundamental frequency of the hydrogen-atom which is 4 times too small, and a value for the ionization-potential of the hydrogen atom which is about half that experimentally found by himself. Besides Thomson's theory apparently gives no indication of an explanation of the laws of the line-spectra, and – making the atom a mechanical system – offers no possibility of evading the well-known difficulties of black-radiation and of specific heat.[83]

Rutherford was less diplomatic. In a letter to the American radiochemist Bertram Boltwood, of Yale University, he characterized the Thomson atom as 'only fitted for a museum of scientific curiosities'.[84] To Arthur Schuster, at the time secretary of the Royal Society, he wrote, no less frankly: 'I believe he [Thomson] knows in his heart that his own atom is not worth a damn and will not do the things it has got to do'.[85] No conflict arose between Bohr and Thomson, who largely cultivated their separate lines of work without bothering too much about the other's theory. Bohr was convinced that Thomson's theory belonged to the past, while his own belonged to the future.

Another of the highly respected physics professors of the old guard, Joseph Larmor of Cambridge University, chose to ignore the Bohr atom. A celebrated pioneer of electron theory, Larmor had given an extensive survey of atomic theory in his Wilde Lecture of 1908, but when the theories of Rutherford and Bohr appeared he remained silent. Only in 1928, in a postscript to a paper of 1921 on non-radiating atoms, did he briefly refer to the Bohr–Rutherford model of the atom.[86] And three years later the 74–year-old physicist wrote lightheartedly about Bohr, 'who, if he were not a Scandinavian, might almost be claimed to be a Cambridge man'.[87]

The case of H. A. Lorentz is somewhat similar. Not feeling at home with atomic and quantum theory, he considered Bohr's theory to be overly speculative. When Peter Debye in 1914 moved from Utrecht to Göttingen, a new professor of theoretical physics had to be found, and Bohr was mentioned as a possible candidate for the vacancy. In a letter to Willem Henri Julius, the professor of experimental physics in Utrecht, Lorentz described Bohr as 'One of the best exponents of the "adventurous" school', but the young Dane was clearly a bit too adventurous to Lorentz's classical taste. 'There must be something in Bohr's theory', Lorentz admitted, referring to its agreement with the hydrogen spectrum. 'He may turn out to be brilliant later on, but . . . I would not venture to recommend him over Dutch candidates at this stage'.[88] Although Lorentz included a reference to Bohr's theory in the second edition of his *Theory of Electrons* of

1915, he showed no interest in the Bohr atom until several years later.[89] This was not an area or a style of physics that appealed to him.

Nor was it an area that appealed to Lord Rayleigh, who never responded to Bohr's theory. According to Robert John Strutt, Lord Rayleigh's son and biographer, in 1913 he asked his father if he had seen Bohr's paper on the hydrogen atom. The third Lord Rayleigh replied, 'Yes, I have looked at it, but I saw it was no use to me. I do not say that discoveries may not be made in that sort of way. I think very likely they may be. But it does not suit me'.[90]

3.6 NICHOLSON'S OPPOSITION

The British atom-building tradition in the style of Thomson did not collapse overnight with the advent of Bohr's new model of atomic structure. It continued for some years, in most cases with the atom builders devising models that incorporated limited features of the quantum theory, as in the works of Thomson and Nicholson. Some of these classical models referred to and were inspired by Bohr's theory, or they were critical responses to it. This was the case with an atomic model proposed in December 1913 by Arthur William Conway, professor of mathematical physics at University College, Dublin.[91] Conway's model was classical, a variant of the Thomson atom, but incorporated some of the features of Bohr's model, including Planck's constant. Its aim was to explain spectral series, or rather illustrate them, without making use of the incomprehensible discontinuous quantum jumps.

Of greater interest is a more elaborate atomic theory proposed a few months later by the Irish physicist William Peddie, a former assistant of P. G. Tait. His model was a 'spherical counterpart of the tubular atom of Sir J. J. Thomson', consisting of a series of negatively charged shells surrounding a positive core and constructed in such a way as to give the desired results. After lengthy calculations Peddie managed to obtain from his model atom Balmer's spectral formula, account for the law of photoelectricity, and come up with a qualitative explanation of radioactivity. His general idea was to derive optical and other phenomena from 'a complicated structure of the atom itself'—and Peddie's spherical atom was indeed complicated. Bohr had admittedly deduced his results in a 'beautifully direct manner', he said, but unfortunately in a way that could not be reconciled with the known laws of dynamics and electromagnetism. As Peddie saw it, for this reason the Bohr atom could not be a model of the real constitution of atoms. He was convinced that the world of atoms could be understood in terms of classical physics: 'It does not seem to me that we are yet under compulsion to forsake the laws of ordinary dynamics in connexion with atomic properties, or the doctrine of a continuous wave-front in æther, or even, apart from magnetic action, the notion of central symmetry in atomic motion'.[92]

As yet another example of a classical alternative to the Bohr atom, consider a work by the American physicist and inventor Albert Cushing Crehore. Immediately following Bohr's first paper on atomic theory in the July 1913 issue of *Philosophical Magazine* there appeared a 60-page (!) paper by Crehore in which he developed and extended the classical Thomson atom into an elaborate theory of molecules, crystals, and more.[93] By February 1915 Crehore had modified the Thomson model into a 'corpuscular-ring gyroscopic theory', in part in an attempt to introduce Planck's constant and take into regard the works of Moseley and Bohr. His theory, which included detailed electron configurations of the elements in the periodic system, built eclectically on features of both the Thomson model and the Bohr–Rutherford model (Figure 3.3). In what he thought was in agreement with Bohr, he assumed, on the one hand, that undisturbed electrons describing circular orbits did not emit radiation. On the other hand, while on the Bohr–Rutherford model beta particles had their origin in the nucleus, in Crehore's theory they might come from any electron in the atomic system. The American physicist used his speculative theory to offer an alternative explanation of X-ray spectra, to account for photoelectricity, to suggest the existence of positive electrons, and to predict an upper limit of atomic weight corresponding to the weight of uranium.

But borrowing a few features from Bohr's theory did not make Crehore accept the theory: 'Although Bohr has in a brilliant manner given an explanation of some of the series of spectral lines, notably those of H and He, yet it may fairly be said that luminous spectra have not been explained by any atomic theory'.[94] Crehore continued to construct electromagnetically based atomic models that avoided the unpalatable

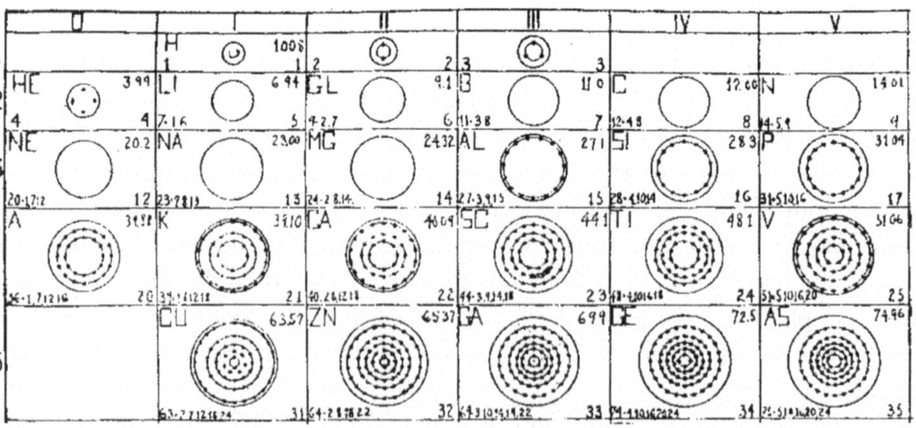

Fig. 3.3. A part of Albert Crehore's atomic periodic system of 1915. Crehore adopted Rydberg's ordinal number, implying two elements (in groups 2 and 3) between hydrogen and helium. The number of electrons in the rings is indicated by the numbers in the lower left-hand corner of the squares. The Rydberg ordinals, two units higher than van den Broek's atomic numbers, are given in the lower right-hand corner. The element of ordinal number 6 is beryllium, then sometimes known by its older name 'glucinium'.

Source: Crehore 1915, p. 323.

quantum jumps. In one of these works, dating from 1921, he summed up his and other conservative physicists' dissatisfaction with the Bohr atom: 'The Bohr model affords no picture of anything that really vibrates with the observed frequencies, since nothing is said about the model during the very time when the radiation is taking place, but merely about its state before and after such radiation'.[95]

At the meeting of the British Association held in 1914 in Melbourne, Australia, Bohr's theory was critically addressed by William Hicks and John Nicholson. Hicks objected that although the theory was 'most ingenious and suggestive', it failed to offer a true explanation, meaning a mechanism for the emission of light. 'It is based on the Rutherford atom, but throws no further light on the structure of the atom itself, as the mechanism of radiation is totally unexplained, and it is this which we are in search of'.[96] This was an objection that Bohr partly acknowledged, for in his Copenhagen address of 1913 he had himself emphasized the lack of a mechanism of radiation. As far as Hicks was concerned, he preferred Conway's electrodynamic alternative, of which he wanted 'more of a similar kind'. Nicholson agreed, but had his own ideas of what the alternative should look like.

Bohr could afford to ignore the alternatives and objections of scientists like Conway, Peddie, Hicks, and Crehore, whose ideas were so clearly out of tune with mainstream physics. The opposition of Nicholson was a different matter, for not only had Nicholson proposed a kind of quantum atomic model before Bohr, his views also enjoyed considerable respect among British physicists and astronomers (whereas he was ignored by continental physicists). There were some similarities between Nicholson's atomic theory and the one proposed by Bohr in 1913, but the two theories nonetheless differed profoundly. While Nicholson's theory presupposed the applicability of classical mechanics and electrodynamics, Bohr's approach severely restricted such applicability and replaced it with non-classical assumptions. First and foremost, there were no quantum jumps in Nicholson's atoms. According to his theory the spectral frequencies were vibration frequencies of the electrons in their circular orbits.

Nicholson's aim in his one-man campaign against Bohr's theory was not primarily to defend his own ring model of atoms. The aim was destructive, to demonstrate irreparable weaknesses in Bohr's theory by examining it from what he thought were its own premises. Contrary to most other critics, he had a deep knowledge of Bohr's atomic theory, which he examined in great technical detail, often greater than that of Bohr himself. His attitude to Bohr's atomic theory wavered somewhat, at times receptive and at other times hostile, but it was basically critical. The many arguments that he launched against Bohr's theory in the period 1913–1915 can be grouped in four classes:

(1) *Emission of X-rays*. According to Nicholson's analysis, two or more coplanar rings of electrons could not exist, neither on the view of the dynamical theory nor on Bohr's theory. Mechanical stability required that either the electrons must move in different planes or they must all lie on the same circle. This implied that Bohr's explanation of Moseley's results of X-rays

was necessarily incorrect. Nicholson objected to the Moseley-Bohr theory because Moseley's formulae did not depend on Z but on $(Z - x)$. It was quite possible, he suggested, to keep Moseley's empirical formulae and interpret them in accordance with the view that there were more elements than given by the Z parameter. Calling Bohr's theory 'so attractive that its retention is desirable' Nicholson nonetheless concluded that 'we must give up the idea of concentric rings in the atom, with X-radiation coming from an inner ring'.[97] Of course, Bohr disagreed and so did Moseley. While Nicholson dismissed the Bohr–Moseley explanation of X-ray spectra, he proposed that a one-ring modification of Bohr's theory might explain the emission of X-rays. However, in that case one would have to abandon the notion of the atomic number as the ordinal number of the periodic system, a notion that was crucial to the Bohr–Moseley theory.

(2) *Atomic number.* It was a recurrent theme in Nicholson's criticism that the hypothesis of the atomic number was in conflict with Bohr's theory. Since Bohr's theory for atoms more complex than helium was founded on the notion that the nuclear charge was the ordinal number for the periodic system, this was a serious charge. Van den Broek responded to Nicholson's arguments, which he thought were ill founded: 'Generally speaking, Bohr's theory is not in disagreement with the atomic number hypothesis'.[98] Willing to accept some version of the atomic number hypothesis, Nicholson denied its essence, that the number defined the place of an element in the periodic system and limited the number of elements in a period. He considered the periodic system less important for atomic theories than astrophysical evidence. That evidence demanded the existence of several elements lighter than helium, and according to the atomic number hypothesis there was no room for these elements.

(3) *Lithium atom.* Since Nicholson had concluded that Bohr atoms could only have a single ring, Bohr's (2,1) model of lithium, as he had presented it in the second part of his trilogy, had to be wrong: 'It is not possible for three electrons and a nucleus to form a lithium atom with a unit valency, after the manner of Bohr's model'.[99] He further argued that Bohr's theory of valency and the structure of complex atoms led to results that were grossly inconsistent with chemical knowledge. For example, lithium should be an inert element, carbon a monovalent element of metallic nature, and nitrogen a divalent metal. So much for Bohr's atomic chemistry!

(4) *Hydrogen and helium.* Bohr's theory was singularly successful when applied to the simplest elements—hydrogen and ionized helium—but according to Nicholson the success was only partially deserved. In detailed analyses he concluded that except for the neutral hydrogen atom the model failed even for simple systems such as He^+, H_2, He and H^-. Even in the case of

> the neutral hydrogen atom it failed to deliver a complete solution. Nicholson summarized: 'Bohr's theory cannot explain any portion of the hydrogen spectrum except the Balmer and Ritz series, and perhaps a Schumann [ultraviolet] series. It also predicts some strong series which are not found'.[100] Of what worth was an atomic theory that was valid, or merely partially valid, only for a single element? Especially with regard to helium he was convinced that Bohr's theory failed to live up to its promises. Having investigated various ways to generalize and modify the theory so as to explain the helium spectra, 'we must conclude that it cannot develop in the manner which its earlier success appeared to foreshadow'[101]

Intending to reply to Nicholson's objections, Bohr drafted a letter to *Nature* and a longer one to *Philosophical Magazine*, but he mailed neither of them. Although 'I admit most readily the importance of the difficulties discussed by Prof. Nicholson', he wrote in the longer draft reply, 'I cannot, on the other hand, feel convinced that the basis for his calculations is sufficiently self-contained to justify his conclusions'.[102] Bohr's published replies came in the form of two papers of 1915, the first on the hydrogen and helium spectra, the second a general development of his theory of atoms and radiation. 'I am unable to agree with Nicholson's conclusions', he stated, apparently unwilling to face these conclusions in detail.[103] He did however take care to repudiate the argument of Nicholson that the 4686 line and the new series discovered by Evans were no evidence for Bohr's theory as they might well be due to hydrogen rather than ionized helium.

Bohr convinced himself that it was not worth entering a dispute with Nicholson, whose premises and way of thinking differed too much from his own to make it worthwhile. Not only did Bohr refrain from entering a public discussion with Nicholson, he also did not communicate with him privately. To H. M. Hansen, his friend and colleague in Copenhagen, he wrote: 'You have probably seen quite a bit of criticism, which has appeared; especially from Nicholson. I do not think it has any foundation. I feel that Nicholson treats the question not as a physical, but as a purely literary one'.[104] He soon came to see the critique from Nicholson, Hicks, and others as insignificant and not worth worrying about. 'I don't think that any of it means anything,' he said in a letter to his brother Harald.[105]

Nicholson's arguments against Bohr's atomic models relied in part on calculations that showed some of Bohr's configurations to be mechanically unstable. Unable or unwilling to counter the mathematical arguments, Bohr decided that they were of little importance because they presupposed that the motion of electrons could be calculated on classical mechanics. Not all physicists ignored them with similar ease. Stability calculations made by the German physicist Ludwig Föppl, a former student of David Hilbert, confirmed some of Nicholson's results that contradicted Bohr's models for atoms more complex than hydrogen.[106] In his famous textbook *Atombau und Spektrallinien* Sommerfeld recognized the force of Nicholson's arguments in the case of the helium atom and other many-electron systems.[107]

Nicholson largely stopped criticizing Bohr's theory after 1915, but he did not significantly change his view on the structure of atoms and the existence of celestial primary atoms. For a while his ideas received cautious support from some British chemists, physicists, and astronomers. For example, Nicholson's work appeared prominently in the series of annual progress reports on radioactivity that Soddy wrote for the Chemical Society. In his progress report of 1917 Soddy echoed Nicholson's sceptical attitude to the Bohr atom: 'In spite of its great initial successes in calculating correctly the magnitude of the Rydberg constant, and in correctly ascribing the Pickering series of lines to helium rather than to hydrogen, Bohr's theory does not seem to have been generally so successful'.[108] Yet, by the end of the decade neither Nicholson's nor others' alternatives were seriously considered by mainstream physicists. Tellingly, while Jeans had covered Nicholson's theory in his reports on quantum theory of 1914 and 1916, in later reviews he dealt exclusively with the theory of Bohr and his school.[109] After World War I the quantum theory of atoms based principally on the works of Bohr and Sommerfeld stood out as the only way to understand atoms and radiation. The continual scepticism in some quarters of British physics, including scattered attempts to establish classical alternatives, only meant that progress happened elsewhere; first and foremost in Germany.

3.7 LIMITED AMERICAN INTEREST

While Bohr's theory made a very considerable impact on physics in the United Kingdom, it was received later and with less interest by American scientists. Among the few early converts was Raymond Birge, a young up-and-coming physicist who became 'a real missionary of the Bohr atom'.[110] As early as in the fall of 1913, while an instructor at Syracuse University, he lectured on Bohr's ideas of atomic structure and later, when he came to the University of California, he brought the Bohr atom to Berkeley. Although the theory was undoubtedly known by many American physicists, it did not make an impression in *Physical Review*, since 1913 the journal of the American Physical Society. The structure of atoms was not what occupied the minds of most American physicists, the large majority of whom worked on experimental rather than theoretical subjects. Until the beginning of 1916, *Physical Review* contained no papers on or references to Bohr's theory of atomic structure and almost no papers that can be classified as atomic theory.[111]

Several of the papers in *Physical Review* and *Astrophysical Journal* dealt with spectroscopy, an area of research that was of equal interest to physicists and astronomers and which American researchers cultivated as actively as their colleagues in Europe. Indeed, astrospectroscopy was something of an American speciality. In a study of the 4686 line and other lines in the spectra of planetary nebulae, William Wright at the Lick Observatory referred to the role of the 4686 line in 'certain theories of the constitution of the atom'. He singled out 'the interesting theory of radiation proposed by Bohr... [which] predicts lines separated by about two angstroms from the members of the Balmer series'.[112] The astronomical interest in atomic theory was also manifest in the presidential address that John Stanley Plaskett gave to the Royal Astronomical Society of Canada in the beginning of 1915. However, Plaskett found Nicholson's theory to be much more promising than Bohr's: 'The question between Nicholson and Bohr as to the constitution of the atom can not be regarded as settled, however, although Nicholson appears to have the last word'.[113]

Readers of *Nature* and *Philosophical Magazine*, which included many American physicists, could hardly avoid to come across Bohr's theory and the Bohr–Rutherford atom, subjects which also appeared in the pages of *Science*, the journal of the American Association of the Advancement of Science. For example, the July 1914 issue of the journal included a survey article by Arthur S. Eve based on a meeting of the Royal Society of Canada on the structure of the atom. Eve, a former assistant (and later biographer) of Rutherford and since 1903 professor of physics in Montreal, presented the ideas of the 'brilliant young Dane, Bohr' whose work 'is remarkable as leading to excellent numerical verification'.[114] He also referred to Bohr's models of water and other molecules, which, although 'somewhat speculative', he found to be 'refreshing'.

Half a year later *Science* brought another survey article which praised the Bohr–Rutherford model of the atom as a great advance, even one that 'will probably remain, suffering but little change in the future'.[115] The author, G. Walter Stewart of the University of Iowa City, recognized the critique of Nicholson but did not find it damaging that Bohr's theory had difficulties with more complex atoms. 'When one contemplates the narrow scope of even this brilliant theory, what a limitless field for research seems ahead!' Not all American physicists were that enthusiastic. More representative may have been the view of Alfred Cole, a professor of physics at Ohio State University, as he expounded it in an address to the American Association for the Advancement of Science at the end of 1914. Cole accepted the nuclear atom, but wavered between the Bohr version of it and the version proposed by Nicholson. Both models were valuable, he thought, but Nicholson's had the advantage that its 'results are obtained by means of established mechanical principles and without the use of such questionable assumptions as the brilliant young Dane cheerfully and confidently makes'.[116]

Nor was Gordon Scott Fulcher of the University of Wisconsin inclined to accept Bohr's theory. In agreement with objections raised by Stark in Germany, he thought the theory was contradicted by Stark's series of experiments with canal rays and their spectra. Fulcher argued that Bohr's 'assumption that the series lines are emitted by the single rotating electron of the hydrogen neutral atom is directly contrary to Stark's experimental result'. The correct interpretation of the experiments was rather that 'The Balmer series is emitted by electrons in the nucleus, vibrating about positions of static rather than dynamic equilibrium'.[117] This was an unorthodox view even among the critics of the Bohr atom, but it reflected Fulcher's adherence to the ideas of Stark.

Just as a few American physicists and astronomers discovered the Bohr atom in about 1915, so too did a few chemists, but they did not particularly like what they found. Together with his doctoral student Ernest D. Wilson, the Chicago physical chemist William Harkins reviewed in the spring of 1915 recent work on the structure of the atom. Like many other scientists of the period, they found it difficult to accept Bohr's postulates, which prevented 'any satisfactory physical picture' of the process leading to the emission or absorption of light. 'Bohr attempts no physical picture or cause of the change of an electron from one steady state of vibration to the next', they complained.[118] Comparing Bohr's theory with the one proposed by Nicholson, they concluded that the latter theory was of 'extreme interest to chemists' and far the more promising of the two alternatives. In Wilson's dissertation of the following year Bohr's theory figured prominently, but 'Bohr's success lies only in the consideration of atoms with one vibrating electron'. Nicholson's alternative had the advantage over Bohr's that it was 'based on calculations made by the classical mechanics', and in addition it was judged superior because of its many brilliant predictions of celestial spectral lines.[119]

The eminent chemist Gilbert Newton Lewis had in 1912 moved from a professorship in physical chemistry at Massachusetts Institute of Technology to become dean of the College of Chemistry at the University of California, Berkeley. Lewis had for a long time nourished an interest in atomic structure (Figure 3.4), but it took until 1916 before he published his ideas of what he called the 'cubical atom'.[120] Since it was essential to this model of the atom that the electrons stayed in fixed positions, he was bound to deny the validity of Bohr's dynamic atom (where fixed electrons were impossible). Concerning Bohr's central postulate that electrons moving in a stationary orbit produce no radiation or other effect, he came up with a curious objection: 'Now this is not only inconsistent with the accepted laws of electromagnetics but, I may add, is logically objectionable, for the state of motion which produces no physical effect whatsoever may better be called a state of rest'.[121] Lewis did not elaborate at the time, but he did so at a symposium of the American Association for the Advancement of Science later the same year, where he amplified his critique of Bohr's theory and its logical weaknesses. Based on a thought experiment, he felt justified to conclude:

> Unless we are willing, under the unslaught of quantum theories, to throw overboard all of the basic principles of physical science, we must conclude that the electron in the Bohr atom not only ceases to obey Coulomb's law, but exerts no influence whatever upon another charged particle at any distance. Yet it is on the basis of Coulomb's law that the equations of Bohr were derived.[122]

In other words, Lewis charged that Bohr's theory was internally inconsistent. Lewis added the objection that on Bohr's theory the revolving electrons would continue their motion even down to the absolute zero of temperature. He apparently believed that at $T = 0$ all motion must cease, including the motion of intra-atomic particles. Some years later he presented a different version of the thought experiment, in which he imagined a metallic wire near a hydrogen atom in its ground state. Due to the orbiting electron an

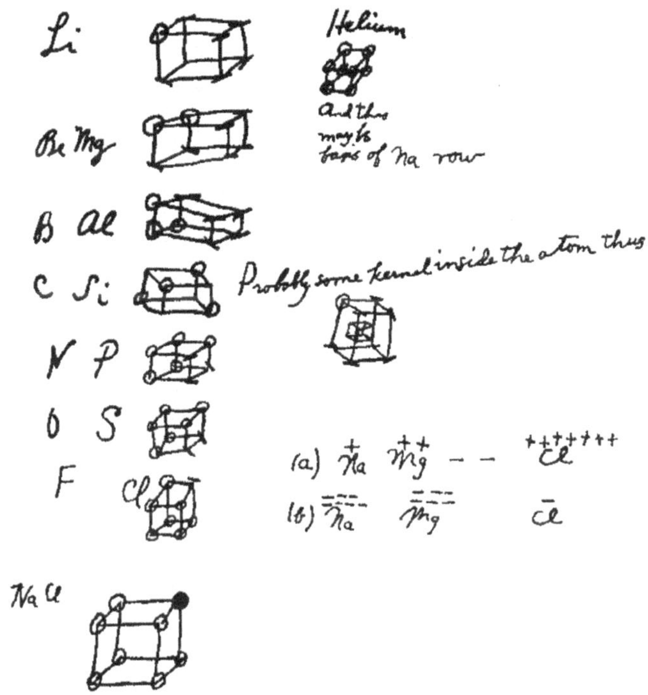

Fig. 3.4. G. N. Lewis's memorandum of 28 March 1902, in which he outlined his idea of atoms consisting of electrons arranged in concentric cubes. Only in 1916 did he publish an improved version of the theory.

Source: Lewis 1923a, p. 29.

alternating current would be produced in the wire, which would therefore get warmer. But, and this is Lewis' paradox, there is no source from which the heat can originate, for the atom cannot be in a state of lower energy (Figure 3.5). As he saw it, Bohr's atom contradicted the fundamental principle of energy conservation.[123]

In spite of his critical attitude to Bohr's theory, Lewis was greatly interested in the ideas of the Danish physicist with whom he wanted to establish connections. He requested the opinion of Rutherford, who replied:

> I regard Bohr as one of the coming men in Mathematical Physics, and I think he has a better grasp of physics that any of the Mathematical people I have come across. He is a man of great originality and, as you know, his work has already attracted wide attention, and I am confident will do so even more in the future.... He is thoroughly *au courant* of all the modern physical problems, and has an extraordinarily wide knowledge of experimental, as well as of theoretical physics. He is a pleasant fellow, speaks English quite well, and is quite a clear and interesting lecturer.[124]

In February 1916 Lewis, acting on behalf of the president of the University of California, invited Bohr to come to Berkeley to give the series of Hitchcock Lectures in natural science and also to remain in residence during the fall semester. Bohr was tempted to

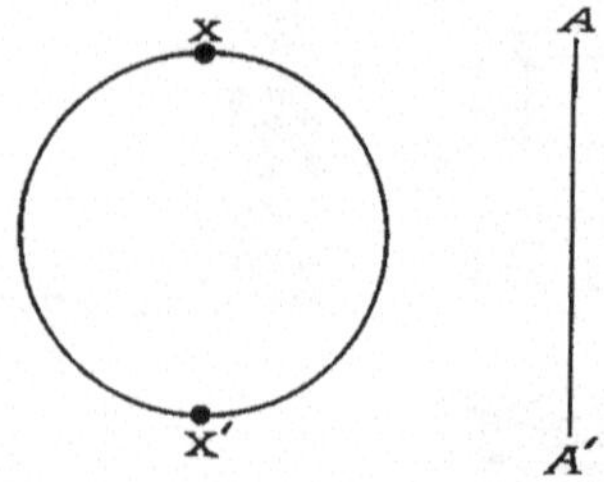

Fig. 3.5. G. N. Lewis's thought experiment of 1923 intended to show how the simple Bohr picture of a hydrogen atom in its ground state is irreconcilable with fundamental physics. AA′ is a small metallic wire near the atom, and the system (atom + wire) is supposed to be isolated from outside influences. When the electron is in position X there will be a flow of positive electricity in the wire toward A, and when it is at X′ there will be a flow toward A′. As the electron revolves in its orbit, it will produce in the wire an alternating current and hence heat will be generated. Lewis suggested that this contradicted the law of energy conservation, since the energy of the atom cannot be lower than the ground state.

Source: Lewis 1923a, p. 51.

accept the offer, but nothing came of it. It seems to have been the prospect of a chair at the University of Copenhagen that made him decline the invitation after it had been impossible to postpone it.[125]

Lewis's theory of valence and atomic structure was to some degree stimulated by a paper that the English chemist Alfred Lauck Parson had published in a series issued by the Smithsonian Institution. After studies in Oxford Parson moved to the United States, where he worked at Harvard and Berkeley and came to know Lewis. In 1915 he published an ambitious and rather speculative 'magneton theory' of the atom (not to be confused with either Weiss's or McLaren's), which he applied to valency, affinity, and a variety of other chemical problems. For the heat of dissociation of hydrogen gas he obtained 135 kcal/mole, a value that deviated more from experiment than the one calculated by Bohr. Parson conceived the electron (or 'magneton') as a tiny ring of negative electricity rotating at great speed, while he took the positive charge to be uniformly distributed in a sphere, like in Thomson's theory. For this reason his model could not account for Rutherford's scattering results. On the other hand, it had the advantage that it afforded 'a rational basis for the assumption made by Bohr, *viz*. that an electron rotating in a stable orbit does not radiate', as stated by William Lewis, a physical chemist at the University of Liverpool.[126] Parson's ring electron was equivalent to a continuous series of charges, and according to classical electrodynamics such a ring should not radiate at all (cf. Section 1.3). Another advantage of Parson's 'toroidal ring model' was that it provided the electron with a magnetic dipole moment and therefore promised a new understanding of the magnetic properties of matter.

For a while Parson's magneton theory was popular among American chemists and physicists, some of whom considered it a preferable alternative to the Bohr–Rutherford model. His work was much closer to the tradition of Thomson than to the theory of Bohr, and he explicitly dismissed the Bohr–Rutherford atom as useless from a chemical point of view. 'Bohr's theory, based upon the conceptions of the nuclear positive charge, gives a interesting treatment of the problem of spectrum series, but its chemical application is very meager indeed'.[127] This was not an unfair characterization. Whether in the version of Thomson, Nicholson, or Bohr, Parson argued that the ring atom 'is experimentally shown to be untenable'.[128] Parson had obviously studied Bohr's papers, including the third and often ignored part of the trilogy, for he referred critically to Bohr's model of the tetrahedral four-valence carbon atom. This part of Bohr's theory, admittedly tentative, failed to impress the chemists, and it certainly did not impress Parson. 'We see there that the theory comes to a complete halt when confronted with the problems of "Chemistry in Space"'.[129] What Parson wanted, and what Lewis and most other chemists wanted, was a static picture of the covalent bond, and that Bohr's theory could not provide. In Chapter 6 we shall return to the use of the Bohr atom in chemistry, including attempts to reconcile the static and the dynamic models of the atom.

By 1916 American scientists were warming up to adopt the new theory of the atom based on the work of Rutherford, Bohr, and Moseley. Robert Millikan, recognized as the leading American physicist at the time, first referred to Bohr's theory in a couple of papers presented in December 1916. As president of the American Physical Society he delivered an address on the new physics of radiation and atoms in which he extolled the 'extraordinary success of the Bohr atom'.[130] According to Millikan's inductivist understanding of Bohr's theory, and of science in general, it was 'guided solely by the known character of the line spectra of hydrogen and helium' and even the postulate of non-radiating stationary orbits was 'merely the statement of the existing *experimental* situation'. The success of the theory, he said, was not least due to 'its adaptability to the explanation of deviations from the behaviour predicted by its most elementary form', as illustrated by the Fowler anomaly and Sommerfeld's recent explanation of the fine structure of the hydrogen spectrum. Millikan was aware of the standard objection that Bohr's atomic theory 'gives us no picture of the mechanism of the production of the frequency', but considered it a strength rather than a weakness. In this regard he likened the theory to the fundamental laws of thermodynamics, which 'are true irrespective of a mechanism'.

Millikan's positive view was only shared by a minority of American physicists and chemists. At the meeting of the American Physical Society of 26–29 December 1916, where he gave his address, several other physicists spoke of atomic structure, but Millikan seems to have been the only one to support the Bohr–Rutherford atom. Many were not aware of this new development in atomic theory. When young Edwin Kemble at Harvard University wrote his dissertation on the application of quantum theory to molecular spectroscopy in 1916–1917, he had not heard of Bohr's theory and consequently did not refer to it.[131] David Webster, a newly appointed professor of physics at Stanford University, criticized in a paper with Leigh Page the Bohr atom from a methodological point of view. Unimpressed by the exact quantitative agreement that Bohr's theory had celebrated in the area of spectroscopy, the two Americans expressed their scepticism by referring to the history of science. The mechanical ether theories had been able to provide quantitative agreement with measurements, and so had Fourier's theory of heat conduction based on the assumption of the heat substance 'caloric'. Might not Bohr's quantum atom turn out to be just a modern version of the ether and the caloric—a chimera?[132]

3.8 THE RECEPTION AMONG GERMAN PHYSICISTS

Bohr stayed in England for most of the period 1912–1915; he was well connected to several leading British physicists, he participated in two of the meetings of the British Association, and with one exception all of his papers appeared in British journals: either *Nature* or *Philosophical Magazine*. Moreover, his theory relied on and was closely related to work of British physicists, in particular Rutherford, Fowler, Barkla, and Moseley. It is therefore natural that his theory of atoms and molecules attracted more and earlier attention in Britain than in Germany, the other of the major powers in physics at the time.

There most likely was another reason, namely that the atom-building tradition was strong and had long roots in the United Kingdom, in contrast to the situation in Germany where this kind of physics was not highly regarded. When Sommerfeld told Bohr in September 1913 that he was 'rather sceptical about atomic models in general', he spoke for a majority of his German colleagues. The difference in attitude was described by Rutherford in a letter to W. H. Bragg of late 1911 in which he said about the first Solvay meeting:

> I was rather struck in Brussels by the fact that the continental people do not seem to be in the least interested to form a physical idea of the basis of Planck's theory. They are quite content to explain everything on a certain assumption, and do not worry their heads about the real cause of the thing. I must, I think, say that the English point of view is much more physical and much to be preferred.[133]

Although Bohr did not attempt to 'form a physical idea of the basis of Planck's theory', his theory clearly belonged to the English tradition and not the continental one.

In the fall of 1913 Harald Bohr stayed in Göttingen where he met and collaborated with Richard Courant, Constantin Carathéodory, David Hilbert, Hermann Weyl, and other leading mathematicians. He wrote back to his brother, then in Copenhagen, that 'People here are still exceedingly interested in your papers, but I have the impression that most of them—except Hilbert, however—and in particular, among the youngest, Born, Madelung, etc., do not dare to believe that they can be objectively right; they found the assumptions too "bold" and "fantastic"'.[134] The somewhat reserved attitude is confirmed by later recollections. Thus, Max Born recalled that Bohr's papers of 1913 'made a deep impression on us and were thoroughly discussed', but also that 'the whole atmosphere of the physics department in Göttingen was, in spite of Debye, not favourable to such revolutionary ideas'.[135] Yet, at the latest by the fall of 1914 Born had become 'an ardent follower of Bohr', preparing a report on the stability of the Bohr atom for Hilbert's seminar in the winter semester of 1914–1915.[136] An early convert to quantum theory, the famous German mathematician met with Bohr but, according to Courant, 'Hilbert could not learn anything from Niels Bohr—it was a problem in itself to communicate mutually with Niels Bohr'.[137] Only later did Hilbert become seriously interested in Bohr's atomic theory, on which he gave lectures in the early 1920s (Section 7.4).

In an interview of 1962 Courant said that the reception in Göttingen was cool and that the eminent spectroscopist and mathematician Carl Runge was particularly antagonistic:

> Carl Runge was between physics and mathematics. He was the great spectroscopist. He knew more about the spectra than anybody else . . . Then Niels came with his model. And I remember that Runge was completely upset. He said, 'Well, such a nice man, and so intelligent. But this man has become completely crazy. This is the shearest nonsense'. It was a violent criticism and opposition.[138]

Courant himself, who had first met Bohr in Cambridge in 1913, immediately found Bohr's theory convincing. He later wrote to Niels Bohr that, 'When I reported these things here in Göttingen, they laughed at me that I should not take such fantasies seriously'.[139] At the Rydberg Centennial Conference held in Lund, Sweden, in 1954 Bohr confirmed what Courant had said about Runge's antagonism. Referring to the explanation of the Pickering–Fowler lines in terms of Bohr's theory, Bohr said: 'I especially recall the warning, given by the latter [Runge] at a colloquium in Göttingen, against such apparently arbitrary use of spectral evidence by theoreticians who did not seem properly to appreciate the beauty and harmony of the general pattern of series spectra, revealed about all by the ingenuity of Rydberg'.[140]

Although Runge conceded that Bohr's theory agreed surprisingly well with spectroscopic data, he considered it to be nothing more than a collection of rules for calculation. It provided no understanding of either atoms or radiation. Runge's dislike of Bohr's theory of spectra did not vanish easily. In a letter of September 1916 Sommerfeld told him that he 'had the impression that you are still somewhat foreign to Bohr's theory'. But Sommerfeld assured him that there was no reason for scepticism: 'One can no longer doubt the absolute correctness of this theory'.[141]

In Zurich, another of the German-speaking centres of physics, Bohr's atomic model was discussed at a colloquium in the autumn of 1913. According to Franz Tank, who attended the colloquium, Max von Laue objected vehemently to the theory: 'That's all nonsense; Maxwell's equations are correct under all circumstances, and an electron orbiting around a positive nucleus is bound to radiate'. Tank further recalled that Einstein, in opposition to von Laue, declared his support for Bohr's model: 'Very remarkable—there must then be something behind it; I do not believe that the derivation of the absolute value of the Rydberg constant is purely fortuitous'.[142] Einstein referred to Bohr's identification $R = 2\pi^2 me^4/h^3c$, which some physicists considered significant and impressive whereas critics tended to see it as a piece of clever numerology.

Only in the summer of 1914 did Bohr meet with German physicists, when he gave talks in Göttingen and Munich in front of Born, Debye, Wien, Sommerfeld, Landé, and others. In a letter to Oseen he wrote: 'I gave a couple of small talks in the seminars in Göttingen and Munich and had many lively discussions. I especially enjoyed talking with Wien and hearing about some experiments going on in his institute'.[143] Alfred Landé, Bohr's junior by three years, recalled of Bohr's seminar in Göttingen:

> He spoke rather poor German with his usual soft voice, and in the front row were all the big wigs. They shook their heads and said, 'If it's not nonsense, at least it doesn't make sense'. I spoke with Max Born after the lecture, and he said to me, 'All this is abolutely queer and incredible, but this Danish physicist looks so like an original genius that I cannot decline that there must be something to it'. This was the attitude.[144]

Whether through formal or informal channels, by the spring of 1914 at the latest Bohr's work was well known and being discussed in German-speaking Europe. The Austrian physicist Arthur E. Haas, who in a work of 1910 had been the first to introduce Planck's constant into the architecture of atoms, had studied Bohr's papers with great interest. He wrote him that 'I . . . shall at the Physical Society of Leipzig this very January render an account of your papers, which of course will also meet with great interest there'.[145] At the same time Bohr's theory and applications of it began to appear frequently and prominently in the programme of the Munich physics colloquia. On 26 January 1914 Paul Epstein reported on Bohr's publications on atomic theory, and four months later Sommerfeld and his former student Wilhelm Lenz discussed Bohr's new work on the Stark effect.[146] As mentioned above, on 15 July Bohr was himself a *Mittwoch* colloquium speaker in Munich.

The trilogy was extensively abstracted in the *Beiblätter* of the *Annalen der Physik* and also in the *Chemische Central-Blatt*. The *Beiblätter* reviewer was Rudolf Seeliger, a young physicist who had taken his doctorate under Sommerfeld and at the time worked at the Physikalisch-Technische Reichsanstalt in Berlin. Seeliger stressed the axiomatic structure of Bohr's theory and its success in explaining the Balmer and Pickering–Fowler series, and also that it was a modification of Rutherford's nuclear model. 'In their last consequences', he said, 'the postulates of Bohr go beyond the assumptions of quantum theory, and they also have rather little connection to the former views of physics; on the other hand, the great heuristic value of Bohr's considerations cannot be belied'.[147] In a detailed and generally positive review in *Naturwissenschaften* of March 1914, Seeliger mentioned the objections of Nicholson, Lindemann, and others. 'One can reasonably ask the question if the postulates of the theory are the only possible ones . . . and if these and the associated deductions are really consistent'. Without answering the question he concluded: 'Even though we may be sceptical with respect to the details, I think we have in Bohr's considerations an important and fundamental advance in the knowledge of the origin of spectral lines and series'.[148]

Another review, even more detailed and positive, appeared the following year in *Physikalische Zeitschrift*, where Eduard Riecke, the 70-year-old professor of experimental physics in Göttingen, gave particular attention to the spectroscopic evidence in favour of Bohr's theory. 'The further development of science may still change much in Bohr's theory, yet it is certain that it has already led to highly valuable information and is of fundamental importance in the area of spectroscopy'.[149] Not only did the spectra of hydrogen and ionized helium agree beautifully with the theory, Riecke also concluded that its importance was of a more general and fundamental kind and not limited to the field of spectroscopy. The positive reviews of Seeliger and Riecke were instrumental in disseminating Bohr's theory among German physicists.

The reception of Bohr's atomic theory in the German physical community in the summer of 1915, two years after it had been introduced, may be judged from the volume on physics that was published in the book series *Die Kultur der Gegenwart* under the editorship of Emil Warburg.[150] The volume, with contributions from leading German and German-speaking physicists, gave a general overview of the state of physics aimed at a general audience. It is evident from the content of the book that atomic structure was not regarded as a subject of high priority. Bohr's theory entered briefly in the chapters on spectrum analysis and magneto-optics, which were written by Franz Exner and Pieter Zeeman, respectively, but only alongside Thomson's model, which was given more attention than Bohr's. None of the chapters dealt with the structure of the atom. Bohr's theory was only given more than brief notice in Wilhelm Wien's chapter on heat radiation, where he emphasized the remarkable reproduction of Balmer's formula that followed from the theory. But Wien, who a decade earlier had helped to pioneer the electromagnetic world view, also pointed out that 'the theory is not yet self-consistent and it contradicts the electromagnetic theory by assuming that the revolving electrons do not emit energy'.[151]

Wien's objection that Bohr's model of the atom contradicted the well-established theory of electromagnetism was common at the time, both in Germany and elsewhere. Oseen, Bohr's Swedish friend and colleague, raised the question in a letter to Bohr of 11 November 1913 in which he congratulated Bohr on his second paper in the trilogy sequel. Now Bohr had developed his theory 'beyond the region of hypotheses and theories and into that of truth itself'. Praise apart, Oseen was curious to know 'how the Maxwell–Lorentz theory should be modified to allow for the existence of an atom of your type'.[152] Oseen (in contrast to Bohr) continued to worry about the problem, and in a detailed analysis in *Physikalische Zeitschrift* he reached the following, unequivocal conclusion: 'Bohr's atom model can in no way be reconciled with the fundamental assumptions of Lorentz's electron theory. We have to make our choice between these two theories. One of them may be correct, but not both of them'.[153] Although greatly attracted by the electron theory based on the Maxwell–Lorentz equations, Oseen refrained from concluding that Bohr's model was in serious trouble. He was careful to stress that his paper should not be considered 'a work of controversy against the theory of Bohr'.

Oseen's analysis was not the last of its kind. Specialists in electrodynamical calculations continued to investigate in great detail the relationship between Bohr's theory and the sacrosanct Maxwell–Lorentz theory of electromagnetism. One of those specialists was George Schott, who earlier had applied his mathematical expertise to shoot down Nagaoka's Saturnian atomic model and now used his weapon against Bohr's postulate of radiation-free stationary states. In 1918 he concluded after lengthy calculations that Bohr's postulate was indeed inconsistent with ordinary electrodynamics and could only be saved by an artificial modification of the theory.[154]

Although German physicists may have been 'exceedingly interested' in Bohr's papers, as Harald Bohr claimed in his letter from Göttingen, for a while the interest did not materialize in scientific papers related to the new theory. Most readers of the 1913–1915 volumes of *Annalen der Physik*, the main journal of the German physics community and at the time coedited by Wien and Planck, would not have perceived that a new revolution in atomic physics was under way or otherwise have come across Bohr and his theory. The journal contained very few papers on quantum and atomic theory, and none at all that dealt with Bohr's new theory of the structure of atoms. Only a couple of papers carried references to Bohr's 1913 papers, among them a paper by H. M. Hansen who in a paper on the inverse Zeeman effect pointed out that some of the spectroscopic details had probably escaped explanation because of the inadequate knowledge of atomic structure. In this context he referred in a note to 'the very important results which N. Bohr has obtained from Rutherford's atomic model'.[155] The near absence of Bohr from the pages of *Annalen der Physik* did not imply a lack of interest from the side of German physicists, as there were other outlets for publication, for example the proceedings (*Verhandlungen* or *Berichte*) of the German Physical Society. Thus, it was in the *Verhandlungen* that Emil Warburg in December 1913 published what was probably the first German research paper relating to Bohr's theory (see the following section).

From a scientific point of view, the most important of the Bohr-related papers published in the *Verhandlungen* were two papers by Walther Kossel, a student of Lenard and Sommerfeld. Appearing in the autumn of 1914 they offered a new and promising analysis of the absorption of X-rays. Kossel's second paper was based on the presupposition that 'as far as systems with one nucleus are concerned, Bohr's model is totally correct and well known'.[156] Kossel, who at the time worked as an assistant at the Technische Hochschule in Munich, only studied Bohr's theory at a rather late date, perhaps in the late spring of 1914. In July he met the Danish physicist in Munich, where Bohr gave a presentation of his work.[157] 'Have you seen a paper by Kossel in Verh. d. phys. Ges. 1914?', Bohr asked H. M. Hansen. 'I think he has got hold of something very important'.[158] And indeed he had, for, as mentioned in Section 3.4, it was only with Kossel's work that Bohr's theory became truly reconciled with the results of Moseley, making it clear that X-rays had their origin in atoms that had lost an electron from an inner ring. Understandably, Bohr very much welcomed the work of Kossel, which he dealt with in some of his papers of 1915.[159]

3.9 THE STARK EFFECT AND OTHER DEVELOPMENTS

In November 1913 Stark announced that he had discovered in experiments with hydrogen canal rays that the lines of the Balmer series H_α and H_β were separated into several components by a strong electric field.[160] Investigating the phenomenon further, he found that it existed for helium as well, and that it followed certain regularities, for example that the separation of the components was roughly proportional to the strength of the electric field. Stark's discovery, which six years later would be rewarded with the Nobel Prize, was an important factor in the increased interest that Bohr's theory attracted in Germany in particular. However, Stark himself did not see the electric effect as positively connected with Bohr's view of atomic structure, which differed profoundly from his own. Concentrating on his extensive experimental work and pursuing his own line of research, he showed little interest in the theoretical discussions concerning Bohr's theory, but he did intervene in the discussion. For example, in May 1914 he confirmed Evans's finding that the 4686 line belonged to the helium spectrum, yet without regarding the identification as supportive of Bohr's prediction.[161] On the contrary, concluding that the line was due to the *doubly* charged helium ion (He^{2+}) he implicitly denied the validity of the Bohr–Rutherford conception of the atom.

Stark's work on the spectra of canal rays convinced him and a few other physicists that experimental data from this area of research were irreconcilable with Bohr's theory and the Bohr–Rutherford nuclear model. In a book of 1914 in which he summarized his work on 'electric spectral analysis' he criticized Bohr's theory in general and confronted it with the canal-rays results and the Stark effect in particular. Although admitting that Bohr's recent theory of the electric effect agreed with some of his experiments, he argued that the agreement was apparent only and in any case annulled by serious disagreements with other of the experimental results:

> The mentioned quantitative agreement between Bohr's theory and observations loses completely its significance in regard of the fact that the theory is unable to reproduce correctly, and not even qualitatively, essential features of the electric splitting of the lines of the H-series . . . Bohr's theory is unacceptable, and that alone for the reason that there is an unsolvable contradiction between it and the observations regarding the splitting of consecutive series terms.[162]

In an extensive paper of 1916 Stark confronted Bohr's theory of the hydrogen atom with his own, very different model, according to which the Balmer spectrum was caused by the H^+ ion and not a neutral hydrogen atom (Starks's H^+ contained several electrons in motion). He proudly emphasized that his model agreed with 'Newtonian dynamics and Maxwell's theory . . . [and] it makes no use of the quantum hypothesis'.[163] The following year Stark launched a frontal attack on the Bohr quantum atom, which to his mind violated ordinary causality: it seemed to allow events, such as quantum jumps from one orbit to another, to be influenced by later events. Stark furthermore concluded that the theory was inconsistent with the experimental detection of H_3^+ ions. However, at the time of the attack—and more would follow—Bohr's theory was so well established that Bohr and most other physicists chose to ignore him.[164]

To return to the electric splitting effect, as early as 5 December 1913 Warburg read a paper to the German Physical Society in which he attempted to explain the Stark effect on the basis of Bohr's theory of the hydrogen atom. Since this theory was new and not generally known to German physicists, he included a condensed account of it.[165] According to Robert Pohl, who attended the meeting in Berlin, Warburg gave a report 'on a very important paper, that was Bohr's paper, . . . He explained . . . that this was a real advance, and I believe that the few hundred listeners at once understood [that] . . . "Planck's *h* proves to be the key for understanding the atom"'.[166] Warburg, who mistakenly seems to have considered Bohr's theory as belonging to the tradition of the electromagnetic matter theories, clearly found the theory appealing. But he was less happy about its foundation in Bohr's postulates, which invited 'weighty misgivings'. Although Warburg obtained a broadening of the spectral lines of the right order of magnitude, he did not succeed in reproducing the distinct line patterns observed by Stark, from which he concluded that Bohr's theory was unable to give more than a partial explanation: 'It does in no way explain it [the Stark effect] completely and for this reason a modification or extension of it is in any case needed'.[167] Bohr disagreed. In a letter to Warburg of early 1914 he remarked that his own as yet unpublished calculations resulted in separate lines with the right separation in wavelength.[168]

Independently of Warburg, the Italian physicist Antonio Garbasso calculated the electric splitting of spectral lines on the basis of what he in a letter to Bohr called his 'marvelous theory of spectral analysis'.[169] He proposed his interpretation of what he called the Stark–Lo Surdo phenomenon in terms of Bohr's theory at a session of the Accademia dei Lincei on 21 December 1913, which most likely was the first reference to Bohr's theory in Italy.[170] However, although Garbasso adopted Bohr's model at an early date, he did it half-heartedly and without following up on his work. On the contrary, shortly afterwards he argued that the Stark effect could be satisfactorily understood on the basis of a revised version of Thomson's classical atomic model! Garbasso's work did not signal the entrance of the quantum theory of atoms into Italian physics, which only occurred after World War I.[171]

While the work of Warburg and Garbasso supported Bohr's theory in a general way, Ernst Gehrcke at the Physikalisch-Technische Reichsanstalt in Berlin disagreed that the Stark effect was best understood on the basis of Bohr's view of the constitution of atoms. In an attempt to establish an alternative to this view, or to translate it into a more acceptable classical theory that avoided Bohr's postulates, Gehrcke derived the Balmer–Bohr formula for hydrogen and suggested an explanation of the Stark effect on the basis of a one-dimensional atomic model. He did not dismiss quantum theory, but preferred to do without it. Contrary to Bohr's theory, 'In my model of emission of light the assumption of energy quanta is admissible but not necessary'.[172] The expression that Gehrcke found for the change in frequency $\Delta\nu$ caused by an electric field differed by a factor of 4/3 from the one published a little later by Bohr. Garbasso's expression for $\Delta\nu$ was twice as large as Bohr's expression.

Bohr's own analysis of the Stark effect appeared in the March issue of *Philosophical Magazine*, which also included a first attempt to explain the Zeeman effect on the basis of the theory. Bohr reasoned that an external electric field F would deform the circular orbits of the electrons but not influence the transition mechanism between stationary states. Restricting his analysis to large quantum numbers and making use of a correspondence argument he found that, theoretically, the hydrogen lines would separate into four components. But in reality only two symmetric components would turn up, for otherwise one could not 'obtain the continuity necessary for a connection with ordinary electrodynamics'.[173] His result for the frequency shift was

$$\Delta\nu = \frac{3}{4\pi^2}\frac{h}{me}F\left(n_2^2 - n_1^2\right),$$

where n_2 and n_1 are the quantum numbers associated with the transition. The formula agreed reasonably well with the observations of Stark, and in any case better than the results obtained by Warburg, Garbasso, and Gehrcke. Realizing that his theory did not cover all the experimental details, Bohr cautiously concluded that 'it seems possible to account for some of the general features of the effect of magnetic and electric fields on spectral lines discovered by Zeeman and Stark'.[174] By the spring of 1914 Bohr was confident that his theory agreed with, or could be developed to agree with, the Stark effect. Enclosing his new paper, in a letter to Stark he wrote:

> My calculations – if right – allow to predict the distances between the outer components of all the hydrogen lines; further they suggest that the effect of the field might be very much smaller for elements of high atomic weight than for such elements as hydrogen, helium and lithium.... I should be very thankfull [sic] if you kindly with a few words would let me know, whether my suggestions might be in conformity with the results of your continuing experiments.[175]

Sommerfeld's growing interest in the Bohr atom was in part indebted to the Stark effect and its relation to Bohr's theory, a subject he dealt with in a Munich colloquium of 27 May 1914.[176] A few days later he wrote in a letter to the French physicist Paul Langevin, this time relating to the Zeeman effect, that 'in the atom an unsuspected number-theoretical symmetry and harmony appears to rule, as from another side Bohr has shown'. Then he expressed his reservations with regard to the state of Bohr's theory: 'Clearly a great deal is true in Bohr's model and yet I think that it must be fundamentally reinterpreted in order to satisfy. In particular, I am presently disturbed that it gives a wrong value for the magneton'.[177] Sommerfeld was warming to the theory, but even after having met with Bohr in Munich, he was not ready to adopt it as a basis for his further research. For example, he continued investigating the Zeeman effect without taking Bohr's model into account.

Only in 1915 did Sommerfeld publish his first research on the new theory, on dispersion theory and on a generalization of Bohr's theory of the hydrogen spectrum. The problem of dispersion, which Bohr had considered from a qualitative point of view in his trilogy, was first treated quantitatively by Peter Debye in a presentation to the Bavarian Academy on 9 January 1915.[178] Using for the hydrogen molecule essentially the same model that Bohr had suggested in 1913, he investigated how the atomic electrons would respond to an incident radiation and a change in its wavelength. His lengthy calculations resulted in a dispersion formula (that is, a formula of the type $\mu = \mu(\lambda)$, where μ is the refraction index) of the same form as the one based on classical physics. However, Debye concluded that the Bohr model for helium, with two electrons moving symmetrically on the same ring, disagreed with dispersion measurements and therefore had to be reconsidered. The success of Sommerfeld's more elaborate theory was also limited to hydrogen.

What she called the 'Bohr–Debye model' of the hydrogen molecule was examined theoretically by Hendrika J. van Leeuwen, a student of Lorentz, in a dissertation from 1915. She concluded that the molecule, if constituted as Bohr and Debye believed, would be mechanically unstable with respect to certain perturbations and that stabilization would only spoil the agreement with dispersion measurements.[179] Van Leeuwen's objections were not the only ones. The following year Clinton Davisson, a former student of Owen Richardson and at the time at the Carnegie Institute of Technology, Pittsburgh, independently calculated the dispersion of light by hydrogen and helium. Unlike Debye and Sommerfeld, his result was disappointing for both elements. 'The calculated values seem quite incapable of accounting for those observed', he concluded, adding that it might be possible that 'this lack of agreement is due to a mistaken conception of the mechanism of dispersion'.[180]

Bohr was pleased that Debye and Sommerfeld were interested in the same questions as himself, but he did not agree with them at all, as he wrote to his brother Harald: 'I look upon the entire problem of dispersion in quite a different way'.[181] The Debye–Sommerfeld theory (as the two theories were collectively known) assumed classical electrodynamics to apply to the perturbations of stationary orbits caused by external radiation, and thus described the interaction between radiation and the orbiting electron in classical terms. This Bohr found to be objectionable. 'If the theory of the hydrogen atom has but the slightest connection with truth', he wrote to Oseen, 'the dispersion (at least in gases) must be a phenomenon of quite a different nature from that assumed by Debye and Sommerfeld. . . . Dispersion cannot, whatever its explanation, be calculated from the motion of the electrons and the usual electrodynamics'.[182]

Debye's work on molecular hydrogen was not the only work in which he developed ideas related to Bohr's theory. At about the same time he investigated the scattering of X-rays on atoms within the framework of the Bohr ring model. Fully realizing that this model of the atom contradicted the laws of electrodynamics, Debye stated that 'The more recent development in our conceptions of atomic structure has forced us to recognize the possibility of electrons in motion which, in spite of very large accelerations, do not emit energy. We have to assume, for instance, the presence of two electrons in a hydrogen molecule, situated opposite one another on a circle 1.05×10^{-8} cm in diameter and revolving with an angular velocity $\omega = 4.21 \times 10^{16}\ \mathrm{sec}^{-1}$.[183] He thought that even if atoms were randomly oriented in space, their constitution in terms of electron rings might be revealed by the X-ray interference patterns they produced. The scattering of X-rays would provide an 'ultramicroscopy of the interior of the atom'. This was the original motivation for what became the celebrated Debye–Scherrer method of X-ray diffraction.[184] Paul Scherrer, who had come to Göttingen in 1913 at the age of 24, recalled of Debye's idea that the regular spacing of electrons on circular orbits should produce diffraction effects:

> One tried hard to become convinced of the reality of Bohr's electron orbits in the atoms in spite of all the hesitations the physicist felt in accepting the hypothesis that the electron on its stationary orbit about the atomic nucleus does not radiate, – a flagrant contradiction to Maxwell's theory. The next job to be done was therefore to find a check on Bohr's hypothesis which worked so simply and directly in the case of spectral emission, by looking for a direct evidence of the reality of the electronic orbits.[185]

Only in 1917 were Debye and Scherrer forced to admit that their method did not allow them to look into the inner of the atom. The work of Debye and Scherrer turned out to be eminently useful in the study of liquids and amorphous solids. Of course, it did not yield 'direct evidence of the reality of the electronic orbits'. As became known only after the advent of quantum mechanics, although it was suspected some time before, orbits in the semiclassical sense of Bohr do not exist.

Notes for Chapter 3

1. Interview by P. Ewald, G. Uhlenbeck and T. S. Kuhn of 29 March 1962 (www.aip.org/history/ohilist/4538_1.html). The French physicist Edmond Bauer recalled similarly that there was no notice taken of the Bohr atom before the war, and that even after the war there still was some resistance 'because it wasn't classical enough'. Interview by T. S. Kuhn and T. Cahan of 8 January 1963 (www.aip.org/history/ohilist/4498_1.html).

2. Bloch 1918, who also included the extensions of Bohr's theory made by Sommerfeld and other German physicists, but not Bohr's correspondence principle. On the reception of quantum theory and the Bohr theory in France, see Kojima 2004.

3. Margenau 1950, p. 311.

4. Ehrenfest to Lorentz, 25 August 1913, as quoted in Klein 1970, p. 278. For later objections of a conceptual nature to Bohr's theory, see Section 4.5.

5. Ehrenfest to Sommerfeld, April–May 1916, in Eckert and Märker 2000, p. 555. The German phrase is 'ganz kanibalischem'.

6. Brillouin 1960, p. v, foreword dated September 1959. The story does not agree with the recollection of Paul Epstein, who spoke on Bohr's theory in Munich on 26 January 1914 (Heilbron 1967, p. 456).

7. Hoyer 1981, p. 603. The reference to the meeting in October was to the second Solvay conference in Brussels, where Sommerfeld and Rutherford were among the invited physicists.

8. See extract of letter from Rutherford to Bohr of 13 May 1913, in Hoyer 1974, p. 173.

9. Lodge 1913, p. 17.

10. 'Physics at the British Association', *Nature* **92** (1913), 304–9, on p. 306. Keller 1983, pp. 173–6 describes the Birmingham meeting and Bohr's role in it.

11. Jeans 1913, p. 379. Bohr later recalled that Jeans's 'lucid exposition was, in fact, the first public expression of serious interest in considerations [Bohr's theory] which outside the Manchester group were generally received with much scepticism'. Bohr, Rutherford Memorial Lecture of 1958, reprinted in Favrholdt 1999, pp. 383–415, on p. 393.

12. Jeans 1913, p. 379.

13. Niels Bohr to Margrethe Bohr, 14 September 1913 (Niels Bohr Archive, Copenhagen). Reproduced with special permission.

14. Ewald 1913, pp. 1298–9. Ewald's report was complementary to and in some respects more detailed than the one in *Nature*. On 19 November 1913 Ewald reported the discussions in Birmingham to a colloquium in Munich where he may have mentioned Bohr's intervention (Heilbron 1967, p. 456).

15. Ewald 1913, p. 1301.

16. Jeans 1914, p. 51. The main section on Bohr's theory appeared on pp. 50–7.

17. Ibid., p. 64.

18. McLaren 1913c.

19. McLaren 1913b, and also McLaren 1913a appearing in the 9 October issue of *Nature*.

20. Nicholson 1913a.

21. McLaren 1913a. For his electrodynamical magneton theory of gravity, see McLaren 1913d.

22. Rutherford and Nuttall 1913, p. 712. A similar reference appeared in Rutherford 1913b. For Rutherford's appreciation of Bohr's theory—positive but not enthusiastic—see Wilson 1983, pp. 318–38.

23. Rutherford 1914a. Also van den Broek referred to Bohr's argument that beta rays have their origin in the atomic nucleus (van den Broek 1914).

24. Rutherford 1914a, p. 498, and similarly in Rutherford 1914b.

25. Rutherford 1914b, p. 351.

26. For the papers presented at the Vienna meeting 21–28 September, see *Physikalische Zeitschrift* **14** (1913), 1073–180. Nancy Greenspan's claim that 'Discussion at the meeting focused on the recent work of twenty-eight-year-old Danish physicist Niels Bohr on the quantum theory of the atom', is unfounded (Greenspan 2005, p. 60).

27. Hevesy to Bohr, 23 September 1913, in Hoyer 1981, p. 531. Hevesy's spelling.

28. Hevesy to Rutherford, 14 October 1913, in Eve 1939, p. 226. That Einstein found Bohr's theory interesting and valuable is supported by Franz Tank's recollection of a colloquium in Zurich at about the same time (see Section 3.8). However, in spite of his sympathy for the theory, Einstein did not refer to it in his publications until 1916.

29. Stachel 2002, p. 369, who cautiously adds that he does not interpret Einstein's letter as an anticipation of Bohr's atomic theory. The letter is reproduced in Kleinert and Schönbeck 1978.

30. Geiger to Bohr, 12 October 1913 (Heilbron 1964, p. 295).

31. Goldschmidt et al. 1921. See also Mehra 1975, pp. 75–94.

32. Thomson 1921, p. 20 and p. 50. Bohr 1913a.

33. Thomson 1913b and Thomson 1913c. On Thomson's research programme on positive rays, see Falconer 1988.

34. Thomson 1913c, p. 121. If Thomson's terminology indicated support of Rutherford's nuclear atom, it was not intended.

35. Bohr interview of 1 November 1962 (Niels Bohr Library & Archives, American Institute of Physics). It had been known since about 1870 that the element palladium is able to absorb large volumes of hydrogen and that the process depends critically on the purity (molecular weight) of the hydrogen gas.

36. Hevesy to Rutherford, 14 October 1913, quoted in Eve 1939, p. 224.

37. Stark 1913a. 'On the basis of our valence hypothesis we should expect the presence of a triatomic hydrogen molecule' (Stark 1915, p. 121).

38. Undated note on 'Spectrum of Hydrogen Isotope', in Hoyer 1981, p. 278. The note was part of a draft paper by Bohr and H. M. Hansen, who apparently made spectroscopic experiments in Copenhagen to detect the isotope effect. The results of these experiments are unknown, but one can assume they were unsuccessful.

39. The isotope effect was first established experimentally in the band spectrum of HCl, where the mass difference of the two isotopes Cl-35 and Cl-37 could be detected. This discovery was made independently by Alfred Loomis in the United States and Adolf Kratzer in Germany (Assmus 1992b, p. 23). In 1932 Harold Urey, Ferdinand Brickwedde, and George Murphy identified the heavy hydrogen isotope (deuterium) by detecting a slightly different wavelength for the spectral lines of hydrogen, $\Delta\lambda = 1.79$ Å for Hα and 1.32 for Hβ. In 1934 Urey was awarded the Nobel Prize in chemistry for the discovery. For a historical and personal account, see Brickwedde 1982.

40. Report of the British Association meeting in Manchester, in *Nature* **96** (1915), 240. It is not generally known that Bohr predicted the isotope effect.

41. 'Note on the Properties of Isotopes and the Theory of the Nucleus Atom', draft of October 1915, in Hoyer 1981, pp. 417–25 (p. 419).

42. Duane and Wendt 1917. The story of triatomic hydrogen is told in Kragh 2010b and in greater detail in Kragh 2011b.

43. Bohr 1919, p. 5, reprinted in Hoyer 1981, pp. 472–88.

44. Wendt and Landauer 1920, who quoted Bohr's paper of the previous year: 'The calculations of Bohr on the basis of Rutherford's atom furnish the only valence hypothesis which satisfactorily accounts for its existence' (p. 945).

45. Tunzelmann 1913 and 1914. Walter 1914.

46. Niels Bohr to Harald Bohr, 30 July 1913, in Rud Nielsen 1972, p. 563.

47. Review of Richardson 1914, in *Nature* **95** (1915), 420–1, signed 'N.B'. According to the 'electromagnetic world view', all of physical reality was of electromagnetic nature and the fundamental laws were those of electrodynamics rather than mechanics. For a description of this view of nature, see Kragh 1999, pp. 105–19. Contrary to many other physicists at the time, the electromagnetic world view left Bohr cold. It is incorrect to describe Bohr and Sommerfeld as 'both once proponents of an all-encompassing electromagnetic view of nature' (Seth 2010, p. 227). The description fits Sommerfeld, but not Bohr.

48. Bohr, 'On the Interaction between Light and Matter', lecture of 13 February 1920, English translation in Rud Nielsen 1976, pp. 227–40 (on p. 232).

49. Richardson 1914, p. 587. Preface dated May 1914.

50. See Knudsen 2001, according to whom Richardson 'gave no indication that he saw a fundamental conflict between this [frequency] postulate and the many applications of Lorentz' electrodynamics in the electron theory of matter' (p. 248).

51. Kaye 1914, p. 18, preface dated February 1914.

52. Ibid., p. 200.

53. Campbell 1914, p. 587.

54. Ibid.

55. Allen 1915b, p. 720. For Allen's unorthodox use of Bohr's theory, see also Allen 1915a.

56. Allen's model was only one of several speculative nuclear models proposed in the second half of the 1910s. For these, see Stuewer 1983.

57. Campbell 1914, p. 587.

58. Spencer 1913. Dawson 1913. The abstracts emphasized the spectroscopic and chemical aspects of the theory, including its picture of the hydrogen molecule.

59. Hevesy 1914, p. 600.

60. Soddy 1914, p. 271, reprinted in Trenn 1975, p. 341. In his report for 1911 Soddy dealt in some detail with Nicholson's atomic model (Trenn 1975, pp. 255–8).

61. *Nature* **96** (1915), p. 240. In a draft from the same time, probably prepared for the Manchester discussion, Bohr dealt in detail with 'The Properties of Isotopes and the Theory of the Nucleus Atom' (Hoyer 1981, pp. 417–25).

62. Fowler and Nicholson 1914, p. 359.

63. Saha 1920, p. 473. The rise of quantum astrophysics is fully described in a series of papers by David DeVorkin and Ralph Kenat (DeVorkin and Kenat 1983a, 1983b; Kenat and DeVorkin 1990).

64. On Barkla and his scientific career, see Stephenson 1967. In 1918 Barkla received the Nobel Prize for his discovery of the characteristic lines, nominated by only a single nominator—Rutherford.

65. The authoritative work on Moseley and the birth of X-ray spectroscopy, including its relation to atomic theory, is Heilbron 1974. See also Heilbron 1966.

66. Moseley 1914a.

67. Moseley to Bohr, 16 November 1913, in Hoyer 1981, pp. 544–6, and Heilbron 1974, pp. 211–13.

68. Bohr to Moseley, 21 November 1913, in Hoyer 1981, pp. 546–7, and Heilbron 1974, pp. 214–15. Only in a paper of 1915 did Bohr publicly dissociate himself from Moseley's idea (Bohr 1915b, p. 412).

69. Moseley to Rutherford, 5 January 1914, in Heilbron 1974, p. 219. The three atomic hypotheses referred to were Thomson's, Nicholson's, and Bohr's.

70. Moseley 1913. He restated the suggestion of a four-electron vibrating ring in Moseley 1914a.

71. Lindemann 1914a, p. 501 and also Lindemann 1914b. For the details of Lindemann's arguments, published in the transactions of the German Physical Society, see Lindemann 1914c. See also Heilbron 1974, pp. 105–8. Nicholson's objections to the X-ray confirmation of Bohr's theory will be discussed in Section 3.6.

72. Bohr 1914b. Moseley 1914a. See also Hoyer 1974, pp. 196–202.

73. Kossel 1914a. Kossel 1914b. For historical context, see Heilbron 1967.

74. Bohr 1915b. As a result of the war there was limited or no exchange of scientific papers between Germany and England. Bohr, from neutral Denmark, did not face the problem.

75. This section builds on Kragh 2011a.

76. Thomson 1913a. Thomson 1921. *Nature* **92** (1913), 305. An extended English abstract of Thomson's Solvay lecture appears in Mehra 1975, pp. 77–81.

77. Thomson 1914. On the application of the Thomson atom to problems of chemistry, see Stranges 1982. These problems included the structure of the H_3 molecule, which Thomson pictured as a triangle of hydrogen atoms, each of which atoms was the origin as well as the termination of a tube of force (Thomson 1914, p. 783).

78. Thomson 1914, p. 781.

79. Thomson 1919, p. 420.

80. Thomson 1936, p. 425.

81. Bohr to Rutherford, 16 October 1913, in Hoyer 1981, pp. 587–9.

82. The unpublished letter is reproduced in Hoyer 1981, p. 268.

83. Hoyer 1981, pp. 588–9.

84. 'J. J. T. . . . knows that I think his atom is only fitted for a museum of scientific curiosities. The idea of a nucleus atom is really working out exceedingly well. You will have seen the work of Bohr and Moseley'. Rutherford to Boltwood, 17 March 1914, in Badash 1969, p. 292.

85. Rutherford to Schuster, 2 February 1914, quoted in Wilson 1983, p. 338.

86. Larmor 1929, pp. 344–72 ('The physical aspect of the atomic theory') and pp. 630–3 ('On non-radiating atoms').

87. Larmor 1931, p. 76.

88. Lorentz to Julius, 14 October 1914, as quoted in van Delft 2007, p. 491. The chair in Utrecht went to Leonard Ornstein, a Dutch physicist.

89. See Nersessian and Cohen 1987, which includes the second edition of *Theory of Electrons* with the reference to Bohr on p. 107. Lorentz 1927, based on a lecture course of 1922, contained discussions of various aspects of the Bohr atom.

90. Strutt 1968, p. 357.

91. Conway 1913. See Heilbron 1964, pp. 299–301 and also Kragh 2011a on which this section relies.

92. Peddie 1914, p. 259.

93. Crehore 1913. At the annual meeting of the American Physical Society in December 1916, Crehore presented two papers on atomic theory. None of them were published, but their titles indicate his stand: 'A criticism of the Rutherford-Bohr atomic hypothesis based upon a theorem of phase-equilibrium of two electrons', and ''h' based upon the classical electrodynamics'. See *Physical Review* **9** (1917), 170.

94. Crehore 1915, p. 310.

95. Crehore 1921, p. 593. The same objection was raised by the respected theoretical chemist Richard Tolman, who found it problematical that 'there is no actual frequency in the atom which corresponds to the frequency ν which is calculated' (Tolman 1922, p. 227). See also Section 4.5.

96. Hicks 1914.

97. Nicholson 1914c, p. 583.

98. Van den Broek 1914, p. 242.

99. Nicholson 1914d, p. 93.

100. Nicholson 1914e, p. 439.

101. Nicholson 1914d, p. 103.

102. The two draft letters are reproduced in Hoyer 1981, pp. 270–1 and pp. 312–16.

103. Bohr 1915b, p. 399. The earlier paper was Bohr 195a.

104. Bohr to Hansen, May 12, 1915, in Hoyer 1981, pp. 517–18. In an earlier letter to Oseen: 'I imagine you understand that I do not place much emphasis on the criticism by Nicholson. His whole point of view is so foreign to me, and by a departure from mechanics I understand something much more radical than he does' (28 September 1914, in Hoyer 1981, p. 562).

105. Niels Bohr to Harald Bohr, 15 April 1915, in Rud Nielsen 1972, p. 579.

106. Föppl 1915.

107. Sommerfeld 1922a, p. 728.

108. Soddy 1917, p. 255. Reprinted in Trenn 1975, pp. 389–416.

109. For example Jeans 1919 and also in the second edition of Jeans 1914, published in 1924.

110. Birge 1956, p. 22, and Sopka 1988, p. 31. Birge's missionary activity did not result in early publications of his dealing with the Bohr atom. Victor Lenzen, who in 1919 was a physics student at the University of California, Berkeley, recalled that there was much opposition to Birge's promotion of Bohr's theory. The physical chemist G. N. Lewis 'was, for a long time, vehemently hostile to the whole dynamical approach, particularly Bohr's'. Interview of 28 June 1962 by T. S. Kuhn, American Institute of Physics, http://www.aip.org/history/ohilist/4739.html.

111. The only sign of interest is a symposium of the American Physical Society of 27 September 1914 on 'Spectroscopic Evidence Regarding Atomic Structure', which included a paper on Nicholson's atomic theory. Bohr's theory may have been mentioned, but the papers of the symposium were not published. See *Physical Review* **5** (1914), 72.

112. Wright 1915, p. 269, read to the National Academy of Sciences on 9 December 1914.

113. Plaskett 1915, p. 47.

114. Eve 1914.

115. Stewart 1914.

116. Cole 1915, p. 77.

117. Fulcher 1915, p. 371. This was one of very few American research papers before 1916 which discussed Bohr's theory. On Stark's critique, see Section 3.9.

118. Harkins and Wilson 1915, p. 1409.

119. Wilson 1916, p. 12 and p. 14, which was probably the first doctoral thesis covering Bohr's atomic theory.

120. Lewis's ideas went back to 1902. For a full historical account of his early conception of atoms and molecules, see Kohler 1971.

121. Lewis 1916, p. 773.

122. Lewis 1917, p. 299, based on a symposium on 'The Structure of Matter' held on 27 December 1916.

123. Lewis 1923a, p. 51. See also Arabatzis 2006, pp. 186–7.

124. Rutherford to Lewis, 13 December 1915, quoted in Peierls 1988, p. 233.

125. Lewis to Bohr, 8 February 1916 (Archive for History of Quantum Physics, Bohr Scientific Correspondence). The honorarium for the lectures amounted to $1000. 'I wrote to Prof. Lewis and said that I hoped to be able to go', Bohr wrote to his brother, 'but could not give a definite answer at once, as I was due to return to my post in Copenhagen in September next' (Niels Bohr to Harald Bohr, 14 March 1916, in Rud Nielsen 1972, p. 585).

126. Lewis 1919, p. 119, who treated Parson's theory as an interesting alternative to Bohr's. Without mentioning Bohr's theory, the Michigan physicist David Webster suggested that Parson's magneton atom could be reconciled with Rutherford's nuclear model (Webster 1918). In addition to Webster, several other American physicists, including Leigh Page and Arthur Compton, expressed interest in the ring electron as the ultimate magnetic particle. These speculations are sometimes claimed to be anticipations of the spinning electron (Jammer 1966, p. 148).

127. Parson 1915, p. 3. On Parson's theory, see Kohler 1971, pp. 364–70, and Stranges 1982, pp. 220–3.

128. Parson 1915, p. 7.

129. Ibid., p. 11.

130. Millikan 1917a, p. 326. The other quotations are from the same paper.

131. See Holton 1988, p. 165, and also Sopka 1988 for a comprehensive account of American responses to the new quantum and atomic theories.

132. Webster and Page 1921. See the quotation in Holton 1988, p. 167.

133. Eve 1939, p. 208.

134. Harald Bohr to Niels Bohr, undated but most likely October 1913, in Rud Nielsen 1972, p. 567. Hilbert would later describe Bohr as the 'Newton of atomic theory' and lecture on the quantum theory of atoms and spectra (Sauer and Majer 2009, p. 509).

135. Born 1978, p. 157.

136. Ibid. On Born's report, see Greenspan 2005, p. 66.

137. Courant 1981, p. 162.

138. Interview with R. Courant, 9 May 1962, by T. S. Kuhn and M. Kac. American Institute of Physics, Niels Bohr Library & Archives. http://www.aip.org/history/ohilist/4562.html.

139. Reid 1976, p. 45.

140. Bohr 1955, as reprinted in Favrholdt 1999, pp. 371–9 (on p. 378).

141. Sommerfeld to Runge, 6 September 1916, in Eckert and Märker 2000, p. 566. Sommerfeld referred to Paschen's recent confirmation of the fine-structure theory, which was often considered strong support for the Bohr–Sommerfeld atom (Section 4.4). On Runge's attitude, see also the description in McCormmach 1982, pp. 73–5.

142. Letter of 11 March 1964 to Max Jammer, as quoted in Jammer 1966, p. 86. According to Friedrich Hund, Otto Stern said in 1914 that he and Laue agreed to give up physics if there was 'anything in this nonsense of Bohr's' (Hund 1967, p. 67).

143. Bohr to Oseen, 28 September 1914, in Hoyer 1981, p. 557.

144. Interview with A. Landé, 5 March 1962, by Thomas S. Kuhn and John Heilbron. American Insitute of Physics, Niels Bohr Library & Archives. http://www.aip.org/history/ohilist/4728_1.html. See also Pais 1991, p. 155.

145. Haas to Bohr, 6 January 1914, in Hoyer 1981, p. 513.

146. Nisio 1973, pp. 59–60. On the Munich *Mittwoch* colloquia 1913–1915, see also Eckert and Märker 2000, pp. 434–9.

147. Seeliger 1914a.

148. Seeliger 1914b, p. 313.

149. Riecke 1915, p. 222. Riecke died the same year, and the paper was published posthumously. In France, Léon Bloch was similarly impressed by the spectroscopic confirmations of Bohr's theory, which in the case of the hydrogen atom 'is very close to reality' (Bloch 1918, p. 171).

150. Warburg 1915.

151. Wien 1915, p. 222. Wien soon recognized the merits of Bohr's theory; witness that he nominated him for a Nobel Prize in 1919 and again in 1920 (Aaserud 2001).

152. Oseen to Bohr, 11 November 1913, in Hoyer 1981, p. 553. Bohr was not interested in the question of a modification of the Maxwell–Lorentz equations. On the other hand, the Norwegian physicist Thorstein Wereide claimed to have derived solutions from a modification of the classical equations that agreed with Bohr's stationarity postulate in the case of circular orbits. To obtain the wanted agreement he made use of the ad hoc hypothesis of an unknown field from the atomic nucleus (Wereide 1917). Wereide explained his theory in a letter to Bohr of 29 September 1916 (Archive for History of Quantum Physics, Bohr Scientific Correspondence).

153. Oseen 1915, p. 404. Stark used Oseen's analysis as ammunition in his critique of the Bohr atom (Stark 1916, p. 76, and in Section 4.5).

154. Schott 1918. One such modification was to conceive one or more rotating electrons as a continuous ring system rather than discrete charges. The unorthodox idea of a 'ring electron' was entertained by several physicists and chemists. Arthur Compton wrote a series of papers on the subject and in England H. Stanley Allen developed his own ideas of a ring electron.

155. Hansen 1914, p. 235.

156. Kossel 1914b, p. 953. On Kossel's X-ray theory, see Section 3.4.

157. On 15 July Bohr and Kossel shared the programme of the university's *Mittwoch* colloquium. See Heilbron 1967, who gives details about Kossel's work and career. In a letter to Bohr of 1921, quoted in Heilbron 1967, Kossel said that he only became acquainted with Bohr's works a short time before they met.

158. Hoyer 1981, p. 517.

159. Bohr 1915a and Bohr 1915b.

160. Stark 1913b. On Stark's discovery see Hermann 1965b, which includes key articles on the Stark effect in the period 1914–1916.

161. Stark 1914a.

162. Stark 1914b, p. 119. Fulcher 1915 agreed with Stark's verdict.

163. Stark 1916, p. 56.

164. Stark 1917. See also Stark 1920, a sharp attack on the atomic theory of Bohr and Sommerfeld. The controversy following Stark's attack is dealt with in Section 4.5.

165. Warburg 1913. This was the first application of Bohr's atom not made by Bohr himself. Moseley's work on X-ray spectroscopy may count as the next one.

166. Interview with Thomas S. Kuhn and John L. Heilbron of 5 March 1962, as quoted in Pais 1991, p. 154. It is doubtful if the German physicists listening to Warburg 'at once understood' the meaning and range of Bohr's theory.

167. Warburg 1913, p. 1266.

168. Bohr to Warburg, 8 January 1914, in Hoyer 1981, p. 608.

169. Garbasso 1914. Garbasso to Bohr, 19 January 1914, in Hoyer 1981, p. 511.

170. The discovery of the electric splitting of spectral lines was made independently by Garbasso's assistant in Florence, Antonino Lo Surdo. See Leone, Paoletti, and Robotti 2004.

171. The first Italian paper, apart from Garbasso's, which cautiously made use of quantum theory and the Bohr atom dates from 1919 and was written by the spectroscopist Rita Brunnetti. See Leone and Robotti 2006.

172. Gehrcke 1914a, p. 839. On his theory of the Stark effect, see Gehrcke 1914b.

173. Bohr 1914a, p. 515. As pointed out in Darrigol 1992, p. 91, this is the first example of a selection rule based on correspondence arguments. See also Nisio 1973, p. 58. On the emergence of selections rules, see Borrelli 2009, who does not mention Bohr's paper.

174. Bohr 1914a, p. 524. On the role of the Stark effect in the reception of Bohr's atomic theory, see Hoyer 1974, pp. 215–41.

175. Bohr to Stark, 31 March 1914 (Archive for History of Quantum Physics, Bohr Scientific Correspondence).

176. For the importance of the Stark effect for Sommerfeld's growing interest in Bohr's theory, see Eckert and Märker 2000, p. 437.

177. Sommerfeld to Langevin, 1 June 1914, in Eckert and Märker 2000, p. 484–5. Sommerfeld mentioned that recent measurements made by Auguste Picard and Pierre Weiss seemed to have confirmed the Langevin–Weiss magneton theory, in contradiction to Bohr's theory.

178. Debye 1915a. For Debye's theory and his model of the hydrogen molecule, see also Wolfke 1916. Sommerfeld's more comprehensive theory appeared in Sommerfeld 1915a. The further development of dispersion theory will be dealt with in Section 8.4.

179. Van Leeuwen 1916, a summary a her 1915 Dutch dissertation.

180. Davisson 1916, p. 27, who was apparently unaware of the works of Debye and Sommerfeld, undoubtedly because of the war.

181. Niels Bohr to Harald Bohr, 10 November 1915, in Rud Nielsen 1972, p. 581.

182. Bohr to Oseen, 20 December 1915, in Hoyer 1981, p. 565. More on dispersion in Section 8.4.

183. Debye 1915b, p. 809.

184. Debye and Scherrer 1916, presented at a meeting of the Göttingen Scientific Society of 4 December 1915.

185. Scherrer 1962, p. 642.

4

The Bohr–Sommerfeld Theory

In spite of its novelty and many radical features, Bohr's atomic theory was generally well received and had by 1918 become accepted by the majority of physicists working in areas of atomic and quantum theory. The theory gained further experimental support through its remarkable interpretation of experiments with electron collisions in gases (Franck–Hertz experiments), and in its reformulation by Sommerfeld it proved able to explain the Stark effect, the normal Zeeman effect, the fine structure of hydrogen, and nearly the whole field of X-ray spectroscopy. The empirical successes depended intimately on theoretical advances of which the generalizations proposed by Sommerfeld and his colleagues in Germany were far the most important. While Bohr's theory had originally been discussed within the British tradition of physics, during the war the centre of gravity in atomic theory changed to Germany, where it would remain for many years. Apart from the Germans, there were also important contributions from Dutch and Swedish physicists. The literature on atomic and quantum physics in Germany and England was markedly different in the decade 1914–1924; in quantity, in style, and not least in quality. Aware of the shift, worried British physicists sought to lure Bohr to England: in 1918 there were plans to create a chair in Manchester for him, and in July 1923 the Royal Society offered him a research professorship.[1] Bohr was flattered, but declined.

The contributions made by the German physicists, and to a lesser extent by Bohr himself, not only changed the national setting of atomic theory, they also considerably changed the picture of the Bohr atom. By the summer of 1920 it did not look very much like the atom introduced seven years earlier. Although it was still based on Bohr's main postulates, it was no longer a planar ring atom governed by a single quantum number. Now it was a many-quantum atom with the electrons moving in elliptical orbits extending into three dimensions rather than being limited to two. Precisely how the electronic orbits were arranged, and how they should be characterized in terms of quantum numbers, remained unsettled questions. This was the case even for the helium atom, the simplest atomic system apart from hydrogen. A few years later the Bohr atom would transmute even more drastically in the direction of a symbolic rather than realistic model, a development which is the subject of Chapter 8.

The progress and consolidation of the Bohr atom in the decade after its birth may be illustrated in various ways. For example, in 1918 Johannes M. Burgers, a student of Ehrenfest's in Leiden, wrote his doctoral dissertation on Bohr's new atomic theory, probably the first dissertation of its kind. Its title was *Het Atommodel van Rutherford-Bohr.* While all of Bohr's works in the early period from 1913 to 1916 had been published in English, in 1921 they appeared collectively in a German translation with a foreword by Bohr, who used the occasion to review the more recent developments in quantum and atomic theory.[2] Although the papers were at the time mainly of historical interest, to German physicists they were more than just of celebratory value. The high regard in which Bohr was held by the German physics community is further illustrated by a special issue of the journal *Die Naturwissenschaften* from the summer of 1923, published as a tribute to 'The First Ten Years of Niels Bohr's Theory of the Structure of Atoms'. The issue included a German translation of Bohr's recent Nobel Prize lecture as well as contributions by Born, Ehrenfest, Kramers, Kratzer, Kossel, Franck, and a few other leading physicists. The introductory article was written by Planck, the dean of German physics, who praised Bohr's theory as 'a unique milestone in the development of theoretical physics'. Planck also used the occasion to thank Bohr on behalf of the German physics community for 'having published a large part of his work in German journals'.[3]

Bohr's theory was expounded not only in scientific books and papers, but from about 1918 onwards also in more popular articles and books intended for a broader public. An early example is *Die Atomtheorie* from 1918, a small book based on a series of lectures given by Leo Graetz, professor of physics in Munich. As Graetz pointed out, the new atomic theory was based on 'arbitrary assumptions', namely the two quantum postulates, which could only be justified by their empirical consequences in spectroscopy and 'by the undeniable fact that in the atomic domain, that is, for the very smallest distances, the discovered laws of mechanics and electrodynamics have no general validity'.[4] Graetz introduced the Bohr atom as a 'an entire solar system' with the electrons whirling around the nucleus 'like the Earth moves around the Sun', but with the important difference that the electrons could only move in discrete orbits given by a whole number. German lay readers could get an updated and more detailed exposition of the Bohr atom from *Die Entwicklung der Atomtheorie*, a book of 1922 written by the physics teacher Paul Kirchberger.[5] From the early 1920s a similar literature, popular or semipopular, appeared in England, the United States and elsewhere. Examples from England included John Sullivan's *Atoms and Electrons* and Bertrand Russell's *The ABC of Atoms*, both published in 1923, and Oliver Lodge's *Atoms and Rays* from 1924.

The Nobel Prize awarded to Bohr in 1922 undoubtedly helped to make him and his theory better known outside the physics community. Congratulating him on the prize, the German spectroscopist Heinrich Konen compared his contributions to atomic physics to Darwin's in biology:[6] no small praise. Bohr had first been nominated in 1916, and in 1919 he received three nominations from highly reputed physicists, namely, Rutherford, Rubens, and Wien. The following year he was nominated by Wien and von Laue, and in 1921 again by von Laue. His luck only changed in 1922, when he was nominated by no less than eleven physicists, including heavyweights like W. L. Bragg, von Laue, Millikan, Planck, Röntgen, and Rutherford.[7] As the Nobel physics committee saw it, his work was connected to Einstein's theory of light and therefore an argument for awarding Einstein the reserved 1921 prize (and without linking the prize to the theory of relativity). As Oseen wrote to Svante Arrhenius, 'I have been pleased to learn that Bohr, both publicly and to me privately, has expressed his happiness that Einstein received the prize and that he received it precisely for that part of his activity which constitutes a preparation for Bohr's'.[8]

Although by far the most prestigious, the Nobel Prize was not the only award with which Bohr was honoured for his atomic theory. In 1921 he received from the Royal Society the Hughes Medal, which included a gift of £1000, and two years later he received the Matteucci Medal from the Italian National Academy of Sciences.[9]

In the same year that Bohr received the Nobel Prize, the first book of a popular nature devoted wholly to his new atomic theory was published, written by Bohr's Dutch collaborator H. A. Kramers and Helge Holst, a Danish physics-trained author and librarian at the Polytechnic College in Copenhagen. The book was written in Danish, but already by the following year it had appeared in an English translation, entitled *The Atom and the Bohr Theory of its Structure* with an introduction by Rutherford. An updated German translation was published in 1925, and it also appeared in a Spanish and a Dutch translation.[10] The successful book gave a semipopular and fairly detailed account of Bohr's theory, emphazing the role of Bohr, Denmark's new national hero. In the preface to the English edition, the two authors were careful to describe the quantum theory of the atom, although an international endeavour, as primarily the brainchild of Bohr: 'The past decade has witnessed an enormous development at the hand of scientists in all parts of the world of Bohr's original conceptions; but through it all Bohr has remained the leading spirit, and the theory which, at the present time, gives the most comprehensive view of atomic structure may, therefore, most properly bear the name of Bohr'.[11]

4.1 CONFIRMATIONS AND EXTENSIONS

Dissatisfied with his position as a docent in Copenhagen, which left him little time for research, in October 1914 Bohr returned to Manchester. Rutherford had arranged that a vacant readership was offered to his protegé, which Bohr happily accepted. He thus spent a second period in Manchester, waiting for a suitable academic position in his native country. This position, a professorship, became a reality in the spring of 1916 and a few months later Bohr returned to Denmark. He would remain as a professor of physics at the University of Copenhagen for nearly forty years (he retired in 1955), during which time he was the unquestioned leader of Danish physics. Shortly after his return Bohr got his first collaborator and student in the shape of the 21-year-old Dutch physicist Hendrik Antonie (Hans) Kramers, who arrived from the Netherlands in 1916 and would stay with him until 1926. In the period from 1915 to 1917 Bohr published relatively little. These were the years in which his quantum theory of atoms was, for the first time, developed and extended independently by other physicists, first and foremost by Sommerfeld and his group in Munich. In the late fall of 1916 Bohr referred in a letter to Rutherford to 'the uninterrupted stream of german [sic] papers on this subject'.[12] By that time Bohr's theory was more than just Bohr's theory.

In a paper that appeared in the September 1915 issue of *Philosophical Magazine* Bohr discussed the essential features and results of his theory. There was a need for such an updated presentation, he thought, 'As the theory has been made a subject of criticism, and as experimental evidence of importance bearing on these questions has been obtained in the meanwhile'.[13] He was thinking of Nicholson, among others. Bohr now organized his theory around six basic assumptions, which he took to be the following:

(1) The existence of stationary states.
(2) The frequency postulate; that is, $h\nu = E_n - E_m$.
(3) The ordinary laws of mechanics hold for the stationary states, but not for transitions between them.
(4) The mean value of the kinetic energy of an electron in a hydrogen atom satisfies $T = \frac{1}{2}nh\omega$, where ω is the frequency of rotation and n an integer.
(5) Electrons in atomic systems revolve in rings and have in their ground state an angular momentum of $h/2\pi$.
(6) An atomic system is stable if the total energy is less than any neighbouring configuration satisfying the above assumption.

Bohr realized that condition (3) implied that his theory was essentially restricted to periodic orbits and that, for non-periodic orbits, 'the general principles applied are no longer a sufficient guidance'. This was a restriction he would soon return to and give his full attention.

Apart from offering a modified presentation of his theory of the hydrogen atom and a new explanation of the Stark effect, Bohr briefly considered the helium atom with its two different systems, known spectroscopically as orthohelium and parhelium. He assumed these were due to different orbital structures, but at the time he did not go further into the matter (see Section 4.7).

The most important remark in the 1915 paper concerned some recent experiments by James Franck and Gustav Hertz on ionization in mercury vapour. In a series of experiments the two Berlin physicists measured what they thought was the ionization potential of gases by bombarding them with accelerated electrons of increasing energy. In experiments of 1914 with mercury vapour they observed that when the electrons were accelerated to an energy of 4.9 eV the electron current dropped suddenly and a characteristic ultraviolet radiation of wavelength $\lambda^* = 2536$ Å was emitted. The phenomenon exhibited a periodic behaviour, and also occurred at small multiples of 4.9 eV (Figure 4.1). Noticing that $hc/\lambda^* = 4.9$ eV, Franck and Hertz interpreted their data as a transfer of the electron energy to cause ionization and emit light in the form of resonance radiation. They were convinced that they had conclusively shown that the ionization energy of mercury was 4.9 eV and also, based on different experiments, that the value for helium was 20.5 eV. Franck and Hertz thought their results were 'completely in agreement with the quantum theory', since according to this theory, as they understood it, 'the vibrations of the electrons in an atom can receive energy only in certain quanta'.[14] It was obviously not Bohr's quantum theory of atoms they were thinking of.

The experiments of Franck and Hertz were not only unrelated to Bohr's theory; at the time they made the experiments they were not even aware of it. In an interview many years later Franck recalled of Bohr's theory: 'We had neither read nor heard about it. We had not read it because we were negligent to read the literature well enough . . . On the other hand, one would think that other people would have told us about it. For instance we had a colloquium at that time in Berlin at which all the important papers were discussed. Nobody discussed Bohr's theory'.[15] Already in his

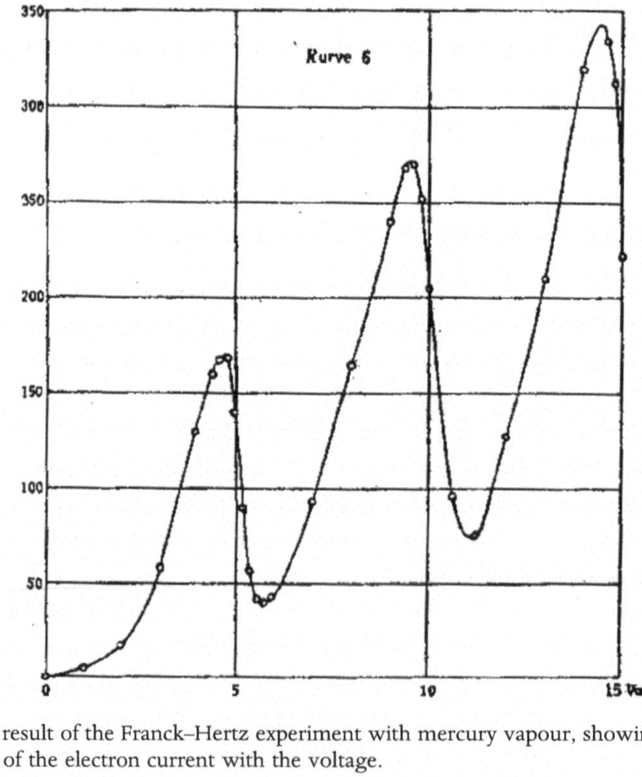

Fig. 4.1. The result of the Franck–Hertz experiment with mercury vapour, showing the characteristic variation of the electron current with the voltage.

Source: Franck and Hertz 1914.

trilogy Bohr had suggested that electron–atom collision experiments of the kind later made in Berlin and elsewhere might yield information about the energy levels of stationary states. The new experiments of Franck and Hertz confirmed him in his belief that the equation $E_n - E_m = h\nu$ might be tested experimentally 'by finding the energy which must be supplied to a system in order to make it emit a certain spectral line', as he wrote to Oseen. In this connection he mentioned 'the beautiful experiments by Franck and Hertz on the ionization of mercury vapours [which] can be interpreted along the same lines'.[16] He consequently proposed an interpretation very different from theirs, namely that the measured value of 4.9 eV constituted the energy difference between two stationary levels in the mercury atom. In his paper of 1915 Bohr phrased his reinterpretation cautiously:

> It seems that their experiment may possibly be consistent with the assumption that this voltage [4.9 V] corresponds only to the transition from the normal state to some other stationary state of the neutral atom. On the present theory we should expect . . . 10.5 volts for the ionization potential instead of 4.9 volts. If the above considerations are correct it will be seen that Franck and hertz's measurements give very strong support to the theory considered in this paper. If, on the other hand, the ionization potential of mercury should prove to be so low as as assumed by Franck and Hertz, it would constitute a serious difficulty . . . since this [mercury] spectrum contain lines of greater frequency than the line 2536.[17]

Bohr obtained the value of 10.5 volts for the ionization potential of mercury from the limit of a spectral series of this element measured by Paschen. This series included a high-frequency line at 1850 Å. However, Franck and Hertz did not agree with Bohr's theory. When they finally referred to it in a paper of 1916, it was briefly and only to reject it.[18] Nor were they alone in interpreting electron collision experiments as a problem for Bohr's atomic theory. For example, the Canadian physicist John McLennan, professor at the University of Toronto, concluded in 1916 that the results of experiments with mercury vapour indicated that 'the theory is invalid'.[19] On the other hand, Bohr's interpretation received support from Robert Millikan, who in an address of 1 December 1916 to the American Physical Society pointed out that according to Bohr the 2536 line should be the longest wavelength in a series, as indeed it was.[20]

Millikan's supportive position was substantiated by experiments made in the spring of 1917 by Bergen Davis and F. S. Goucher at Columbia University.[21] Applying a modified version of the Franck–Hertz apparatus, the two American physicists confirmed Bohr's picture of what happened in collisions between electrons and mercury atoms. As Bohr expressed it in an address in Copenhagen of February 1920, the Davis–Goucher experiment showed a 'perfect agreement' with his theory.[22] In fact, in collaboration with the Manchester physicist Walther Makower, in 1915 or 1916 Bohr had himself done experiments with an apparatus of essentially the same design as the one used at Columbia University. However, the complicated apparatus broke down and Bohr and Makower never succeeded in getting data from it. In a later letter to Makower, he wrote: 'I have just in these days had much opportunity of thinking of our experiments on ionisation in Manchester, which unfortunately came to so suddenly [sic] an end through the diminutive but destructive fire in the cotonwool packing'.[23]

Franck and Hertz belatedly came to the same conclusion as Bohr with regard to their experiments, as they reported in a review article of 1919 on 'the confirmation of Bohr's theory' by investigations of inelastic collisions between slow electrons and gas molecules.[24] Six years later they received the Nobel Prize for their series of works, a prize which was awarded as much to Bohr's theory as it was to the experiments themselves. This was made clear by Oseen in his presentation speech, where he emphasized that although the experiments of Franck and Hertz, were valuable by themselves, 'even more important at the present time is the general finding that Bohr's hypotheses concerning the different states of the atom and the connexion between these states and radiation, have been shown to agree completely with reality'.[25] One can safely speculate that had it not been for Bohr's theory, Franck and Hertz would not have been honoured with the Nobel Prize.

The application of his atomic theory to new areas of experimental physics was important to Bohr, but he was no less interested in generalizing the theory, systematically exploring its relations to classical theory, and presenting it in a form logically consistent with existing quantum theory. 'I have thought quite a bit about the general foundation of the quantum theory', he confided in a letter to Oseen of late 1915.[26] During the next couple of months he worked hard to put his thoughts on paper, which resulted in yet another article submitted to *Philosophical Magazine*, this time on 'The application of the quantum theory to periodic systems'. It was intended to appear in the April issue, but Bohr withdrew it shortly before publication. The explanation appears in another letter to Oseen: 'I had already had proof of my paper... when Sommerfeld's new and exceedingly interesting paper on the structure of spectral lines appeared. This paper has quite changed the present state of the Quantum theory, and I have therefore postponed the publication of my paper for the present'.[27] In fact, Bohr's paper was never published as intended. Two days after having written to Oseen, he addressed Sommerfeld:

> I thank you so much for your most interesting and beautiful papers. I do not think I have ever enjoyed the reading of anything more than I enjoyed the study of them, and I need not say that not only I but everybody here has taken the greatest interest in your important and beautiful results... I decided at once to postpone the publication [of my own paper] and to consider it all again in view of all, for which your papers have opened my eyes.[28]

In spite of the fact that Bohr's original paper never reached publication as planned, it included several innovative thoughts that a couple of years later would be incorporated in some of his other papers. Indeed, it has been called 'the most famous unpublished paper in the history of physics'.[29]

Bohr's overall aim was to harmonize his theory with the general quantum theory—but which theory? In 1911–1912 Planck had formulated a new version of his original quantum hypothesis according to which a vibrating atomic system would absorb energy continuously and thus exist permanently in an energy state of any value up to a critical one given by $h\nu$ or multiples of it. Only when it reached this value would energy be emitted in the form of discrete quanta, a process governed by a probability law.[30] In spite of the emission of discrete quanta, the oscillators did not possess intrinsically discontinuous energies. It followed from Planck's 'second theory' that a collection of oscillators at zero temperature would not have zero energy but a finite average energy given by ½ $h\nu$. As Bohr pointed out, Planck's second theory was inconsistent with the basic assumption that an atomic system can exist permanently only in a series of discrete stationary states and that both emission and absorption of radiation energy occur by a transition between two such states. In Bohr's theory of 1916 he regained the zero-point energy for a harmonic oscillator, but it 'has here an origin quite distinct from that in Planck's theory'. According to Bohr, the probability of a periodic system of more than one degree of freedom being in the ground state $n = 0$ was zero, implying that at zero temperature all systems of this kind must be in a $n = 1$ state.

Another noteworthy feature of the unpublished work was Bohr's adoption of the so-called adiabatic invariance principle proposed by Ehrenfest in 1913.[31] According to this principle or hypothesis, only certain 'adiabatically invariant' quantities can be subjected to quantization, namely such quantities that have the same value before and after they are affected by a slowly changing action. According to Ehrenfest, 'If a system be affected in a reversible adiabatic way, allowed motions are transformed into allowed motions'.[32] Another way of expressing this idea is that ordinary mechanics applies to slow deformations of a quantized system.

Ehrenfest's hypothesis was inspired by Planck and Einstein, not by Bohr's atomic theory (which he disliked at the time). However, when Bohr came across it he instantly realized its relevance for his own work. He described the hypothesis as follows: 'It allows us by varying the external conditions to obtain a continuous transformation through possible states from a stationary state of any periodic system to the state corresponding with the same value of n of any other such system containing the same number of moving particles'.[33] As he saw it, the hypothesis functioned as a guarantee of the stability of the stationary states: the mechanical laws were not only valid under constant external conditions, but also during very slow ('adiabatic') changes of them. This was an understanding of the hypothesis that to some extent differed from Ehrenfest's, as Bohr pointed out in his first letter to the Dutch physicist:

> As far as I understand, however, I consider the problem from a point of view which differs from yours, and I have therefore not made use of the same terminology as in your original papers. In my opinion the condition of the continuous transformability of motion in the stationary states may be considered as a direct consequence of the necessary stability of these states and to my eyes the main problem consists therefore in the justification of the application of ordinary 'mechanics' in calculating the effect of a continuous transformation of the system. As it appears to me it is hardly possible to base this justification entirely on thermodunamical considerations, but it seems naturally suggested from the agreement with experiments obtained by calculating the motion in the stationary states themselves by means of ordinary mechanics.[34]

In his works of 1918 and the following years Bohr underlined his different understanding of the hypothesis by using the term 'principle of mechanical transformability' rather than 'principle of adiabatic invariants', where the latter expression refers to the thermodynamical origin of Ehrenfest's hypothesis.

Bohr pointed out that stationary states could be characterized by the adiabatic invariant $\bar{T}/\omega$, where $\bar{T}$ is the average value of the kinetic energy over a complete period and ω the frequency of the periodic motion:

$$\frac{\bar{T}}{\omega} = \frac{1}{2} nh.$$

This replaced his earlier quantization condition $T = ½nh\omega$. As an example of a simple periodic system Bohr considered in some detail a model of diatomic molecules consisting of two atoms rotating round their centre of gravity. This model he compared with the earlier theory of his compatriot Niels Bjerrum and spectroscopic observations of the absorption spectra of diatomic gases. By means of a correspondence argument he suggested that transitions would occur between neighbouring stationary states n and $n + 1$. From this kind of selection rule he found for the radiation frequency the quantized values

$$\nu = \frac{h}{8\pi^2 J}(2n+1),$$

where J is the inertial moment of the molecule. It follows that the constant difference in frequency is given by

$$\Delta\nu = \frac{h}{4\pi^2 J}.$$

Bjerrum, who had quantized the energy of the rotating molecule, but without using Bohr's frequency condition, obtained for $\Delta\nu$ a value twice as large. Because the moments of inertia of molecules were not known independently, at the time it was not possible to determine which of the two predictions was the best one.[35]

4.2 SOMMERFELD'S APPROACH TO ATOMIC THEORY

Max Planck had no interest in atomic models, but in late 1915 he nonetheless took up Bohr's theory of atoms and spectra at a few occasions. His work in this area was, as he expressed it in a letter to Sommerfeld, 'just a little excursion into for me unfamiliar territory'.[36] In one of these little excursions he derived the Bohr–Balmer formula in a way that he considered more satisfactory than the original one of Bohr. Like several other physicists at the time, he objected to Bohr's discontinuous quantum jumps. 'I hate discontinuity of energy even more than discontinuity of emission', as he wrote to Ehrenfest.[37] In the paper on the emission of spectral lines, he explained the difference between his and Bohr's views: 'The idea developed here is different from Bohr's in so far that here the emission process is not necessarily connected with a jump of the oscillating electron from one stationary orbit to another stationary orbit, but instead it can be followed without some significant change in the elliptic orbit'.[38] In Planck's theory the frequencies of the spectral lines were assumed to be equal to frequencies of rotation characterizing atomic electrons, which means that he rejected both of Bohr's basic postulates. Apart from this excursion into the unfamiliar territory of atomic models, Planck decided to leave the further development to his younger colleague in Munich.

Sommerfeld innovatively developed Bohr's theory of the hydrogen atom (and other one-electron atoms) in two important works of 1915–1916, one dealing with a two-quantum extension of the theory and the other with the fine structure of the spectrum caused by the theory of relativity. His extensions of Bohr's atomic theory were informally presented in his lecture course in the winter semester 1914–1915, but only presented to the Bavarian Academy of Sciences at meetings of late 1915 and early 1916.[39] The 'exceedingly interesting' generalization, as Bohr called it, included as a key feature that the electron of the hydrogen atom moved in an elliptic rather than a circular orbit. This was a feature that Bohr had previously suggested, but without developing the idea and without introducing a quantum number related to the eccentricity of the elliptic orbit. Sommerfeld started his first paper by praising Bohr's theory:

> The theory of the Balmer hydrogen spectrum appears at first glance to have been brought to a conclusion through the wonderful investigations of N. Bohr. Bohr could explain not only the general form of the series law, but even the numerical value of the constants detailed therein, and their refinement when the motion of the nucleus is taken into consideration.[40]

Yet Sommerfeld also pointed out, as Nicholson and others had done previously, that the Bohr model was essentially a model for the hydrogen atom: 'One might even say that the capability of the Bohr theory is for the time limited to this hydrogen series and to the hydrogenic series (ionized helium, X-ray spectra, series ends of the visible spectra)'.[41] In his alternative formulation Sommerfeld made use of two quantum numbers instead of one, and, applying a different and more general method to Bohr's, derived the generalized Balmer formula.

Sommerfeld's method was based on the postulate that for a periodic system of f degrees of freedom the stationary states can be determined by quantum conditions in the form of the action integrals $\int p_i dq_i = n_i h$. Here q_i is a coordinate and p_i its corresponding generalized momentum; n_i is a positive integer and the index i attains whole numbers from 1 to f. For the case of an electron moving elliptically about a positive nucleus he used the quantum condition for the angular momentum

$$\oint p_\phi d\phi = nh,$$

where the integration is taken over a whole period. Since $p_\phi = ma^2\omega = nh/2\pi$ for a circular orbit, this expresses the quantization of angular momentum introduced by Nicholson and Bohr. But for an elliptic orbit this condition alone will not do, because it gives frequencies for the spectral lines that vary continuously with the continuous change of the eccentricity between zero and one and therefore fails to reproduce the discrete lines. Sommerfeld showed that the energy of a stationary state n governed only by the angular quantum condition would vary as

$$W_n = -\frac{2\pi^2 me^4}{h^2}\frac{1-\epsilon^2}{n^2},$$

where ϵ is the eccentricity of the orbit. He therefore needed to quantize the eccentricity as well, which he did by applying an analogous quantum condition to the radial momentum,

$$\oint p_r dr = n'h,$$

where r is the varying distance of the electron from the nucleus and n' a new quantum number. He considered the quantization rules to be the 'unproved and... perhaps unprovable foundation of the quantum theory'. For the ratio between the minor semiaxis b to the major semiaxis a (which expresses the eccentricity ϵ) he derived

$$\frac{b}{a} = \frac{n}{n+n'}.$$

Thus, an elliptical orbit with a given major axis might have different shapes depending on the value of n'/n.

Having found from his two quantum conditions the energy of a stationary level in terms of n and n', Sommerfeld could use Bohr's frequency condition to calculate the frequency of radiation emitted by a transition from state (m, m') to state (n, n'). With Rydberg's constant expressed as $R = 2\pi^2 me^4/h^3$, the result was

$$\nu = R\left[\frac{1}{(n+n')^2} - \frac{1}{(m+m')^2}\right],$$

which agreed with the sharp lines observed spectroscopically. He commented that the result was 'surprising in the highest degree and of striking certainty' and further noted that 'the energy is uniquely determined by the sum of the quanta of action, which we may distribute arbitrarily among the azimuthal and radial coordinates'.[42] Since the energy depends only on the quantity $(n + n')$, there is no simple and unique relationship between energy and stationary orbit, as there was in Bohr's original theory. For each value of n' there will be n orbits that are less eccentric the closer the values of the two quantum numbers (Figure 4.2). For $n' = 0$ the orbit will always be a circle. For example, the Balmer state given by $(n + n') = 3$ can be obtained in four different ways, corresponding to a circular orbit, two elliptical orbits, and a straight line:

$$n' = 0, n = 3, b = a$$

$$n' = 1, n = 2, b = 2a/3$$

$$n' = 2, n = 1, b = a/3$$

$$n' = 3, n = 0, b = 0$$

Sommerfeld excluded the last possibility, sometimes known as a 'pendulum orbit', because it pictures an electron oscillating through the atomic nucleus in a straight line. This he judged to be unphysical (*unwirklich*) and conflicting with the stability of atoms. The symbols n and n' for the quantum numbers are those used by Sommerfeld. In the later literature it became common to denote the azimuthal (or subordinate) quantum number as k and keep n for the principal quantum number, meaning that

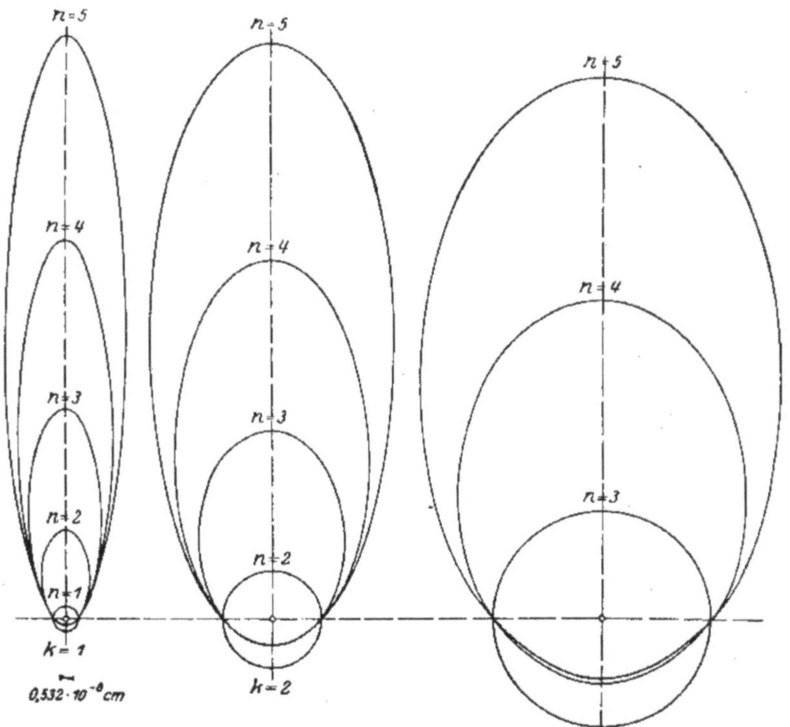

Fig. 4.2. Electron orbits of the hydrogen atom with different values of the n and k quantum numbers.

Source: Grotrian 1928, vol. 1, p. 15.

$$n = n' + k$$

In a paper of 1923, Nicholson argued that pendulum orbits were dynamically possible and that electrons might perhaps penetrate the nucleus. His paper, picturing the electron and the nucleus as 'states of strain in the aether', was probably ignored by Sommerfeld and most other quantum physicists.[43] However, the pendulum orbits $k = 0$ remained problematic in the sense that the reason to rule them out was physical rather than based in theory.

Sommerfeld also considered the three-dimensional case, in which connection he introduced the concept of space quantization (*Richtungsquantelung*) as given by the condition

$$\oint p_\psi d\psi = mh$$

The quantity ψ is an angle in a plane perpendicular to some direction z (as defined by an external field) that forms an angle α with the perpendicular to the plane of the electron orbit (Figure 4.3). Denoting the projection of the orbital angular momentum M on the z-direction M_z, the quantization condition can be written $M_z = mh/2\pi$. What Sommerfeld called the 'equatorial quantum number' m can take on $2k + 1$ integral values from k to $-k$, and the allowed inclinations of the orbit are given by

$$\cos\ \alpha = \frac{M_z}{M} = \frac{m}{k}$$

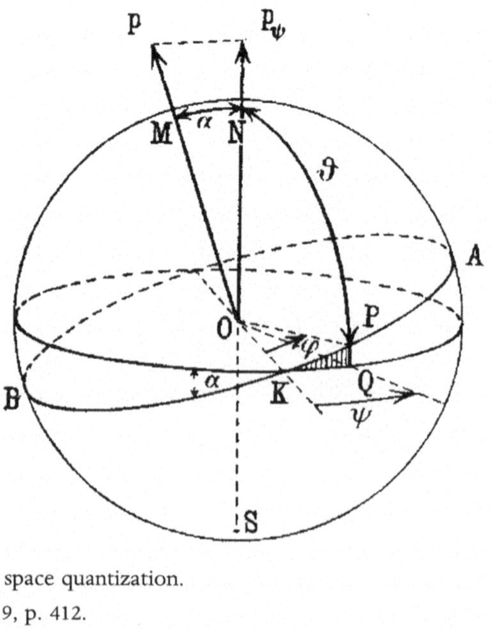

Fig. 4.3. Sommerfeld's space quantization.
Source: Sommerfeld 1919, p. 412.

Sommerfeld thus concluded that only certain orientations of the electron orbit relative to the direction of an applied magnetic field were possible, a result of obvious relevance to the study of the Zeeman effect. 'This space quantization of the Kepler orbits is undoubtedly among the striking results of the quantum theory', he wrote in 1919. 'In the simplicity of its derivation and results it gives us almost the impression of witchcraft'.[44] With the discovery of the Stern–Gerlach effect two years later (Section 8.5), the witchcraft was confirmed.

Returning to the two-dimensional hydrogen atom, in Sommerfeld's picture of it a spectral line did not represent a single quantum transition, but a superposition of several transitions. In the case of the H_α line, for example, where the initial state is $(m + m') = 3$ and the final one is $(n + n') = 2$, there are six possible transitions; for the H_β line there are eight transitions. Sommerfeld supposed that not all of the transitions would actually occur, but that some of them were ruled out by what he called a 'quantum inequality' (*Quantenungleichung*) the nature of which he was at the time uncertain. Nonetheless, his quantum inequality represented the first step in the emergence of the concept of selection rules. His theory of the Balmer spectrum was novel and differed from Bohr's in so far as the electron orbits were now ellipses of differerent eccentricities. However, Sommerfeld's different and richer picture did not lead to any additional energy levels and therefore also not to additional spectral lines. From an experimental point of view it did not go beyond Bohr's picture and was, in this respect, somewhat disappointing. On the other hand, Sommerfeld's second extension of Bohr's theory (Section 4.4) was far from disappointing from an empirical point of view.

Apart from being published in the *Sitzungsberichte* of the Bavarian Academy, the content of Sommerfeld's two works on the hydrogen atom were also expounded in an exceedingly long article appearing in the September and October 1916 issues of the *Annalen der Physik*. Moreover, the new theory appeared prominently in Sommerfeld's most influential monograph, *Atombau und Spektrallinien*, which was published in 1919 and subsequently appeared in several expanded editions. An English translation of the third edition appeared in 1923. In this famous book, often described as the 'Bible' of atomic theory and quantum spectroscopy, Sommerfeld was careful to keep the mathematics at a fairly elementary level (by his standards) and to connect the theoretical results, whenever possible, with experimental data.[45] The book was also an exposition of atomic theory which, especially in the first edition, emphasized the action-angle approach followed in Munich, at the expense of the methods cultivated by Bohr and his colleagues. Indeed, although Born found the book to be 'marvelous' he also thought it was at places disturbingly 'local patriotic' and that it failed to appreciate 'Bohr's formulation [which] is also very beautiful'.[46] Whatever the differences between Sommerfeld and Bohr, the latter greatly appreciated the papers coming from Munich. As soon as he had digested them, he stopped the publication of his new paper on periodic systems, as mentioned in Section 4.1. He initially tried to save the paper by extending it with Sommerfeld's new ideas, but apparently concluded that the extended version was not publishable.[47]

Atombau had a major effect on the physics community, not only as a textbook but also as a convincing argument for the Bohr–Sommerfeld theory of atomic structure, a theory which many physicists continued to find strange. In October 1921 W. Lawrence Bragg reported to his father that he had studied the second edition of *Atombau*: 'I have been reading Sommerfeld's book. It is very good, and not a bit abstruse except in certain chapters at the end. He is convinced of the truth of Bohr's atomic model, and pictures the atoms as planet-systems. The numerical checks he get, the explanation of the "fine-structure" etc., is very convincing, but do you really think these orbits have any existence? I can't believe it'.[48]

Sommerfeld was not the first physicist to formulate a general quantum condition applicable to all systems governed by the quantum theory. Such a condition, in the form $\int p_i dq_i = n_i h$, was suggested in the spring of 1915 by William Wilson, a physicist at King's College, London. The theory, Wilson said, 'while formally distinct from Bohr's theory, leads to the same results when applied to the Rutherford type of atom in which an electron travels in a circular orbit round a positively charged nucleus'.[49] However, the English physicist rested content with having expressed Bohr's theory in a new and more general way and did not apply his dynamical formulation to the calculation of atomic spectra or to other problems of physics.

Independently of Wilson, at about the same time the Japanese physicist Jun Ishiwara presented a paper with a proposal somewhat similar to the one of Wilson and to the one proposed a little later by Sommerfeld.[50] As Ishiwara demonstrated in his paper, his formulation of the quantum conditions yielded the results of Bohr's theory of atomic structure. His work may have been the first published reference to Bohr's theory by a non-Western physicist. However, Ishiwara's quantum condition led him to results that differed from Bohr's by a factor of two. He was to some extent influenced by Nicholson's view of atomic constitution and probably for this reason he found it natural to assume that the neutral hydrogen atom contained two electrons rather than one. At any rate, the papers of Wilson and Ishiwara are of more interest from a priority point of view than from a scientific point of view. They exerted no significant influence on the further development of the quantum theory of atoms.

4.3 SOMMERFELD'S THEORY AT WORK

As mentioned in Section 3.9, in 1914 Bohr had suggested a first explanation of the Stark effect on the basis of his atomic theory, but without being able to account for the effect quantitatively. In view of the fact that the electric splitting of spectral lines could not be explained on classical grounds, a full explanation in terms of quantum theory was desirable and considered an important test case for the new theory. Sommerfeld occupied himself with the problem, realizing that his phase-integral formulation of quantum theory had to comprise a theory of the Stark effect. Rather than trying to solve the difficult problem himself he handed it over to his former student Paul Epstein, a Polish-born Russian citizen who in 1914 had obtained his doctorate under Sommerfeld. At the same time that Epstein was struggling with the theory of the Stark effect it was also investigated by the eminent physicist and astronomer Karl Schwarzschild, since 1909 director of the Astrophysical Observatory at Potsdam. Schwarzschild was familiar with the Stark effect, for in 1914 he had proposed a theory for it along classical lines.[51] He now decided to use the machinery of the Bohr–Sommerfeld quantum theory in combination with the powerful mathematical methods from celestial mechanics that he, a theoretical astronomer, knew so well. Epstein followed largely (and independently) a similar approach, which combined the Hamilton–Jacobi formalism of celestial mechanics and the phase-integral formalism of quantum theory used by Sommerfeld.

The two theorists completed their complex calculations almost simultaneously and in the spring of 1916 they arrived at theories that were substantially the same.[52] To account for the Stark effect, a problem that involved three degrees of freedom, they had to make use of three quantum numbers, n_1, n_2, and n_3. Abbreviating $n = n_1 + n_2 + n_3$ and denoting the strength of the electric field F, Epstein obtained for the energy of a hydrogenic atom the expression

$$W = -\frac{2\pi^2 me^4 Z^2}{n^2 h^2} - \frac{3h^2 F}{8\pi^2 meZ} n(n_1 - n_2)$$

The first term is Bohr's expression for the energy of the atom (H or He^+) in the absence of the field and the second term gives the separation of the frequencies in the presence of an electric field. Making use of selection rules proposed by Sommerfeld he succeeded in giving a complete description of the experimentally known Stark effect—not only the frequency separations but also the polarizations of the lines. The term 'selection principle' (*Auswahlprinzip*) appeared for the first time in Epstein's paper.

In the case of the Balmer spectrum the agreement between theory and experiment turned out to be practically perfect, even though it was not possible at the time to calculate the intensities of the lines. Well aware of the significance of his results, Epstein concluded: 'We believe that the reported results prove the correctness [*Richtigkeit*] of Bohr's atomic model with such striking evidence that even our conservative colleagues cannot deny its cogency. It seems that the potentialities of quantum theory as applied to this model are almost miraculous and far from being exhausted'.[53] Schwarzschild, too, saw the theory of the Stark effect as a remarkable proof of the essential truth of the Bohr–Sommerfeld theory.

The importance of the work of Epstein and Schwarzschild was not restricted to the Stark effect of hydrogen. It was only with these papers, and with Schwarzschild's in particular, that Hamilton–Jacobi dynamics and the so-called action-angle technique was introduced into quantum theory as an integral part of Sommerfeld's quantization procedure.[54] Einstein found the methods introduced by Epstein and Schwarzschild to be of fundamental importance and in 1917 he lectured on them before the German

Physical Society. In fact, Einstein's lecture was more than just a discussion of existing theory. He used the occasion to suggest what he modestly called 'a little modification of the Sommerfeld–Epstein condition', a new version of the quantization rule that was far from as unimportant as Einstein apparently thought.[55] Although he did not spell it out very clearly, he realized that Sommerfeld's quantization condition was inapplicable to many realistic problems of what came to be known as the 'chaotic' type. Einstein's method was only fully appreciated in the 1970s, when it was applied in the new subfield known as quantum chaos.

Generally speaking, the mathematical techniques that physicists adopted from astronomy proved very important in the further development of quantum theory, especially as worked out by Bohr, Born, Pauli, and Heisenberg. Referring to a seminar in Göttingen under Born on the methods of celestial mechanics, Heisenberg wrote to Pauli: 'We study here with all our might Poincaré, for in him one can after all already find the quantum theory of all atomic models'.[56] Born suggested that the use of astronomical methods strengthened belief in the unity of the laws of nature. 'If one succeeds in understanding the structure of the atom by means of perturbation calculations', he wrote, 'then the similarity of cosmic and atomic processes has been demonstrated and it provides an indication of the unity of the processes and laws of the world'.[57]

Whereas, in the case of the Stark effect, quantum theory was able to solve a problem outside the reach of classical physics, the case of the magnetic separation of spectral lines was different. This effect, discovered by Pieter Zeeman at the University of Leiden in a famous experiment of 1896, for which we was awarded the Nobel Prize, was immediately explained by Lorentz in terms of the then-new electron theory.[58] Assuming the existence of elastically bound electrons (or 'ions') of mass m in the atoms, Lorentz deduced that as a result of the magnetic field the vibration frequency ν_0 of an electron would split up into a triplet of frequencies for a field that was perpendicular to the path of light. If H denotes the strength of the magnetic field, he obtained

$$\nu = \nu_0 \pm \frac{e}{m}\frac{H}{4\pi c}.$$

Lorentz's theory was singularly important because it showed that electrons had to be negatively charged and with an e/m ratio of a magnitude 1,000 times that of the hydrogen ion (see Section 1.2).

The separation into triplets is known as the 'normal' Zeeman effect, but it soon turned out that in most cases a spectral line placed in a weak magnetic field splits into a larger number of lines, multiplets following a more complicated pattern. This 'anomalous' Zeeman effect was first reported by the Irish physicist Thomas Preston, who in 1897 noticed that the sodium doublet lines D_1 and D_1 were split into a quadruplet and a sextuplet, respectively.[59] Preston's effect was only anomalous in the sense that it could not, in contrast to the triplet effect, be explained on the basis of electron theory. Bohr was at an early stage aware of the challenge from the Zeeman effect; that is, the need to reproduce it on the basis of his atomic theory. His 1914 paper on the Stark effect contained a discussion of the magnetic effect, which he at the time thought required a modification of the frequency condition $E_m - E_n = h\nu$. His attempt to explain the Zeeman effect did not go very far and neither did an early attempt by the Austrian physicist Karl Herzfeld to apply Bohr's theory to the effect.[60]

A quantum understanding of the Zeeman effect was only advanced in 1916, on the basis of the quantum-theoretical formalism established by Sommerfeld and others. Sommerfeld himself, and independently his former student Debye, at the time a professor in Göttingen, developed theories of the normal Zeeman effect that were based on essentially the same methods. They were directly inspired by and used the technical resources of the Epstein–Schwarzschild theory of the Stark effect. The two papers by Debye and Sommerfeld appeared consecutively in the *Physikalische Zeitschrift* in October 1916.[61] In both cases the result was a magnetic splitting of the form $\Delta\nu = \Delta p$ $(eH/4\pi mc)$, where p is a magnetic quantum number restricted to changes $\Delta p = 0$ or $\Delta p = \pm 1$. In this way the classical formula of Lorentz was regained. Although Planck's constant does not appear in the formula, it was there implicitly: it just happened to cancel out in the derivation. Commenting on his work in *Atombau*, Sommerfeld said: 'In the present state the quantum theory of the Zeeman effect achieves just as much as Lorentz's theory, but not more: it can account for the normal triplet including its polarisation, but hitherto it has not been able to explain the complicated Zeeman types'.[62] The complicated or anomalous Zeeman effect would soon become a major headache for the quantum theorists.

In a letter to Bohr, Sommerfeld informed his colleague in Copenhagen about his recent work: 'I have just finished an article on the Zeeman effect in which the normal triplet, at least, is derived according to your general approach $h\nu = W_2 - W_1$ (contrary to your expectations)'.[63] Sommerfeld also calculated the Zeeman effect of hydrogen by taking into account the theory of relativity, which according to his theory resulted in a doubling of the Balmer lines. He was surprised to find that the relativistic correction did not affect the result at all, the reason for his surprise being that the hydrogen doublets, in contrast to the singlet lines, ought to exhibit an anomalous Zeeman effect. To Sommerfeld this was 'highly suspicious' and it even caused him to describe, in a letter to Ehrenfest, the quantum theory of the Zeeman effect as 'false'.[64]

The problem was further complicated by the experimental finding that the form of the Zeeman effect depended on the strength of the magnetic field. The so-called Paschen–Back effect, discovered in 1912–1913 by Paschen and his collaborator Ernst Back at the University of Tübingen, was particularly confusing. What they found was that with increasing magnetic field strength nearly all the anomalous Zeeman components coalesced into a normal Zeeman–Lorentz triplet pattern. The effect found by Paschen and Back was not in itself a problem, but it was a problem that they also found it in the H_α line, where, according to Sommerfeld's theory, the doublet should split into a normal triplet for any strength of the magnetic field. The possible existence of a

Paschen–Back effect in hydrogen worried Sommerfeld, who realized that if the effect were real the theory was in trouble. According to experiments made by Peter Erochin at the University of Göttingen the effect was indeed real, but for a while Sommerfeld convinced himself that the evidence was inconclusive. Only in 1922, when the evidence could no longer be ignored, did the Paschen–Back effect in hydrogen turn into a recognized if not very important anomaly for the Bohr–Sommerfeld theory of atoms.[65]

In Sommerfeld's extension of Bohr's theory X-ray spectroscopy played no motivating role. Nonetheless, by means of his two-quantum theory he was able to account successfully for new discoveries in this nascent branch of atomic physics. Generally speaking, Sommerfeld's theory of the atom quickly became closely connected with X-ray spectroscopy, a field he covered extensively in the various editions of his *Atombau*. Kossel's explanation of X-rays as originating in electron transitions from an outer to an inner ring (Section 3.4) received support from the discovery, made by the Swedish physicist Ivar Malmer at the University of Lund, that the K_α line consisted of two closely adjacent lines. To Kossel, this indicated a doublet structure of the K-level. Kossel reported on Malmer's discovery and his interpretation of it at a colloquium in Munich on 22 December 1915, which caused Sommerfeld to look for an explanation of it. According to Sommerfeld, the doublet was due to the two possible quantum states of what he thought was the single L-electron described by $n = 2$, one state corresponding to a circular orbit ($k = 2$) and the other to an ellipse ($k = 1$).[66] He pictured the L electron in the heavier elements like the one in the hydrogen atom, except that it would be screened from the nuclear charge because of the effect of the other atomic electrons. The effective nuclear charge would not be Ze but $(Z - s)e$, where s is a screening factor. From his relativistic theory of the hydrogen atom (see next section) he obtained for the difference in frequency between the two L doublet levels the approximate result:

$$\Delta\nu_L \cong \frac{R\alpha^2}{2^4}(z - s)^4,$$

where α is the fine-structure constant, $\alpha = 2\pi e^2/hc \cong 0.007$, and R is the Rydberg constant as given by Bohr's theory. The formula fitted experimental data over a wide range of the periodic system, from $Z = 41$ to $Z = 92$, for $s = 3.5$. As Sommerfeld emphasized, this strongly indicated that the elements shared the same electron structure for their innermost regions. The L doublets showed a systematic increase in the value $\Delta\nu/(Z - s)^4$, but this was only to be expected from Sommerfeld's theory (Figure 4.4). Including higher approximations, the theory predicted the quantity to vary as

$$\frac{\Delta\nu}{(Z - s)^4} = \frac{R\alpha^2}{2^4}\left\{1 + \frac{5\alpha^2}{2}\frac{(Z - s)^2}{2^2} + \frac{53\alpha^4}{8}\frac{(Z - s)^4}{2^4}\right\},$$

which accounted for the experimentally determined variation. Thus, as the Lund physicist Manne Siegbahn expressed it, using the entire relativity correction, 'a full and numerically complete explanation of the doublets in the L series is obtained'.[67]

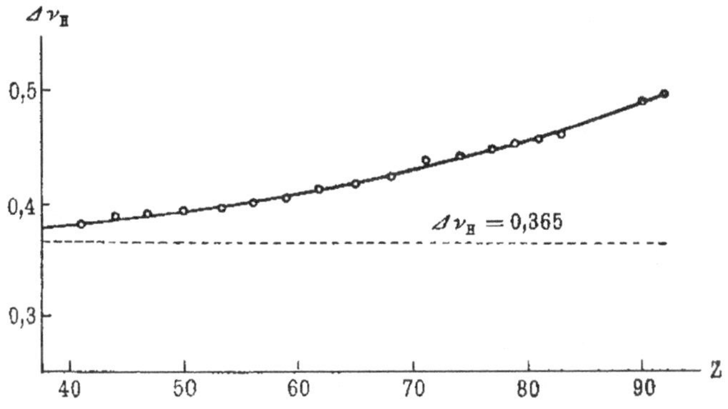

Fig. 4.4. The variation of the X-ray L doublets with the atomic number, agreeing with Sommerfeld's theory.
Source: Sommerfeld 1922a, p. 608.

Although successful, Sommerfeld's explanation of X-ray spectra was far from complete. For example, it was unable to account for the value of s, and there were other difficulties. Still, it was a remarkable advance and was seen as such by contemporary physicists. Having studied the first version of Sommerfeld's theory as it appeared in the *Sitzungsberichte* of the Bavarian academy, Bohr expressed his admiration for it in a letter to Richardson:

> I am very enthusiastic indeed over Sommerfeld's paper, it is much brilliant and contains almost innumerable illuminating remarks, he has f[or] inst[ance] even attached a much interesting explanation of the components of the K and L lines in the high frequency spectra. This appears to be quite analogous (even quantitatively) to the doubling of the hydrogen lines, and is only so much bigger, because the doubling effect, which is a relativity one, is proportional to the 4th power of the nucleus charge.[68]

4.4 A REVELATION: HYDROGEN'S FINE STRUCTURE

The most spectacular success of Sommerfeld's theory was his incorporation of the special theory of relativity in atomic structure and the resulting explanation of the fine structure of the spectra of hydrogen and ionized helium. In 1887 Albert Michelson and his collaborator, the chemist Edward Morley, reported that in a highly resolving spectroscope the H_α line appears as a doublet with a separation in wave number $\Delta\nu \cong 0.3\ \text{cm}^{-1}$. The corresponding separation in wavelength is given by $\Delta\lambda = -\Delta\nu/\nu^2$. By 1913 the discovery was confirmed by other measurements and it was further established that also the H_β line exhibited a 'fine structure' of a similar magnitude.

Bohr was apparently unaware of hydrogen's fine structure, which he did not mention in his trilogy.[69] In fact, the phenomenon contradicted his original theory of the hydrogen atom, where the Balmer lines necessarily had to be singlets. There seemed to be no way in which a doublet structure could be accommodated within the model. By the following year he became aware of it, but for a brief time he tended to assign it to the electric field in the discharge tube, that is, to a Stark effect.[70] Together with H. M. Hansen he performed experiments in Copenhagen to test the electric hypothesis, only to conclude that it was wrong. This he reported in a letter to Rutherford in which he suggested an alternative explanation:

> If the hydrogen lines should be true doublets, this fact seems to me strongly to indicate a complex structure of the hydrogen nucleus, as from quite general reasons it seems difficult to account for a doubling by help of a system of a single electron rotating round a point charge. The dimensions of the nucleus might still be extremely small, thus a calculation indicates that the presence in the nucleus of a dipol of unite [sic] charges with a distance apart of about 10^{-13} cm might explain the distance between the components observed.[71]

By 1915 Bohr had come to the conclusion that the fine structure was neither an electric nor a nuclear effect, but that it was due to the relativistic variation of the mass of the electron. He suggested that the doubling of the lines could be explained if the electron orbited in an ellipse that slowly rotated around the central charge: 'The frequency of this rotation of the orbit will depend on the degree of excentricity. For very small alterations from the circular orbit the ratio between the frequency of rotation of the orbit and the frequency of revolution of the electron is given by $2\pi^2e^4/n^2c^2h^2$, which for $n = 2$ is of the same order of magnitude as the doubling of the hydrogen lines observed'.[72] However, Bohr did not follow up the idea, which in a qualitative sense formed the basis of Sommerfeld's detailed and much more advanced theory.

Sommerfeld's basic idea was, not unlike Bohr's, to include Einstein's formulae $m = m_0(1 - v^2/c^2)^{-1/2}$ and $E = mc^2$ in the analysis of an elliptically moving electron; unlike Bohr, he did it by means of the new quantization conditions relating to the variables p_r and p_ϕ.[73] To solve the relativistic Kepler problem was mathematically demanding but not an impossible problem for a mathematical physicist of Sommerfeld's calibre. It was just the kind of work that suited his style of mathematical physics. After lengthy calculations he arrived at a picture in which the ellipse slowly rotated around the nucleus of atomic number Z (Figure 4.5). The two frequencies, the one of the revolving electron and the other of the precession, corresponded to a doubling of the stationary state. For the energy of an electron in a state with principal quantum number n and azimuthal quantum number k he obtained the exact formula

$$W(n,k) = m_0c^2\left\{1 + \frac{\alpha^2 Z^2}{\left[(n-k) + \sqrt{k^2 - \alpha^2 Z^2}\,\right]^2}\right\}^{-1/2} - 1$$

Expanding the square root and omitting terms in higher orders of α^2 resulted in the more manageable fine-structure formula

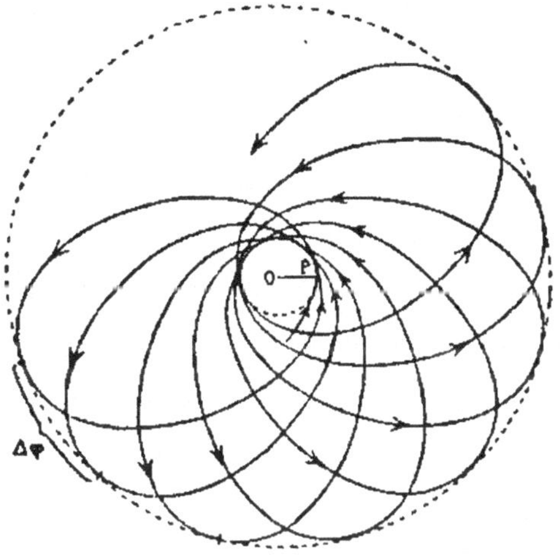

Fig. 4.5. Sommerfeld's picture of the fine structure in hydrogen, showing the perihelion of the elliptically moving electron precessing around the nucleus.
Source: Sommerfeld 1922a, p. 571.

$$W(n,k) = -\frac{RhcZ^2}{n^2}\left\{1 + \frac{\alpha^2 Z^2}{n^2}\left(\frac{n}{k} - \frac{3}{4}\right)\right\}.$$

The first term is the ordinary Bohr value for the energy, and the second term is the relativistic correction. The quantity α, a measure of this correction, is the famous fine-structure constant defined as the dimensionless combination of natural constants

$$\alpha \equiv \frac{2\pi e^2}{hc} \cong \frac{1}{137}.$$

Although it was only with Sommerfeld's theory that the ratio became known as the fine-structure constant and turned into a spectroscopic number, something like it had been considered previously.[74] One example is Jeans' suggestion made at the 1913 meeting of the British Association, namely that $h/2\pi = (4\pi e)^2/c$ (Section 3.1). It is also worth pointing out that Bohr's expression in his revised Balmer formula of 1915 included a value corresponding to $\alpha/8$.

By using Bohr's frequency condition, Sommerfeld found the frequencies of the transitions between the three $n = 3$ levels and the two $n = 2$ levels and thereby reproduced the fine structure. As a test of his theory he paid particular attention to the difference between the hydrogen sublevels (2,2) and (2,1) which, if expressed as a wave number ($1/\lambda = \nu/c$), should be

$$\Delta\nu_\alpha = \frac{R\alpha^2}{2^4} = 0.365\ \mathrm{cm}^{-1}$$

Now the fine structure does not appear as six distinct lines, corresponding to the six possible transitions in Sommerfeld's scheme, but as a doublet, which is the composite result of overlapping lines. Sommerfeld realized that, to be directly testable, his theory had to be supplied with rules of intensity. Unfortunately the theory did not predict intensities, but only frequencies, and this constituted a problem. In his work of 1916 he was therefore forced to introduce arbitrary assumptions about the intensities, for example that transitions in which k decreased had very low intensity. He suggested a general intensity rule that gave the intensity of a transition relative to that of the most probable transition, which he took to be between circular orbits. However, this and other intensity and selection rules were educated guesswork rather than rules derived from quantum theory.

In any case, with these assumptions Sommerfeld obtained a fine-structure splitting that could be compared with the experiments that Paschen made in his spectroscopic laboratory in Tübingen. Paschen had earlier performed precision experiments on the 4686 'Pickering–Fowler line,' which according to Bohr's theory had its origin in a transition from $n = 4$ to $n = 3$ in ionized helium, and so he was well prepared. In his tests of Sommerfeld's fine-structure theory he focused on the same line, which had several advantages, one of them being that the doublet separation was much larger than in hydrogen. (Because $Z = 2$ for helium and the separation increases with the fourth power of Z, the separation is sixteen times as large.) Another advantage was that the greater mass of the helium atom caused a reduction of the Doppler broadening of the lines, which varies inversely with the square of the mass of the atom.

After a series of delicate experiments, on 21 May 1916 Paschen could inform Sommerfeld that 'My measurements are now finished and they agree everywhere most beautifully with your fine structures'.[75] This was no exaggeration, for he concluded that he had obtained a value for $\Delta\nu_a$ that agreed precisely with Sommerfeld's theoretically derived value (Figure 4.6). In reality the agreement between theory and experiment was not so complete as claimed by Paschen and Sommerfeld. For various reasons, and especially because of the questionable assumptions used for the theoretical intensities, the harmony was less impressive than the majority of physicists was led to believe. At any rate, Sommerfeld's fine-structure theory was generally considered to be excellently and unambiguosuly confirmed by experiment. Because the theory rested on the foundation provided by Bohr, the experiments were also taken as strong support of his theory of atomic structure. Epstein said that he was converted to Bohr's theory only 'after Sommerfeld in his theory of the fine structure of the hydrogen lines achieved such a striking agreement with the experiment'.[76]

In spite of his early interest, as reported by Hevesy in the autumn of 1913 (Section 3.1), it seems that Einstein only 'discovered' Bohr's theory in 1916, mainly in connection with his own work on the emission and absorption of radiation. It is also only from that time that Bohr's name begins to enter Einstein's correspondence. In a paper from the summer of 1916, Einstein referred to 'Bohr's theory of the spectra' and derived from statistical considerations Bohr's frequency condition.[77] The theory of the fine structure

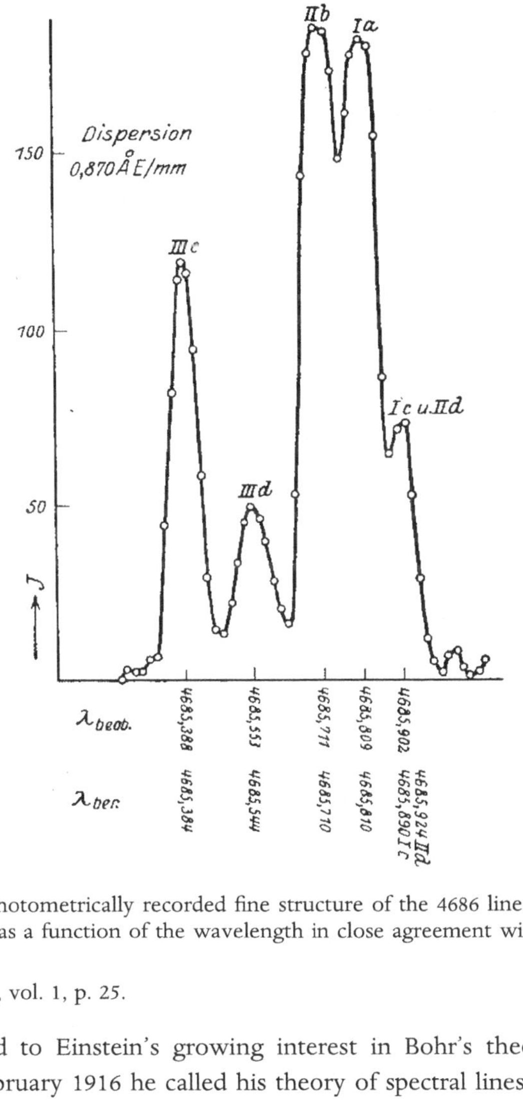

Fig. 4.6. Paschen's photometrically recorded fine structure of the 4686 line in He^+, showing the intensity of the lines as a function of the wavelength in close agreement with Sommerfeld's fine-structure formula.

Source: Grotrian 1928, vol. 1, p. 25.

greatly contributed to Einstein's growing interest in Bohr's theory. In a letter to Sommerfeld of February 1916 he called his theory of spectral lines 'a revelation', and in a later letter he wrote: 'Your investigation of the spectra belongs among my most beautiful experiences in physics. Only through it does Bohr's idea become completely convincing. If I only knew what little bolts the Lord had used for it!'[78] Although Einstein recognized that the quantum theory was highly successful in accounting for one-electron atoms, he was also keenly aware of its lack of unity and a satisfactory logical foundation. How could such a patchy theory explain simple atomic systems so perfectly? In a letter to Born of 1919 he wrote about the quantum theory: 'One really ought to be ashamed of its success, because it has been obtained in accordance with the Jesuit maxim: "Let not thy left hand know what thy right hand doeth" '.[79]

Like Einstein, Planck considered Sommerfeld's theory something of a revelation. In his Nobel lecture of 1920 he called attention to the impressive agreement between

Paschen's measurements and what he called Sommerfeld's magical formula (*Zauberformel*). Comparing Sommerfeld's work to the French astronomer Urbain Leverrier's sensational discovery in 1846 of the planet Neptune, he attributed to the theory the highest possible praise. Bohr's theory was the culmination of yet another piece of magic. 'The quantum theory attains its most strong foundation in Niels Bohr's arguments for and developments of the atomic theory', Planck said. 'For it was allotted to this theory to discover in the quantum of action the long sought key to the entrance door of the wonderland of spectroscopy, which since the discovery of spectral analysis obstinately defied all attempts to open it'.[80]

Much of the interest in Sommerfeld's fine-structure theory was not related to atomic structure but rather to its use of Einstein's theory of relativity, which was seen as controversial in some parts of the physics community. The confirmation of the fine-structure theory was often taken as proof of Einstein's theory, or at least of the relativistic formula of mass variation. Young Wolfgang Pauli, who at the time was a student of Sommerfeld's, was among those who considered the fine structure the definitive confirmation of Einstein's formula. In his remarkable monograph on the theory of relativity, written at the tender age of 21 years, Pauli concluded that the theory of spectra 'leads to a complete confirmation of the relativistic formula, which can thus be considered experimentally verified'.[81] The work of Sommerfeld on the fine structure of hydrogen was widely seen as final and in need of no modification. Although several non-relativistic alternatives—and even a few based on the general theory of relativity—were proposed, they attracted little interest. The perceived success had the effect that mainstream physicists passed over a critical examination of the Bohr–Sommerfeld theory of hydrogen. This theory, it turned out with the emergence of spin quantum mechanics in early 1926, gave the right result but built on a quite wrong foundation. As first shown by Heisenberg and Jordan, the fine structure is the combined result of relativity theory and the electron's spin.

At the end of the old quantum theory, clouds began to gather on the fine-structure heaven. Gerhard Hansen, a physicist at the University of Jena, found a new line (IIc) in the hydrogen H_α spectrum, which corresponded to the $\Delta k = 0$ transition $(3,1) \rightarrow (2,1)$ forbidden by the Bohr–Sommerfeld theory.[82] Since the IIc line could not be explained away as the result of a Stark effect, it constituted an anomaly. However, this blemish only became widely known after the introduction of spin, and then it came to be seen as confirmatory evidence rather than an anomaly.

4.5 VOICES OF DISSENT

The atomic theory proposed by Bohr was generally successful and its scientific successes multiplied with Sommerfeld's extensions from the years 1915 to 1917. Yet doubts remained, both with regards to the confirmation of the theory and, not least, its conceptual foundation. Ludwik Silberstein, a Polish-born physicist living in England and a former student of Planck, pointed out in a review of 1920 that 'the weak side of the theory consists in the heavy sacrifices it requires at the very outset', the chief of the sacrifices being the denial of Maxwellian electrodynamics in the stationary orbits. Yet, in spite of the sacrifices, and 'notwithstanding its somewhat magically arithmetical character', Silberstein valued the theory's impressive predictive power and agreement with experiment higher than its weaknesses.[83]

While many physicists agreed, not all did. Physicists of a more conservative inclination found it hard to accept the fundamental message of Bohr's theory; that the frequency of emitted light does not pre-exist in the atom in the form of electrons vibrating or rotating with the observed frequency. It was equally hard for them to accept stationary states with non-radiating revolving electrons so blatantly defying the laws of electrodynamics, and no more understandable was the picture of electrons jumping acausally and discontinuously from one stationary state to another. When Bohr was nominated for the 1919 Nobel Prize, the physics committee turned down the proposal with the argument that 'the assumption on which Bohr's atomic model had been built stands in contradiction to physical laws that are still indispensable'[84]—which was indeed an essential message of Bohr's theory! Not only did it violate classical laws of physics, a physical mechanism for what happened in the atom was also conspicuously missing. We have seen how Rutherford at an early date objected to the element of acausality built into Bohr's model, how Wien and Oseen worried about its disagreement with electrodynamics, and how J. J. Thomson found it to be 'arithmetical' rather than 'physical'.

Objections against the theory, either in its original version or in the more advanced version argued by Sommerfeld, continued during World War I and the years thereafter.[85] Ehrenfest was full of praise for Sommerfeld's generalization of the quantum theory of atoms, and especially of Epstein's successful use of it, but qualified his praise because of its basis in Bohr's theory: 'Even though I consider it *horrible* that this success will help the preliminary, but still completely *monstrous*, Bohr model on to new triumphs, I nevertheless heartily wish physics at Munich further success along this path!'[86] The American physicist Merle Tuve encountered Bohr's theory while a student at the University of Minnesota. He later recalled that not only was it 'quite irrational and absurd' from the viewpoint of classical physics, it also seemed to be methodologically flawed: 'Various mathematical formalisms were devised which simply "described" atomic states and transitions, but the same arbitrary avoidance of detailed processes, for example, descriptions of the actual *process* of transition, were inherent in all these formulations'.[87]

A few physicists attempted to derive the mysterious Bohr quantization rules on a classical basis, either from electrodynamics or hydrodynamics: Thorstein Wereide in Norway, Leigh Page in the United States, and Marcel Brillouin in France. Bohr's atomic theory disseminated slowly to the French physics community, where it became widely known only after World War I and even then without attracting a great deal of interest.

In fact, Brillouin's paper of 1919 was one of the first research papers on the subject, but this contribution from Paris was in a tradition completely different from the one followed in Copenhagen and Munich. The aim of Brillouin, a distinguished professor at the Collège de France, was to offer a dynamical explanation of the quantum rules in terms of a modified classical theory that on a continuum basis could reproduce the discrete features of Bohr's theory. 'It appears', he wrote, 'that one is able to formulate a *dynamical* hypothesis with the necessary qualities to represent the essential properties of the Bohr atom'.[88] In order to reproduce the discrete features of Bohr's theory from a continuum theory, Brillouin had to assume that the velocity of light in the vicinity of atoms greatly exceeded the ordinary value in a vacuum! Physicists outside France ignored or were unaware of Brillouin's theory.

A look at the German scene will give a further impression of the mixed reception that the Bohr–Sommerfeld theory faced in some quarters of the physics community. For example, Arthur Szavassi, a physicist at the German Technical High School in Brno, Moravia, looked for the 'true nature' of Bohr's theory, which he found was empirically convincing but incomprehensible in its foundation. He offered a demystification of the enigmatic features of the theory, which, he thought, were 'really of a statistical nature due to the cooperation of many atoms'.[89] Planck, as we have seen, denied in 1915 the validity of Bohr's basic postulates.

Woldemar Voigt, the highly esteemed professor of mathematical physics in Göttingen, had difficulties with Bohr's theory. In a letter to Sommerfeld, congratulating him on the award of the Helmholtz Prize for his work on the quantum theory of spectra, he wrote: 'It seems to me to follow from the results of Bohr's theory that the electron can hardly be as simple as in the Lorentz–Einstein theory. There must be *something* in it which . . . forces the electron into this or that defined orbit after it has been thrown out of another one. Perhaps this "something"—say an internal rotation—is also the cause why it *does not* radiate in these special orbits *but does so* in the transition from the one to the other. If so, this "something" must be intimately linked to the "elementary quanta". I wonder if you think along similar lines?'[90] The 67-year-old Voigt was a classical physicist, but some of his reservations were shared by younger physicists as well. In his letter to Stark, part of which was quoted above, Epstein said:

> Generally I fully agree with you that in its present state Bohr's theory requires a readjustment of our thinking which involves different logical hardnesses. As to myself, I have found this internal transformation very difficult; apart from the problem to which you refer, that the emission frequency of the electron is determined by its properties already before it has reached its final orbit, what offended me particularly was the circumstance that the electron during its motion in the stationary orbit does not radiate.[91]

The following year, in a review article in *Die Naturwissenschaften*, Epstein pointed out in public that although the Bohr theory of the atom was undoubtedly correct in its main features, it was not without problems. 'Why is there no radiation in the case of motion in a stationary orbit, and what are the circumstances when the electron jumps from one stationary orbit to another?' he asked. Epstein thought that Bohr's frequency condition had 'a teleological character which flatly offends the scientist'.[92]

The objection mentioned by Epstein, Stark, and other physicists, namely that an electron in an excited state somehow seems to 'know' the lower state to which it transits, seems to have been widespread. In March 1917 Hendrik Kramers was lecturing in Stockholm on Bohr's theory, and he reported to Bohr in Copenhagen that in the lively discussion following the lecture, 'there were many who criticized the frequency condition $h\nu = E' - E$; the electron would need an information bureau to be able to calculate the frequency to be emitted'. His evasive answer to the objection seems to have been that, 'fortunately, the formula is only formal, and one cannot deduce anything about the mechanism of the radiation, etc'.[93] Following another lecture visit in December 1918, Kramers told Bohr that the Swedish physicists were no longer quite as sceptical. Oseen still thought 'that it was all mysterious, but he cannot deny that all the reasoning is sound, and that it is a fruitful mysticism'.[94] In an address of 1924 Sommerfeld referred to the same problem, lightheartedly suggesting that perhaps the mechanism in quantum transitions was teleological rather than causal. 'It appears to me', he said, 'that such a teleological reversal of causality is less opposed to quantum theory than classical theory'.[95]

The fine-structure theory and its confirmation by Paschen and other experimentalists gave rise to a public debate concerning the validity of the Bohr–Sommerfeld model. Parts of this debate were of a technical nature, relating to the reliability of Paschen's measurements, while other parts of it were of a conceptual or, in a few cases, even ideological nature.

The experimental data produced in Tübingen in favour of Sommerfeld's theory were some years later challenged by precision experiments made at the Physikalisch-Technische Reichsanstalt near Berlin. Ernst Gehrcke and his collaborators found for the doublet separation a lower value that disagreed with Sommerfeld's prediction, and they laid the blame on Sommerfeld's use of Einsteinian relativity. According to Gehrcke and some other conservative physicists, the experimentally determined fine structure implied that Einstein's mass variation formula was wrong and had to be replaced with an older one proposed by Max Abraham on the basis of the classical theory of the rigid electron. Gehrcke had from an early date been opposed to Bohr's theory of the atom, which in 1914 he translated into a more classical model (Section 3.9). In later works he developed a visualizable atomic model based on a corpuscular ether, which had the advantage (in contrast to Bohr's) that it provided 'the physical reasons that determine spectral phenomena and not merely a formal principle from which the laws can be derived'. His model allegedly explained 'what *really* goes on when spectral lines are produced'.[96]

The most sustained criticism came from Stark, one the leaders of the conservative faction of German physicists and a bitter enemy of the new quantum theory of atoms (see also Section 3.9). In his critique of Bohr's theory from 1917 Stark primarily focused on Bohr's arguments that the H_2^+ ion had to be unstable and dissociate into a H^+ ion and a neutral hydrogen atom. As he correctly pointed out, Bohr's model of molecular hydrogen was inconsistent with experiments. Not only did Stark's own experiments with canal rays prove the existence of H_2^+, so did Dempster's more recent experiments with positive ions. In addition to the experimental problems, he emphasized that Bohr's theory was plagued by problems of a conceptual and logical nature. As was he echoing Rutherford's letter to Bohr of 20 March 1913, Stark asked: 'How can the electron, when it is leaving its initial orbit, already know its final orbit; or, by which considerations does it choose its final orbit, so that already before it has attained it [it knows] the definite frequency it has to emit to reach the chosen final orbit?'[97] This question was not only raised by enemies of the theory, but also by some of its friends. In their semipopular account of Bohr's theory from 1922, Kramers and Holst stated the paradox of the apparently teleological behaviour of the electron as follows:

> On the whole it is very difficult to understand how a hydrogen atom, where the electron makes a transition from orbit 6 to orbit 4, can during the entire transition emit a radiation with a frequency different from that when the electron goes from orbit 6 to orbit 5 . . . Even from the very beginning the electron seems to arrange its conduct according to the goal of its motion and also according to future events. But such a gift is wont to be the privilege of thinking beings that can anticipate certain future occurrences. The inanimate objects of physics should observe causal laws in a more direct manner, i.e., allow their conduct to be determined by their previous states and the contemporaneous influences on them.[98]

While Stark considered the paradox a great problem for Bohr's theory, Kramers and Holst saw it as a stimulating challenge to our thinking about the subatomic world, indicating that it might be 'impossible to obtain a consistent picture of atomic processes in space and time'.

Stark used a considerable part of his Nobel Prize lecture of 1919 to argue against Bohr's atomic theory, which he was unable to accept 'because in its provisions it postulates suppositions which contradict, not only Maxwell's theory, but the very spirit of physics'.[99] Although he admitted that the theory agreed impressively with the simple spectra, he found it objectionable from both a conceptual and empirical point of view. Yet he also realized that the significance of his own Stark effect was much enhanced by Bohr's theory, which indirectly was a contributing cause for the Nobel Prize. Bohr had congratulated Stark on the prize, and in a letter of thanks the German physicist wrote: 'Your words of appreciation are even more honourable to me because they come from a man who has given such a fruitful theory to science. Moreover, it is precisely by way of your theory that the phenomenon discovered by me has been clearly understood'.[100] This private appreciation of Bohr's theory was not to be found in Stark's published works.

The following year Stark intensified his attack, warning Sommerfeld in advance: 'In the next issue of Jahrbuch der Elektronik there will appear my critique of Bohr's theory from the standpoint of an experimental physicist. It shows that your theory of fine structure does not agree with experience'.[101] In his article in *Jahrbuch* (of which he was the editor) Stark repeated and amplified some of his old objections, including that the frequency of the emitted radiation was mysteriously determined before the electron reached its final state. He granted that Bohr's theory had scored some empirical successes, of which he mentioned the derivation of the Balmer series and the Rydberg constant, the main features of the Stark effect, and the normal Zeeman effect. But the prize for the successes was much too high, for it rested on the introduction of several enigmatic postulates, including the notion of stationary states and the frequency condition. To this should be added, Stark said, the experimental facts that flatly contradicted Bohr's theory: certain continuous spectra, the H_2^+ ion and other constituents in canal rays, the Paschen–Back effect in hydrogen, the fine-structure spectra, and the asymmetric intensity of the lines in the Stark spectrum. On balance Stark thought that there was little merit in Bohr's theory and that consequently it deserved to be rejected.

Bohr was aware of Stark's attack, but had no wish to be dragged into a polemics with the quarrelsome and insistent German physicist. After Sommerfeld had informed him that he would reply to Stark's 'quite reckless' criticism, Bohr wrote to him: 'I was pleased to hear that you have written a rejoinder to Stark's critique. As you assumed, I had not intended a reply'.[102] Bohr was puzzled that Stark thought that the electric effect named after him posed a problem for quantum theory. As Bohr saw it, in this area theory and experiment stood in a beautiful symbiotic relationship and Stark should rather be grateful to quantum theory for having endowed his discovery with a higher status. This position—that Stark's experiments represented the strongest possible support for the quantum atom—was also what he said in the only paper in which he referred to Stark's anti-Bohr article.[103]

In his reply to Stark, Sommerfeld carefully and diplomatically answered the criticism to the best of his ability. Only when it came to the question of the apparently predetermined emission frequency was he vague and had to admit that he had no good answer. Perhaps, he speculated, one might have to rely on the teleological principle of least action in order to understand the emission process. 'Who can rule out that one day there will be found, also in the theory of emission of light, a formulation that makes the anticipation of final states dispensable?' Sommerfeld complained that Stark's short list of the accomplishments of Bohr's theory systematically underrated the power of the theory. At the very least, he said, one should include areas from optical spectroscopy, the whole field of X-ray spectroscopy, and the Franck–Hertz experiments. Based on these and other accomplishments, Sommerfeld found it justified to call Bohr's theory 'the greatest progress of all times in the understanding of the atom'. Moreover, 'one should not forget that Bohr's theory is only eight years old', and nor should one forget that 'Rutherford's nuclear theory of the atom has become vital and fruitful only by Bohr's quantum-theoretical contribution'.[104]

The Viennese physicist Adolf Smekal joined Sommerfeld in repudiating Stark's anti-Bohr article.[105] His reply to 'Stark's causality objection to [Bohr's] frequency condition' was no less vague than Sommerfeld's, and no more convincing. With respect to Stark's experimentally based criticism, Smekal claimed that it was either due to misunderstandings or to a failure to take into account the unknown effects of interatomic electric fields. He implied that when these effects had been understood the anomalies would disappear.

Besides repeating his old objections, Stark included a criticism, based on Gehrcke's new and 'very reliable' measurements, of the experimental support of Sommerfeld's fine-structure theory. To Stark's mind there was no doubt that the lower value of the doublet separation obtained at the Physikalisch-Technische Reichsanstalt disproved Sommerfeld's quantum theory of spectra. He could have chosen to place the guilt on the theory of relativity, but at the time he thought that the relativistic ('Lorentz–Einstein') mass formula was convincingly confirmed by experience. In his persistent crusade against the 'dogmatic spirit' of modern physics, Stark often emphasized the case of the fine structure. He maintained that Sommerfeld and his associates manipulated the experimental data in order to force them to agree with theory. The following excerpt from a booklet of 1922 is representative:

> I claim that in several of the cases where Sommerfeld believes he sees lines in the observations, there are none at all, and the separation of the components of the hydrogen and helium lines deviates from the theoretical value... It is indeed regrettable that a theorist distorts the facts when he looks at them through the spectacles of his theory, although in the long run it can do no harm. The dogma creates more serious damage, however, if the experimentally working physicist takes on the spectacles of the theorist and more or less consciously makes it his task to confirm the prevailing theory.[106]

The attempts of Stark and other conservative German physicists to change the course of physics did not succeed. The opposition against the Bohr–Sommerfeld quantum theory of atoms was weak and scattered, and the scientific reputation of the Nobel laureate Stark was rapidly declining. By the early 1920s the opposition was ignored by Sommerfeld and most other mainstream physicists. While Bohr never cared about Stark's critique, he did care about the views of Sommerfeld, who was not only an esteemed colleague but also, in a real sense, a scientific rival. Sommerfeld's style of physics was not Bohr's style, and the two physicists differed in their views of how to develop the quantum theory, in particular about the role of the correspondence principle. However, their disagreements were not of a fundamental nature and did not involve their personal relationship. Bohr and Sommerfeld had first met shortly before the war and would only meet again after the war had come to an end and imperial Germany had changed into the Weimar Republic. Meanwhile they had stayed in contact by means of letters and exchange of scientific papers.

In September 1919 Sommerfeld attended a conference in Lund, Sweden, invited by the leading Swedish X-ray spectroscopist Manne Siegbahn. Among the participants was Bohr, who gave an address in English on 'Atomic Physics and the Problem of Radiation'. After Sommerfeld had given lectures in Lund and also in Stockholm, he spent three happy days with Bohr in Copenhagen. Having returned from his Scandinavian journey, he wrote to his daughter: 'Bohr is just like Einstein, only better washed and much neater. He stayed with me for the entire three days and was most attentive. The stay in Kop. [Copenhagen] was really wonderful . . .'[107] From that time onwards Bohr's connections with German physicists intensified. While his earlier scientific publications had all been in English, he now published some of his most important work in German and generally established a close relationship with the strong German physics community. In the spring of 1920 he was in Berlin, invited by the German Physical Society, where he met many of the important physicists, including Planck, Born, Landé, Kossel, and—for the first time—Einstein. With the establishment the following year of an institute of theoretical physics in Copenhagen, headed by Bohr and created for him, the connections with German physicists grew even closer.

4.6 X-RAY ATOMS

The model for atoms with more than a few electrons was usually conceived as a structure of concentric rings of the 'Saturnian' kind that Bohr had discussed in the second part of his trilogy. Chemical facts and the interpretation of the characteristic X-rays seemed to support the simple ring model. This was what Kossel argued in an article of 1916 in the *Annalen der Physik*, more than 130 pages long, in which he connected the appearance of X-ray series with the periods of the periodic system of the elements in greater detail than earlier. In building up electron structures, he said that 'The next electron, which appears in the heavier element, should always be added at the periphery [and] in such a manner that the observed periodicity results'. He elaborated:

> This leads to the conclusion that the electrons, which are added further, should be put into concentric rings or shells, on each of which . . . only a certain number of electrons – namely, eight in our case – should be arranged. As soon as one ring or shell is completed, a new one has to be started for the next element; the number of electrons, which are most easily accessible, and lie at the outermost periphery, increases again from element to element and, therefore, in the formation of each new shell the chemical periodicity is repeated.[108]

Although the ring or shell picture of the Bohr atom can still, nearly a century later, be met in elementary textbooks, among quantum physicists it did not survive for long. It formed the basis of the first attempts to derive electron configurations of the elements on the basis of X-ray spectroscopy, but a few years after Kossel's article it was realized to be not only inadequate, but wrong.

In February 1919 Sommerfeld told Bohr that he found it 'rather difficult to make progress in the Röntgen spectra'. There was, he said, 'too much that is hypothetical about the positions and populations of the rings'.[109] X-ray spectra and their use in atomic theory had been of great importance to Sommerfeld since his theory of 1916, and over the next few years he and other physicists endeavoured to extend it into a theory of the atomic structure of higher atoms. A major goal of this line of research was to determine the population numbers in the various rings; that is, to find the number of electrons in the rings surrounding the atomic nucleus. This research programme was obviously of great importance to both physicists and chemists, for it promised to yield an understanding of the electron configurations of atoms and thus provide an answer to a problem that Bohr had addressed in his theory of 1913 (and which J. J. Thomson had addressed even earlier). If the programme succeeded, X-rays and atomic theory would turn out to be the keys that unlocked the secrets of the periodic system and perhaps reduced it to atomic physics. The X-ray spectroscopic approach to atomic structure was for a couple of years pursued with great determination by Sommerfeld, Debye, and several other physicists, whereas Bohr in Copenhagen merely looked on from the sidelines. When X-ray spectra became important to his own work on atomic structure, as they did in the early 1920s, it was in a different way to the one followed previously.

From studies of the intensity of X-rays reflected from crystals, Arthur Compton at Princeton University found in 1917 that he obtained reasonable agreement with experimental data if he assumed the innermost ring of atoms to contain four electrons and the one outside it six electrons.[110] In the case of calcium he suggested population numbers of 4, 6, 8, 2 and for oxygen the more reasonable numbers 2, 6 (Figure 4.7). Compton's work did not rely specifically on Bohr's theory, but was consistent with it. A more systematic approach, based on the frequencies of X-rays and not on their intensities, was proposed the same year by Debye, who made explicit use of the theoretical conceptions of Bohr, Sommerfeld, and Kossel. Debye reasoned that the frequency due to an electron transition to the innermost K ring could be expressed as the energy difference between two rings, the energy depending on the number p of electrons in the K ring. From the energy difference followed the frequency as a function of the atomic number, and by fitting the resulting $\nu(Z)$ function to the measured K_α frequencies for elements between $Z = 11$ (sodium) and $Z = 60$ (neodymium) he found good agreement for $p = 3$. Debye thus pictured the first electron ring as three symmetrically arranged electrons rotating around the nucleus. 'From this ring one electron can be removed and be brought on a circular orbit associated with two quanta', he wrote. 'The two remaining electrons then come closer to the nucleus and describe, at an angular distance of 180° from each other, a new circular orbit around the nucleus. The transition of the three electrons from the second state to the first state creates the K_α line'.[111]

A similar but more sophisticated approach was followed by the Polish physicist Jan Kroo, who considered an atom with $(p - 1)$ electrons in the K ring and q in its L ring.[112] An electron would pass from K to L, leaving the two rings with p and $(q - 1)$ electrons, respectively. From this picture he obtained an expression in which both p and q appeared:

$$\frac{\nu(K_\alpha)}{R} = p(Z - s_p)^2 - (p-1)(Z - s_{p-1})^2 + (q-1)\frac{(Z - p - s_{q-1})^2}{2^2} - q\frac{(Z - p + 1 - s_q)^2}{2^2}$$

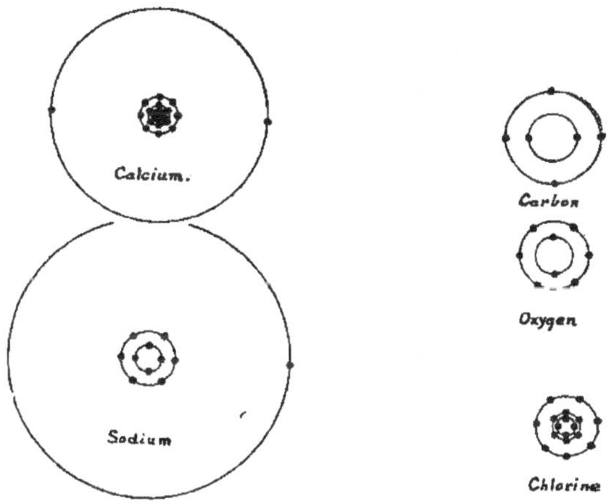

Fig. 4.7. Arthur Compton's atomic models for five elements based on X-rays reflected by crystals. *Source*: Compton 1917, p. 51.

The *s*-quantities are screening factors due to the repulsions from the other electrons. Planck's constant does not appear explicitly in the expression, but only implicitly in Rydberg's constant *R*. In fact, the only way in which quantum theory entered the derivation was through the quantization of the angular momentum, much as Bohr had found in 1913. From a comparison of his formula with experimental data Kroo concluded that in the normal state $p = 3$ and $q = 9$. When an electron was lost by ionization, the atom would have $p = 2$, and after the emission of a K_α ray it would end in a state with $p = 3$ and $q = 8$. Referring to Kroo's result in the first edition of *Atombau*, Sommerfeld commented that the population numbers showed 'a remarkable agreement' with the periodic system.[113] His curious argument was that they were only one unit greater than the number of elements in the first two periods of the periodic system (he did not suggest how to account for the difference of one).

Another industrious contributor to the approach initiated by Debye was the Norwegian physicist Lars Vegard, a newly appointed professor at the University of Oslo and a specialist in the physics of the aurora borealis. In a series of works from 1917–1919, published in both German and English journals, he dealt extensively with atomic models derived from X-ray spectroscopic data. In November 1917 he concluded that his results agreed well with experiments if elements with $Z > 9$ contained one K ring with quantum number $n = 1$ containing three electrons, two closely spaced L rings with $n = 2$ containing seven and eight electrons, respectively, and one M ring with $n = 3$ containing nine or ten electrons.[114] Two years later he suggested that the best data indicated an M ring with twelve electrons (Figure 4.8). Vegard believed that there was an '*l* ring' just outside the L ring and that it had the same quantum number, $n = 2$. For the light elements from lithium to fluorine he argued that they had an internal K ring

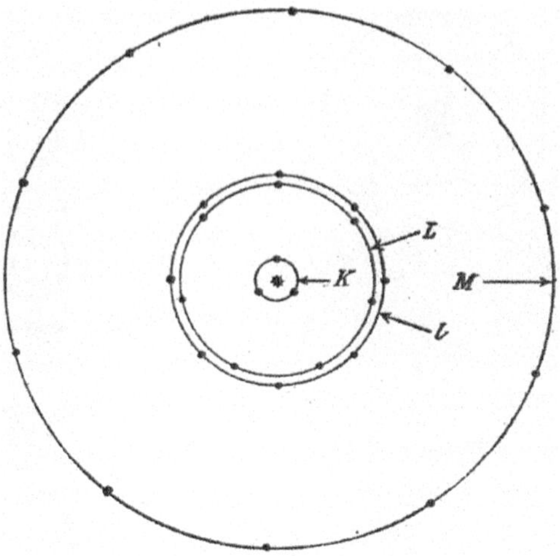

Fig. 4.8. Lars Vegard's atomic model of manganese, atomic number 25.
Source: Vegard 1918, p. 317.

with two electrons. New electrons would be added to the L ring, from Li = (2,1) to F = (2,7), while the next electron would fill up the K ring, meaning that Ne = (3,7). In the period starting with sodium the *l* ring would be built up, so that argon was assigned the electron structure (3,7,8). The next inert gas, krypton, was similarly characterized by an outermost ring containing eight electrons, the structure being Ar = (3,7,8,10,8).

Whatever the population numbers, it is worth noticing that Vegard based his system on the assumption that the quantum numbers of the rings in the *normal* (unexcited) atoms increase by one unit as one moves outward from the nucleus. 'I have succeeded', he wrote to Bohr, 'to obtain a most striking agreement with experimental data on the basis of the hypothesis of increasing quantum numbers'.[115] Vegard's hypothesis implied that all elements belonging to the same period have the same value of the principal quantum number n. 'If at all we shall be able to proceed further in the direction pointed out by Bohr', he said, 'we can hardly avoid the assumption that systems of electrons exist in the normal atom with quant numbers greater than 1'.[116] That is, contrary to Bohr's original atom, which in its normal state was characterized by $n = 1$, Vegard's was a many-quantum atom. The idea of atoms with rings $n > 1$, which also appeared in Debye's work of 1917, was quickly adopted by Sommerfeld and other atomic physicists. In a letter to Manne Siegbahn, Sommerfeld indicated his low regard of the Norwegian physicist: 'I have sent for publication in Physik. Ztschr. a note in which I attempt to determine the filling of the ring with electrons from the Röntgen spectrum, in a manner similar to that employed (though perhaps rather haphazardly) by Vegard. In doing so I attempt above all to grasp the interaction between the various rings'.[117]

Debye, Sommerfeld, Kroo, and Vegard all agreed that the K ring contained three electrons, such that, for example, chlorine was assigned the structure (3,7,7) and phosphorus (3,7,5). Of course, these structures did not easily comply with the periodic system and other chemical knowledge. Nonetheless, for a few years they were widely accepted by the physicists. 'Aren't there three electrons in the K ring?', a somewhat surprised Sommerfeld asked in 1919, when he realized that this might not be the case and that the ring atom might have to be abandoned.[118]

On the basis of X-ray data and his version of Bohr's ring atom Vegard attempted to provide all the atoms with population numbers, thus to account for the entire periodic system in terms of atomic theory. Bohr had made a similar attempt in 1913, but more limited and tentative. Vegard's project, as he presented it in two large papers in *Philosophical Magazine*, was much more ambitious.[119] When it came to the higher atoms his population numbers were little more than guesswork. 'We have more or less to grope in the darkness and feel our way forward', he admitted, and this he did by considerations of the same kind that had guided earlier atom-builders. These considerations included the valences of the elements, the variation of the electropositive properties along the periodic system, and Meyer's curve of the atomic volumes of the elements. As to the rare earth metals he assumed they were characterized by the addition of a new ring with a total of fourteen electrons and with $n = 4$, the new ring being placed inside the outermost ring to which he ascribed four electrons throughout the group of the rare earths.

'I think I have succeeded on the basis of the hypothesis of increasing quantum numbers in obtaining very striking agreement with the experimental data', an optimistic Vegard wrote to Bohr in the fall of 1918.[120] But his optimism was short-lived, as the entire approach on which his work was based soon turned out to be a blind alley. This was demonstrated primarily by Smekal and his German collaborator Fritz Reiche, who in a critical analysis of the approach showed that it was unable to discriminate between, for example, population numbers (3,7) and (2,8) for the K and L rings. They also pointed out that disturbances from one ring to another would destroy the results obtained by Debye and Vegard.[121] 'It seems to me that the goal of determining the number of electrons in each ring still lies far in the future', Sommerfeld admitted in a letter to Bohr.[122] Six months later he was ready to put the blame on the ring atom, as Reiche and Smekal had suggested. Their analysis left little hope for modifications on the basis of the planar ring atom, which made Smekal conclude that this kind of model was inadequate.

Abandoning the simple ring atom was not a great sacrifice to Bohr. On Christmas day 1919 he wrote to Richardson: 'I am quite prepared, or rather more than prepared, to give up all ideas of electronic arrangements in "rings"'.[123] What should replace the ring atom, he did not yet know, but he would soon come up with an alternative. Bohr publicly renounced his earlier belief in the ring atom a few months later, in the lecture he gave in Berlin on 27 April 1920. At the end of this lecture he briefly pointed out that many of the assumptions of his 1913 theory would have to be changed because of 'the lack of agreement of the theory with experiments'. In particular, 'It appears no longer possible to justify the assumption that in the normal states the electron move in orbits of special geometrical simplicity, like "electron rings".... Considerations... force us to look about for possibilities of more complicated motions'.[124]

4.7 THE BOHR–SOMMERFELD ATOM AROUND 1920

According to Sommerfeld, there were two kinds of L-orbits, either circular ($n = k = 2$) or elliptical ($n = 2$, $k = 1$). It was no problem to place several equally spaced electrons on a circle, but it could not be done on an ellipse, for then they would experience different Coulomb forces during their revolutions and the orbit would be unstable. As a possible solution Sommerfeld placed each of the electrons on its own ellipse, with the position of the electrons arranged in such a way that at any moment each of them would be at a corner of a regular polygon. In 1918 he described the basic idea of what he called an *Ellipsenverein*, a union of ellipses, as follows: 'If the actual positions of the n electrons are connected by straight lines, there results a regular n-sided polygon, whose area increases or decreases if the n electrons approach their aphelion or perihelion, respectively, which they do by following exactly the same rhythm'.[125] The L-electrons would either move in a ring or in their carefully coordinated ellipses, but the atom could not accommodate both types of motion.

The picture of electrons moving rhythmically on ellipses (Figure 4.9) evidently appealed to him, and not only for technical reasons. Some might find the picture artificial, even contrived, but not Sommerfeld. In a footnote he said that he saw 'the artful interlocking of the n electronic paths' as 'not unnatural' but rather as a sign of the 'high harmony of motion that must rule within the atom'. The *Ellipsenverein* appeared

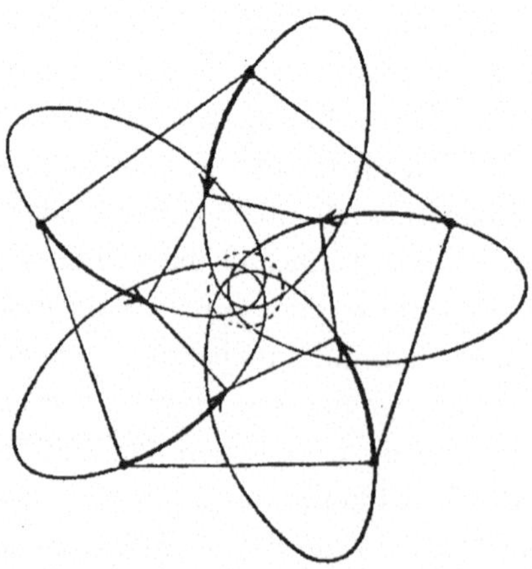

Fig. 4.9. Sommerfeld's *Ellipsenverein* for five electrons moving symmetrically around the nucleus. The position of the electrons at any point of their elliptical orbits spans a pentagon pulsating in size as the electrons synchronously move about in their orbits.

Source: Sommerfeld 1919, p. 367.

prominently in the various editions of *Atombau*, followed by similar comments highlighting the aesthetic quality of the model. Yet, the aesthetically pleasing features did not mean that the model was scientifically satisfactory. Sommerfeld recognized that it faced several problems, including that it was not possible to derive the screening factor of $s = 3.6$ using the model. A more serious objection, pointed out by the Dutch physicist Johannes Burgers, was that the group of elliptically moving electrons would penetrate into the region of the circular K-orbit, causing a change in the Coulomb force experienced by them. This would be the case in the planar *Ellipsenverein*, but perhaps this just indicated that the configurations of electrons had to be extended from two to three dimensions. Sommerfeld thought that this might well be the case, and he was not the only one.[126]

One unlikely source for the spatial atom was the theoretical investigation of the compressibility of crystal lattices, a branch of physics in which Born was an expert. In the fall of 1918, Born and his pupil Landé studied the properties of simple crystals on the basis of the Bohr–Sommerfeld ring model of atoms. They arrived at the conclusion that planar electron orbits were unable to explain the properties and that atoms therefore had to be viewed as spatial structures. Born later recalled (obviously with hindsight) that his work with Landé led him to doubt orbital models or at least take them less seriously: 'From that moment on, my endeavour was not to find confirmations of the orbital theory but to provide arguments for its insufficiency'.[127]

As to Landé, the failure of the planar atom led him to propose complicated spatial models which, in some of the versions, were three-dimensional analogues to Sommerfeld's *Ellipsenvereins*. He referred to this class of models, which he developed 1919–1920 in a series of papers with the common title 'Dynamik der räumlichen Atomstruktur', as a *Polyederverein*. These models were governed by polyhedral symmetries of which the cubical symmetry resulted in a *Würfelatom* with eight electrons describing small circles near the corners of a cube (Figure 4.10).[128] For large values of the effective atomic number ($Z - s$) his theory resulted in more complicated arrangements of moving

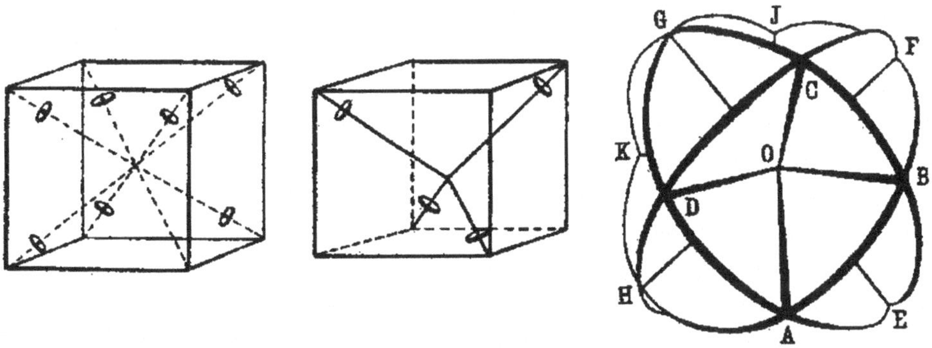

Fig. 4.10. Landé's cubical atom with the electrons moving in small circles near the corners of a cube.

Source: Landé 1920, p. 382.

electrons. With these cubical atoms he thought it possible to build up electron structures of some the elements and their ions. He even applied his theory to chemical problems, including the formation of water from oxygen and hydrogen, but the results were not empirically convincing.

Landé's *Würfelatom* was consistent with, and received indirect support from, X-ray experiments made in 1917 by the American physicist Albert Hull, at the Research Laboratory of the General Electric Company. To explain the results obtained with iron crystals, Hull suggested that the electrons in an iron atom were situated at the corners of three concentric cubes and that they had 'fixed positions in the atom, about which they move in small orbits'.[129]

Although Landé's cubical atoms did not last for long, they are of interest because they signified the break with the planar atom and were for a time received favourably, even enthusiastically. 'I find your cubical arrangement much more agreeable than the pancake atom, especially if you now can quantize it', Sommerfeld wrote to Landé in March 1919, and half a year later he described the cubical atom as a 'salvation' of the troubled theory of X-ray spectra.[130] Bohr expressed his 'extraordinary' interest in Landé's theory and invited him to Copenhagen, where he gave a lecture on the spatial atomic models in October 1920.[131] However, although interested in the cubical atom, Bohr did not believe in it. At any rate, his own ideas about atomic structure went in a different direction, and in March 1921 he explicitly rejected Landé's model as incompatible with his own. The electrons, he said, 'are not at every moment arranged in configurations of simple axial or polyhedral symmetry as in Sommerfeld's and Landé's work'.[132]

Among chemists it was no novelty that the atom had three dimensions, and some of the chemists thought of the spatial structures as cubical. One of them was Alfred Parson, who in 1915 imagined ring-formed electrons to be placed at the corners of cubes (see Section 3.7). It was indeed natural for chemists to think of cubical atoms, for then the number eight, as appearing in Abegg's valence rule and the periodicity of the elements, might attain a physical representation. Ideas of this kind inspired G. N. Lewis to formulate a model of cubical electron atoms as early as 1902, but he only published his theory fourteen years later.[133] By picturing the outer electrons of an atom to be placed at the corners of a cube he offered a structural model for the single and double (but not triple) bonds in molecules, namely as pairs of shared electrons. Since it was essential to Lewis's model that the electrons were stable, it was fundamentally inconsistent with Bohr's theory of the atom.

Another cubical atom, in some respects an extension of Lewis's, was proposed by Langmuir in 1919. Suggesting that the position of electrons 'probably correspond fairly closely to the corners of a cube', he constructed elaborate molecular models on this basis.[134] Like Lewis's model, the one of Langmuir was static and both models rested on the assumption of intra-atomic forces for which there were no physical justification. In reality the *Würfelatom*, based on the Bohr–Sommerfeld quantum theory, had little in common with the atoms of Lewis and Langmuir except for the shared assumption of cubically arranged electrons. Although Landé and Born were aware of the ideas of the American chemists, they did not take them seriously. We shall return to the Lewis–Langmuir theory in Chapter 6.

One of the subjects in which both Landé and Bohr had a focal interest was the structure of the helium atom, a vexed problem in the old quantum theory. The spectrum of helium was puzzling because it consisted of two spectral terms that did not combine; that is, there are no transitions from the energy levels of the one to those of the other. Originally, shortly after helium was discovered in terrestrial sources in 1895, several scientists suggested that the gas was composed of two elements.[135] Runge and Paschen tentatively assigned an atomic weight of about 3 to the lighter, unknown constituent and about 5 to ordinary helium or 'orthohelium'. The lighter constituent was known as 'parhelium'. Rydberg even assigned it a chemical symbol (Pa, later used for protactinium), and Lockyer suggested the name 'asterium' for the enigmatic element. It took a few years before it was recognized that parhelium is fictitious and exists only in the form of a particular spectral state.

Bohr's original model of 1913 pictured the helium atom as two electrons revolving in the same one-quantum orbit around the nucleus. It resulted in an ionization energy (about 27 eV, by 1920 revised to 28.8 eV) that was too high and it also failed to reproduce dispersion data. Neither did the calculated ionization energy for He^+ (82.9 eV) agree with the measured value. Understandably, the situation worried Bohr, who from the autumn of 1916 began reconsidering the helium atom in collaboration with his new assistant, Hendrik Kramers. 'I have used all my spare time in the last months to make a serious attempt to solve the problem of the ordinary helium-spectrum, working together with a young Dutch physicist Dr Kramers', Bohr told Rutherford. 'I think really that at last I have got the clue to the problem'.[136] But Bohr's optimism was premature. It would take a long time until his and Kramers's extensive calculations produced results that were satisfactory and could be compared with experiments. And then the comparison turned out to be a disappointment.

In a joint paper of 1919 Sommerfeld and Kossel announced what they called a spectroscopic displacement law (*Verschiebungssatz*), namely that the spark spectrum of an element has the same character as the arc spectrum of the element preceding it in the periodic system.[137] For example, the series in the spark spectra of the alkali elements would be similar in type to those of the arc spectra of the noble gases. On Bohr's interpretation that spark spectra had their origin in ionized atoms and arc spectra in neutral atoms, this was a regularity to be expected from the Bohr–Kossel shell model of the elements. The regularity was hardly a surprise to Bohr, who in a letter to Sommerfeld related it to the problem of the structure of the helium atom:

> It has also pleased me very much to read what you and Kossel have written about the displacement law. I have recognized the significance of this law for a long time, and, as early as two years ago, we began experiments in the physical laboratory of Copenhagen in the hope of observing the spark spektrum of lithium, for the special purpose of testing the

> consequences of a theory of the ordinary helium spectrum which had been worked out already in the autumn of 1916 in collaboration with Dr. Kramers.[138]

Bohr's point was that since the structure of Li^+ was very similar to helium, the spark spectrum of lithium might act as a test case for models of the helium atom. Although the experiments that Bohr and H. M. Hansen conducted in 1917 failed to produce the spark spectrum, he did not forget about the idea. About 1922, when the helium crisis became clearly recognized, he assigned to one of his assistants, Sven Werner, the task of investigating lithium's spark spectrum.[139]

The first notable advance in the helium problem was made by Landé, who in papers from 1919 and 1920 developed models of the spectra and structure of helium based on the general methods of quantum theory formulated by Sommerfeld and others.[140] He basically investigated models in which an outer electron—the valence electron or *Leuchtelektron*—moved in an orbit around an inner orbit carrying the other electron, the two planes being either coplanar or differently oriented. In *Atombau* Sommerfeld referred to this model as a 'double star model' (*Doppelstern*).[141] Landé argued that the coplanar structure corresponded to the normal state ortho-helium, while the spectrum of parhelium had its origin in atoms where the orbits were crossed. From an examination of spectroscopic data Franck and Reiche, on the other hand, argued that the normal state of helium was given by the one-quantum crossed structure.[142] This was also what Landé ended up with and what many physicists, if not all, accepted by about 1921. There was however considerable disagreement and neither Landé's final model nor other models resulted in an ionization energy close to the one observed. The exception was the 'oscillating model' advocated by Langmuir, from which he obtained an ionization energy of 25.62 eV, but this model did not satisfy the ordinary quantization conditions.[143]

While the experimental value of helium's ionization energy had been uncertain at the time of Bohr's trilogy, in 1919 the value was narrowed down to 25.4 ± 0.25 eV, as determined by Franck and Knipping by means of the electron-collision method. In 1922 Franck improved the value to 24.6 eV and Lyman, using an optical method, found 24.5 eV. Detailed calculations of a modified Landé model made by John H. Van Vleck at Harvard University resulted in an unacceptably low ionization energy of 20.7 eV. He concluded that the helium atom was an unsolved mystery:

> The conventional quantum theory of atomic structure does not appear able to account for the properties of even such a simple element as helium, and to escape from this dilemma some radical modification in the ordinary conceptions of the quantum theory or of the electron may be necessary.[144]

Bohr and Kramers continued their calculations on helium for several years. Although the methods they employed were quite different from those used by Landé, they arrived at largely the same conclusions, namely that parhelium was the ground state and corresponded to the crossed orbits, while orthohelium corresponded to the

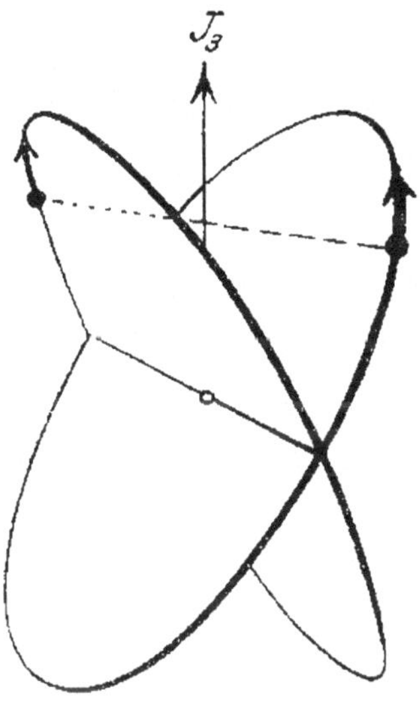

Fig. 4.11. The model of the normal helium atom favoured by Bohr, Kemble, and Kramers. The symbol J_3 denotes the total angular momentum.
Source: Born 1925, p. 331.

coplanar orbits. The two Copenhagen physicists excluded the Landé coplanar state and also the single-ring state of 1913 as candidates for the normal state of helium, and thus ended up with the case of two intersecting orbits (Figure 4.11).

Bohr intended to publish his work with Kramers, but for various reasons the paper never materialized. However, he dealt with the subject in correspondence and addresses, and included a detailed treatment of it in an unpublished manuscript probably intended for the 1921 Solvay Congress.[145] (Bohr was invited to the congress, but overwork and stress prevented him from participating.) Some parts of the collaborative work in Copenhagen survived in the form of a detailed paper Kramers published in early 1923, in which he examined a crossed model with two equivalent one-quantum orbits inclined at an angle of 60°. For this model of helium in its normal state he found an ionization energy close to the one calculated by Van Vleck and equally far from the experimental value. In spite of the disappointing result he maintained that the model was probably correct and instead concluded, not unlike Van Vleck, that 'already in this simple case—mechanics is no longer valid'.[146] This was just the first round in the story of the helium anomaly; we shall return to the second round in Section 6.5.

Notes for Chapter 4

1. Peierls 1988. In 1921 Bohr was appointed a honorary member of the Royal Institution and five years later he was elected a foreign member of the Royal Society.
2. Bohr 1921b, translated by Hugo Stintzing, a German physical chemist.

3. Planck 1923, p. 537. The special issue of *Die Naturwissenschaften* carried the title 'Die ersten Zehn Jahre der Theorie von Niels Bohr über den Bau der Atome'.

4. Graetz 1918, p. 78. The book must have sold well, for in 1922 it came out in a fourth printing and the same year it appeared in a Russian translation.

5. Kirchberger 1922.

6. Konen to Bohr, 16 December 1922, in Rud Nielsen 1977, p. 702.

7. On Bohr and the Nobel Prize, see Pais 1991, pp. 210–17, and Aaserud 2001. Bohr was also nominated for a chemistry prize, in 1920 by Fritz Weigert, a physical chemist at the University of Leipzig.

8. Oseen to Arrhenius, 20 November 1922, quoted in Aaserud 2001, p. 287. Both Oseen and Arrhenius served on the Nobel physics committee.

9. The Hughes Medal is named after the British physicist and engineer David E. Hughes (1831–1900), who is best known for his inventions related to telephone technology. The Matteucci Medal is named after the Italian physicist Carlo Matteucci (1811–1868), who did important work in bioelectricity in particular.

10. Kramers and Holst 1923, translated by the American physicist Robert B. Lindsay and his wife, Rachel T. Lindsay, who spent the years 1922–1923 in Copenhagen. The German translation was Kramers and Holst 1925, translated by Fritz Arndt, a chemist at the University of Breslau. The title of the Spanish translation of 1925 was *El Átomo y su Estructura según la Teoría de N. Bohr*, and the one of the Dutch translation of 1927, *De Bouw der Atomen*. In the Danish original of 1922 Holst appeared as the first author and Kramers as the second. A revised and expanded Danish edition was published in 1929, at a time when Kramers had left Denmark. He was replaced by another of Bohr's young collaborators, the Swedish physicist Oskar Klein.

11. Kramers and Holst 1923, p. viii. Dresden 1987, p. 134, calls the book 'a truly missionary venture to spread the gospel according to Bohr'. Missionary it was, yet the authors did not hide the provisional and incomplete nature of Bohr's theory.

12. Bohr to Rutherford, 29 November 1916, in Hoyer 1981, p. 595.

13. Bohr 1915b, p. 395.

14. Franck and Hertz 1914, p. 512. English translation in ter Haar 1967, pp. 160–6. Sources related to the Franck–Hertz experiment are compiled in Hermann 1967. For a historical and philosophical perspective, see Hon 1989.

15. Quoted in Holton 1961, p. 808.

16. Bohr to Oseen, 28 September 1914, in Hoyer 1981, pp. 562–3.

17. Bohr 1915b, p. 411.

18. Franck and Hertz 1916, p. 438.

19. McLennan and Keys 1916, p. 607.

20. Millikan 1917b, which was one of the first American expressions of support for Bohr's theory.

21. Davis and Goucher 1917, which also included an analysis of electron collisions with hydrogen and a comparison of these data with Bohr's theory. Hendrik van der Bijl, a South-African physicist working at the Western Electric Research Laboratory in New York, provided further support for Bohr's interpretation of the Franck–Hertz experiments, showing that the alleged ionization effect was in fact a secondary photoelectric effect (van der Bijl 1917).

22. Lecture of 13 February 1920 to the Royal Danish Academy of Sciences and Letters on 'The Interaction Between Light and Matter'. Translation in Rud Nielsen 1976, pp. 227–40.

23. Bohr to Makower, 21 February 1920 (Archive for History of Quantum Physics, Bohr Scientific Correspondence). Bohr referred to his failed experiment with Makower in his Copenhagen address: 'Our apparatus was broken up by a mishap, and, under the war conditions in England, it was unfortunately impossible to get the apparatus, which was complicated and made of fused quartz, repaired again'.

24. Franck and Hertz 1919.

25. http://www.nobel.se. According to Oseen, the main result of the Franck–Hertz experiments was that Bohr's two basic postulates 'are no more hypotheses but experimentally proved facts'.

26. Bohr to Oseen, 20 December 1915, in Hoyer 1981, p. 568.

27. Bohr to Oseen, 17 March 1916, in Hoyer 1981, p. 572. A week later, in a letter to H. M. Hansen, Bohr described Sommerfeld's paper as 'most beautiful and extremely important', adding that it constituted 'a tremendous progress in the whole quantum theory' (24 March 1916, Archive for History of Quantum Physics, Bohr Scientific Correspondence).

28. Bohr to Sommerfeld, 19 March 1916, in Hoyer 1981, pp. 603–4. Sommerfeld replied in a letter of 20 August (Eckert and Märker 2000, p. 565). The proofs of Bohr's paper to the *Philosophical Magazine* are reproduced in Hoyer 1981, pp. 431–61. In 1921, at a time when it was no longer of much scientific interest, it appeared in a German translation together with other of Bohr's early works (Bohr 1921b, pp. 123–51). The collection included a historically valuable introduction by Bohr in which he accounted for his withdrawal of the paper in March 1916.

29. Gearhart 2010, p. 146.

30. Planck expounded his latest version of the second theory in Planck 1913, which was the source used by Bohr. In his trilogy of 1913 he referred to earlier versions. For an account of Planck's second theory and its fate until it was abandoned about 1920, see Kuhn 1978, pp. 235–354. See also Needell 1980 for the various versions of Planck's theory. The success of the Bohr theory was a major reason why the second theory disappeared from the literature of physics.

31. For a detailed historical analysis of Ehrenfest's adiabatic principle, a name coined by Einstein in 1914, see Navarro and Pérez 2006. See also Klein 1970, pp. 264–92.

32. Ehrenfest 1917, p. 501, reprinted in Van der Waerden 1967, pp. 79–93.

33. Hoyer 1981, p. 436.

34. Bohr to Ehrenfest, 18 May 1918, in Rud Nielsen 1976, p. 12. Bohr's comments related to his new memoir on quantum theory (Bohr 1918a), a copy of which he had just mailed to Ehrenfest.

35. On Bjerrum's theory of diatomic molecules and its relation to Bohr's, see Fujisaki 1983 and Asmuss 1992a. Bohr valued the works of Bjerrum, to which he frequently referred. In his letter to Oseen of 20 December 1915, he wrote about the experimental reality of quantum theory that 'it is perhaps most evident from Bjerrum's beautiful theory' (Hoyer 1981, p. 568). Molecular spectroscopy in the old quantum theory is dealt with in Section 6.3.

36. Planck to Sommerfeld, 30 January 1916, in Eckert and Märker 2000, p. 445.

37. Planck to Ehrenfest, 23 May 1915, quoted in Kuhn 1978, p. 253.

38. Planck 1915, p. 913.

39. Historical works on Sommerfeld's theory include Nisio 1973, Mehra and Rechenberg 1982a, pp. 213–30, Arabatzis 2006, pp. 148–62, and Seth 2010, pp. 162–73. For a philosophical perspective and comparison between the theories of Bohr and Sommerfeld, see Hettema and Kuipers 1995.

40. Sommerfeld 1915b, p. 425.

41. Ibid.

42. Ibid., p. 439.

43. Nicholson 1923, which also included some unorthodox ideas about isotopes and nuclear structure. On the pendulum orbits, see further Lindsay 1927, who pointed out an error in Nicholson's reasoning and suggested that the 'distasteful' passage of an electron through the nucleus might be avoided if a short-range nuclear repulsive force was introduced.

44. Sommerfeld 1919, p. 143.

45. On Sommerfeld's *Atombau*, see Seth 2010 and Eckert and Märker 2004, pp. 41–3. The description of *Atombau* as the Bible of atomic theory is not a later construction. It was used in the early 1920s by Born, Paschen and Weyl, among others. The book was received enthusiastically by German physicists, who agreed that it was an 'excellent opus', as Karl Bergwitz summarized a review in *Physikalische Zeitschrift* **21** (1920), 223–4.

46. Born to Sommerfeld, 5 March 1920, in Eckert and Märker 2004, p. 75. Born found that Sommerfeld presented the state of atomic physics too optimistically, without emphasizing the doubts and problems sufficiently. Readers might easily get the impression 'that everything is in order; but that is often not the case, for example the molecular models of H_2 etc'.

47. Bohr wrote in 1916 a draft version intended as an addition to his paper, see Hoyer 1981, pp. 463–70. It consisted mainly in a summary account of Sommerfeld's theory and included no new ideas.

48. W. L. Bragg to W. H. Bragg, 17 October 1921, as quoted in Jenkin 2008, p. 412.

49. Wilson 1915, p. 795.

50. Ishiwara 1915. For Ishiwara and his work on the quantum theory, see Mehra and Rechenberg 1982a, pp. 210–11 and Nisio 2000. Ishiwara, who spent the years 1912–1913 in Europe, intended to attend the Birmingham meeting of the British Association where Bohr's theory was first discussed. However, he only arrived in England in October and thus missed the opportunity to meet Bohr. In his unpublished paper of 1916 Bohr referred to the works of Wilson and Ishiwara (Hoyer 1981, pp. 451–2). He pointed out that Ishiwara's theory was inconsistent with the usual one-electron model of neutral hydrogen.

51. Schwarzschild 1914. For Bohr's interest, see Bohr to Schwarzschild, 23 February 1914, in Hoyer 1981, pp. 600–1.

52. Epstein 1916a (reprinted in Hermann 1965b, pp. 145–76) and Epstein 1916b. Schwarzschild 1916.

53. Epstein 1916b, p. 150 and similarly in Epstein 1916a, p. 520.

54. Darrigol 1992, pp. 104–18. See also Shore 2003 for the links between quantum theory and the Hamilton–Jacobi equation and other methods derived from celestial mechanics.

55. Einstein 1917a, reprinted in Kox et al. 1996, pp. 556–66. Bohr probably knew about Einstein's work, but he never referred to it, possibly because he found it irrelevant to his own work in atomic theory. For a full account, see Bergia and Navarro 2000. The relevance of Einstein's insight for a quantum theory of chaotic systems is discussed in Stone 2005.

56. Heisenberg to Pauli, 12 December 1922, in Hermann et al. 1979, p. 74, and similarly Heisenberg to Sommerfeld, 28 October 1922, in Eckert and Märker 2004, p. 131. The work studied so intensively in Göttingen was Henri Poincaré, *Les Méthodes Nouvelles de la Mécanique Céleste* (Paris: Gauthier-Villars, 1892–1893).

57. Born 1923, p. 538.

58. See, for example, Robotti and Pastorino 1998.

59. For the discovery and early history of the anomalous Zeeman effect, see Weaire and O'Connor 1987 and Mehra and Rechenberg 1982a, pp. 445–52.

60. Bohr 1914a, pp. 513–18. Herzfeld 1914. Bohr to Herzfeld, 2 March 1914, in Hoyer 1981, p. 526.

61. Debye 1916. Sommerfeld 1916a.

62. Sommerfeld 1922a, p. 374.

63. Sommerfeld to Bohr, 20 August 1916, in Eckert and Märker 2000, p. 565. The remark related to Bohr's suggestion of 1914 that the frequency condition needed modification to be able to accomodate the Zeeman effect.

64. Sommerfeld 1916a, p. 501. Sommerfeld to Ehrenfest, 16 November 1916, in Eckert and Märker 2000, p. 572.

65. For the Paschen–Back effect and its role in the old quantum theory, see Jensen 1984 and Robotti 1992, which include references to the primary literature. We shall return to the anomaly in Section 6.6.

66. Sommerfeld 1916b. For historical analysis, see Heilbron 1967.

67. Siegbahn 1925, p. 169. Sommerfeld 1922a, p. 610.

68. Bohr to Richardson, 3 March 1916, quoted in Heilbron 1967, p. 470.

69. Pais 1991, p. 185, finds it plausible that Bohr knew of the fine structure at the time he wrote the trilogy, but gives no reasons. I, on the other hand, find it most unlikely that Bohr, who was meticulous in citing relevant experimental knowledge, was aware of it. The fine structure was not common knowledge at the time and did not appear in, for example, Stark's *Prinzipien der Atomdynamik* (Stark 1911), which Bohr may have used in 1913. Nor was it mentioned in Christiansen's textbook in physics (Christiansen 1910). Moreover, none of the early spectroscopic critics of Bohr's theory referred to the fine structure as an argument against his theory, which I take as evidence that the doublet nature of the hydrogen spectrum was not well known and also not known to Bohr.

70. See Section 2.5 and Bohr's letter to Fowler of 28 April 1914, in Hoyer 1981, pp. 506–7. The published version of the hypothesis appeared in Bohr 1914a, p. 521.

71. Bohr to Rutherford, 20 May 1914, in Hoyer 1981, p. 593. He briefly mentioned the nuclear hypothesis in Bohr 1915b, p. 335. Five years later the Polish-Italian physicist Ludwik Silberstein proposed a nuclear fine-structure theory based on the hypothesis that the shape of the atomic nucleus is not spherical (Silberstein 1920a).

72. Bohr 1915b, p. 335.

73. Sommerfeld 1915b and, in more detail, Sommerfeld 1916c. All editions of *Atombau* contained an extensive treatment of the fine-structure problem, which in this way became widely known.

74. For these early considerations, see Kragh 2003b.

75. Paschen to Sommerfeld, in Eckert and Märker 2000, p. 559. The experiments on the fine structure of hydrogen and helium and their relations to Sommerfeld's theory, are detailed in Kragh 1985a and Robotti 1986.

76. Epstein to Stark, 11 December 1917, as quoted in Benz 1975, p. 97.

77. Einstein 1916. See also the comments in Kox et al. 1996, p. 147.

78. Einstein to Sommerfeld, 8 February 1916 and 3 August 1916, in Eckert and Märker 2000, p. 525 and p. 563.

79. Einstein to Born, 4 June 1919, in Born 1971, p. 10.

80. Planck 1922, p. 161.

81. Pauli 1958, p. 83, originally published as volume V 19 of *Encyclopädie der Mathematischen Wissenschaften* (Leipzig: Teubner, 1921). For further details on the historical relationship between the fine-structure theory and the special theory of relativity, see Kragh 1985.

82. Hansen 1925, who had already noticed the possibility of the forbidden transition the year before. See Robotti 1986.

83. Silberstein 1920b, p. 2.

84. Quoted in Aaserud 2001, p. 283.

85. For more on these objections, see Kragh 2011c.

86. Ehrenfest to Sommerfeld, April–May 1916, in Eckert and Märker 2000, p. 555. Emphases added. See also Klein 1970, p. 286.

87. Quoted in Holton 1970, p. 1037.

88. Brillouin 1919, p. 1319. This paper was followed by a couple of other papers in which the French physicist applied his theory to the spectrum of hydrogen. Brillouin's work influenced to some extent the thinking of Louis de Broglie in his development of a wave mechanics of atoms. Indeed, de Broglie later called Brillouin 'a veritable precursor of wave mechanics' (Kojima 2004, p. 774). See also Darrigol 1993, p. 327.

89. Szavassi 1918, p. 505.

90. Voigt to Sommerfeld, 11 February 1917, in Eckert and Märker 2000, p. 573.

91. Epstein to Stark, 11 December 1917, as quoted in Benz 1975, p. 97. The letter related to Stark's recent critique of Bohr's theory (Stark 1917).

92. Epstein 1918, p. 252.

93. Kramers to Bohr, 12 March 1917, in Rud Nielsen 1976, p. 654.

94. Kramers to Bohr, 27 January 1919, quoted in Robertson 1979, p. 51.

95. Sommerfeld 1924a, p. 1049, an address given at the meeting in Innsbruck of the Society of German Scientists and Physicians.

96. Gehrcke 1920, p. 172. For his earlier atomic model, see Gehrcke 1914a and Section 3.9.

97. Stark 1917, p. 114.

98. Kramers and Holst 1923, pp. 136–7, who stressed that atomic processes did not violate the principle of causality.

99. http://nobelprize.org/nobel_prizes/physics/laureates/1919/stark-lecture.html.

100. Stark to Bohr, 22 November 1919 (Archive for History of Quantum Physics, Bohr Scientific Correspondence).

101. Stark to Sommerfeld, 9 December 1920, in Eckert and Märker 2004, p. 89. For Stark's anti-Bohr article, see Stark 1920. Sommerfeld urged Stark to stop his polemics, maintaining that 'Paschen's experiments and my theory of them are certain beyond any doubt [*hieb- und stichfest*]'. Sommerfeld to Stark, 12 December 1920, ibid. p. 91.

102. Sommerfeld to Bohr, 18 February 1921, and Bohr to Sommerfeld, 22 February 1921, in Eckert and Märker 2004, p. 94.

103. Bohr 1923d, p. 294. This was the Seventh Guthrie Lecture, given to the Physical Society of London on 24 March 1922.

104. Sommerfeld 1921, p. 510 and p. 501.

105. Smekal 1921, a summary of a lecture given in Vienna on 7 March 1921.

106. Stark 1922, pp. 22–3.

107. Letter to Margarethe Sommerfeld, 24 September, in Eckert and Märker 2004, pp. 16–17.

108. Kossel 1916, p. 237. As indicated by its title, the aim of Kossel's memoir was to derive molecular structures from Bohr's theory. Although he succeeded in accounting for the ionic bond, his attempt to explain the covalent bond on the same basis was unsuccessful.

109. Sommerfeld to Bohr, 5 February 1919, in Eckert and Märker 2004, p. 47.

110. Compton 1917. Dated 19 June 1916, it was one of the first American papers that cited Bohr's atomic theory. Bohr's 1913c structure for calcium was 8, 8, 2, 2.

111. Debye 1917, p. 277. The method employed by Debye and other physicists in the same tradition of the ring atom is explained in Heilbron 1967 and Nisio 1969. For the role of X-ray spectroscopy in atomic models, see also Mehra and Rechenberg 1982a, pp. 319–32.

112. Kroo 1918.

113. Sommerfeld 1919, p. 265. He admitted that the results based on X-ray spectroscopy were 'more of a potential than actual significance'. For Sommerfeld's own and very important works in the area, see Sommerfeld 1918 and also Sommerfeld to Siegbahn, 4 June 1918, in Eckert and Märke 2000, pp. 597–9. See also the various editions of *Atombau*.

114. Vegard 1917.

115. Vegard to Bohr, 16 September 1918. Bohr Scientific Correspondence, Archive for History of Quantum Physics. In 1917 Bohr still thought in terms of the one-quantum atom, interpreting the L electron as an excited K electron and not as a constituent of the atom in its normal state (Heilbron 1967, p. 484).

116. Vegard 1918, p. 317.

117. Sommerfeld to Siegbahn, 8 May 1918, in Eckert and Märker 2000, p. 592. Sommerfeld 1918.

118. Sommerfeld to Landé, 17 March 1919, as quoted in Heilbron 1967, p. 478. Graetz (1918, p. 85) commented that Debye's structures were 'completely illusory'.

119. Vegard 1918 and Vegard 1919.

120. Vegard to Bohr, 16 September 1918, quoted in Heilbron 1967, p. 476.

121. Reiche and Smekal 1918.

122. Sommerfeld to Bohr, 18 May 1918, in Eckert and Märker 2000, p. 595.

123. Bohr to Richardson, 25 December 1919, as quoted in Heilbron 1967, p. 478.

124. Bohr 1920. English translation in Bohr 1922a, pp. 20–60, reproduced in Rud Nielsen 1976, pp. 241–82, quotation on p. 282.

125. Sommerfeld 1918, p. 297. As Sommerfeld mentioned, a similar idea had earlier been suggested by Nicholson in his critical analysis of the Bohr atom (Nicholson 1914f). It was also advocated by Vegard, who imagined that 'the elliptic axes are arranged radially and with equal angular intervals, and that . . . at any moment the electrons will be evenly distributed on the circumference of a circle, the radius of which undergoes periodic changes as time passes' (Vegard 1919, p. 259). Heilbron 1967 deals with the history of the *Ellipsenverein*.

126. Sommerfeld 1919, p. 368.

127. Born 1978, p. 183.

128. Landé 1919a and Landé 1920. See the description in Heilbron 1964, pp. 371–80.

129. Hull 1917, p. 87, who did not refer to Bohr's theory or other theories of atomic constitution.

130. Sommerfeld to Landé, 17 March 1919 and 3 September 1919, as quoted in Heilbron 1967, p. 479.

131. See Forman 1970, p. 165. The invitation came after Landé had himself suggested coming to Copenhagen to speak of his theory of the cubical atom. Landé to Bohr, 25 July 1920, in Rud Nielsen 1977, p. 720.

132. Bohr 1921c, p. 106. On this important paper, see Section 7.1.

133. Lewis 1916. Lewis 1923a, reprinted 1966 in a Dover Book edition. For historical details, see Kohler 1971. Stranges 1982 refers to other suggestions of cubical atoms in the period. While Lewis in his further work kept to the static atom and the shared electron pair hypothesis, he abandoned the cubic arrangement of the electrons.

134. Langmuir 1919, p. 872. See also Kohler 1974. Langmuir cited the X-ray investigations of Hull, who was his colleague at the General Electric laboratory, as evidence for his theory. For more on Langmuir's ideas of the constitution of atoms, see Section 6.2.

135. See Kragh 2009, which includes references to the literature.

136. Bohr to Rutherford, 29 November 1916, in Hoyer 1981, p. 595.

137. Kossel and Sommerfeld 1919. As they noted, the law has a formal similarity to the radioactive displacement law dating from 1912. See also Sommerfeld 1922a, pp. 456–66.

138. Bohr to Sommerfeld, 27 July 1919, in Rud Nielsen 1976, p. 689. See further Bohr's Copenhagen address of December 1920 (Rud Nielsen 1977, p. 53).

139. Werner's experiments were only completed in 1925 (Werner 1925), at a time when German and American physicists had already published results on the lithium spark spectrum. As expected, they found two term systems corresponding to orthohelium and parhelium.

140. Landé 1919b. For the details of Landé's helium theory and further references to the literature, see Heilbron 1964, pp. 376–80, and Mehra and Rechenberg 1982a, pp. 399–405.

141. Sommerfeld 1919, p. 70 and p. 290.

142. Franck and Reiche 1920, which was criticized in Kemble 1921.

143. Langmuir 1921c.

144. Van Vleck 1922, p. 847.

145. Unpublished manuscript on 'Constitution of Atoms', Section 2: 'Helium'. Reproduced in Rud Nielsen 1977, pp. 99–174. See also Darrigol 1992, pp. 161–5.

146. Kramers 1923a, p. 341. See also Dresden 1987, pp. 119–22.

5

A Magic Wand

During the years from 1917 to 1921 Bohr was not only busy with his research on atomic theory but also with preparations and plans for a new institute devoted to fundamental physics at the University of Copenhagen. In his proposal to the Faculty of Science of April 1917 he stressed how research contributions to the new quantum theory of matter and radiation necessitated an interaction between theory and experiment closer than in traditional physics. Theory alone would not do. One is now faced, he said, 'with having one or another consequence of the theory tested experimentally, before the choice between the various possibilities presented can be made, and therefore it is as mentioned necessary that the practitioners of the subject have the opportunity to carry out and guide scientific experiments in direct connection with the theoretical investigations'.[1] The institute came to carry the name *Universitetets Institut for Teoretisk Fysik*, meaning The University's Institute for Theoretical Physics, with the term 'theoretical' to be understood as 'fundamental' rather than opposite to experimental. Financed by a combination of public and private money, including a substantial grant from the Carlsberg Foundation, the institute—'Bohr's Institute' as it was known from the very beginning—was officially inaugurated on 3 March 1921. (It changed its name to *Niels Bohr Instituttet* only in 1965, three years after Bohr's death.)

Apart from Bohr, the professor and director of the institute, its first staff included Kramers, the Danish experimentalists Jens C. G. Jacobsen and Hans M. Hansen, and James Franck, who had come from Göttingen to help getting the instruments working. They were joined by Hevesy, whom Bohr had known since their stay in Manchester with Rutherford, and also by the young Swedish physicist Oskar Klein, who first had met Bohr in May 1918. Trained in physical chemistry as a student of the eminent Swedish chemist Svante Arrhenius, under the guidance of Kramers Klein swiftly changed to theoretical physics and became one of Bohr's close associates. Svein Rosseland from Norway was another young Scandinavian in the earliest phase of Bohr's institute and another of his assistants who would evolve into a leading scientist, in his case in astrophysics. In a letter of support for Bohr's application to the Carlsberg Foundation, Sommerfeld wrote in 1919: 'The Institute of Mr Bohr should not only serve the up and coming generation of Denmark, it will also be an international place of work for foreign talent whose own countries are no longer in a position to make available the golden freedom of scientific work'.[2] And this is what happened. Sommerfeld was primarily thinking of Germany, and during the early phase of Bohr's institute it was indeed young German physicists who mostly benefited from it and contributed to its growing status in atomic and quantum physics.

If there is one word that characterizes Bohr's scientific work in the period it is the 'correspondence principle', which he first enunciated in an important work of 1918 and continued to develop and apply over the following years. This famous principle or analogy between quantum theory and classical physics not only became the beacon that guided Bohr's own research, it was also a method that dominated much of quantum theory until the emergence of quantum mechanics in the autumn of 1925. Although it was to some extent also adopted by German and other foreign physicists, it was a speciality of Bohr and his collaborators, of whom Kramers was particularly an adept. If Bohr were the philosopher and architect of the correspondence principle—and he was—Kramers was its computer. The apparently intuitive way Bohr used the principle and the consequences he drew from it was not matched and rarely fully understood outside Copenhagen. Physicists ocassionally referred to it as Bohr's 'magic wand', sometimes as an expression of admiration and at other times with a critical connotation. Sommerfeld, whose voice was of the greatest importance in the period, belonged to the latter group. Without referring specifically to the correspondence principle, Einstein thought that Bohr approached the problems of atomic physics 'emotionally'. Whether logically justified or not, under the guidance of the correspondence principle quantum physics made considerable progress in the early 1920s and paved the way for the kind of non-mechanical and unvisualizable mathematical model of the atom that Heisenberg first presented in a historic paper received by *Zeitschrift für Physik* on 29 August 1925. Heisenberg's pioneering work was as enigmatic as it was fruitful, and this was not its only debt to the correspondence principle.

5.1 QUANTUM THEORY OF LINE SPECTRA

Although Bohr published nothing in the years 1916 and 1917, he was far from idle or willing to leave his brainchild—the quantum theory of atoms—in the hands of the productive and innovative German physicists. He found it necessary to provide the theory with a consistent and more satisfactory basis, and this he worked hard to do. Only at the end of 1917 was he able to complete the first part of this basis, which appeared in three sequels, the first two in 1918 and the third belatedly in 1922. Published in the transactions (*Skrifter*) of the Royal Danish Academy of Sciences, the common title of the lengthy and complex treatise was 'On the Quantum Theory of Line-Spectra'. At the unusually young age of 32, Bohr had recently been elected a member of the prestigious society, which at the time comprised only 68 national members (in addition there were 124 foreign members). In 1939 he became the society's president, an office he held until his death in 1962. He would also publish many of his later papers on quantum, atomic and nuclear physics in the transactions of the academy rather than submitting them to the normal physics journals, and he would induce his assistants and collaborators to follow the same publication strategy. The decision to use a local publication series undoubtedly implied a more limited circulation in the physics community, but Bohr had his reasons, as he told Rutherford at the end of 1917:

> This last term I have used all my spare time to complete a long paper on the general principles of the quantum theory. On account, however, of the great difficulties of postal communication I have thought it advicable [sic] first to publish it here in the transactions of the Royal Danish Academy, where I could publish it in English, and where I had the special advantage to be able to start the printing before the end of the long paper was quite finished . . . As soon as I have got a corrected proof of it all I look forward to send it to you. I hope very much that the Phil. Mag. will publish it later although it has first appeared here.[3]

On the other hand, the disadvantage was that far from all physicists had access to the paper. In Bohr's correspondence from the period there are several letters from physicists requesting him to send them reprints of the paper, which they understood was important but had not seen. One of them was Erwin Schrödinger, recently appointed professor at the University of Breslau. 'May I add rather an impudent prayer?' he wrote in February 1921. 'Your two important communications in the Danish Akademy from 1918 are so very difficult to procure for us in Germany. At Breslau we have but *one* copy, I know of, that is the one of Mr Ladenburg. Would it be possible to you, to lett me have a copy?'[4] Bohr sent him the requested reprint.

In his letter to Rutherford of December 1917, Bohr wrote about his forthcoming work, which he evidently considered a breakthrough in the rational understanding of the quantum atom. 'It seems now really possible to a certain degree to overlook the theory with all its different applications from a uniform point of view', he wrote, briefly referring to what had made this possible, namely, a new 'analogy between this [quantum] theory and the ordinary theory of electrodynamics'. By using this analogy—the correspondence principle—it had proved possible 'to obtain an immediate interpretation of the apparently capricious law which governs the intensities and polarisations of the large number of components into which each of the hydrogen lines are split up in the presence of an electric field, and of which hitherto no explanation has been offered'. Bohr felt he was back on track: 'At the present I am myself most optimistic as regards the future of the theory'.

Although intended as a single treatise, in reality 'On the Quantum Theory of Line-Spectra' consisted of three separate papers of which the third one dealt mainly with the spectra of higher elements. While written in the spring of 1918, when this part was eventually published it included an appendix in which Bohr discussed more recent developments, theoretical as well as experimental. Originally he had planned the treatise to have four parts, where the final one was to cover the theory of 'the constitution of atoms and molecules' (just as in the 1913 trilogy), but this part remained unpublished. He only wrote some fragmentary drafts that mainly dealt with the hydrogen molecule, dispersion in gases, and molecular spectra; that is, the same subjects he had treated on earlier occasions.[5] Generally there was a great deal of repetition in Bohr's publications from the period.

While the trilogy of 1913 was clearly structured and fairly easy to follow, the new trilogy of 1918–1922 was anything but. It was, as Bohr said with understatement to Ehrenfest, 'somewhat unmanageable'.[6] Both the structure and style of the memoir—Bohr did not bother to split up sections into smaller bits or to limit the length of his sentences—made it something of a nightmare for the reader. Important as it was, it was definitely not a lucid work. On the other hand, it expressed what Bohr wanted to express and is a good example of what Pauli once called 'the subtlety of Bohr's style'. Part of this subtlety was that Bohr 'knew well what he wished *not* to say when he strove in long sentences to express himself in his scientific papers'.[7]

Thus, in his original formulation of the first postulate Bohr referred to the validity of ordinary mechanics in the stationary states of an atomic system (Section 2.4). It is instructive to compare this formulation with the one he presented in the beginning of his 1918 memoir, where classical mechanics no longer appears in the postulate but is relegated to separate discussion. His carefully worded formulation was this: 'That an atomic system can, and can only, exist permanently in a certain series of states corresponding to a discontinuous series of values for its energy, and that consequently any change of the system, including emission and absorption of electromagnetic radiation, must take place by a complete transition between two such states. These states will be denoted as the 'stationary states' of the system'.[8]

The first part of the treatise dealt primarily with the general quantum theory and its application to systems of one degree of freedom and to the so-called *multiply-periodic systems* first discussed in quantum theory by Epstein and Schwarzschild. The motion of a system of this kind is compounded of a number of motions, each of which separately is periodic, but which have different periods. The coordinates of such dynamical systems can be resolved in Fourier series, and the systems return asymptotically to their initial configurations after a long time. For this reason they were called 'conditionally periodic' (*bedingt periodisch*) in the German literature, a name that Bohr adopted in his Copenhagen memoir.[9] The elliptically revolving electron in a hydrogen atom represents a simple multiply-periodic system with two periods, one being the period of revolution and the other the slow rotation of the major axis of the ellipse. The important point is that physicists agreed that quantum theory only applied to motions that are multiply periodic in terms of separated coordinates.

The second part, subtitled 'On the hydrogen spectrum', was the longest and most important, as it was here he applied the correspondence principle introduced in the first part (although not yet by that name). In this second part Bohr dealt at length with the general mechanical theory of perturbations, which he innovatively combined with the correspondence principle in such a way that it provided a means for deriving the spectrum of an atomic system perturbed by, for example, an external electric field.

The theory of perturbations and similar methods borrowed from celestial mechanics had earlier been introduced into quantum theory by Epstein and Schwarzschild, but Bohr's method was more direct and empirically promising.[10]

Bohr never found it easy to put his thoughts on paper. Writing the memoir on his new approach to atomic theory was a most demanding task, which took many rewritings and nearly exhausted his strength. This is what he told Richardson in a letter from the summer of 1918 in which he described the state of quantum theory as a 'terrible riddle':

> I know that you understand how things happen and how my life from the scientific point of view passes of[f] in periods of overhappiness and despair, of feeling vigorous and overworked, of starting papers and not getting them published, because all the time I am gradually changing my views about this terrible riddle which the quantum theory is. In the last years things have in this respect of unsuccessful literary occupation come to a climax (I hope so at any rate) and I have been rewriting and rewriting again the paper on the principles of the theory of which we spoke already in England.[11]

While some scientists first think and then write, for Bohr the writing process was an integral part of his cognitive process. It was a process he used to obtain as much clarity as possible, and for this reason it was a difficult one. Heisenberg vividly described Bohr's tortuous way of producing a paper:

> Bohr would always change the sentences again and again. He could have filled half a page with a few sentences and then everything was crossed out and changed again. And even when the whole paper was almost finished – say, ten pages or so – the next day everything would be changed over again. So it was a continuous process of improvement, change and discussions with others... The final text of Bohr's paper was so subtle and he would think about half an hour whether in a certain case he would use indikativ or the konjunktiv and so on.[12]

In the period from about 1917 to 1922 Bohr was repeatedly overworked and occasionally stressed, not only because of teaching obligations and the responsibilities of planning his new institute but also because of his slow and methodical style of working. As he confided to Sommerfeld, 'I suffer so much from the difficulties of getting the papers in a satisfactory form, and from an unfortunate inclination to make all results appear in a systematic order'.[13]

As soon as the first part of his Copenhagen memoir appeared in print, namely in late April 1918, Bohr sent copies of it to several of his colleagues, among them Epstein, Schrödinger, Debye, Sommerfeld, Ehrenfest, and the London physicist William E. Williams. Among those who requested and received copies was also the Swedish mathematician Ivar Fredholm, professor of mechanics and mathematical physics in Stockholm, who was known in particular for his work on integral equations. Fredholm at the time took an interest in atomic and quantum theory.[14]

Although Sommerfeld was rather reserved (see Section 5.4), the initial response was generally positive. Epstein found the new correspondence approach to be 'very convincing' and of 'amazing simplicity', as he wrote in a letter to Bohr of May 1918. 'It actually seems that the discrepancy between quantum and classical treatments is not at all so great as has heretofore been assumed', he commented. Debye, too, expressed great interest in the memoir: 'I exaggerate not in the least when I say that it has aroused universal interest. Especially your rule for computing intensities is evidently of the greatest importance'.[15] In spite of such enthusiastic responses, Bohr himself came to feel that his new series of papers in atomic theory had not been properly understood and appreciated. He was frustrated that they had not had a greater impact, as he told Sommerfeld in a letter of 1922—the same year that he was awarded the Nobel Prize for his contributions to atomic theory:

> I have often felt myself scientifically very lonesome, under the impression that my efforts to develop the principles of the quantum theory systematically to the best of my ability have been received with very little understanding. For me, it is not a matter of didactic trifles but of a serious attempt to achieve such an inner coherence that one can attain a secure basis for the further development. I understand quite well how little the matters are clarified as yet, and how helpless I am at expressing my thoughts in easily accessible form.[16]

In his paper of 1918 Bohr stressed even more strongly than before that the transition between stationary states could not be understood on the basis of either classical mechanics or classical electrodynamics. To fix the stationary states of an atomic system one would have to specify the energy difference between the states, or correspondingly the frequencies according to the quantum rule $E' - E'' = h\nu$, and this presented what he called a 'fundamental difficulty'. The difficulty was that 'we have assumed that the direct transition between two such states cannot be described by ordinary mechanics, while on the other hand we possess no means of defining an energy difference between two states if there exists no possibility for a continuous mechanical connection between them'.[17]

To solve the problem, or rather to bypass it, he made use of his own version of Ehrenfest's adiabatic principle or what he called the principle of mechanical transformability of the stationary states, which allowed him to 'transform mechanically the stationary states of a given system into those of another'. Starting from the simple case of the harmonic oscillator he argued that if the forces acting on the system were changed very slowly, the system would gradually evolve into some other system rather than exhibit discontinuous quantum transitions. The invariants that belonged to the former system would keep the same values for the later one. As he pointed out, this was an interpretation of the principle not to be found in Ehrenfest's paper of 1914 but still one that 'might in a more direct way indicate the content of the principle and the limits of its applicability'.[18] Ehrenfest, who had originally disliked the new atomic theory, came to recognize the advantage of Bohr's understanding of the adiabatic principle. In the 1923 special issue of *Die Naturwissenschaften* dedicated to Bohr and his atomic theory, he said that 'in the hands of Bohr it [the adiabatic principle] has become a wonderfully sharp and flexible instrument' with applications he had not himself thought of. At the time a Bohr devotee, he doubted that it was possible to give 'a deeper and more concise formulation of the adiabatic principle than the one Bohr offers'.[19]

Apart from relying on an innovative use of Ehrenfest's principle, Bohr's new formulation of his theory was directly inspired by Einstein's recent theory of the emission and absorption of radiation (see Section 5.2). The use of Ehrenfest's and Einstein's ideas enabled him to arrive at what he called 'a rational basis for the determination of the a-priori probability of the different stationary states of a given atomic system'.[20] This rational basis was what would soon be known as the correspondence principle, and it made it possible 'to throw some light on the outstanding difficulties by trying to trace the analogy between the quantum theory and the ordinary theory of radiation as closely as possible'.[21] By following this route he would not arrive at a complete theory, but he was confident that he would extend the theory progressively and increase its explanatory power. While in his earlier work Bohr had wavered somewhat with regard to the number of postulates forming the basis of his theory—recall that in 1915 he operated with six postulates or basic assumptions—from 1918 onwards he concluded that there were only two really fundamental postulates, the one being the postulate of stationary states and the other the frequency condition. As to the other assumptions he considered them to be provisional and of a more limited significance.

Apart from his great but difficult work in the transactions of the Royal Danish Academy, in the period 1918–1922 Bohr published several other papers in which he discussed his correspondence approach to atomic theory and applied it to a variety of problems. For example, on 27 April 1920 he gave a lecture in Berlin to the German Physical Society, which was subsequently published under the title 'Über die Serienspektra der Elemente'.[22] As mentioned above, in Berlin he met many of the leading German physicists, including Planck, Born, Landé, Haber, and Einstein, and he also had the opportunity to discuss his ideas with the younger people at a *bonzenfreie* (bigwig-free) colloquium. In this meeting, with partipants including, among others, Lise Meitner, James Franck, and Gustav Hertz, full professors were not allowed.[23]

Bohr was also invited to give a report to the third Solvay congress which would convene in Brussels in the first week of April 1921. Although unable to participate because of poor health, he prepared the requested report dealing with applications of quantum theory to atomic problems. In his absence it was presented by Ehrenfest who, acting as his emissary, also gave a brief talk on the correspondence principle.[24] At the Solvay meeting Bohr's atomic theory appeared in quite a different context, namely in Heike Kamerlingh Onnes's address on superconductivity and its possible explanation in terms of the Bohr–Rutherford model of the atom. According to Kamerlingh Onnes superconductivity was clearly a non-classical phenomenon that could be understood, if understood at all, on the basis of Bohr's quantum atom. Unfortunately, neither he nor

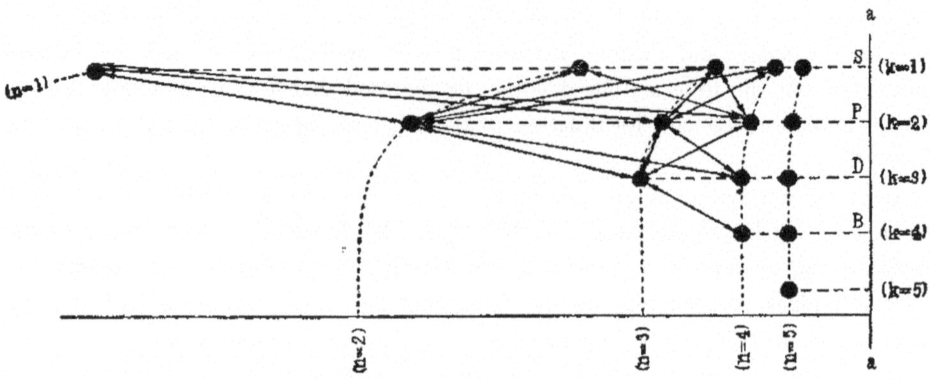

Fig. 5.1. Bohr's diagram of the series spectrum of sodium, as used in his lecture in Berlin in April 1920.

Source: Bohr 1922a, p. 30.

others were able to tell how. 'When the atoms of the Rutherford–Bohr model combine to form a metal, what becomes of their peripheral electrons? Do they lose all or part of their kinetic energy?' he asked, unable to provide an answer.[25]

The lecture Bohr gave in Berlin in April 1920 is of interest not only because he used the term 'correspondence principle' for the first time, but also because he introduced a new graphical representation of 'the apparent capriciousness in the occurrence of lines predicted by the combination principle'.[26] Spectroscopists had earlier used graphical representations of frequencies or wave numbers to visualize the bewildering regularities of spectra, but Bohr's diagram, as he illustrated it with the series spectrum of sodium, was novel by relating to the frequencies of the spectral *terms* and not to the observed lines (Figure 5.1). By connecting with lines the energy levels between which transitions could take place, he presented a bridge between atomic theory and spectroscopic data. A few months later Walter Grotrian, at the Astrophysical Observatory in Potsdam, derived from Ritz's combination principle the energy levels involved in the spectral lines of neon, representing his results in a similar graphical way and acknowledging Bohr's priority.[27] This kind of diagram became widely used in spectroscopy, and is known as either a 'Grotrian diagram' or sometimes a 'Bohr–Grotrian diagram'.

5.2 THE CORRESPONDENCE PRINCIPLE

In a general sense, correspondence arguments had been known many years before Bohr elevated them to a formal principle of correspondence in quantum theory.[28] For example, it was recognized that the equations of Einstein's special theory of relativity, or, for that matter, those of Lorentz's classical electron theory, had to contain the equations of classical mechanics in the limit $v/c \rightarrow 0$ or, formally, $c \rightarrow \infty$. Again, the field equations of general relativity must yield Newton's law of gravitation for weak gravitational fields. Likewise, in the limit $h \rightarrow 0$ the results of quantum theory must converge toward those obtained from classical physics, as was the case with Planck's radiation law in relation to the Rayleigh–Jeans law based on classical considerations. It was known that for small values of $h\nu/kT$ the Planck law would pass over into the classical radiation law.

However, the elementary recognition that the results of classical and quantum physics converge in the limit $h \rightarrow 0$ (or $h\nu \rightarrow 0$), sometimes known as 'Planck's correspondence principle', does not capture what the correspondence principle, properly understood, is about. Léon Rosenfeld recalled that he once suggested to Bohr that this was the first germ of the correspondence principle, but that Bohr disagreed: 'He said emphatically that "it is not the correspondence argument. The requirement that the quantum theory should go over to the classical description for low modes of frequency, is not at all a principle. It is an obvious requirement for the theory"'.[29] A better candidate for a pre-Bohr correspondence principle may be the reasoning applied by Planck in deriving the blackbody radiation law from his second quantum theory in 1911. According to Max Jammer, Planck's argument 'was probably the earliest instance in quantum theory of applying what more than ten years later became known as the "correspondence principle"'.[30]

The way that Bohr first applied correspondence arguments was rather different, as he argued that when energy intervals $\Delta E = h\Delta\nu$ between stationary states became arbitrarily small the emitted frequencies would coincide with the frequencies expected from classical theory. He used such reasoning in 1913 as one of several ways to derive the energy levels of the hydrogen atom (Section 2.4), and the following year he reasoned similarly in his analysis of the Stark effect (Section 3.9). Correspondence arguments of this kind also appeared in the manuscript he withdrew from *Philosophical Magazine* of 1916. Bohr pointed out that the frequency of vibration would not be the same for two successive stationary states of a system, and for this reason alone one could not expect a simple relation between the mechanical frequency ω and the radiation frequency ν. Such a relation, he said, 'can only be expected in the limit where n is very great and the ratio between the two frequencies of vibration in the two states differs only very little from unity'.[31] In this case n can be considered a continuous variable and Bohr assumed, without any proper justification, that the equation $dE/dn = h\omega$ would hold for any value of n. For a simple harmonic oscillator, as given by the linear equation $x(t) = a\cos(\omega t)$, a direct application of the frequency condition would lead to

$$\nu = \frac{E' - E''}{h} = (n' - n'')\omega. \tag{1}$$

Thus it yields not only the ground frequency $\nu = \omega$ but also the higher frequencies $\nu = 2\omega, 3\omega, \ldots$, in disagreement with Planck's fundamental assumption. According to Bohr, the analogy to classical electrodynamics would exclude the overtones or higher harmonics and thus restore the agreement. Bohr's limited correspondence arguments in 1916 and in his earlier work were concerned with frequencies only and he did not extend them to a general formal analogy between quantum theory and classical electrodynamics. He provided such an extension in his paper of 1918 and on several later occasions, for example in his lecture of 1920 to the German Physical Society and in his Solvay address of the following year.

In his 1918 memoir Bohr amplified his earlier remarks about frequencies, firstly relating them to a simple periodic atomic system of one degree of freedom. In classical physics the displacement x of a charged particle in such a system—which might be an electron revolving circularly around an atomic nucleus—can be resolved in a trigonometric series of the form

$$x(t) = \sum_{\kappa} C_{\kappa} \cos 2\pi(\kappa\omega t + c_{\kappa}), \quad \kappa = 1, 2, 3, \ldots \tag{2}$$

where C_{τ} and c_{τ} are constants. The system will emit a series of spectral lines corresponding to the ground frequency ω and its higher harmonics 2ω, 3ω, etc., and it will emit all the lines at the same time. Moreover, the intensity of the lines are given by the square of the amplitudes, $I \sim |C_{\tau}|^2$. As Bohr pointed out, in quantum theory the picture is very different, for here the emitted frequencies do not reflect in any direct way the motion of the particle and they do not occur simultaneously. All the same, 'in the limit where n is large there exists a close relation between the ordinary theory of radiation and the theory of spectra'. This insight, sometimes called Bohr's correspondence theorem, should be distinguished from his correspondence principle or postulate, which concerns the intensities and polarizations of the emitted lines. Bohr stated the principle as follows:

> We must further claim that a relation, as that just proved for the frequencies [equation 1], will, in the limit of large n, hold also for the intensities of the different lines in the spectrum. Since now on ordinary electrodynamics the intensities of the radiations corresponding to different values of τ are directly determined from the coefficients C_{τ} in [equation 2], we must therefore expect that for large values of n these coefficients will on the quantum theory determine the probability of spontaneous transition from a given stationary state for which $n = n'$ to a neighbouring state for which $n = n'' = n' - \tau$. . . . We may expect that also for small values of n the amplitude of the harmonic vibrations corresponding to a given value of τ will in some way give a measure for the probability of a transition between two states for which $n' - n''$ is equal to τ.[32]

According to Bohr, then, an electron in an orbit characterized by its periodic properties would have a certain probability of passing to another stationary state. In this probabilistic sense its final destination would be determined, and that without ascribing to the electron any 'free will'. The correspondence principle can be seen as a kind of reply to the frequently raised objection that an electron in a certain excited state would need to 'know' the state to which it was destined to pass (Section 4.5).

Importantly, Bohr realized that his principle or analogy might function as a selection principle. In continuation of the arguments in the quotation above he said: 'Thus in general there will be a certain probability of an atomic system in a stationary state to pass spontaneously to any other state of smaller energy, but if for all motions of a given system the coefficients C in [equation 2] are zero for certain values of τ, we are led to expect that no transition will be possible, for which $n' - n''$ is equal to one of these values'. Applied to the possible electronic orbits in an atom he phrased the correspondence selection principle as follows: 'Among the processes that are conceivable and that according to the quantum theory might occur in the atom, we shall reject those whose occurrence cannot be regarded as consistent with a correspondence of the required nature'.[33] While Bohr in 1918 did not speak of the correspondence principle, but instead of a 'formal analogy', he did so in his Berlin lecture of 1920, where he introduced it as 'a far-reaching *correspondence* between the various types of possible transitions between the stationary states on the one hand and the various harmonic components of the motion on the other hand'.[34]

Neither in his 1918 memoir nor in his Berlin lecture did Bohr enunciate a proper definition of the correspondence principle. The closest he came to such a definition was perhaps in a paper of 1921, in which he said of the correspondence principle that it grew out of 'endeavors to attain a simple asymptotic agreement between the spectrum and the motion of an atomic system in the limiting region in which the stationary states differ only little from one another'. The principle, he continued, assumes that 'the occurrence of any transition process connected with emission or absorption of radiation depends on the presence in the motion of the system of certain corresponding harmonic components of vibration'.[35]

It was an important part of Bohr's correspondence principle, conceived as a method of calculation, that he could associate the amplitude coefficients C_τ with the radiation probabilities recently introduced by Einstein in a series of three important papers from 1916–1917 (written while he was developing his new general theory of relativity into a theory of cosmology!). Although Einstein had known about Bohr's atomic theory since 1913, it was only in connection with his radiation theory that it became important to him and also only then that he first referred to it in public: 'Now that Bohr's theory of spectra has obtained its great results it seems beyond doubt that the basic ideas of quantum theory must be retained'.[36] The element of Einstein's theory that attracted Bohr's attention was his idea that the emission of radiation occurred spontaneously and probabilistically, thereby apparently defying the laws of classical physics. Einstein considered a system (or 'molecule', as he called it) that might exist in a quantum state n and a higher excited state m. For the probability that the system decays from m to n in the time interval dt, and thereby emits radiation of frequency ν, he wrote

$$dW = A_{mn}dt, \tag{3}$$

where the inverse of the probability coefficient A_{mn} signifies the lifetime of the system in the excited state. Indeed, it follows that the number of atoms in the state m will diminish in the same way as radioactive atoms do according to the Rutherford–Soddy decay law, namely as

$$N_m(t) = N_m(0)\,exp\,(-A_{mn}t).$$

The similarity of the emission of light to radioactive processes was explicitly noted by Einstein, who on the one hand found that it supported his theory. But on the other hand, it also left 'the duration and direction of the elementary processes to "chance"', and this he felt uneasy about.[37] Whatever the issue of chance or indeterminism, he showed in a very general way that his theory led to Planck's radiation law and also, by a comparison with Wien's displacement law, to the relation $E_m - E_n = h\nu$. This, he commented, 'is, of course, the second main hypothesis of Bohr's theory of spectra of which we can now state after Sommerfeld's and Epstein's extensions that it belongs to those parts of our science which are certain'.[38]

Einstein believed his derivation of Planck's blackbody formula lent strong support to the existence of the light quanta (or photons) that he had first proposed in 1905. 'There is no emission in spherical waves', he concluded in 1917, a view which was diametrically opposed to the one held by Bohr, who adamantly resisted Einstein's light-quantum hypothesis. This was not only because because it introduced 'insuperable difficulties' when applied to the phenomena of interference, but also because it excluded in principle a 'rational definition' of the conception of a frequency. His conclusion: 'The hypothesis of light-quanta, therefore, is not suitable for giving a picture of the processes, in which the whole of the phenomena can be arranged, which are considered in the application of the quantum theory'.[39] Bohr would continue to battle the light quantum in the years to come, in large measure because he felt it contradicted the basis of the correspondence principle. In view of his later theory involving merely statistical conservation of energy and momentum (the BKS theory of 1924, see Section 8.3), it is of interest to note that as early as 1917 or 1918 he was ready to consider violation of strict energy conservation in relation to the photoelectric effect. In a manuscript on 'Principles of the Quantum Theory' he wrote:

> It would seem that any theory capable of an explanation of the photoelectric effect as well as the interference phenomena must involve a departure from the ordinary theorem of conservation of energy as regards the interaction between radiation and matter. This view has been expressed by several authors and an interesting attempt to build up a theory on this basis has been made by Nernst.[40]

In a lecture to the Royal Danish Academy of Sciences of February 1920 Bohr said of Einstein's theory that it 'looks very much like Newton's, and it can no more than that give any sort of explanation of the interference phenomena'.[41] While Bohr rejected the light-quantum hypothesis he found it more appealing that, as he understood Einstein, an atomic system 'will start *spontaneously* to pass to the stationary state of smaller energy',[42] and also that Einstein's theory directly supported his frequency condition of 1913. While Einstein almost certainly did not identify the 'spontaneous' emission of radiation with an uncaused emission, Bohr's view was more radical. In 1922 he was ready to entertain an acausal interpretation of the radiative emission process:

> According to the postulates of the quantum theory, we not only abandon any such immediate connection between the motion of the atom and the result of the process of emission or absorption, but we are even compelled to depart so far from the ordinary descriptions of nature as to assume that the result of such a process depends actually on the final, as well as the initial state . . . We do not seek a *cause* for the occurrence of radiative processes, but we simply assume that they are governed by *the laws of probability*.[43]

It is relevant to recall that Stark and other critics had questioned how the electron in an excited state could 'know' its final state and what caused it to make the transition to this particular state (Section 4.5). Bohr's view, as stated in the quotation, provided a kind of answer to the question, but it was hardly an answer that his critics would have found satisfactory. At any rate, he cautiously avoided equating 'spontaneous' with 'acausal'. Defining radioactive and radiative spontaneous processes as such processes that occur 'without any assignable external stimulation', he left room for causes internal or not yet known.[44]

The connection that Bohr saw between the correspondence principle and Einstein's probability coefficients of equation 3 was that the latter could be expressed by the Fourier coefficients or amplitudes appearing in equation 2. Including numerical coefficients the connection can, for the component τ, be written in the form

$$A_{mn} \cong \frac{16\pi^4 e^2}{3c^3} \nu^3 C_\kappa^2,$$

where $\tau = m - n$. In this way it should be possible to obtain the intensity of spectral lines, irrespective of the values m and n of the quantum states from which they had their origin. Bohr indicated in a general and qualitative way how this could be done, in particular in relation to the lines of the Stark effect and the Zeeman effect, but did not engage in the complicated calculations necessary for obtaining quantitative results that could be compared with experiments. Such calculations were first made by Kramers in his dissertation of 1919, a document that will be considered below.

In the second part of his memoir, Bohr indicated that arguments based on the correspondence principle implied that the stationary states of an atom are not sharply defined ($\Delta E = 0$) but have an energy that will vary over a small interval ($E \pm \Delta E$). As he made clear in a later work, from this it follows that discrete spectral lines must have a natural line width that is independent of the external conditions that cause, for example, the temperature-dependent Doppler broadening of the lines due to the motion of atoms or molecules.[45] In his lectures in Göttingen in 1921 he explained that from a correspondence point of view one would have to take into account the radiation in the stationary states and expect that the duration of the emission process was of the same order of magnitude as that calculated by classical theory. 'In view of the fact that the values of E' and E'' are not determined exactly, because of our neglect of this radiation, the question arises as to the accuracy of the equation $E' - E'' = h\nu$', he said. 'If we ask ourselves what we are to understand by ν, we must assume that in reality there exists no strictly monochromatic radiation'.[46] Thus Bohr's correspondence principle led to a conceptual change in the very foundation of his original atomic theory, the two fundamental postulates.

In the years following 1918 Bohr continued to reflect on the correspondence principle and formulate it in different versions, not all of which were very clear or, for that matter, mutually consistent.[47] Nevertheless, logic and clarity apart, they were empirically fruitful. Bohr realized that the use of classical conceptions in formulating a law of quantum theory was problematic from a logical point of view, and that the originally used expression for the correspondence principle—'a formal analogy between the quantum theory and the classical theory'—might not be appropriate. In a paper completed in the autumn of 1922 he remarked that 'Such an expression might cause misunderstanding, since, in fact . . . the Correspondence Principle must be regarded purely as a law of the quantum theory, which can in no way diminish the contrast between the postulates and electrodynamic theory'. This was an important point, in fact so important to him that one page later he felt it necessary to repeat and sharpen the message:

> In the limiting region of large quantum numbers there is in no wise a question of a gradual diminution of the difference between the description by the quantum theory of the phenomena of radiation and the ideas of classical electrodynamics, but only of an asymptotic agreement of the statistical results.[48]

When Bohr sometimes referred to the principle as a 'law', as he did in the quoted remark, it was because he considered it as a selection principle that holds universally, irrespective of the value of the quantum number. Yet he also insisted that quantum theory and classical physics were irreconcilable and for this reason he tended to conceive the correspondence principle, which after all bridges the two in a certain sense, as primarily of *heuristic* value. In the concluding part of his lecture in Berlin he said that 'in spite of the fundamental differences between these [quantum] points of view and the ordinary conceptions of the phenomena of radiation, it still appears possible on the basis of the general correspondence between the spectrum and the motion in the atom to employ these conceptions in a certain sense as guides in the investigation of the spectra'.[49]

It is worth emphasizing that some of Bohr's later statements concerning the correspondence principle did not reflect the way he thought about it in the early 1920s, before the emergence of quantum mechanics. Some of these later statements were clearly influenced by the new mechanics and his complementarity interpretation of it in which the correspondence between the quantum region and the classical region was an important element. For example, in his Como lecture of 1927, in which he first introduced the complementarity principle, he spoke of 'a farreaching correspondence between the consequences of the classical theory and those of the quantum theory'.[50] He sometimes seemed to believe that the correspondence principle was an element integrated in the complementarity principle; even a derivative of it. However, this was obviously not how he had conceived the principle of correspondence just a few years earlier, before the era of quantum mechanics. Although Bohr's understanding of the principle changed with the emergence of quantum mechanics, there is no doubt that the correspondence principle shaped his interpretation of this theory and was very important for his 1927 formulation of the complementarity principle.

It was only after the Great War that Einstein and Bohr got an opportunity to meet personally, an encounter that took place when Bohr gave his lecture to the German Physical Society in the spring of 1920. The meeting between the two great physicists seems to have been successful. Shortly after Bohr's return to Copenhagen, Einstein penned his first letter to him:

> The magnificent gift from the neutral world, where milk and honey still flow, gives me a welcome occasion to write to you. Not often in life has a person, by his mere presence, given me such a joy as you. I understand now why Ehrenfest is so fond of you. I am now studying your great papers, and in doing so – when I get stuck somewhere – I have the pleasure of seeing your youthful face before me, smiling and explaining. I have learned much from you, especially also how you approach scientific matters emotionally . . . I am looking forward to our talks in Copenhagen.[51]

Invited by the Danish Astronomical Society, Einstein used the occasion of a visit to Oslo to visit Copenhagen on his way back to Berlin. In Copenhagen he gave a lecture on the theory of relativity on 25 June 1920 and spent much of his time in conversation with Bohr. About his travel he said in a letter to Lorentz: 'The journey to Kristiania [Oslo] was really beautiful, but the most beautiful was however the time I spent with Bohr in Copenhagen. He is a highly intelligent and excellent man'.[52] Later the same year Einstein met with Bohr in Berlin, where Bohr told him about his new ideas concerning the atomic structure of the chemical elements. Einstein was impressed by the advances Bohr had made. 'His intuition is much to admire', he said to Sommerfeld.[53] Much later, in his autobiographical notes written in 1946, Einstein characterized Bohr's atomic theory as a 'miracle' and an expression of 'the highest form of musicality in the sphere of thought'. It was, he said, due to Bohr's 'unique instinct and tact' that he had been able 'to discover the major laws of the spectral lines and of the electron-shells of the atoms together with their significance for chemistry'.[54]

5.3 SELECTION AND INTENSITY RULES

In the popular book of 1923 coauthored by Helge Holst, Kramers wrote about the correspondence principle that 'It is difficult to explain in what it consists, because it cannot be expressed in exact quantitative laws, and it is, on this account, also difficult to apply'.[55] Nonetheless, it could be and was in fact applied as a tool of theoretical physics for a variety of purposes, and Kramers himself was an acknowledged master of the art of making mathematical sense of the correspondence philosophy.

In 1918, when completing his great work on the quantum theory of line spectra, Bohr wrote to Ehrenfest about his former student Kramers that 'he is at present on his own account preparing for publication two papers: on the intensity of the components of the fine structure and the Stark effect of the hydrogen lines, and on the simultaneous effect on these lines of the relativity modifications and of an external electric field'.[56] Bohr suggested the mentioned topics as a subject for Kramers' doctoral dissertation, which in 1919 appeared in the same journal as Bohr's memoir: the transactions of the Royal Danish Academy. In the second part of his memoir Bohr referred to Kramers, 'who on my proposal has kindly undertaken to examine the resolution of the motion of the electron in its constituent harmonic vibrations more closely'.[57] The dissertation of 1919 was the first detailed and quantitatively successful application of the correspondence principle and it was also the work that established Kramers' position in the physics community. He received his doctorate from the University of Leiden in the spring of 1919, in the presence of Bohr, among others.

Kramers' dissertation rested solidly on Bohr's theory and the 'intimate formal connection' that Bohr had shown to exist between the quantum theory and classical electrodynamics. Only in a paper of the following year, an extension of the work conducted for his thesis, did Kramers refer to Bohr's *Korrespondenzprinzip*, possibly the first time that the term was used by someone other than Bohr himself.[58] Following his master and idol, he argued in the dissertation that not only would the formal connection hold for the frequencies, but 'also the intensities and polarisations of the spectral lines emitted in the region of large *n*'s will asymptotically be the same as the intensities and polarisations of the corresponding lines which on ordinary electrodynamics would be emitted by the atom'.[59]

In the calculations of a transition between two stationary states m and n by means of a Fourier resolution of the classical orbit it is not obvious which of the orbits to use. Should it be the orbital frequencies ω of the initial state or those of the final one? For large values of the quantum numbers, where the two orbits are infinitesimally close, it does not matter, but it does matter in the case of small quantum numbers. Picking up on an idea of Bohr's, Kramers suggest considering (in the case of one dimension) not only the states given by the action integrals $J_{\mathrm{m}} = mh$ and $J_{\mathrm{n}} = nh$, but also a sequence of states given by

$$J(\lambda) = [n + \lambda(m - n)]h,$$

where λ is a parameter that varies continuously between 0 and 1. He could then express the radiation frequency as the average over the continuum of 'intermediate states' between the two discrete lines: 'The frequency ν of the radiation emitted during the transition under consideration is equal to the mean value, taken over all states from $\lambda = 0$ to $\lambda = 1$, of the frequency $(n_1' - n_1'')\omega_1 + \ldots (n_s' - n_s'')\omega_1$ which appears in the motion of the electron when this motion... is resolved in its constituent harmonic components'.[60] The intermediate states had a classical meaning only, but by using them Kramers could write the Bohr frequency of the emitted line as

$$\nu = \frac{1}{h}\int_{\lambda=0}^{\lambda=1} \omega(\lambda)dJ = \int_0^1 (m - n)\omega(\lambda)d\lambda,$$

where $\omega(\lambda)$ denotes the vibrational or orbital frequencies of the infinite number of intermediate states. The equation, a version of the correspondence theorem for frequencies, is valid regardless the magnitude of the quantum numbers and their changes. As Kramers pointed out, for large quantum numbers $(m - n)\omega$ can be considered a constant and therefore placed outside the integration sign, 'so that the frequency ν of the emitted radiation approaches asymptotically to the frequency... present in the motion of the system'. The result is the ordinary correspondence formula

$$\nu = (m - n)\omega.$$

Kramers used a similar averaging method in his use of the correspondence principle for computing intensities, which he did by considering the Einstein *A* coefficients as an average of the *C* amplitudes appearing in the classical Fourier series. In his dissertation he investigated in particular the components of the Stark spectrum and the fine-structure components, in both cases for hydrogen and helium. He also included a section on the lines produced by the normal Zeeman effect. It will be possible, he said, 'to form an idea of the relative intensities with which the different components of the Stark effect will appear, by comparing the intensity of each component with the values of the squares of the amplitudes of the corresponding harmonic vibrations occurring in the motion of the system in the initial state and in the final state and in the mechanically possible states lying "between" these states'.[61]

Carrying out this programme, which required very extensive calculations, Kramers arrived at theoretical values for the relative intensities of the Stark lines that he modestly described as 'convincing'. In fact the agreement between theory and experiment was nearly perfect, which evidently supported Bohr's correspondence principle and its idea of connecting the *C* amplitudes with the Einstein *A* coefficients (Figure 5.2). In a review published for the National Research Council, the Princeton physicist Edwin

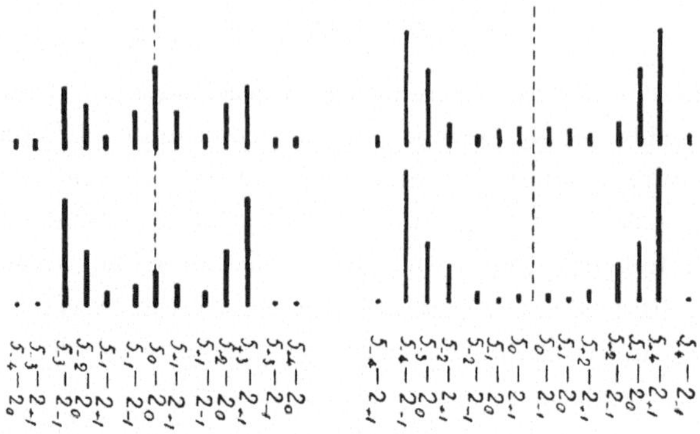

Fig. 5.2. The Stark effect of the H_γ line in hydrogen as displayed by Bohr in his Nobel lecture of 1922. The two pictures above are schematic representations of Stark's experimental results for the components observed parallel (left) and perpendicular (right) to the direction of the electric field. The height of the lines is a measure of their intensity and the dotted line in the middle indicates the H_γ line in the absence of an electric field. The pictures below show the theoretical results as calculated by Kramers.

Source: Bohr 1923b, p. 38.

Adams included a detailed account of Kramers' 'valuable memoir', which he praised as a most important contribution to quantum and atomic theory. In the view of Adams, the results obtained by Kramers 'afford most convincing evidence for the value of the principles which Bohr has introduced in the quantum theory'.[62] Kramers' theory not only accounted for all known facts of the Stark effect, it also led to predictions of the relative intensities of the components of the weak H_ϵ line corresponding to a quantum transition from $n = 7$ to $n = 2$. These components were not known at the time, but later experiments confirmed that they agreed with Kramers' calculations.[63]

As another important case Kramers investigated the components of the fine-structure lines. As described in Section 4.4, in 1916 Sommerfeld had successfully accounted for the experimentally known fine structure of the Balmer lines in the hydrogen spectrum by extending Bohr's atomic theory into the domain of special relativity theory. His theory was in excellent agreement with Paschen's spectroscopic measurements, but only by introducing somewhat arbitrary assumptions about the intensities of the lines. By 1919, in the first edition of *Atombau*, Sommerfeld had incorporated in his theory the work of Adalbert Rubinowicz (see p. 208) and thereby produced a picture of the fine structure of the H_α line where the observed doublet was the composite result of three adjacent spectral lines. Sommerfeld's way of estimating the number and intensity of the lines lacked a solid theoretical foundation, for which reason Kramers made alternative calculations based on the methods of the correspondence principle. Moreover, he investigated in detail the effect of an electric field as it existed in the tubes of Paschen's experiments, concluding that as the field strength increased the fine-structure splitting would gradually change into a pattern dominated by the Stark effect.[64]

According to Kramers, the predicted doublet separation in the hydrogen spectrum, as usual in terms of the wave number, would be $\Delta H_\alpha = 0.328\ cm^{-1}$ in the absence of an electric field and $0.341\ cm^{-1}$ in the presence of a field of strength 100 V/cm. He examined in particular the lines at 4686 Å and 3203 Å in the He^+ spectrum, originating in the transitions $(n = 4) \rightarrow (n = 3)$ and $(n = 5) \rightarrow (n = 3)$, respectively, and obtained a very good agreement with Paschen's data. The agreement relied to a large extent on Kramers' analysis of the action of the perturbative electric field, which provided his theory with sufficient flexibility to also cover the results of other measurements.

In the third edition of *Atombau* Sommerfeld adopted Kramers' correspondence-based analysis, concluding that now the fine structure of one-electron atoms was perfectly understood: 'The exemplary agreement with experiment, which Kramers has obtained in particular in the case of Paschen's He^+ observations and also in the case of the Stark effect . . . confirms convincingly the fertility of Bohr's correspondence principle'.[65] Kramers' calculations explained the lines observed by Paschen, including the IIId component due to the $(4, 1) \rightarrow (3, 1)$ transition, which has $\Delta k = 0$ but could be explained as an effect of an external electric field. However, according to the same reasoning there should also appear a fairly intensive component (IIIb) due to the transition $(4, 3) \rightarrow (3, 1)$, and this component was in fact missing in the experiments. In principle, the missing IIIb component constituted a problem for Sommerfeld's (and Kramers') theory.

Strangely, neither Sommerfeld, Kramers nor other experts in the fine-structure theory pointed out that the missing line was inconsistent with Paschen's measurements. The anomaly of the missing line went unnoticed for more than five years until it was brought to light in the summer of 1925 by Kramers' two compatriots Samuel Goudsmit and George Uhlenbeck, of spin fame.[66] In their new classification scheme, IIId came out naturally and there was no room for the IIIb component. Some of Kramers' results were revised and extended by the American physicist Raymond Birge, at the University of California, who clarified the complex relationship between experiment and theory in fine-structure spectroscopy. Concluding that the Bohr–Kramers theory was most satisfactory, Birge argued that Paschen's measurements lent support to Bohr's 'selective principle' based on the idea that 'the actual orbit of the electron, in the initial and in the final state, can be considered as the sum of a number of harmonic vibrations'.[67]

The notion of selection rules, or what Birge called selective principles, originated in Munich, not in Copenhagen, and principally with the works of Sommerfeld, Epstein, and Rubinowicz.[68] In connection with his theory of the two-quantum relativistic hydrogen atom Sommerfeld realized the necessity of an *Aufwahlprinzip* that would reduce the number of possible transitions between stationary states, but at first he could only suggest rules of an ad hoc character. The situation changed in the spring of 1918, when Adalbert Rubinowicz, a young Polish physicist working in Sommerfeld's group in Munich, used complex arguments based on the conservation of energy and angular momentum to argue that in radiative transitions from one state to another the azimuthal quantum number k could at most change by one unit. In other words, allowed transitions were restricted to

$$\Delta k = 0 \text{ and } \Delta k = \pm 1.$$

Rubinowicz associated the first case with a linear polarization of the emitted light and the second case with circular polarization. In *Atombau* Sommerfeld praised Rubinowicz's derivation as being obtained by 'a remarkably rigorous manner of deduction, reminiscent of the incontrovertible logic of numerical inferences'.[69] In this respect, and because Rubinowicz's rules had a solid basis in fundamental physics, they were superior to Sommerfeld's earlier *Quantenungleichungen*. Admitting as much, Sommerfeld reported the good news to Bohr:

> By comparing energy and angular momentum, Rubinowicz finds a condition for the azimuthal quantum number: it can at most change by one unit (0, ± 1). In this way the superfluous components in the Zeeman effect are cut off and my 'quantization equations' justified. You arrive at the same results by your interesting comparison between classical and quantum-theoretical emission for high quantum numbers. Your method is perhaps more far-reaching, but the view indicated above seems physically more instructive to me.[70]

Bohr found Rubinowicz's deduction interesting and referred to it in a footnote added to the second part of his 1918 memoir, but he did not share either Sommerfeld's enthusiasm or his view that it was 'physically more instructive' than his own. Based on his analogy or correspondence principle he had independently derived the same result, although with the difference that transitions with $\Delta k = 0$ were not generally allowed (Figure 5.3). In his Berlin lecture of 1920 he contrasted the Munich method with the Copenhagen correspondence method as follows:

Fig. 5.3. Bohr lecturing in the USA 1923, possibly one of the Silliman lectures he gave at Yale University between 6 and 13 November. Based on the correspondence principle, he demonstrates the selection rule for the azimuthal quantum number to be $\Delta k = \pm 1$.

Courtesy of the Niels Bohr Archive, Copenhagen.

> The conclusions which we can draw by means of the correspondence principle are of a more detailed character than can be obtained solely from a consideration of the conservation of angular momentum. For example, in the case of the hydrogen atom perturbed by a central force we can only conclude that *k* can not change by more than unity, while the correspondence principle requires that *k* shall vary by unity for every possible transition and that its value cannot remain unchanged. Further, this principle . . . also enables us to draw conclusions about the relative probabilities of the various possible types of transitions from the values of the amplitudes of the harmonic components.[71]

On the basis of the correspondence principle Bohr could also account for the changes in the so-called inner quantum number *j* that Sommerfeld had introduced phenomenologically in 1920 in addition to the quantum numbers *n* and *k*. The inner quantum number described multiplet structures in spectroscopy—that is, sublevels of the (*n*, *k*) orbits—but had no clear physical meaning except that it was assumed to arise from 'a complication of the motion of the outer electron due to a small departure from central symmetry of the inner system, which causes the plane of the orbit of the outer electron to undergo a slow precession round an axis coinciding with the axis of angular momentum of the atom'.[72] The selection rule for the *j* quantum number was found, both in Munich and in Copenhagen, to be $\Delta j = \pm 1, 0$.

Spectroscopy proved that Bohr's selection rule for the azimuthal quantum number gave better results than the one of Rubinowicz and Sommerfeld, although in practice the difference was of little importance. The reason was that although $\Delta k = 0$ was allowed according to Rubinowicz's theory, he and Sommerfeld argued that such transitions would in reality be inactive.[73] Whether in Rubinowicz's or Bohr's version, the selection rule seemed to be violated by certain lines that appeared in the helium spectrum and also in the spectrum of potassium. In 1922 three American spectroscopists at the National Bureau of Standards, Washington DC, argued that their experiments showed transitions of $\Delta k = 2$ and that the result could not be explained by a perturbing effect due to the electric field. When confronted with the apparent anomaly Bohr suggested a way of avoiding it, namely that the formation of ions would explain it as an electric perturbation effect. To his mind the experiments did not pose a threat to the correspondence principle. As he saw it, they 'may not be considered as presenting the theory with serious difficulties, but rather as providing a means of investigating the conditions under which the spectra are emitted'.[74]

5.4 ATTITUDES TO THE CORRESPONDENCE PRINCIPLE

As indicated above, while Kramers and Ehrenfest were early and enthusiastic converts to the new programme in quantum theory based on the correspondence principle, Sommerfeld's response was more restrained. It was not that he opposed the principle and Bohr's use of it, but rather that he was insensitive to it and felt that it was not of a sufficiently fundamental nature. He thought it rested on a rather shaky foundation and much preferred one that was rigorous and mathematically closed in the traditional sense he valued. While admitting the correspondence principle as heuristically useful—he could not possibly deny that—Sommerfeld never became convinced that it could serve as a reliable foundation of quantum theory. Indeed, in the first edition of *Atombau* he characterized it as an analogy principle that was 'somewhat foreign' to this theory and grafted on it rather than arising from it in a natural manner.[75] When he described it as a 'magic wand' it was not an expression of enthusiasm but rather of guarded distrust. It just was not Sommerfeld's style of physics. Contrasting his own and Rubinowicz's approach to that of Bohr, he wrote that 'Bohr has found a magic wand in his analogy principle (he himself regards it as a "formal" principle), which without clearing up the conceptual difficulties allows him to make the results of the classical wave theory directly useful for the quantum theory'.[76]

The magic wand metaphor was used by some other physicists as well, either at the time or in retrospective writings. In an article of 1935 in the Danish journal *Fysisk Tidsskrift*, Kramers expressed how many of his colleagues perceived Bohr's use of the principle: 'In the beginning the correspondence principle appeared to the physicists as a somewhat mystical magic wand, which did not act outside Copenhagen'.[77]

In a letter to Bohr of 11 November 1920, Sommerfeld expressed his sustained uneasiness about the correspondence principle:

> In the appendixes to my book you can see that I have gone to some pains to treat your correspondence principle better than in the first edition . . . Nonetheless, I must confess that your principle, the origin of which is foreign to the quantum theory, is still a source of distress to me, however much I recognize that through it a most important connection between quantum theory and classical electrodynamics is revealed.[78]

Bohr's atomic theory was openly and deliberately somewhat ill-defined in a sense that did not square with Sommerfeld's ideas of what a fundamental theory of physics should look like. Rather than proving the consistency of his theory in the usual formal-mathematical way, Bohr argued for it heuristically: he emphasized that the many applications of the theory provided sufficient reason to believe that it was pragmatically consistent. Speaking of the 'correspondence viewpoint' he said in 1921 that it had 'proved fruitful in ever new realms of application, without thereby bringing us a step closer to a solution of the difficulties touched upon; at least only in so far as, with each extension of the application of the quantum theory, we perceive the nature of this riddle in ever clearer light'. He then went on: 'This is connected with the circumstance that this viewpoint is not at all a closed formal one, but may rather be regarded as a description of certain general features of the radiation processes'.[79]

In spite of his reservations Sommerfeld could not help being impressed by the results that Bohr, the undisciplined Copenhagen magician, had produced by means of his magic wand. In a couple of papers from 1922, coauthored with his bright young student

Heisenberg, he himself applied Bohr's principle to solve problems of spectral physics. The third edition of *Atombau* contained a full and basically sympathetic account of the correspondence principle, which he now praised as a superior and most fertile method of physics. Yet Sommerfeld's praise was largely confined to the undeniable heuristic power of the principle.

> Bohr's method is not only of greater consequence [than the one of Rubinowicz] in the question of intensity, but also leads to sharper and more definite results as regards the question of polarization. In the matter of method the principle of correspondence has the great advantage that it postulates that Maxwell's theory be generally valid for long waves (Hertzian oscillations of wireless telegraphy), and that it does not throw overboard the many useful results, which the classical theory gives for optical waves and Röntgen rays. In the matter of atomic models we must now recognize the complete superiority of the correspondence principle. For here it seems that Bohr, by applying classical mechanics and electrodynamics to obtain definite statements about the periodic system and the atomic shells, has succeeded to advance where other roads have been inaccessible.[80]

Although Sommerfeld never fully came to terms with the correspondence principle, he learned to appreciate it as an indispensable approach, if not the only approach, to the problems of quantum and atomic theory. In a paper of 1924 read to the 88th meeting of the Society of German Scientists and Physicians he expressed once again his sympathetic yet critical attitude: 'The magic of the correspondence principle has proved itself generally through the selection rules of the quantum numbers and in the series and band spectra', he admitted. But then: 'Nonetheless I cannot view it as ultimately satisfying on account of its mixing of quantum-theoretical and classical viewpoints'.[81] He made it clear that the ideal to strive toward was an axiomatic-deductive theory of the type known from classical mechanics and the theory of general relativity. Quantum theory might not be rife for such a formulation, but this he did not consider an excuse for not developing it in this direction. And Bohr's correspondence approach was not, after all, a step in this desirable direction.

Upon receiving the fourth edition of *Atombau* from Sommerfeld, Bohr politely expressed his admiration for the work but also mentioned some of his reservations with regard to specific issues. Realizing that his own work and general attitude to atomic physics had a reputation of being 'philosophical', even deliberately obscure, he closed the letter by stressing that he was no less interested in clarity and empirical data than the Munich professor: 'I should not like you to get the impression that my inclination to pursue obscure, and undoubtedly often false analogies makes me blind to that part of the formation of our conceptions that lies in the unveiling of the systematic of the facts. Even if I were blind, I would only need to glance at your book to be healed'.[82]

Apart from discussing the correspondence principle in his 1920 Berlin lecture and (in absentia) at the Solvay meeting of the following year, Bohr also spread the gospel in a series of lectures in Göttingen of 1922 and included it in his Nobel lecture later the same year. Both in Göttingen and in Stockholm the constitution of higher atoms in terms of quantum levels was the main subject, as will be discussed in Chapter 7. As Bohr saw it at the time, the correspondence principle played an important role in the building up of atomic structures and the explanation of the periodic system of the elements based on these structures.

Invited by Franck and Born to lecture in Göttingen, in June 1922 Bohr gave a series of seven lectures sponsored by the Wolfkehl Foundation. In what informally became known as the Bohr festival (*Bohr Festspiele*) he exposed his ideas to a large group of physicists, many of them young and coming leaders of the quantum mechanics that still belonged to the future (Figure 5.4). Some of them had attended his lecture in Berlin two years earlier, but most had not. The audience in Göttingen included experienced physicists and mathematicians such as Franck, Born, Hilbert, Sommerfeld, and Runge,

Fig. 5.4. The 'Bohr festival' in Göttingen, June 1923. Max Born is sitting on the chair, and behind him stand, from the left, Carl Wilhelm Oseen, Niels Bohr, James Franck, and Oskar Klein. Courtesy of the Niels Bohr Archive, Copenhagen.

but Friedrich Hund, Werner Heisenberg, Wolfgang Pauli, Pascual Jordan, and other younger practitioners also listened to his lectures. Heisenberg, who was then 21 years old, later spoke of Bohr's 'long stream of thoughts, of which only the beginning was clear, and whose end disappeared in the darkness of a philosophical attitude that was very stimulating to me'.[83] He came to believe, at least in retrospect, that 'Bohr was primarily a philosopher, not a physicist'.[84]

The correspondence principle and its manifold uses occupied much of the first three lectures Bohr gave in Göttingen, where he basically covered the same ground as in his Solvay address and on other occasions. Referring to the connection between the classical amplitudes C_τ and Einstein's probability coefficients A_{mn} for quantum transitions, he said that 'This raises the serious question of whether we must rest content with statements of probabilities for individual processes'. His tentative answer was that 'As matters stand at present, we are so far from being able to give a real description of these processes that we may well assume that Einstein's mode of treatment may actually be the most appropriate'. He refrained from advocating a stronger version of acausality in the subatomic domain. Bohr did not offer a strict definition of the correspondence principle to his audience in Göttingen, but he did offer an illustration:

> We may regard H_β as the octave of H_α, since H_β corresponds to a quantum jump of 2 and H_α to a quantum jump of 1. It is true that H_β does not have twice the frequency of H_α, but it corresponds to the octave. This relationship we call the the 'correspondence principle'. To each transition there corresponds a harmonic component of the mechanical motion.[85]

After having dealt in some detail with the Stark effect and the fine-structure spectrum, in the third lecture he pointed out that the transition of the fine structure to the usual Stark effect by a gradual increase of the electric field, such as investigated theoretically by Kramers, still lacked experimental confirmation. But this did not worry him at all, for he was confident that future experiments would verify the predictions:

> The quantum theory yields very many details of the phenomenon to be expected. Even if we really should not be unprepared to find that the quantum theory is false, it would surprise us very much if such a detailed picture obtained from the quantum theory should not be valid; for our belief in the formal reality of the quantum conditions is so strong that we should wonder very much if experiments were to give a different answer than what is demanded by the theory.[86]

The correspondence principle and its wide range of uses was a contributing cause why Bohr was awarded the Nobel Prize for 1922. It featured prominently in Svante Arrhenius's presentation speech, where the Swedish physical chemist (himself a Nobel laureate of 1903) emphasized the value of Bohr's principle and its role in 'the determination of the tracks of electrons which are possible within atoms that are heavier than the atom of hydrogen'. Arrhenius illustrated the use of the correspondence principle by referring to the structure of the helium atom:

> It occurs in two different modifications: one is called parhelium, and the other is called ortho-helium – these were supposed at first to be different substances. The principle of correspondence states that the two electrons in parhelium in the tracks of rest [stationary orbits] run along two circles, which form an angle of 60° to one another. In ortho-helium, on the other hand, the tracks of the two electrons lie in the same plane, the one being circular, while the other is elliptical.[87]

In his Nobel lecture delivered in Stockholm on 11 December 1922 Bohr gave a popular account of the correspondence principle, emphasizing its empirical power (Figure 5.2). Rather than focusing on helium, he paid particular attention to Kramers' calculations as proof that 'the theory reproduces completely the main feature of the experimental results, and in the light of the correspondence principle we can say that the Stark effect reflects down to the smallest details the action of the electric field on the orbit of the electron in the hydrogen atom'. In addition, he referred to how fruitful the correspondence approach had proved in the areas of atomic and molecular spectroscopy.[88]

Several physicists developed the Bohr–Kramers approach in order to use it for electrons moving in orbits around a nucleus and thereby to calculate the intensity of spectral lines. Whereas Bohr and Kramers had not themselves worked out harmonic representations of electron orbits, this is what Frank Hoyt, a young American physicist working at Bohr's institute, attempted to do. Impressed by what he called 'the uniform success of Bohr's correspondence principle', Hoyt applied it to the intensities of X-ray lines and the Balmer lines of the hydrogen spectrum.[89] In a series of papers from 1923–1925 he calculated the *C* amplitude coefficients of electron orbits, including the so-called penetrating orbits corresponding to $\Delta k = 0$, and tried to estimate transition probabilities or relative intensities of the resulting lines. This he did by averaging over the intensities that would be emitted by the intermediate states on the basis of classical theory. Hoyt thus tried to work out quantitatively what Bohr had grasped more intuitively, but despite laborious calculations the results were meagre. They varied from one averaging method to another, and none of the methods agreed even approximately with the experimentally determined intensities.[90]

Richard Tolman, professor of physical chemistry and mathematical physics at the recently established California Institute of Technology, was not all that impressed by Bohr's atomic theory. Yet he followed the same approach as Hoyt, only focusing on the much simpler system of the original Bohr hydrogen atom. He found by one method of averaging a zero probability that an electron in its ground state would fall into the nucleus, while the other methods led to an infinite probability.[91] Although Tolman speculated that his result might have astrophysical consequences, neither his nor Hoyt's work produced results of any scientific significance. They were soon recognized as blind alleys and in retrospect they merely served to illustrate the limitations of the correspondence principle. Bohr assured Hoyt that 'the problem with which you have been occupied lately is one which will turn out to be fruitful for the discussion of the origine [sic] of spectra', but it seems that he had little confidence in the Hoyt–Tolman approach.[92] For the sake of completeness it should be noted that the use of the correspondence principle was not limited to spectra and orbital electrons. In a few cases it was also applied to the atomic nucleus, as Svein Rosseland did in his quantum theory of radioactivity, dating from 1923.[93] Characteristically, Rosseland worked at Bohr's institute.

Although the correspondence principle belonged to the semiclassical Bohr–Sommerfeld theory it has been said, not without justification, that 'there was rarely in the history a comprehensive theory which owed so much to one principle as quantum mechanics owed to Bohr's correspondence principle'.[94] Physicists in the era before quantum mechanics did not know that, of course, but they generally appreciated the value of the principle and considered it an important part of the quantum theory of atoms. Almost all comprehensive reviews of quantum theory, whether published as articles or monographs, contained careful descriptions of the correspondence principle and its role in atomic theory. This was the case with Sommerfeld's *Atombau* and it was also the case with Born's *Atommechanik* based on a series of lectures in Göttingen in the winter semester of 1923–1924.[95]

In France, Bohr's atomic theory was carefully reviewed by Léon Brillouin, who in a monograph of 1922 included a discussion of the correspondence principle and its applications. Together with the related adiabatic principle, he considered the correspondence principle to be 'a guide of extraordinary fecundity'.[96] His compatriot Louis de Broglie agreed, although not without a stint of criticism. In his important doctoral thesis of 1924, he wrote: 'This idea of correspondence, which seems still imprecise and rather elastic, will have to guide the bold researchers who wish to set up a new electromagnetic theory that agrees better with quantum phenomena than the present theory does'.[97] According to de Broglie, the correspondence theorem of frequencies could be interpreted in accordance with his pet hypothesis of phase waves associated with the electron orbits.

While neither Sommerfeld, Born, nor Brillouin expressed much enthusiasm about Bohr's 'rather elastic' principle, John Van Vleck presented it as a cornerstone of quantum physics in his extensive report on *Quantum Principles and Line Spectra*, which was completed in August 1925 but only published the following year. Van Vleck dealt in great detail with the correspondence principle in its various versions, including what he called the 'principle of continuous formation'. According to this principle, 'the atom must be capable of formation by the gradual shrinking of the orbit of the captured electron without abrupt or complicated alterations in the essential character of the motion'.[98]

Pauli's book-length review of quantum theory in *Handbuch der Physik* appeared, like Van Vleck's, only in 1926, at a time when the new quantum mechanics had made much of the older quantum theory obsolete. In his careful account of the correspondence principle, Pauli remained faithful to Bohr's ideas. For example, he emphasized with Bohr that the correspondence principle was 'purely a principle of the quantum theory' and that the asymptotic agreement between quantum and classical theory in the limit of high quantum numbers in no way implied 'a diminution of the contrast' between the two theories.[99] Moreover, he left no doubt that the correspondence principle provided a more general and satisfactory basis for quantum theory than other alternatives: 'As regards its physical content', he said, 'the correspondence principle does not, contrary to the adiabatic principle, depend directly on the validity of classical mechanics for the motion of an atomic system in stationary states; it has proved fruitful also in numerous applications of more complicated atoms, where the quantum theory of periodical systems obviously fails. From this one may perhaps hope that the further development of quantum theory leads to a generalization and quantitative sharpening of the correspondence principle'.[100]

While Pauli was generally in sympathy with the correspondence principle, he was quite critical of Bohr's use of it to establish electron configurations for the chemical elements and thereby to account for the structure of the periodic system (Section 7.6). He was similarly critical with regard to the possibilities of calculating the intensity of spectral lines on the basis of Bohr's principle. In a letter to Heisenberg of early 1925 he referred ironically to the *Imperialismus des Korrespondenzprinzips* (the 'imperialism of the correspondence principle').[101]

Specialists in quantum theory disagreed over whether or not the imperialism of the correspondence principle extended to the realm of the intensities of multiplet components, a branch of spectroscopy that attracted great attention in the early 1920s. The Dutch physicists Leonard Ornstein, Herman Burger, and Hendrik Dorgelo established in a series of papers several empirical rules for the intensities of multiplet components, for example that the sum of the intensities of the components of a multiplet associated with the same initial state was proportional to the value of the inner quantum number j of the state. The question was whether this and other rules could be explained from the quantum theory of the atom.[102] More specifically, could they be accounted for on the basis of Bohr's correspondence principle? In a detailed review of the problem in the fourth edition of *Atombau* Sommerfeld concluded that this was not the case, or perhaps that is was only possible by introducing auxiliary hypotheses of an artificial nature. 'The problem does not lie in the areas of analytical mechanics and classical electrodynamics, but in the arithmetical and discontinuous quantum theory', he wrote. 'It is just from the experiences with the intensity questions that we base our conviction that the correspondence principle cannot be the final formulation of the quantum problems'.[103]

Heisenberg begged to differ from his esteemed professor. In his earlier development of the so-called core model of the atom, he relied strongly on the correspondence principle, which he rated very highly. In September 1921 he thought that his core model was sanctioned by Bohr's principle, the virtues of which he extolled in a letter to Landé: 'The fact that the entire model interpretation of a process can be calculated from purely empirical material is another brilliant achievement of the Bohr correspondence principle, which I am beginning to want to consider *as important as the entire quantum theory*'.[104] Three years later, after having arrived in Copenhagen for a stay at Bohr's institute, he wrote to Pauli about the multiplet problem:

> Together with Bohr I have once more carefully considered the question and we have arrived at the conclusion that, against Sommerfeld's opinion, the sum rules do not elude understanding in terms of the correspondence principle; on the contrary, they are *strict consequences* of the correspondence principle and really the most beautiful example that the correspondence principle sometimes permits the drawing of unambiguous conclusions.[105]

With Bohr he had found a way to account for the intensities of the multiplet lines, and 'We are very happy with this interpretation, for now the attacks on the correspondence principle are entirely refuted'. Pauli was not convinced and tended to side with Sommerfeld, which made Heisenberg make the somewhat trivial comment that the entire question of course depended on what one meant with the correspondence principle:

> If by correspondence principle one means, as you do, the false statement that one can come from averaging the classical intensities to those of quantum theory, then *you* are right that one *cannot* come to Ornstein's rule by means of the correspondence principle; however, if one understands a meaningful logical connection to the classical theory, then *I* am right.[106]

Clearly, the correspondence principle could be and in fact was understood in different ways. In the spring of 1925 Heisenberg succeeded in justifying the Ornstein–Burger–Dorgelo rules theoretically, which was a triumph of the old quantum theory and its foundation in the correspondence principle. The triumph forced Heisenberg to sharpen the arguments based on the correspondence principle and thus contributed to the process that a few months later led him to quantum mechanics.

5.5 CORRESPONDENCE ARGUMENTS IN QUANTUM MECHANICS

It is generally agreed that the correspondence principle and arguments related to it played a crucial role in the formation of the new quantum mechanics that appeared in the late summer of 1925. Among the important steps on the road were papers on dispersion and radiation theory by Rudolf Ladenburg, John Slater, Kramers, Van Vleck, Bohr, and Heisenberg himself (Section 8.4). All this work relied on or was connected significantly with the correspondence principle. For example, in 1924 Van Vleck published in the *Physical Review* a detailed analysis of the radiation problem. It was in two parts, the first of which, entitled 'Some Extensions of the Correspondence Principle', included a correspondence theory of absorption. Van Vleck's paper has been called 'the most sophisticated application of the correspondence principle in the old quantum theory'.[107] Another of the important papers on dispersion theory, written jointly by Kramers and Heisenberg and published in March 1925, took as its starting point 'Bohr's correspondence principle [which] gives us a significant hint on how to describe the reaction of the atom to the radiation field, using classical concepts'.[108]

The role of the correspondence principle in the formation of quantum mechanics is undisputed. However, it is often assumed that in late 1925, after the new mechanics had secured a satisfactory formulation with the matrix-mechanical theory of Born, Heisenberg, and Jordan, and independently with Paul Dirac's alternative theory, there was no longer any need for it. After all, now it could be replaced by the much more satisfactory arguments based directly on the fundamental theory of quantum mechanics. Yet Bohr did not think that the correspondence principle had become obsolete because of the quantum revolution. According to him the principle survived the revolution, living on in the new theory of Heisenberg and his associates.

In a semi-historical review of developments in quantum and atomic theory from the late autumn of 1925, Bohr dealt with the correspondence principle—which 'expresses the tendency to utilise in the systematic development of the quantum theory every feature of the classical theories in a rational transcription appropriate to the fundamental contrast between the [quantum] postulates and the classical theories'—not as were it a tool of the past but as one that belonged to the present. Referring to Heisenberg's new theory, he stated that 'the whole apparatus of the quantum mechanics can be regarded as a precise formulation of the tendencies embodied in the correspondence principle'.[109] In other words, Bohr considered the principle to be alive and well because it had been fully incorporated into quantum mechanics. At about the same time Born, Heisenberg, and Jordan completed a comprehensive work that established the new matrix or quantum mechanics on a firm basis. In this important work, often known as the *Dreimännerarbeit*, the authors emphasized that the formal similarity to classical theory should be understood in the sense that 'the new theory . . . can itself be regarded as an exact formulation of Bohr's correspondence considerations'.[110] Dirac too referred to the correspondence principle, but he reinterpreted it in a way that was foreign to its original meaning: 'The correspondence between the quantum and classical theories lies not so much in the limiting agreement when $h \rightarrow 0$ as in the fact that the mathematical operations on the two theories obey in many cases the same laws'.[111]

We may get another impression of how important Bohr continued to believe the correspondence principle to be from a letter he wrote to Oseen in late January 1926. Looking back on the stormy development of quantum physics over the last half year he expressed renewed optimism with regard to the prospect of a full understanding of atomic structure. The most recent progress was the incorporation of the electron's spin within the framework of the new quantum mechanics, which he characteristically saw as a vindication of the correspondence principle, which until then had failed 'in such a capricious manner that the confidence in all conclusions that could be drawn from the consideration of mechanical models was deeply shaken'. But now the spinning electron had given 'a qualitative understanding of all details in complete agreement with the correspondence principle'.[112]

To the extent that Bohr's correspondence principle lived on in the post-1925 era it was mainly in alternative versions of radiation theory, as temporary substitutes for quantum electrodynamics or as a general principle of a more qualitative nature.

Heisenberg considered the correspondence principle to be one of two sources for a mathematically formulated quantum theory, the other source being empirical data. Describing the quantum transitions between stationary states, in 1930 he cautioned that the traditional principle was 'far from satisfactory . . . and easily leads to false conclusions'. Instead he proposed a kind of generalized correspondence principle, which postulated 'a detailed analogy between the quantum theory and the classical theory appropriate to the mental picture employed'. This analogy, he said, 'does not merely serve as a guide to the discovery of formal laws', but also, and even more importantly, 'furnishes the interpretation of the laws that are found in terms of the mental picture used'.[113]

The most sophisticated and consistent use of correspondence arguments was made by Oskar Klein in 1927, when he attempted to formulate a quantum theory of radiation based on a direct connection of Maxwell's equations with the relativistic (and five-dimensional) version of wave mechanics he had recently proposed. Applying a method inspired by Bohr's calculation of the Einstein probability coefficients of spontaneous emission, Klein found expressions for the charge and current densities and evaluated these by means of correspondence rules to obtain the fields emitted by perturbed atoms.[114] In this way he succeeded in giving a correspondence interpretation of wave mechanics and explanations of, among other things, the photoelectric effect, dispersion, the Compton effect, and the selection rules for the Zeeman effect.

Klein's correspondence approach, with its unmistakable imprint of the Copenhagen spirit, was held in great esteem at Bohr's institute, where the attitude toward quantum electrodynamics was somewhat sceptical. The Swedish physicist Ivar Waller, who spent the spring of 1927 at the institute, worked on radiation theory along the lines indicated by his compatriot Klein, although without explicitly referring to the correspondence principle.[115] In 1931 the young Danish physicist Christian Møller formulated a relativistic scattering theory that used Klein's correspondence approach but not the methods of quantum electrodynamics.[116] In this way he avoided the troublesome divergence problems that plagued quantum electrodynamics and was able to solve problems that could not easily be treated by this theory. Also Léon Rosenfeld, who was a specialist in quantum electrodynamics, turned to the correspondence approach after he came to Copenhagen in 1931.[117] As one of the few non-Copenhageners who retained an interest in the correspondence principle, Van Vleck applied the correspondence principle to problems of quantum electrodynamics, in his case to formulate an alternative version of the Jordan–Dirac transformation theory.[118]

Although the 'correspondence principle' appeared prominently in several publications in the decade following 1925, in most cases it was in versions that had little in common with Bohr's original principle in whatever its formulation. Klein's theory of 1927 was an exception, whereas Møller's scattering theory merely related a classical concept (the four-current distribution in Maxwellian electrodynamics) to a corresponding concept in quantum mechanics (the matrix elements). References to the correspondence principle by name rarely implied use of Bohr's principle in its authentic version and sometimes the authors only used the principle in a rhetorical sense. Another noteworthy fact about the correspondence approach in the 1920s was that it was mostly cultivated by physicists at Bohr's institute in Copenhagen or who had close associations to it. Physicists without a Copenhagen connection felt no need to appeal to or reconsider the correspondence principle.

Notes for Chapter 5

1. Letter of 18 April 1917, as translated in Robertson 1979, p. 21.

2. Ibid., p. 35.

3. Bohr to Rutherford, 27 December 1917, in Rud Nielsen 1976, p. 683.

4. Schrödinger to Bohr, 7 February 1921, in Rud Nielsen 1977, p. 737, Schrödinger's spelling. Irving Langmuir was another scientist unable to obtain a copy of Bohr's memoir (Bohr to Langmuir, 3 December 1920, in Rud Nielsen 1977, p. 726).

5. The drafts to the fourth part are reproduced in Rud Nielsen 1976, pp. 185–200. According to this source Bohr was discouraged by Sommerfeld from publishing this last part of the memoir. The two relevant letters are Kramers to Sommerfeld of 24 May 1922 and Sommerfeld's reply of 1 June 1922.

6. Bohr to Ehrenfest, 18 May 1918, in Rud Nielsen 1976, p. 11.

7. Pauli 1945, p. 99.

8. Bohr 1918a, p. 5. Subsequent formulations of the first postulate also avoided reference to the validity of classical mechanics (e.g. Bohr 1924, p. 2).

9. A clearer presentation can be found in Bohr 1924, where he spoke of the systems as 'multiply' rather than 'conditionally' periodic.

10. Bohr relied on one of the standard textbooks of celestial mechanics, the two-volume *Die Mechanik des Himmels* (Leipzig: Veit, 1902–1907) written by the Swedish astronomer Carl Charlier. The same source was used by Epstein, Schwarzschild, and Sommerfeld. In the 1962 interview by Kuhn et al. (Niels Bohr Library & Archives), Bohr recalled also using a book by Boltzmann, probably *Vorlesungen über die Principe der Mechanik* (Leipzig: Barth, 1904).

11. Bohr to Richardson, 15 August 1918, in Rud Nielsen 1976, p. 14. From neutral Denmark Bohr also commented on the war, which was still going on: 'Although the war is felt very much in allmost [sic] everything the scientific life here is not much disturbed. In fact people find it rather more necessary than usual to concentrate on things not too much connected with material life'.

12. Interview with W. Heisenberg by T. S. Kuhn on 19 February 1963. American Institute of Physics, Niels Bohr Library & Archives. http://www.aip.org/history/ohilist/4661_6.html.

13. Bohr to Sommerfeld, 27 July 1919, in Rud Nielsen 1976, p. 17.

14. Grandin 1999, p. 107.

15. Epstein to Bohr, 14 May 1918, and Debye to Bohr, 6 June 1918, in Rud Nielsen 1976, p. 637 and p. 607.

16. Bohr to Sommerfeld, 30 April 1922, in Rud Nielsen 1976, p. 691. Bohr thanked Sommerfeld for sending him the third edition of *Atombau und Spektrallinien* and expressed his satisfaction that in this edition Sommerfeld's attitude to the correspondence principle was markedly more positive than in the earlier edition (Section 5.4).

17. Bohr 1918a, p. 9. For an insightsful analysis of Bohr's memoir, see Darrigol 1992, pp. 115–49. The first part is reproduced in van der Waerden 1967, pp. 95–138. A German translation of the full memoir, made by the Göttingen physicist Paul Hertz, appeared in 1923 as *Über die Quantentheorie der Linienspektren* (Braunschweig: Vieweg & Sohn).

18. Bohr 1918a, p. 8.

19. Ehrenfest 1923, p. 543.

20. Bohr 1918a, p. 9.

21. Ibid., p. 4. A few years later he said about the quantum-classical correspondence that 'the present theory of spectra is in a certain sense to be regarded as a rational generalization of the ordinary theory of radiation' (Bohr 1920, p. 427).

22. Bohr 1920, with an English translation in Bohr 1922a. On this and other of his works from the early 1920s, see also chapters 6 and 7.

23. Meitner 1964. Interview with Franck, conducted by Thomas S. Kuhn and Maria Goeppert Mayer 10 July 1962 (American Institute of Physics, Niels Bohr Library & Archives. www.aip.org/history/ohlist/4609_1.html).

24. Bohr's English manuscript on which the French translation in the Solvay proceedings was based is reproduced in Rud Nielsen 1976, pp. 364–80, which also includes Ehrenfest's address on the correspondence principle and the discussion at the meeting. Among those who discussed the addresses of Bohr and Ehrenfest were Langevin, Lorentz, W. L. Bragg and Rutherford. See also Mehra 1975, pp. 109–12.

25. Quoted in Dahl 1992, p. 104, who also refers to Bohr's unsuccessful and unpublished attempt of 1932 to explain superconductivity in terms of atomic theory.

26. Bohr 1922a, p. 30.

27. Grotrian 1920. See also Grotrian 1928 for a large collection of energy level diagrams for atoms with one to three valence electrons.

28. Some physicists and philosophers have suggested correspondence principles of such a general kind that the principle tends to lose any specific meaning. Thus, according to Fadner 1985 the correspondence principle dates back to Newton's *Principia* (!) and was used by a dozen physicists before Bohr. See also the appendix.

29. Rosenfeld 1973, p. 253. 'Nevertheless', Rosenfeld insisted, 'that was the first beginning of it'.

30. Jammer 1966, p. 50. On the relationship between Planck's second theory and Bohr's correspondence principle, see Whitaker 1985.

31. Hoyer 1981, p. 444.

32. Bohr 1918a, pp. 15–16.

33. Bohr, manuscript entitled 'Atomernes Bygning' (Niels Bohr Archive, Copenhagen), as quoted in Kragh 1979, p. 156. On the use of the correspondence principle in atomic structure, see Chapter 7.

34. Bohr 1920, p. 427. The term *Korrespondenzprinzip* appeared on p. 444.

35. Bohr 1921a, p. 2.

36. Einstein 1916, p. 318.

37. Einstein 1917b, p. 127. For the historical relationship between radioactivity and quantum probabilities, see Van Brakel 1985.

38. Einstein 1917b, p. 128. English translation in ter Haar 1967, pp. 167–83. Bohr summarized the theory in Bohr 1918a, pp. 6–8. Einstein also introduced the probability of radiation induced by

an external radiation field, given by coefficients B_{mn}, but in the present context this is less important. In modern literature the *A* and *B* quantities are often known as the Einstein coefficients.

39. Bohr 1924, p. 35. This paper was an English translation, by the American physicist L. F. Curtiss, of Bohr 1923a.

40. Manuscript quoted in Stolzenburg 1984, p. 15. For Nernst's idea about energy nonconservation, see Nernst 1916 and Nernst 1922.

41. Bohr, 'On the Interaction Between Light and Matter', in Rud Nielsen 1976, pp. 227–40, on p. 234. Bohr's emphasis.

42. Bohr 1918a, p. 7. Einstein did not use the expression 'spontaneous emission', but just called it *Ausstrahlung* (emission) and for induced radiation he used the term *Einstrahlung* (which may perhaps be translated as immission).

43. Bohr 1924, pp. 20–1.

44. Ibid., p. 21.

45. Bohr 1918b, p. 94 and more clearly in Bohr 1924, pp. 31–2. Sommerfeld and Heisenberg (1922) used the correspondence principle to calculate the spectral width of the hydrogen lines. See also Pauli 1926a, pp. 68–9. The impossibility of spectral lines with a zero broadening was later shown to be a consequence of Heisenberg's energy–time uncertainty relation $\Delta E \Delta t \geq h/2\pi$, but the result has its origin in the old Bohr quantum theory.

46. Rud Nielsen 1977, p. 361.

47. On Bohr's various versions of the correspondence principle and other physicists' conceptions of them, see Bokulich 2010. The historical literature on the correspondence principle is considerable. It includes Jammer 1966, pp. 109–18, Darrigol 1992, pp. 137–49, and Petruccioli 1993, pp. 78–110. See also Meyer-Abich 1965, which is a more philosophically oriented analysis. A clear discussion from a modern perspective of the technical aspects of the correspondence principle is given in Fedak and Prentis 2002. For a comprehensive contemporary review, see Buchwald 1923.

48. Bohr 1924, p. 22 and p. 23.

49. Bohr 1920, p. 469 (English translation in Bohr 1922a, p. 60).

50. Bohr 1928, p. 576, reproduced in Kalckar 1985 (on p. 124). On Bohr's changing views of the correspondence principle before and after quantum mechanics, see Meyer-Abich 1965, pp. 75–7.

51. Einstein to Bohr, 2 May 1920, in Rud Nielsen 1976, p. 22. 'The magnificent gift' refers to a packet of butter which Bohr sent to Einstein after he had returned to Copenhagen. For Einstein's growing interest in Bohr and his theories, see also Einstein to Ehrenfest, 18 November 1919, in Buchwald et al. 2004, p. 228. On Bohr and Einstein, see Pais 1991, pp. 227–30. In January 1920 Bohr had proposed Einstein as a foreign member of the Royal Danish Academy of Sciences and Letters, an invitation that Einstein accepted.

52. Einstein to Lorentz, 4 August 1920, in Buchwald et al. 2006, p. 364. See also Bohr to Einstein, 24 June 1920, in Rud Nielsen 1976, pp. 634–5.

53. Einstein to Sommerfeld, December 1920, in Buchwald et al. 2006, p. 532.

54. Einstein 1949, p. 47.

55. Kramers and Holst 1923, p. 139.

56. Bohr to Ehrenfest, 18 May 1918, in Rud Nielsen 1976, p. 11.

57. Bohr 1918b, p. 69.

58. Kramers 1920, on p. 199.

59. Kramers 1919, p. 44. On Kramers and his early works dealing with the Bohr theory and the correspondence principle, see Dresden 1987, pp. 107–10 and ter Haar 1998, pp. 11–16.

60. Kramers 1919, p. 43.

61. Kramers 1919, pp. 47–8. See also Bohr 1924, p. 24.

62. Adams 1923, p. 88.

63. Foster 1924.

64. Kramers 1919, pp. 75–83, and in greater detail in Kramers 1920.

65. Sommerfeld 1922a, p. 599.

66. Goudsmit and Uhlenbeck 1925. See also Kragh 1985a, Robotti 1986, and Jammer 1966, p. 95.

67. Birge 1921, p. 598, who did not speak of the 'correspondence principle'.

68. On the emergence of selection rules in quantum theory and spectroscopy, see Borrelli 2009.

69. Sommerfeld 1919, p. 393. Rubinowicz 1918. The essence of of Rubinowicz's lengthy and complex argument is summarized in Darrigol 1992, p. 140.

70. Sommerfeld to Bohr, 18 May 1918, in Rud Nielsen 1976, p. 686. The letter was written a few days before Rubinowicz submitted his paper to *Physikalische Zeitschrift* and shortly after Sommerfeld had received (and obviously studied) the first part of Bohr's work on the quantum theory of line spectra.

71. Bohr 1920, p. 459. Rubinowicz had ample opportunity to discuss the matters with Bohr in 1920, when he spent the months from May to August at Bohr's new and still unfinished institute in Copenhagen.

72. Bohr 1922b, p. 1113. There is more about the inner quantum number in chapters 7 and 8.

73. Sommerfeld 1922a, p. 327. Pauli later commented that Sommerfeld's conclusion that $\Delta k = 0$ transitions were inactive rested on 'somewhat artificial additional assumptions, whereas the correspondence principle leads directly to the result' (Pauli 1926a, p. 56).

74. Bohr 1922b, p. 1116. The possibility of $\Delta k = 2$ transitions in the presence of perturbative fields had been recognized by Bohr and Kramers in their work of 1918–1919. The experimental anomaly was reported in Foote et al. 1922. The disagreement evolved into a minor dispute in which some experimenters thought to have disproved the Rubinowicz–Bohr–Sommerfeld selection rule while others concluded in favour of it. See Mehra and Rechenberg 1982a, pp. 655–6.

75. Sommerfeld 1919, p. 401. For the different views of Bohr and Sommerfeld—or between Copenhagen and Munich—with regard to foundational issues in quantum theory, and the correspondence principle in particular, see for example Darrigol 1992, pp. 137–49, and Seth 2010, pp. 231–5. Seth speaks of the 'Bohr–Sommerfeld controversy' (p. 336), which I think is to exaggerate the significance of the disagreement between the two physicists.

76. Sommerfeld 1919, p. 403. The characterization of the correspondence principle as a magic wand (*Zauberstab*) also appeared in the later editions of *Atombau*.

77. Kramers 1935, p. 87.

78. Sommerfeld to Bohr, 11 November 1920, in Rud Nielsen 1976, p. 690. The last part of the quotation refers to Bohr's use of the correspondence principle to build up structures of atoms, a subject that will be dealt with in Section 7.1.

79. Bohr 1921a, p. 9, as translated in Rud Nielsen 1976, p. 356. On Bohr's view of pragmatic consistency, see Darrigol 1997.

80. Sommerfeld 1922a, p. 339.

81. Sommerfeld 1924a, p. 1048. The meeting took place in Innsbruck between 21 and 27 September.

82. Bohr to Sommerfeld, 21 November 1924, in Stolzenburg 1984, p. 38.

83. Heisenberg 1969, p. 45. Hund, five years older than Heisenberg, said that 'The glamour that surrounded this event cannot be communicated in words today; for us it was as brilliant as the Händel-Festival of those days in Göttingen' (interview of 25 June 1963, quoted in Mehra and Rechenberg 1982a, p. 345). On Bohr's lectures in Göttingen, see Mehra and Rechenberg 1982a, pp. 344–58. The manuscripts of the lectures are reproduced in English translation in Rud Nielsen 1977, pp. 341–420.

84. Heisenberg 1967, p. 95.

85. First lecture of June 12, Rud Nielsen 1977, p. 348.

86. Third lecture of June 14, Rud Nielsen 1977, p. 371.

87. http://www.nobelprize.org/nobel_prizes/physics/laureates/1922/press.html.

88. Bohr 1923b, p. 38.

89. Hoyt 1923 and Hoyt 1924. He stayed at Bohr's institute from October 1922 to September 1924.

90. Hoyt 1925.

91. Tolman 1925. A specialist in the theory of relativity, Tolman had an interest in astrophysics and would soon emerge as a leading theoretical cosmologist. He speculated that if an electron fell into the hydrogen nucleus, it would 'presumably lead to the destruction of the atom with the production of an amount of radiant energy equivalent to its mass' (p. 136). That is, he proposed electron–proton annihilation as a source of stellar energy, an idea that had been suggested by Eddington as early as 1917 and was also entertained by Jeans and a few other astronomers.

92. Bohr to Hoyt, 26 September 1924 (Archive for History of Quantum Physics, Bohr Scientific Correspondence).

93. Rosseland 1923a.

94. Jammer 1966, p. 118.

95. Born 1925, pp. 67–71 and pp. 114–23. Born paid particular attention to Kramers's theory of the Stark effect, which he covered in detail (pp. 242–52). A large part of *Atommechanik* was actually written by Friedrich Hund, who at the time served as Born's asistant in Göttingen. The only major review of quantum and atomic theory that failed to mention the correspondence principle may have been an excellent series of papers in the *Bell System Technical Journal* written by the American physicist Karl Darrow (1925–1926).

96. Brillouin 1922, p. 121.

97. De Broglie 1924, p. 75. On de Broglie's interpretation of the correspondence principle, see Darrigol 1993.

98. Van Vleck 1926, p. 97. A similar principle laid behind Bohr's earlier use of the correspondence principle in his theory of the structure of atoms (Section 7.2).

99. Pauli 1926a, p. 45.

100. Ibid., p. 49.

101. Pauli to Heisenberg, 28 February 1925, in Hermann et al. 1979, p. 212.

102. Darrigol 1992, pp. 235–46.

103. Sommerfeld 1924b, p. 658. Preface dated October 1924.

104. Heisenberg to Landé, 29 October 1921, quoted in Cassidy 1979, p. 211. It soon turned out that Heisenberg's core model was incompatible with Bohr's construction principle and its basis in the correspondence principle (Section 8.2).

105. Heisenberg to Pauli, 30 September 1924, in Hermann et al. 1979, p. 162.

106. Heisenberg to Pauli, 8 October 1924, in Hermann et al. 1979, p. 167.

107. Van Vleck 1924. Reprinted in van der Waerden 1967, pp. 203–22. Duncan and Janssen 2007, p. 665.

108. Kramers and Heisenberg 1925, p. 682. English translation in van der Waerden 1967, pp. 223–52.

109. Bohr 1925a, p. 849 and p. 852. A German version of the paper appeared in *Die Naturwissenschaften* **14** (1926), 1–10.

110. Born et al. 1926, p. 558 (van der Waerden 1967, p. 322).

111. Dirac 1925, p. 649. Although Dirac made use of correspondence arguments in his 1925 paper, in general he did not appreciate the correspondence principle.

112. Bohr to Oseen, 29 January 1926, in Stolzenburg 1984, pp. 238–9. Bohr was referring to a still unpublished paper in which Heisenberg and Jordan explained the fine structure of the hydrogen spectrum in terms of a combination of spin and relativity corrections.

113. Heisenberg 1930, p. 82 and p. 105. The book was based on lectures he gave at the University of Chicago in the spring of 1929.

114. Klein 1927. Pauli was critical to Klein's theory, see Pauli to Bohr, 29 March 1927, in Hermann et al. 1979, p. 389.

115. Waller 1927.

116. Møller 1931. On Møller's works in the period, see Kragh 1992.

117. Rosenfeld 1931.

118. Van Vleck 1928. For a lucid summary of correspondence arguments up to the late 1920s, see de Broglie 1938.

6

Molecules and Other Failures

During the early years of the 1920s, Bohr's atomic model and its extension by Sommerfeld and others enjoyed a nearly paradigmatic status within atomic and molecular physics. The theory was developed in particular by physicists belonging to institutions in Copenhagen, Göttingen, Munich, and Leiden, between which there was close collaboration and a fruitful exchange of ideas. While the German (or Germanic) dominance was marked, during this period American physicists entered seriously into the field for the first time, making important contributions of both an experimental and a theoretical nature. Remarkably, British physicists were nearly invisible in the endeavour to develop the quantum theory of atoms. They were observers of the development, not contributors, and when they contributed, it was often with works of a somewhat unorthodox kind within the older British tradition of mechanical and electrodynamical models. While *Philosophical Magazine* was the flagship of atomic models before World War I, by the early 1920s the journal had lost its significance and had replaced by the new *Zeitschrift für Physik*, founded in 1920 as a complement to the *Verhandlungen* of the German Physical Society, as the favourite journal of the quantum physicists.

A work by Edmund T. Whittaker, an esteemed mathematician at the University of Edinburgh and later a historian of ether and quantum physics, may illustrate how the British style in model-building (Section 3.3) was still alive in the 1920s. Using only the ordinary laws of electrodynamics, Whittaker undertook to construct an atomic theory in which quantum phenomena were explained in terms of 'magnetic currents'. His theory was considered important in Britain and the United States, but not elsewhere. In 1923 Bohr politely referred to Whittaker's ideas and his 'ingenious... mechanism which reproduces the characteristic features of the quantum theory', but without hiding that he had no faith in them at all. 'It must be emphasized', he said, 'that they are scarcely suited, from the nature of the case, to throw light on the actual applications in the present state of the theory'.[1] What he meant was that Whittaker's theory was nonsense.

The Bohr–Sommerfeld theory had celebrated great victories, but it soon became clear that it also faced great difficulties. Bohr's original and most ambitious project of establishing a common theory for atoms and molecules—one which would be of equal significance to physics and chemistry—turned out to be a failure. From the point of view of the chemists, the Bohr atom had little to offer since it did not provide a model of the covalent bond that was chemically useful. Molecules were not outside the reach of the theory, but success was largely limited to molecular spectroscopy. Analysis of the simple cases of H_2 and H_2^+ proved the inadequacy of Bohr's theory, and the helium atom was an even more serious case. Not only did quantum models of this atom disagree with experiments, in 1923 it was proved that *none* of the helium models allowed by classical mechanics and quantum theory fitted the bill. In addition to these failures there were a number of other phenomena that either contradicted the Bohr–Sommerfeld theory or could not be explained by it. The last section of this chapter examines the Paschen–Back effect, the Ramsauer effect, the crossed-field problem, and superconductivity. Another and far more serious anomaly, the anomalous Zeeman effect, will be considered in Chapter 8, where the anomalies will also be evaluated in a general way.

6.1 THE BOHR ATOM AND THE CHEMISTS

As we have seen in previous chapters, chemical considerations played an important part in Bohr's atomic theory of 1913, and he continued to use chemical facts as evidence and inspiration in his further development of the theory. However, during the first years of the Bohr atom, most chemists either ignored or responded cooly to his ideas of atomic and molecular structure. From about 1920 many textbooks in inorganic and physical chemistry contained sections on the new atomic theory, although in most cases it was mentioned only briefly and without connecting it to the chemical properties of matter.[2]

As mentioned in Section 2.6, Ernst H. Buchner, a physical chemist at the University of Amsterdam and a student of Jacobus van't Hoff, was an early exception. In a letter to *Nature* from the summer of 1915 he suggested that Bohr's theory might be useful in understanding the well known similarity between radicals and chemical elements.[3] For example, ammonium compounds are analogous to alkaline compounds, indicating a structural similarity between NH_4 and Na or K. Noting that while a structure for NH_3 can be constructed in agreement with Bohr's ideas, the radical NH_4 cannot, Buchner assumed that in NH_4 the electrons would rotate in four rings (4, 4, 2, 1) around the five nuclei joined in the centre of the system. In that case one would have a nice analogy between NH_4 and the alkali atoms. He suggested explaining the chemical resemblance between cyanogen $(CN)_2$ and the halogens on the same basis, assigning CN the structure 13 (4, 4, 4, 1) by analogy with fluorine's 9 (4, 4, 1) and chlorine's 17 (8, 4, 4, 1). Bohr found Buchner's chemical speculations interesting, but also pointed ou that 'on the theory under consideration it would hardly be permissible to assume that the nuclei of radicals or compounds are so close together that they will act upon all the electrons as a single nucleus'.[4] To illustrate his point he discussed the possible configuration of Thomson's triatomic hydrogen molecule, which he suggested might have a structure similar to the radicals discussed by Buchner. However, Bohr decided not to publish his note and only returned to the question of H_3 in 1919 (Section 3.2).

Physical chemists had for a long time played an important role in the development of quantum theory, but mostly in areas that did not involve the structure of atoms (such as photochemistry, statistical mechanics, and gas theory). By 1920 Bohr's theory and its application to atomic and especially molecular spectroscopy had become part of advanced courses in physical chemistry. For example, the Liverpool chemist William Lewis included a detailed review of the theory, based on Bohr's sequence of papers from 1913, in his textbook *A System of Physical Chemistry*.[5]

Although chemists were not uninterested in atomic theory, few considered the Bohr atom to be particularly relevant to their science and for this reason they paid little attention to it. They simply did not find it very helpful to the sort of problems they wanted to solve, of which the vexing problem of the nature of the chemical bond in molecules stood out as the most important.[6] In 1923, at a time when the Bohr atom was firmly established as the unrivalled theory of atomic structure, Walther Kossel suggested that it would only be a matter of time before valence phenomena were fully explained by electrostatic forces in accordance with Bohr's theory. But this was more wishful thinking than based on fact. Throughout the reign of the Bohr atom it was plainly unable to account for valency in a way that satisfied most chemists. Even the simple hydrogen molecule seemed outside its reach. The French chemist Alexandre Lepape represented the voice of the chemical community better than the physicist Kossel: 'Just as active as the Bohr atom is in the hands of the physicist, it is inert in the hands of the chemists'.[7]

In spite of his interest in and knowledge of chemistry, Bohr's mind was never genuinely chemical. For him chemistry was basically a resource to be exploited for the purpose of theoretical physics, not a science on par with physics. In 1919, in a lecture given to the Chemical Society in Copenhagen, he reflected on the relationship between the two sister sciences, diplomatically stating that, 'Of course, the separation of physics and chemistry is of a purely opportune character'. Referring to the relatively new discipline of physical chemistry, a science established in the late 1880s, he said:

> For many chemical problems, it has turned out to be possible and necessary to adopt viewpoints and apply methods borrowed from physics. But, as far as the results are concerned, the two fields exist as separate disciplines; one deals with what might be called the general phenomena, while the other deals with the properties of the different elements . . . Whether it is necessary to say that chemistry becomes part of physics, as one might be inclined to from a certain point of view, or if one should perhaps say the opposite, is not easy to decide; for, one might also say that, from the most modern point of view, the problem of general physical properties hardly exists, but everything depends on the properties of the chemical elements having atomic numbers from 1 to 100.[8]

In a talk a few months later to the Royal Danish Academy of Sciences he adopted a more reductionist point of view, now emphasizing the lead of atomic physics: 'Since . . . a possibility has been opened up of interpreting chemical experiences with the aid of considerations originating in the so-called physical phenomena, a connection between physics and chemistry has been created which does not correspond to anything conceived of before'.[9]

Bohr was not alone in his cautious physicalism, which was expressed less cautiously by some of his German colleagues who tended to see chemistry as a somewhat immature science in comparison with physics. According to Sommerfeld, quantum theory would in the future lead to 'a complete topology of the atom and, in addition, a mathematical chemistry'. Whereas a mathematical physics had existed for a century, 'It is only now that one can begin to conceive an effective mathematical chemistry . . . which can explain the as yet obscure concept of valency and at least in typical cases is able to predict the occurrence of chemical reactions'.[10] His pupil Kossel held that the covalent bond was of electrostatic nature and 'a purely physical problem'. It was, he said, 'up to us physicists to interest chemists in this conception'.[11] But the majority of chemists had no wish to be taught about chemical matters by scientists outside their own field.

Nor was it only German physicists who tended to see Bohr's atomic theory as a whole new chapter in the history of the relationship between physics and chemistry. In a semipopular book on atomic structure, Lodge wrote lyrically about 'The brilliant attempts at further analysis of the atoms of all the chemical elements, so as to deduce their properties—the full beauty of atomic astronomy which is now unfolding before the eyes of enthusiatic experts'. He concluded that, 'we are living in the dawn of a kind of atomic astronomy which looks as if it were going to do for Chemistry what Newton did for the Solar System'.[12]

Born, who had worked extensively with problems of crystal lattices and chemical physics, had little respect for classical laboratory chemistry, a science he tended to see as inferior to physics because it was not mathematically based. 'We realize that we have not yet penetrated far into the vast territory of chemistry', he wrote in 1920, 'yet we have travelled far enough to see before us in the distance the passes that must be traversed before physics can impose her laws upon her neighbour science'.[13] Although Bohr did not use such imperialist rhetoric, he agreed in the reductionist goal of deriving all chemical properties from calculations of atomic physics. Of course, chemists found such reductionism an unwarranted provocation. Niels Bjerrum, Bohr's friend and former teacher of inorganic chemistry, said in an address of 1922 that 'In the view of Bohr a complete mathematization is possible', but it was a view he found neither realistic nor desirable: 'We may be convinced that even when the electron theory has reached perfection the chemical formulae of the nineteenth century will still continue to be the ideal instrument of stating the composition of substances and of understanding their interactions'.[14]

Apart from his unsuccessful work of 1919 on triatomic hydrogen Bohr did not develop his atomic theory in chemical directions until the early 1920s, when he turned to the periodic system of the elements (Chapter 7). The first person to systematically exploit the potential of the new theory for chemistry was Kossel, a physicist with no training in chemistry. In his unusually long article in *Annalen der Physik* from 1916 he extended Bohr's ring structure model to higher elements and used it to account for the ionic bond and various properties of heteropolar compounds.[15] His table of the chemical elements gave, for the first time, the correct atomic numbers for all the known elements from hydrogen to uranium. Moreover, he provided population numbers for the shells of the lighter elements (up to $Z = 25$) that improved on those tentatively proposed by Bohr in 1913. For example, while Bohr had proposed (8, 2, 2) and (8, 8, 2, 2) for magnesium and calcium, respectively, Kossel argued that the two elements were filled with electrons according to (2, 8, 2) and (2, 8, 8, 2).

When Kossel turned to the covalent or 'homopolar' compounds he was less successful. He had to admit that representing, *à la* Bohr, the bond in binary molecules by a rotating ring of electrons placed between the two nuclei was not a very satisfactory model. Kossel attached special attention to rings with eight electrons, but not to the shared electron pair on which Lewis based his atomic theory from about the same time. Although Kossel thought that most diatomic molecules were kept together by an octet of electrons, in other cases he chose a different number. For example, he populated the ring binding two nitrogen atoms into a N_2 molecule with all the ten valence electrons, and in the case of O_2 with twelve electrons (Figure 6.1).

Kossel's version of Bohrian atomic chemistry did not appeal to the chemists, who could easily point out serious shortcomings. For example, Kossel's theory was unable to account for the tetrahedral structure of methane and many other organic molecules because it was faithfully based on Bohr's original planar atomic model. Nor could it account for the magnetic properties. Kossel made a vain attempt to explain the phenomena of stereochemistry by appealing to electrostatic distortions of the rings, but his explanation was far from convincing.

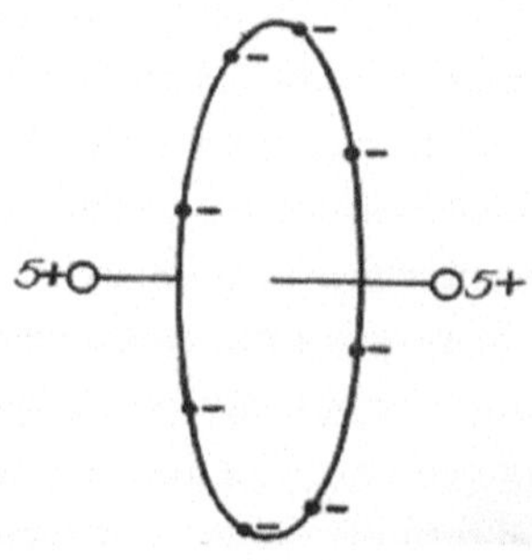

Fig. 6.1. Kossel's ring model of 1916 of the nitrogen molecule, with eight electrons on a ring common to the two atoms.

Source: Kossel 1916, p. 359.

By 1920 Bohr was ready to abandon the ring atom, which made him sceptical about Kossel's theory. His dissatisfaction was not rooted in considerations of chemical inadequacy, but in quantum-physical considerations of a more fundamental nature, which led him to a picture of spatially structured elliptical orbits. For a short time he believed that the new picture made sense of Kossel's valence theory, as he indicated to him in a letter of July 1921:

> When I have kept back for so long with regard to your valence theory it was because I thought to find great difficulties in bringing it into detailed agreement with the picture of atomic structure. But now I believe to recognize how it is possible to make the valence theory appear in the most beautiful harmony with the principles on which the quantum theory is based. I look much forward being able to send you my full account.[16]

However, when Bohr published his revised atomic theory he was silent about valency and also about the harmony between this theory and the models proposed by Kossel.

If the physicists' models did not live up the chemists' expectations, so the chemists' models did not live up to the expectations of the physicists. The kind of atomic models that chemists found useful violated the standards for acceptable models that Bohr and his colleagues in quantum theory found necessary. According to them a satisfactory atomic model would have to include a detailed picture of electrons arranged in various orbits or quantum states. The picture should allow for quantitative predictions and hence be formulated mathematically. In order to be acceptable it was essential that the model was in agreement with the principles of quantum theory and, more generally, with the laws of established physics. This implied that the angular momentum of every electron was $h/2\pi$, that the model atom was dynamically stable, and that it was kept together by electrostatic forces only. An atomic model that did not violate these and related theoretical criteria would be judged empirically on its ability to account for the facts of spectral physics and, to a lesser extent, other facts such as electron–atom collisions, dispersion, and the periodic system. Of course, it counted to the advantage of a model if it could predict novel facts and these were actually found to be in agreement with the prediction.

The chemists did not feel obliged to accept these criteria for atom-building. Their atomic models were primarily designed to meet the needs of chemistry, in particular the covalent bond, the spatial structure of molecules, isomerism, crystals, coordination compounds, and the periodicity of the elements. Although their models showed a great deal of dissimilarity they had some essential traits in common that distinguished them from those proposed by the quantum physicists. First of all, the chemists felt justified in disregarding the quantum theory and free to introduce features that lacked physical justification. Their models were typically of a pictorial, qualitative sort in which the detailed arrangement of electrons did not enter or was not considerered important. Moreover, they showed little interest in whether their models were mechanically (or electrodynamically) stable or not; they considered it legitimate to introduce new intra-atomic forces not known experimentally if these could solve or illuminate chemical problems. In his important paper of 1916 G. N. Lewis presented the chemical tradition as follows:

> We must first of all, from a study of chemical phenomena, learn the structure and the arrangement of the atoms, and if we find it necessary to alter the law of force acting between charged particles at small distances, even to the extent of changing the sign of that force, it will not be the first time in the history of science that an increase in the range of observational material has required a modification of generalisations based upon a smaller field of observation.[17]

Seven years later, in his monograph entitled *Valence*, Lewis characterized the quantum theory as 'the entering wedge of scientific bolshevism'.[18] More generally, he found the 'convergent' method of the chemists to differ radically from the 'divergent' one adopted by the theoretical physicists, yet being no less valuable. While the latter insisted on explanations and deductions from models, chemists successfully based their generalizations on numerous experimental data. In his Silliman Lectures of 1926, published as *The Anatomy of Science*, he spoke of 'the snobbery of science', one aspect of which was the desire of some chemists to imitate the methods of the physicists. But, he warned: 'It is a common fault of mankind to refuse to recognize the existence of a phenomenon unless some mechanism has been devised or, as we say, some explanation is offered'.[19]

Lewis's position was representative of a large part of the chemical community, but not all of it. Other chemists, more impressed by the successes of atomic physics, believed that the role of chemistry in atomic theory was mainly to provide the physicists with a fresh stimulus based on the requirements of a related science. This was the message of James Walker, professor of chemistry in Edinburgh, who in his presidential address of 1922 to the London Chemical Society dealt with the role of the physicist in the development of chemical theory. Describing the physicist as 'the scientific *conquistador* of to-day' he willingly admitted that a large part of chemistry owed a debt to physics. As to Bohr's atomic theory, he expressed reservation: 'This type of atomic model satisfies the physicist, but it offers to the chemist no adequate account of valence'. Walker suggested that this was an area in which chemistry might repay its debt to atomic and quantum physics: 'The fact that no atomic model can be regarded as satisfactory by the chemist unless it accounts in a simple way for the phenomena of valency may guide the physicist to altogether new conceptions', he said.[20] But his insistence of the fruitfulness of valency to physical atomic theory fell on deaf ears.

As an illustration of the different conceptions held by chemists and physicists one may consider the ambitious atomic and molecular theory developed by Irving Langmuir, a physical chemist and future recipient of the Nobel Prize in chemistry. Langmuir's theory, which to some extent was an extension of Lewis's, assumed atoms to be cubically structured in such a way that they could form compounds in agreement with chemical knowledge. By imaginative use of his cubical structures, which he assigned to all the elements of the periodic table, he built up models of both organic and inorganic compounds (Figure 6.2).

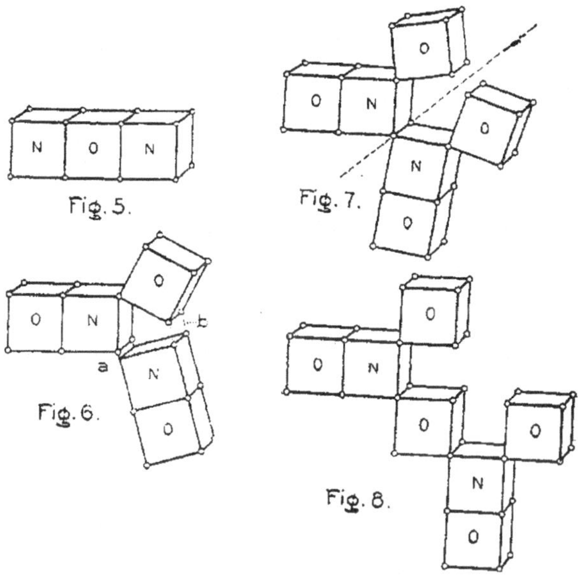

Fig. 6.2. Langmuir's cubical atoms making up various molecules of nitrogen oxides (N_2O, N_2O_3, N_2O_4, N_2O_5).
Source: Langmuir 1919, p. 898.

One of the very few chemists who had contacts to Bohr and his colleagues in Germany, Langmuir corresponded with Bohr and in November 1921 he visited him in Copenhagen. He continued to Göttingen, from where Born reported to Einstein: 'We liked Langmuir very much; he knows lots of physics'.[21] Langmuir wanted to base his views on the quantum theory, which was unusual for chemists. He admired Bohr and worked on problems, such as the structure of the helium atom, that were considered highly interesting by the physicists. Almost a parody of Bohr's 1913 theory, Langmuir based his own atomic theory of 1919 on no less than eleven postulates, proudly emphasizing its deductive nature. He paid lip service to the ideals of the quantum theorists, and perhaps to Sommerfeld and the German physicists in particular, by claiming that his theory promised 'a new chemistry, a deductive chemistry, one in which we can reason out chemical relationships without falling back on chemical intuition'.[22] But his vision of a deductive chemistry rested on premises for which there was no justification except that they led to the wanted results. As Edward Andrade, professor of physics at the Artillery College, Woolwich, dryly observed, 'the postulates have very much an ad hoc character, often paying little attention to any previously established laws of electromagnetism'.[23]

Langmuir's atom shared two of the features that characterized most of the chemical models, Lewis's included, namely, the existence of static electrons and non-Coulombian forces. In order to justify the static electrons physically he invented an ad hoc 'quantum force' designed to counterbalance the Coulomb attraction between an electron and the nucleus. He believed such a step was justified, because then 'we are not troubled by mysterious quantum conditions'.[24] The quantum force—Andrade called it 'hypermysterious'—was taken to be given by

$$F_q = \frac{1}{mr^3}\left(\frac{nh}{2\pi}\right)^2,$$

where m is the electron mass and n is an integer corresponding to the principal quantum number of the Bohr–Sommerfeld theory. With this assumption it was easy to calculate the equilibrium energy, which turned out to be identical to the energy of the stationary states in Bohr's theory. Of course, the identity was no accident, for it was secured by the expression chosen for the quantum force. However, from a physical point of view Langmuir's hydrogen atom differed radically from Bohr's, for in the theory of the American chemist there were neither stationary states nor quantum jumps in the sense of Bohr, only positions of equilibrium: 'According to the present theory, if the quantum number of an electron in a stable position decreases by one unit, the electron is no longer stable but falls towards its new position of equilibrium, and oscillates about this position. It then radiates its energy of oscillation according to the usual laws of electrodynamics'.[25]

By means of the frequency condition $E' - E'' = h\nu$ Langmuir could reproduce the generalized Balmer spectrum in the same form that Bohr had presented in 1913. But what about the other successes of Bohr' theory, such as the nuclear mass correction to the hydrogen spectrum and the fine structure of the Balmer lines? No problem, according to Langmuir. In the first case he just changed the expression for the quantum force by replacing the mass of the electron m with the reduced mass $mM/(m + M)$, where M is the mass of the nucleus. As to the fine structure, Sommerfeld's explanation in terms of relativity theory would seem to require a swiftly moving electron and hence contradict the static atom. Langmuir countered the objection by still another modification of the quantum force, this time to

$$F_q = \frac{1}{mr^3}\left(\frac{h}{2\pi}\right)^2\left[(n_a + n_r)^2 - \alpha^2 Z^2\left(\frac{n_r}{n_a} + \frac{1}{4}\right)\right]$$

where $\alpha = 2\pi e^2/ch$ is Sommerfeld's fine-structure constant, Z is the atomic number, and n_a and n_r correspond to the angular and radial quantum numbers of the Bohr–Sommerfeld theory. Voilà, the fine structure is reconciled with the static atom! (Even the imaginative Langmuir was unable to produce an expression for the quantum force that corresponded to Sommerfeld's exact formula for the fine-structure energy levels.)

The quantum physicists could not accept this approach, which they considered methodologically illegitimate and contrary to established quantum theory—which it was. Consequently Langmuir's atom became a target for occasional criticism, although few physicists bothered to respond to it in public. Sommerfeld chose to ignore the quantum force and relegated Langmuir's theory to a brief footnote in his *Atombau*. Bohr showed more interest, or more politeness. 'I must confess doubt in the reality of the interesting models of hydrogen molecules and helium atoms, proposed in your recent papers', he wrote in a letter to Langmuir of December 1920. He continued: 'In difference to the model of a hydrogen molecule, proposed in my first papers, I cannot imagine any way, in which models of the type, proposed by you, might be formed by a natural process. For the rest I do not believe at all in my own models for the hydrogen molecule and the helium atom, but think that the motion of the electron in these structures is of a more complex type, which do not at present allow of any simple mechanical treatment'.[26] In public he emphasized in a letter to *Nature*

> the fundamental difference between the picture of atomic constitution indicated in this letter and that developed by Langmuir on the basis of the assumption of stationary or oscillating electrons in the atom . . . In Langmuir's theory the stability of the configuration of the electrons is considered rather as a postulated property of the atom for which no detailed *a priori* interpretation is offered.[27]

In less diplomatic language Andrade dismissed Langmuir's atom as hopelessly contrived: 'It is scarcely necessary to insist on the artificiality of this picture', he quipped. 'The electrons in the Langmuir atom have, in fact, so few of the known properties of electrons that it is not immediately clear why they are called electrons at all'.[28] While Langmuir advertised his theory as deductive chemistry, Bohr and his fellow quantum theorists viewed it quite differently. Noting that 'Langmuir's work has attracted much attention among chemists', in an address of October 1921 Bohr pointed out that 'Such a *descriptive theory* is sharply differentiated from one where an attempt is made to explain the specific properties of the elements with the aid of general laws applying to the interaction between the particles in each atom'.[29] With the latter kind of theory he thought of his own quantum theory of atomic structure.

The chemists' dissatisfaction with the Bohr atom, and more generally with 'the so-called quantum theory', was given voice in an address Richard Tolman gave at the end of 1921. Pretending to represent the chemical point of view, he stated it as 'extreme hostility to the physicists, with their absurd atom, like a pan-cake of rotating electrons, an attitude which is only slightly modified by a pious wish that somehow the vitamine 'h' ought to find its way into the vital organs of their own, entirely satisfactory, cubical atom'.[30] Tolman had no problem with the mathematics of the Bohr–Sommerfeld atomic theory, but regarded as physics he found it to be unnecessarily arbitrary and conflicting with classical dynamics and the wave theory of light. Quantum theory was indeed interesting, even indispensable. Yet, two decades after Planck had introduced it, the theory 'must be regarded as still an experiment'. As to the unsatisfactory nature of the physicists' atom, he commented:

> This atom was constructed by the physicists, like a solar system . . . partly because they were entirely unfamiliar with the actual facts concerning the behavior of atoms in chemical combination. No chemist would be willing to think of a carbon atom as a positive nucleus with rings of electrons rotating around it in a single plane. The carbon atom must have

> tetrahedral properties, and in general I feel that the cubical atom of Lewis and Langmuir must be regarded as representing chemical facts better than anything proposed by the physicists.[31]

Robert Millikan, Tolman's colleague at the California Institute of Technology, had for a long time been a loyal supporter of the Bohr atom. In his Faraday Lecture of 1924 he took the opposite position to Tolman, criticizing American chemists for imagining 'the electrons sitting around on dry good boxes at every corner, ready to shake hands with, or hold on to, similar loafer electrons in other atoms'. This picture he contrasted with the one of the physicist, who 'has preferred to think of them as leading more active lives, rotating with enormous speeds in orbits, and occasionally flying out of these orbits, for one reason or another'. As Millikan saw it, Bohr's dynamical theory of the atom—'one of the established truths of modern physics'—was far better and more fertile than the view of the chemists. He called the Bohr atom 'about as much hypothetical to-day as is the theory of the rotation of the earth upon its axis or of the planets around the sun'.[32]

6.2 STATIC VERSUS DYNAMIC ATOM

The atom of the quantum physicists was essentially dynamic; that is, the electrons had to perform orbital motions in order to be in a mechanically stable configuration. This was as essential to the original Bohr atom as it was to its later developments. When Bohr abandoned the ring atom in 1920, neither he nor other quantum physicists dreamt of dispensing with its dynamical nature. The Bohr atom was modified in various ways by Sommerfeld, Born, Landé, Schrödinger, and Bohr himself, but none of the modifications were in the direction of the static atom favoured by Lewis, Langmuir, and most other chemists. The cubical atom (*Würfelatom*) advocated by Landé and Born shared some features with the one of the chemists, yet the similarity was superficial. Although the electrons in this kind of atomic model did not revolve around the nucleus, they did revolve rather than being at rest. Nevertheless, Lewis and Langmuir took the superficial similarity as evidence that the physicists' dynamic atom and the static one of the chemists were not irreconcilable.[33] They were not alone in proposing a reconciliation of the two different pictures of atomic constitution.

The chemists were in need of an atom with fixed electrons, but refrained from simply dismissing the Bohr atom, which at the time had gained a far too authoritative status to be either dismissed or ignored. Atoms supposedly do not distinguish between the domains of physics and chemistry, so a conception of the atom common to both domains seemed desirable, even necessary. Could the static and the dynamic atom not somehow be united? Norman Campbell considered Bohr's theory to be 'beyond doubt', and since he found the theory of Lewis and Langmuir to be 'extremely plausible', it followed that there could not be any real inconsistency between the two theories.[34] The question had to be answered affirmatively. According to Campbell, the apparent disagreement between the static and the dynamic atom was just that—apparent. He held it to be a manifestation of the formal principles of quantum theory, and in particular of the correspondence principle. In a response to Campbell, Bohr denied that the correspondence principle was of a formal nature only and maintained that it did not justify the dualistic view that 'the pictures of atomic constitution used in explanations of different phenomena may have a totally different aspect, and nevertheless refer to the same reality'.[35]

Many chemists, and also a few physicists, thought that the desired unification could be achieved. During the early 1920s there were several attempts to devise intermediate models, typically by interpreting Bohr's theory in such a way that it corresponded to a static atom. In most of these proposals the covalent bond was pictured as one or two electrons orbiting elliptically around the two nuclei rather than circulating between them (Figure 6.3). For example, this was the view of Campbell, according to whom, 'with this principle as a guide, it is merely a matter of linguistic translation to interpret on the basis of a dynamic atom the conclusions which have been reached on the basis of the static atom'.[36] Lewis agreed that the Bohr atom was not intrinsically in conflict with the model favoured by himself and Langmuir. His linguistic translation was this:

> Bohr has removed every essential element of conflict between the views of the physicist and the chemist. If we regard as the important thing the orbit as a whole, and not the position of the electron within the orbit, and if each electron is assigned an independent orbit, then we may think of each electron orbit as having a fixed position in space. The average position of the electron in the orbit may be called the position of the electron and will correspond entirely to that fixed position which was assigned in the theory of the static atom.[37]

Several other scientists, including Nevil Sidgwick in Oxford, A. E. Oxley in London, and Carl Knorr and Kasimir Fajans in Munich, argued for a reconciliation along similar lines.

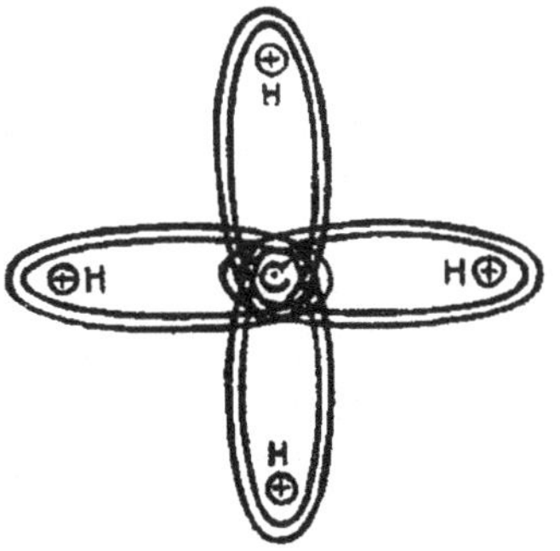

Fig. 6.3. An attempt to harmonize the Bohr atom with the chemical bond. In this model of the methane molecule, proposed by the German chemist Carl Knorr, the bond consists in a pair of electrons moving in elliptic orbits around both the carbon and the hydrogen atom.

Source: C. A. Knorr, ‚Eigenschaften chemischer Verbindungen und die Anordnung der Elektronenbahnen in ihren Molekülen', *Zeitschrift für anorganische Chemie* **129** (1923), 109–140.

It was widely believed among British and American chemical authors that the trend was towards a 'Rutherford–Bohr–Lewis–Langmuir atom', as one chemist called the supposed hybrid model of the future.[38] The popular chemical models were rivals of a sort to the Bohr theory, but it may be more appropriate to describe them as complementary rather than rival models. After all, since they only addressed a small part of the domain of the Bohr theory—basically the covalent bond in molecules—they were not rivals in a proper sense. In any case, Bohr disagreed that a reconciliation of the kind proposed by Campbell was possible. He planned to write an answer to *Nature*, but the paper remained an unfinished draft.[39]

The Cambridge physicist Ralph Fowler was one of the few mainstream physicists who adopted the chemists' point of view, although not without some reservations. Recognizing that the Bohr–Sommerfeld theory was unable to account for the covalent bond, he praised Lewis's shared-electron hypothesis as one which 'accounts in an entirely satisfactory way for a large body of chemical facts, and carries with it a conviction that it represents some real physical feature in the molecule'. For this reason he considered it important to establish a bridge between the models of Bohr and Lewis:

> In order to connect it [Lewis's theory] with recent physical views in which the electron must be thought of as moving round the nucleus in orbits obeying (at least to some extent) the laws of dynamics and characterized by certain definite energies and definite quantum numbers the shared electrons must be thought of as moving in orbits which are common to both nuclei . . . It is unfortunately very difficult to gain any clear conception of the nature of these shared orbits, or of the energy of the electrons in them.[40]

Indeed, for the reconciliation of physical and chemical atomic theories was largely an illusion, at least until the coming of quantum mechanics. While Bohr and his German colleagues chose to ignore the proposals of Campbell, Fowler, Knorr, and others, the British chemist John Main Smith responded in a way of which they might have approved. Main Smith, who expected that 'future advances in the knowledge of the intimate structure of atoms will be made largely in chemistry', pointed out that the ideas of Lewis, Langmuir, and their followers 'have no mathematical or dynamic necessity consequent on the electron configurations of the suggested Bohr atoms, but are extraneous assumptions arising from the absence in Bohr's theory of any provision for combination between atoms without ionization'.[41] However, Main Smith was equally critical of Bohr's atomic chemistry as he was to the Lewis–Langmuir chemical atomic theory. Bohr's scheme of electron configurations, he argued in 1924, 'is not compatible with the detailed properties of specific elements', nor could it explain the existence of carbon monoxide and several other molecules.[42] Contrary to Lewis, he thought (as Stark had proposed a decade earlier) that in most cases the chemical bond was identical with *one* shared electron. For evidence of such a chemical bond he referred to, among other examples, triatomic hydrogen and the H_2^+ molecule.

The quantum theorists' lack of interest in the chemical atom did not imply a lack of interest in molecules, as illustrated by the large amount of work devoted to molecular spectroscopy and the structure of the hydrogen molecule. Born and his colleagues and students in Göttingen, including Franck, Heisenberg, and Erich Hückel, made extensive calculations on the formation and structure of simple molecules on the basis of quantum theory, but their efforts bore little fruit and none that could arouse the interest of the chemists. In a letter to Einstein, Born looked forward to the day when 'the question of homopolar [covalent] binding forces between atoms has been solved from Bohr's point of view'. Although not doubting that the problem was solvable, he realized that it was a difficult one: 'Unfortunately every attempt to clarify the concept fails. I am fairly sure though that in reality it must all be very different from what we think now'.[43] Born was right. It took until 1927 before the chemical bond was understood on the basis of quantum mechanics, an accomplishment that primarily relied on the works of Walther Heitler, Fritz London, and Friedrich Hund.[44]

Born's elaborate calculations of molecules, made collaboratively with his student Heisenberg, were an orgy in mathematics but with almost no connection to empirical reality. Defining a molecule as 'a mass system consisting of nuclei and electrons which are held together by electrostatic forces', the two physicists applied the heavy machinery of perturbation theory to analyze in a general way the motion of the system.[45] They made no attempt to apply their elaborate theory to real molecules. As an exasperated Heisenberg wrote to Pauli, 'The work on molecules that I did with Born ... contains bracket symbols [*Klammersymbole*] with up to 8 indices and will probably be read by no one'.[46] Certainly, it was not read by the chemists.

6.3 MOLECULAR SPECTROSCOPY

In his Nobel lecture of 11 december 1922 Bohr mostly dealt with the atomic theory of the periodic system, but he also found time to mention the band spectra of molecules and their relations to the correspondence principle.[47] 'The first to apply the [quantum] postulates to this problem was Schwarzschild', he said, 'but the important work of Heurlinger especially has thrown much light on the origin and structure of band spectra'. Bohr continued:

> The considerations employed here can be traced back directly to ... Bjerrum's theory of the influence of molecular rotation on the infra-red absorption lines of gases. It is true we no longer think that the rotation is reflected in the spectra in the way claimed by classical electrodynamics, but rather that the line components are due to transitions between stationary states which differ as regards rotational motion. That the phenomenon retains its essential features, however, is a typical consequence of the correspondence principle.[48]

To understand the relevance of molecular spectra for Bohr's theory we recall that in 1912 Niels Bjerrum quantized the rotational energy of a molecular dipole with a moment of inertia J as $E = J(2\pi\nu)^2 = mh\nu$, or

$$\nu = m\frac{h}{2\pi^2}\frac{1}{J},$$

where m is an integer. The radiation frequency appearing in Bjerrum's equation was the same as the rotational frequency, implying that the mechanism of radiation was assumed to follow classical electrodynamics. Corresponding to the two directions of rotation, the emitted or absorbed frequencies would be given by a superposition of the rotational frequencies on the vibrational frequency: $\nu = \nu_{\text{vib}} \pm \nu_{\text{rot}}$. Bjerrum's theory of vibration–rotation bands quickly won general acceptance and was the following year verified by measurements made in Berlin by Eva von Bahr, a Swedish physicist working in Heinrich Rubens's laboratory. Her work was widely recognized as important. In 1915 Bohr suggested that 'Eva von Bahr's papers almost seem to offer direct proof of the quantum laws or at least of the impossibility of treating the rotation of molecules with anything resembling ordinary mechanics'.[49]

As mentioned by Bohr, Schwarzschild was the first to apply the Bohr–Sommerfeld quantum theory to molecular spectra, as he did in his important work of 1916, which is probably better known for its pioneering treatment of the Stark effect.[50] Applying Sommerfeld's quantization rule to the rotating molecule and interpreting the radiation from it as arising from quantum transitions following Bohr's frequency condition, Schwarzschild obtained for the frequencies in the visible bands

$$\nu = \frac{1}{h}(E_2 - E_1) + \frac{h}{8\pi^2 J}(m_2^2 - m_1^2).$$

The first term represents the electronic energies and the second the contribution from the quantized rotation. The quantities m_1 and m_2 are integral quantum numbers. This expression, which was restricted to band spectra in the optical region, agreed with the an empirical formula that the French physicist Henri Deslandres had found in 1887. According to Dealandres's rule, the frequencies of lines in the band spectrum follow an expression of the form $\nu = Am^2 + Bn^2 + C$, m and n being integers. When it came to the infrared region, Schwarzschild thought, in agreement with Bjerrum, that the frequencies were mechanical, directly reflecting the rotational frequencies of the molecules. This was also the opinion of Edwin Kemble at Harvard University, one of the American specialists in molecular spectroscopy.

Thus while Bohr's frequency condition governed one kind of band spectrum, it seemed not to work for another kind. The lack of unity was widely seen as unsatisfactory. As mentioned in Section 4.1, in his withdrawn paper of 1916 Bohr discussed Bjerrum's theory of infrared spectra, arguing that it needed to be changed in accordance with the fundamental postulate relating radiation energy to jumps between quantum states. He was not at that time aware of Schwarzschild's analysis, which had not yet appeared in the proceedings of the Prussian Academy of Sciences.

In the years after 1918 molecular spectroscopy evolved into a major research field, cultivated in particular by German and American physicists and physical chemists.

With the authoritative status that Bohr's atomic theory had achieved, it became important to base a theory of the band spectra fully on this theory, meaning that all lines in whatever range of the spectrum had to represent a transition between quantized energy states. Work within this framework was vigorously pursued by German physicists such as Wilhelm Lenz, Fritz Reiche, and Adolf Kratzer. It was a major research area for the Sommerfeld school and appeared prominently in the various editions of *Atombau*. Although it was not equally important to Bohr and his associates in Copenhagen, it was of importance to Bohr's general ideas of atomic structure, not least because spectra of molecules could serve as a test case for the use of the correspondence principle outside the atomic domain. Bohr followed the progress in molecular spectroscopy at a distance, yet with keen interest. In his drafts for the unpublished Part IV of the Copenhagen memoir he dealt critically with Bjerrum's theory and mentioned the more recent one of Schwarzschild. He argued by means of the correspondence principle that the rotational quantum number can only change by one unit, leading to rotational lines separated in frequency by the quantity $\Delta\nu = (2m + 1)h/8\pi J$.[51]

A full integration of Bohr's theory into the theory of band spectra occurred in the years 1920–1922, when the frequency condition was applied to combined electronic, vibrational, and rotational transitions, primarily by Reiche and Kratzer in Germany, Heurlinger in Sweden, and Kemble in the United States. Through this work it became understood that the energy of a molecule can be divided into three spectral ranges: in the far infrared rotation spectra dominate, while in the near infrared one observes vibration–rotation spectra; the visible and ultraviolet parts of the spectrum are characterized by lines originating in electronic transitions. In a paper of 1920 Reiche argued that the frequencies of vibration–rotation spectra were given by

$$\nu = \nu_{vib} \pm (m + \frac{1}{2})\frac{h}{4\pi^2 J},$$

where $m = 0, 1, 2, \ldots$ However, it then follows that the lines in a band are evenly spaced with $\Delta\nu = h/4\pi^2 J$, and this implication was contradicted by precision experiments on the absorption in HCl and HBr made by Elmer Imes at the University of Michigan (Figure 6.4). Imes's measurements showed very clearly a gap in the centre of the pattern of lines that could not easily be explained theoretically.[52]

Reiche's theory was based on three elements, namely the standard rotational quantization, Bohr's frequency condition, and 'Bohr's analogy principle'; that is, the correspondence principle that Bohr had recently introduced but not yet named. When Reiche realized the discrepancy with Imes's data he suggested changing the first of the elements. By changing the formula for the energy of a rotator from

$$E_{rot} = m^2 \frac{h^2}{8\pi^2 J} \quad \text{to} \quad E_{rot} = (m + \frac{1}{2})^2 \frac{h^2}{8\pi^2 J},$$

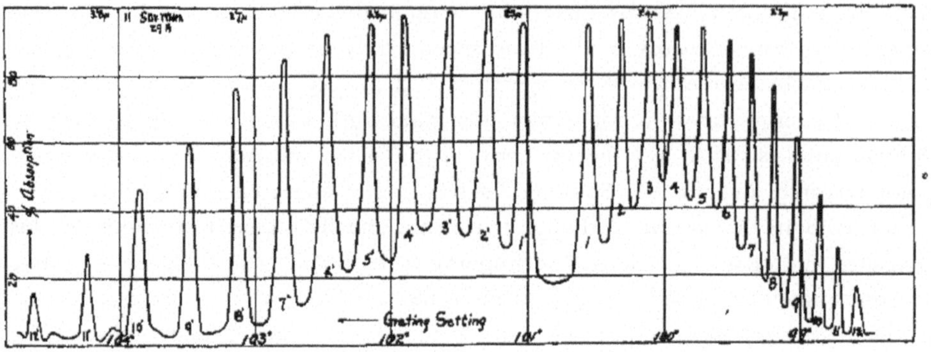

Fig. 6.4. Elmer Imes's absorption spectrum of HCl at $\lambda = 3.46\ \mu$, showing the gap in the pattern of lines that forced physicists to consider half-integral quantum numbers.
Source: Imes 1919, p. 260.

he could account for the observed gap. According to Reiche, this suggestion of rotational 'half-quanta' was first suggested by Einstein, 'with whom I have often had the opportunity to discuss these matters ... [and who mentioned] a possible way to change the rotational quantization so as to annul the contradiction with observations'.[53] Although the half-quanta were theoretically controversial they seemed to be necessary, and they were incorporated into an influential and more elaborate theory of the band spectra that Kratzer published in 1923.[54] Realizing half-quanta to be 'revolutionary', the British spectroscopist William Curtis argued that the bands of helium formed two non-combining classes, one of which involved half quantum numbers. While one of the classes corresponded to angular momenta given by $mh/2\pi$, the other corresponded to helium atoms having angular momenta $(m + \frac{1}{2})h/2\pi$.[55]

Reiche's theory and the results associated with it were independently and a little earlier derived by Torsten Heurlinger, a young Swedish physicist at the University of Lund.[56] Like Reiche, Heurlinger was acquainted with and made use of the analogy or correspondence principle, namely to argue that only transitions satisfying $\Delta m = \pm 1$ would occur in the rotational spectrum. He had already applied the correspondence principle to rotating molecules in a limited way in a paper of early 1919, where he referred to 'Bohr's general principle' (*einem allgemeinen Satze von Bohr*).[57] It is possible that he had studied Bohr's memoir in the fall of 1918, or he may have learned about it from his compatriot Oskar Klein. In September 1919, at the age of 25, Heurlinger participated in the Lund conference organized by Manne Siegbahn, where he gave a talk on his theory of band spectra and had the opportunity to discuss it with Bohr, Sommerfeld, Bjerrum, and other of the attendees. Because Heurlinger's paper appeared earlier than Reiche's, he was unaware of the gap in the band spectrum detected by Imes. Although illness forced Heurlinger to stop what looked like a promising career in physics, his few papers on molecular spectra were important and highly regarded by contemporary physicists. In a letter to Kemble the American spectroscopist Raymond Birge described Heurlinger's dissertation of 1918 as 'the most important paper on band spectra ever written'.[58]

The half-integral quantum numbers suggested by Reiche also appeared in the theories of the anomalous Zeeman effect developed by Landé and Heisenberg (see Section 8.1) and in some of the attempts to explain the helium atom, but in both cases in an atomic context. They also turned up in a theory of rotational molecular spectra proposed by Kramers and Pauli in early 1923. Six months earlier Bohr wrote to Kramers:

> I myself had a strenuous but also very interesting time in Göttingen . . . Pauli is coming up here in October to stay for about half a year. He is an excellent man in all respects . . . I talked a little with Pauli about band spectra, and it appears to me that your computations might be of value even for such simple compounds as HCl. Pauli was of the opinion that the spectra indicate that the axis of angular momentum makes an angle with the molecular axis; however, he had not been able to solve the mathematical problem himself.[59]

In agreement with Kratzer the two young physicists at the new Copenhagen institute argued that the electronic part of the angular momentum was one-half and that a diatomic molecule could only exist in stationary states if it rotated. 'No rotation-free stationary state of the molecule exists', they wrote about the HCl molecule.[60] In the scheme of Kramers and Pauli the intense lines on both sides of the fundamental vibration frequency were produced by quantum jumps between the rotational states given by $m = \frac{1}{2}$ and $m = \frac{3}{2}$. Like all other works on band spectra from the period, their theory relied to some extent on the correspondence principle.

By 1924 the theory of band spectra was in good shape and in general agreement with Bohr's theory of the constitution of atoms and molecules. The theory was widely seen as a success for the correspondence principle. However, minor details remained unexplained and so did the meaning of the mysterious half-quanta which was much discussed in the small community of quantum physicists. Kemble considered it a problem that Kratzer's theory implied that the lines symmetrically placed about the gap should have different intensities, while experiments proved that they had nearly the same intensity. He therefore attempted to 'harmonize the Kratzer half integral quantum numbers, for which there is strong evidence, with the observed symmetry in the intensities of the two branches of the HCl band'.[61] This he did by an advanced use of Bohr's correspondence principle.

The existence of a zero-point energy, as originally proposed by Planck in his second quantum theory of 1911, was controversial whether or not it was associated with Planck's theory. Einstein and his collaborator Otto Stern supported it for a brief period of time, but Einstein soon decided that the hypothetical zero-point energy was 'as dead as a doornail' [*mausetot*], as he wrote to Ehrenfest in the fall of 1913.[62] It was not quite dead, though, for it continued to attract attention in theories of magnetism and specific heat, and even entered cosmology in the form of Nernst's hypothesis of the cosmic vacuum—or world ether, as he thought it—being a huge energy reservoir. Nernst speculated that fluctuations in the energy density of the vacuum might form configurations out of which atoms would be created.[63] With the new studies of band spectra in the early 1920s, the concept of zero-point energy became respectable among molecular physicists. Yet it was only in the fall of 1924 that half-quanta were firmly established in molecular spectroscopy. In a study of the spectrum of boron monoxide (BO), Robert Mulliken, a young physical chemist at Harvard University, concluded that observations could only be understood on the assumption of quantum numbers with a minimum value of one half. In a preliminary announcement of his results in *Nature*, he wrote:

> It is probable that the minimum vibrational energy of BO (and doubtless of other) molecules is ½ quantum. In the case of molecular rotational energy, the necessity of using half quanta is already well established. Analogous relations appear in line spectra; e.g. Heisenberg has successfully used half-integral radial and azimuthal quantum numbers in explaining the structure and Zeeman effect of doublets and triplets.[64]

When Mulliken's full report appeared in *Physical Review* in March 1925, his work was widely considered a final confirmation of the zero-point energy, although it did not arouse much attention among the quantum theorists.

Molecular spectroscopy was the most important hunting ground for physicists investigating the quantum theory of molecules, but it was not the only one. Another was the electric and magnetic properties of molecules, an area of research cultivated in particular by the Sommerfeld school in Munich. According to the classical theory, as developed by Debye in a work of 1912, the dielectric constant ϵ for a diatomic polar molecule such as HCl can be written as

$$\frac{1}{4\pi}(\epsilon - 1) = N\alpha + C\frac{N\mu^2}{3kT},$$

where α is the polarization coefficient, μ the electric dipole moment, C a pure number, and N the number of molecules per unit volume. The classically calculated value for the quantity C was $C = \frac{3}{2}$. In a paper of 1921, Pauli compared the classical theory with the one based on the Bohr–Sommerfeld quantum theory.[65] He arrived at the same expression as the classical one, but with a value $C = 1.54$, corresponding to an electric moment 2.15 times larger than in the classical theory. In principle the case provided a means of testing the quantum theory, but this was in principle only: unfortunately the relevant experimental data were not available. The best data were obtained in 1924 by Charles T. Zahn at Princeton University, yet they were not good enough to yield an unambiguous result. 'The measurement of the dielectric constant of dipole gases at different temperatures offered a possibility of distinguishing between the two theories', Zahn concluded, but 'it is impossible to choose between the quantum theory and the classical theory'.[66]

In early 1926 the young American physical chemist Linus Pauling reconsidered the case. Using half-integral quantum numbers he found an even larger deviation from the classical theory, namely $C = 4.57$. As was only pointed out later, on the old quantum theory the discrepancy with the classical $C = \frac{3}{2}$ value persisted regardless of the temperature, while from the correspondence principle one would expect an asymptotic connection at high temperatures.[67] While Pauling did not use the new quantum mechanics, later the same year this was done in work by Ralph Kronig and, independently, Pauli and Lucie Mensing.[68] Remarkably, the three physicists found that quantum mechanics restored the classical value of $C = \frac{3}{2}$.

6.4 THE SIMPLEST MOLECULES

In a letter to Bohr of February 1919, Sommerfeld expressed his hope that his colleague in Copenhagen would come up with an approach to dispersion theory superior to the one of Debye and himself. 'I would also not object', he continued, 'if you could replace the H_2 model, which after all contains so many contradictions, with something better'.[69] Bohr realized that his model of molecular hydrogen was far from perfect, but at the time he had nothing better to offer. His old model of 1913 was known to be wrong, or at least seriously inadequate, as shown by its inability to reproduce the experimentally determined heat of formation of hydrogen gas (Section 2.6). Moreover, the same kind of model predicted the H_2^+ ion to be highly unstable, which disagreed with the experiments on positive rays carried out by J. J. Thomson, Arthur Dempster, and others. The poor empirical record of Bohr's model of the hydrogen molecule was also shown by the ionization process

$$H_2 \rightarrow H + H^+ + e^-$$

for which the model predicted an energy of 16.24 eV; this failed to turn up experimentally. Only in the case of dispersion was the Bohr model, or rather the Bohr–Debye model, of the hydrogen molecule moderately successful, and even in this case only for a while.

The final value that Langmuir obtained for the heat of formation, namely 84 kcal/mole, was too large to accommodate the value calculated by Bohr, which was about 62 kcal/mole. Langmuir's determination was complicated and indirect, but it was substantiated by Nernst and also by electron–collision experiments performed by James Franck and his collaborators in Göttingen. Franck's alternative method resulted in a dissociation energy of (81.3 ± 5.7) kcal/mole.[70] The discrepancy was considered from a theoretical perspective by Planck, who in a paper of 1919 calculated the heat of formation on the basis of the two existing versions of quantum theory (his 'first' and 'second' theory). In none of the cases did his elaborate calculations result in values close to the measured one.[71] Whereas Planck recovered Bohr's value of 62 kcal/mole on the ordinary or first quantum theory, his calculations based on the second theory resulted in a value of 140 kcal/mole.

Bohr was of course aware of the unsatisfactory state of his molecular model. In the unpublished Part IV of his memoir 'On the Quantum Theory of Line Spectra' he reconsidered in purely qualitative terms the formation of a hydrogen molecule from two hydrogen atoms. He started with two atoms far apart and then imagined, much like he had done in Part III of his trilogy, that they slowly approached each other as a result of external forces acting on the nuclei (Figure 6.5). By considerations of this kind he found that it was possible to 'form systems in which several electrons rotate in the same circular orbits at equal angular intervals without at any moment during the process to change the angular momentum of any of the electrons'. If two hydrogen atoms were in this way brought into the closest possible contact it would result in a system corresponding to a helium atom, or strictly speaking a hypothetical helium-2 isotope. Bohr found this way of thinking to be relevant for the electron structure of both molecules and more complex atoms:

> By proceeding in this way and performing corresponding processes with the systems already obtained we may step by step form more complex systems which as far as the position and motion of the charged particles are concerned may be regarded as representing atoms and molecules of higher atomic numbers.[72]

In his lecture delivered to the Physical Society in Copenhagen in December 1920 Bohr made it clear that although he considered the general picture of formation of the hydrogen molecule to be satisfactory, the simple model could not be maintained. Not only did it fail to produce the correct heat of formation and ionization potential, it also disagreed with measurements of the specific heat at high temperatures. In addition, the

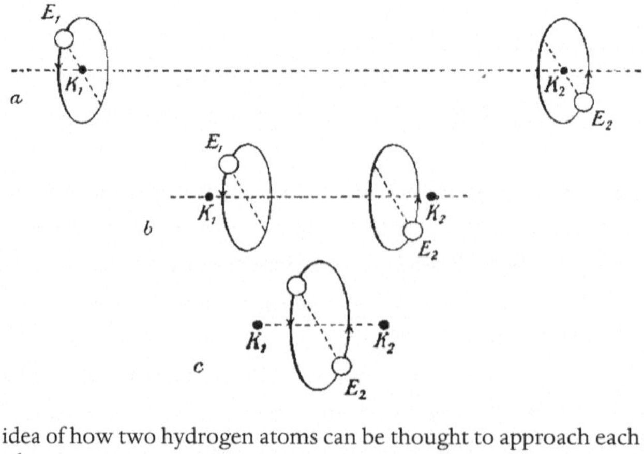

Fig. 6.5. Bohr's idea of how two hydrogen atoms can be thought to approach each other and form a hydrogen molecule.

Source: Kramers and Holst 1923, p. 181.

model was unable to account for the magnetic properties of molecular hydrogen. Bohr's hydrogen molecule would have a magnetic moment and thus be paramagnetic, whereas experiments implied that it was in fact diamagnetic, with zero angular momentum along the nuclear axis. The first to notice the discrepancy may have been the American physicist Jacob Kunz, at the University of Illinois, who in 1918 calculated the magnetic susceptibility of a hydrogen gas on the basis of Bohr's model. His value was much higher than the one established experimentally.[73] Kotaro Honda and his collaborators at Sendai University, Japan, attempted to account for the discrepancy by assuming the molecule rotated with constant frequency about an axis perpendicular to the line joining the two nuclei. Although they could in this way obtain a diamagnetic susceptibility of the right order of magnitude, their work attracted little attention and was not seen as a solution to the magnetic problem of the Bohr hydrogen molecule.[74]

Problems of magnetism were not limited to the hydrogen molecule, but turned up also for atoms. On Bohr's view, the electron structures of the sodium ion and the chlorine ion (K^+ and Cl^-, respectively) were almost identical to that of the argon atom, for which reason they were expected to have the same magnetic properties. However, experiments made by the Japanese physicist Také Soné showed that the diamagnetic susceptibility of argon was about ten times that of the two ions. According to Pauli, this implied a value of the inertial moment of argon which was much too large: 'quite impossible from a model point of view and from the view of the kinetic theory of gases!'.[75] Pauli saw no way to explain the discrepancy.

In his Copenhagen lecture Bohr expressed some confidence in a new model of the hydrogen molecule proposed by Lenz, according to whom the two electrons moved in opposite directions but in the same plane as the nuclei.[76] In his careful analysis of 1919 Lenz concluded that the Bohr–Debye model was irreparably wrong and needed to be replaced by a better one. His proposed substitute model possessed no magnetic moment and had a greater mechanical stability than Bohr's, but it was never developed in such mathematical detail that it could be compared with experiments. Lenz not only considered the hydrogen molecule, but also the helium atom, which he pictured as two electrons moving in different orbits around the nucleus. As he noted, when one of the electrons was close to the nucleus and the other far from it, the helium atom would be approximately equivalent to a hydrogen atom. If so, he suggested, two helium atoms might unite to form an unstable helium molecule of constitution He_2.

The problem of the magnetic properties of hydrogen also motivated A. E. Oxley, a British physicist associated with University College, London, to propose a molecular model related to Langmuir's, which he considered superior to Bohr's. According to Oxley, the two electrons rotated in very small orbits about fixed equilibrium points in such a way that the system would be diamagnetic. 'There is little possibility of reconciling such small orbital motions with the coplanar one of Bohr', Oxley commented in 1920. When he became aware of the Landé–Born cubical atom, he suggested that his own qualitative model with nearly fixed electrons received support from the work done in Germany.[77] It did not. Oxley's speculations, which did not rely on either quantum theory or dynamical considerations, were ignored outside Great Britain.

In 1922 a very different kind of model was proposed by Born, who followed the same line of thinking that Bohr had adopted in his Copenhagen lecture: to consider two distant hydrogen atoms and letting their nuclei approach each other so slowly that the adiabatic principle applied. In this way he arrived at four possible models, of which he singled out one as the most satisfactory from a physical point of view. In this model the two electrons moved synchronously but in opposite directions in planes that formed angles of $\pm 60^\circ$ with the axis between the nuclei. Like Bohr, Born found it an attractive feature that the hydrogen molecule could be regarded as an early stage in the formation of a helium-like atom: 'If one lets the nuclei gradually approach each other, one finally arrives at exactly that model of parhelium in the ground state which has recently been given by Bohr and which, for many reasons, must be considered as the most probable configuration of helium'.[78]

However, Born's favoured model was quickly shot down by his doctoral student Lothar Nordheim, who examined models of the hydrogen molecule as part of his doctoral dissertation in Göttingen.[79] Contrary to the expectations of Bohr and Born, he proved that a stable and empirically satisfactory model of the molecule could not be obtained by adiabatical changes of the distance between the two hydrogen nuclei. According to Nordheim's systematic study, all energetically stable configurations would be dynamically unstable (Figure 6.6). Among the possible models allowed by quantum theory the original one of Bohr was the 'relatively best', but it was nonetheless unacceptable from both an empirical and a dynamical point of view. Many physicists took Nordheim's analysis as proof that a precise model of the hydrogen molecule was outside the reach of the Bohr–Sommerfeld quantum theory of atoms. If this were the case one might consider models outside this theory, as a few physicists did. For example, Stanley Allen found it worthwhile to introduce a 'quantum force' of the same kind that Langmuir had considered. On this basis he constructed a model in which the two electrons were at rest relative to the nuclei, the four particles being situated at the corners of a square.[80] Allen's model did not result in an improved dissociation energy and his main motive for proposing it seems to have been his belief in static atoms. His model shared the fate of Oxley's model of the hydrogen molecule: it was ignored by Bohr and the German physicists.

Sommerfeld dealt extensively and critically with Bohr's model of the hydrogen molecule in his *Atombau*. He made it clear that the model was wrong and that a better one was not yet at hand. Reiche arrived at the same verdict in his 1921 review of quantum theory: 'Bohr's model seems not to correspond to reality: The configuration of the nuclei and the electrons must obviously be quite different'.[81] But how?

If quantum theory seemed unable to explain the hydrogen molecule, a system of four particles, perhaps it might be more successful in the even simpler case of three particles as manifested in the H_2^+ ion, the simplest possible molecule. As mentioned in Section

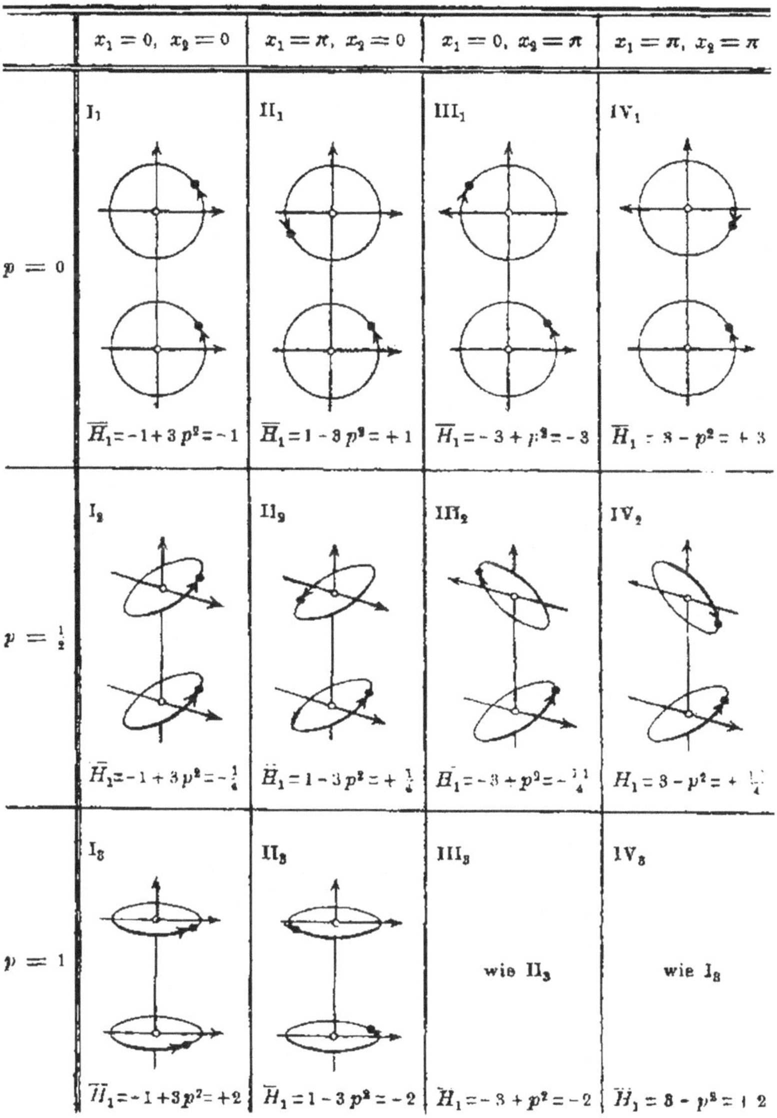

Fig. 6.6. Nordheim's classification of possible models for the hydrogen molecule.
Source: Nordheim 1923, p. 81.

2.6, in his trilogy of 1913 Bohr briefly dealt with the positive hydrogen molecule ion, which he found to be problematic. As early as February 1913 he realized that Thomson's identification of H_2^+ in positive rays disagreed with his still unpublished theory of atomic constitution, as he reported in a letter to Oseen.[82] He said the same in the letter to Rutherford of 1 July 1913 that accompanied the manuscript of the third part of his treatise:

> I have for a few days been in a terrible doubt as to the validity of the foundation of the theory. The reason was, that I had not before sufficiently realized how impossible it was on the basis of the assumptions made to construct a stable system representing a positively charged hydrogen-molecule, such as observed by J. J. Thomson in experiments on positive rays . . . It seems however now to me, that there is a possibility to escape from the difficulty in question by assuming that the conditions of equilibrium and stability of a stationary system cannot be applied on the systems in the experiments in question on account of the different circumstances under which the systems are formed.[83]

In his 1919 paper on triatomic hydrogen Bohr basically repeated what he had said about H_2^+ six years earlier. The same year Sommerfeld took a closer look at the atomic system in the first edition of *Atombau*. His model was the same as Bohr's, a single electron revolving in a circle of radius a placed symmetrically between two hydrogen nuclei separated by a distance $2b$. For the ratio a/b he found

$$\left(\frac{a}{b}\right)^2 = \sqrt[3]{16}-1 \cong 1.52$$

Since the corresponding ratio for the neutral molecule is $(a/b)^2 = 3$, 'the H_2^+ ion is considerably longer than the H_2 molecule'.[84] More importantly, Sommerfeld calculated for the total energy of the ion $W_+ = -0.88\ Rh$, where R is Rydberg's constant. This is greater than the energy of a hydrogen atom and a hydrogen nucleus (with zero kinetic energy), which is given by $W_H = -Rh$, and for this reason the molecule ion should decay according to $H_2^+ \rightarrow H + H^+$. However, ionization experiments on neutral H_2 indicated that $W_+ < W_H$, and Sommerfeld therefore concluded that Bohr's model of H_2^+ was no better than his model of H_2. A full discussion of the H_2^+ system was presented a few years later by Pauli and independently by the Dutch physicist Karel Niessen, in both cases as parts of their doctoral dissertations.[85]

Pauli's analysis of the H_2^+ ion did not presuppose a particular electronic orbit but considered the totality of mechanical orbits and then applied principles of a general kind to select the orbits that were allowed by quantum theory. Among those general principles the most important was a new stability criterion, namely that the electron orbits had to be stable in the sense of ordinary mechanics. This condition was not part of Bohr's original view of the constitution of atoms, where motions in the normal state did not necessarily have to be mechanically stable. (This was one of the issues of disagreement between Bohr and Nicholson.) Apart from making use of Bohr's ordinary correspondence principle, Pauli introduced what he called a 'mechanical correspondence principle', which he thought of as analogous to the principle valid for radiation processes. Referring to non-radiative transitions between two quantum states caused by a colliding electron, he said: 'Ordinary mechanics must (at least approximately) yield a measure of the relative frequency of the various quantum-energetically possible inelastic collisions'.[86] According to Pauli, this agreement between classical and quantum theory was of a statistical nature.

Bohr found the terminology of Pauli's principle to be unfortunate because it did not really reflect the essential message of the correspondence principle. As pointed out in Section 5.2, Bohr insisted that the correspondence principle was a law of quantum theory, whereas Pauli's criterion suggested an agreement between the classical description of collision processes and the ones obtained from quantum theory. Bohr's comment on Pauli's mechanical principle throws some light on how he understood the correspondence principle:

> He [Pauli] emphasizes the formal applicability of the classical laws in the limiting region of large quantum numbers and concludes from this that the application of the classical laws, even for smaller quantum numbers, in a formal respect gives a certain approximation. From this conclusion, several interesting inferences are drawn. If, however, they are designated as a mechanical correspondence principle, this is a mode of expression which deviates essentially from the conceptions presented in the text... The law for the appearance of radiative processes referred to as the Correspondence Principle is to be regarded as a typical law of the quantum theory... and, in itself, is not concerned in principle with the question of the degree of approximation of the applicability of the laws of radiation of the classical theory.[87]

Pauli's systematic and general analysis of the H_2^+ system yielded the result that only spatial orbits were possible; that is, orbits in which the electron moves in a plane that rotates around the axis connecting the nuclei. He ended up with a symmetrical spatial orbit as the only one that satisfied his general criteria. The same result was obtained by Niessen. In the ground state of the Pauli–Niessen H_2^+ model the electron moved on an ellipsoid, the ends of which were cut off (Figure 6.7). For the energy and the distance between the nuclei Pauli calculated

$$W_+ = -0.52\,Rh = 7.0\text{eV} \text{ and } a_+ = 5.54\,a_0 = 2.95 \times 10^{-8}\text{cm},$$

where a_0 denotes the Bohr radius of the hydrogen atom. Although the system thus did not represent the state of lowest energy, Pauli argued that it was 'metastable' and would not decay spontaneously into this state.

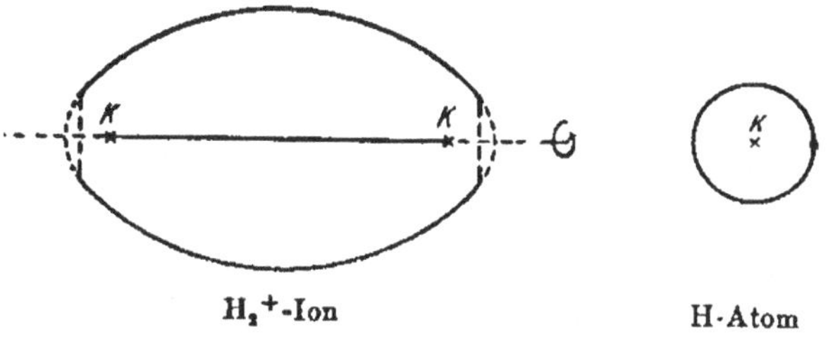

Fig. 6.7. The Pauli–Niessen model of the ground state of the positive hydrogen molecule ion, compared to the ground state of the hydrogen atom.

Source: Pauli 1922, p. 234.

Comparing the Pauli–Niessen model with experiments proved difficult because there was no direct determination of the ionization energy of the H_2^+ ion ($H_2^+ \rightarrow 2H^+ + e^-$). One sought instead to find the corresponding energy $I(H_2)$ for the neutral molecule; that is, for the process $H_2 \rightarrow H_2^+ + e^-$, and compare it with experiments. This was done by the energy equation

$$I(H_2) = D(H_2) + 2I(H) - I(H_2),$$

where $D(H_2)$ refers to the dissociation energy of H_2 into two hydrogen atoms. Out of this came a semi-theoretical value $I(H_2)$ = 23.7 eV, which could be compared with measurements. However, the interpretation of the experimental data was far from straightforward.[88] For a while measurements appeared to be consistent with Pauli's value of 23.7 eV, but in 1924 it became clear that his model disagreed with experiments. In early 1925 two American chemists, T. R. Hogness and E. G. Lunn, confirmed earlier results that showed the prediction from Pauli's theory to be wrong: 'The several theoretical models for H_2^+ give values for the ionization potential of H_2 higher than the observed 16.0 volts. The Bohr model, considered untenable for other reasons, leads to the value 17.85 volts (2.20 *Rh* − 0.88 *Rh*), the Pauli model to 23.7 volts, and the various models of Niessen to values 23.5–28.8 volts'.[89]

Given the general nature of Pauli's theory, the disagreement with experiment was more than just the failure of a particular model of the H_2^+ ion. It could be interpreted as a failure of existing quantum theory itself, as indeed it was retrospectively. But in the years 1922–1924 the Pauli–Niessen model did not attract a great deal of attention and its disagreement with experiment was not seen as a sign of crisis. Remarkably, Pauli himself never quoted his early work on the constitution of the hydrogen molecule ion. Born discussed it in his lectures on *Atommechanik* held in the winter semester 1923–1924, but without considering it very important. Having noted the discrepancy between theory and experiment he merely commented that 'at present it is quite unclear where the reason for the failure of the theory is to be sought'.[90]

6.5 AN ATOMIC CATASTROPHE

Much more serious than the hydrogen molecule anomalies were the failed attempts to construct a model of the helium atom that conformed in detail with experimental data. In continuation of what has been said in Section 4.7, by 1922 the case of the helium atom was unresolved if not yet considered unresolvable. In Bohr's unpublished manuscript for the 1921 Solvay conference he argued that the motions of the two interacting electron orbits would be nearly periodic only in two cases: 'In the first of these cases the electrons move through all times in one and the same plane, and in the second the planes of the orbits form with a high degree of approximation constantly a right angle with each other, the major axis of the inner orbit remaining perpendicular to the plane of the other'.[91] In the case of coplanar orbits he reported for the energy in the stationary states the result

$$E(n,k) = \frac{-2\pi^2 e^4 m}{h^2\left(n \pm \frac{9}{32}\frac{1}{k^2}\right)^2},$$

where n is the principal quantum number of the outer electron and k its azimuthal quantum number. For the minus sign in the denominator he found an approximate agreement with the spectrum of ortho-helium. He further reported that Kramers had calculated the energy of parhelium, where the planes of the orbits were supposed to be inclined by an angle of 60°. The result was $W_{\mathrm{He}} = -5.85\ W_{\mathrm{H}}$, corresponding to an ionization energy of 25.2 eV. Since this was close to the observed value, Bohr optimistically considered it a 'most interesting support' for his and Kramers' view of the constitution of the ground state of the helium atom. However, the agreement did not survive for long. In the paper that Kramers finally published in 1923, the energy was found to be $W_{\mathrm{He}} = -5.525\ W_{\mathrm{H}}$, or an ionization energy of $1.525\ W_{\mathrm{H}} = 20.63$ eV.[92] A parhelium model somewhat similar to the one considered by Bohr and Kramers was proposed by Edwin Kemble in 1921, but without calculations of the energy.[93] In the literature of the period the crossed-orbit model was sometimes referred to as the 'Bohr–Kemble model' or the 'Bohr–Kramers–Kemble model'.

Landé, Bohr and Kramers were far from the only physicists who occupied themselves with the difficult helium problem. On the contrary, in the early 1920s it was investigated by a dozen or more physicists, both from the United States and Europe. Among the American contributors were Kemble, Millikan, Langmuir, and Van Vleck; among the German physicists who took up the challenge were Epstein, Reiche, Born, and Sommerfeld, as well as Sommerfeld's highly talented students Pauli and Heisenberg. Late in 1922 Epstein wrote to Bohr from the newly established California Institute of Technology, telling him that by using a method different from Van Vleck's he had 'for some time ago' calculated the ionization energy of the symmetric helium atom to 21 eV. He thought the discrepancy might perhaps be due to the assumption of symmetrically moving electrons.[94]

One of the problems of Bohr's model (or the Bohr–Kemble model) for parhelium was that the total angular momentum in the normal state was $h/2\pi$, in apparent contradiction to the experimentally established fact that helium is diamagnetic, that is, it has no magnetic moment. In March 1922 Heisenberg wrote to Pauli in Göttingen: 'Bohr can only explain the diamagnetism by assuming that in a magnetic field He is always oriented in such a way that its moment about the axis of the field is 0. This doesn't please me very much, but of course one cannot doubt that Bohr is right'.[95] Heisenberg's phrase, reflecting the great authority of Bohr in matters of atomic theory, is remarkable: *Of course* one cannot doubt that Bohr is right!

The magnetic behaviour was a major reason why Sommerfeld proposed a model of his own rather than accepting Bohr's model. In a paper of 1923 he considered a model with coplanar orbits in which the electrons moved in opposite directions about the nucleus, thus resulting in a zero net angular momentum and therefore also a zero magnetic moment.[96] At the end of his paper, Sommerfeld referred to new calculations of Heisenberg, 'who has solved in detail the dynamical problem of our model for unexcited helium, and has found that the ionization potential is about 24.5 volts, which is correct'. Sommerfeld further advocated his helium model in the fourth edition of *Atombau*, although he admitted that his preference was to some extent based on his 'subjective conviction'. Realizing that none of the several proposed models had proved stable and at the same time produced agreements with experiments, he declared that from now on he would no longer speak of a *helium model* but only of the *helium problem*.[97]

On 30 April 1922 Born wrote Einstein a letter in which he said that he and Pauli, who had come from Munich to Göttingen to work with him, were writing a joint paper on the perturbation methods in quantum theory. 'We now really understand Bohr's ideas, at least in part', he said, adding: 'We have also started to do calculations for ortho-helium (two coplanar electrons) and were able to confirm Bohr's old claim that the inner electron moves around fast on an elliptic orbit whose major axis points towards the slowly moving outer electron'.[98] Einstein replied: 'I greatly admire the sure instinct which guides all of Bohr's work. It is good that you should be working on helium'.[99] Although Pauli worked hard on the helium problem in the summer of 1922, nothing came out of his efforts. Born, on the other hand, went on to collaborate with Heisenberg in producing an important paper that not only analyzed the helium case in great detail but also demonstrated more clearly than previously the limitations of the Bohr–Sommerfeld atomic theory.

Young Heisenberg took an interest in the helium problem even before engaging in the collaboration with Born. In the fall of 1922 he wrote to Sommerfeld that although Bohr's model of helium was probably wrong, 'for the sake of beauty in physics' he hoped that Bohr's more general ideas of the constitution of elements might still turn out to be true. Apparently, they were considered too beautiful to be wrong. 'One must hope that most of them will somehow remain intact', as he expressed it.[100]

Shortly afterwards, in another letter to Sommerfeld, he reported the results of his initial calculations of a planar model of the kind preferred by Sommerfeld; that is, two electrons moving in different ellipses and in opposite directions. 'I obtain an ionization potential of 24.6 volts with a possible error of about $\pm 2/3$ volts, but a probable error of $\pm 1/3$ volts. This agrees perfectly with the value of 24.5 volts measured spectroscopically by Lyman'.[101] In obtaining the result, Heisenberg relied on the general theory of perturbation methods that Born and Pauli had recently formulated. For a time he was optimistic, believing that he could solve the helium problem by introducing half-integral quantum numbers in the sense that the azimuthal quantum number of an individual electron would be $k = \pm\, h/4\pi$. However, he faced opposition from Bohr and Pauli in Copenhagen, and in his later work with Born the half-integral quantum numbers of the angular momenta were excluded. According to what he heard from

Pauli in Copenhagen, 'Bohr in any case wants to allow mechanics to be no longer valid, that is, only to a certain approximation. He does not yet believe in half quanta and helium'.[102]

Although Heisenberg's calculations of the normal helium atom were known to physicists in Göttingen, Munich, and Copenhagen, they were never published in full. His collaboration with Born primarily dealt with the excited states of the atom, which offered the advantage that they could be treated by the powerful methods of perturbation theory. In an excited state the outer electron moves in an orbit somewhat similar to the one of a hydrogen atom except that it is perturbed by the system consisting of the atomic nucleus and the inner electron. By means of perturbation calculations it should be possible to analyze the system in detail and thereby decide if any of the orbits allowed by the existing quantum theory agreed with experimental knowledge. This was the primary purpose of his and Born's work. In early 1923 Heisenberg sketched the research programme in a letter to Bohr: 'The other work of which I wanted to write you is a general investigation of all mechanically allowed orbits of excited helium. If in the end the experimentally found terms are not included, then one knows that the mechanics is wrong'.[103]

The work that Born and Heisenberg completed in Göttingen was a mathematically rigorous and completely general investigation of all possible types of electronic motion in excited helium. The aim was to derive theoretical spectral terms and, by comparing them to the terms inferred from the observed spectral lines, draw conclusions with regard to the viability of quantum atomic theory. This is what Heisenberg told Pauli in February 1923, at a time when the investigation was not yet completed but nonetheless suggested an answer, and not the answer he had hoped for:

> I must also tell you about a second catastrophe. Born and I have now calculated the most general model of excited helium in all details . . . and we find that the energy certainly comes out wrong. It was rather easy to have an overview of all possible types of motion: there are, as was to be expected, only three classes of motion possible in quantum theory (Bohr's ortho-helium and parhelium types, besides the ortho-helium in which the two electrons move in opposite directions) in addition to a couple of quantum solutions in which the inner electron goes through the nucleus (or does so at least for $k \to \infty$).[104]

The way that Born and Heisenberg obtained testable results from their general theory was to calculate the so-called Rydberg correction δ, which was a measure of the departure of a spectral term from the hydrogenic term having the same quantum numbers. They wrote the total energy of an excited atom of the hydrogenic type as

$$W = -RhZ^2 - Rh\frac{(Z-1)^2}{(n+\delta)^2},$$

where $Rh = W_H$ is the energy of the hydrogen atom in its ground state, n is the principal quantum number, and $Z = 2$ in the case of helium. After having expressed the δ corrections for the three possible classes of quantum orbits in terms of the azimuthal quantum number they could compare them with the Rydberg corrections obtained experimentally from ortho- and parhelium (Table 6.1). The verdict was brief and unequivocal: 'The result of our investigation is totally negative'.[105] In conclusion they stated that there were only two ways out of the problem. Either the quantum conditions were wrong, and one would, for example, have to consider non-integral values for the azimuthal quantum numbers; or the motion of the electrons in the stationary states did not follow the laws of mechanics. Sometime before the paper was submitted to *Zeitschrift für Physik*, Heisenberg suggested to Pauli that, in spite of their different attitudes, 'we are both convinced that all present He models are just as wrong as the entire atomic physics'.[106]

Confessing that 'I don't seem to get any closer to the great mystery of the quanta, in spite of all my efforts', Born also reported the failure of the helium calculations to Einstein: 'We have been looking at perturbation theory (Poincaré's) to determine whether it is possible to obtain the observed term values from Bohr's models by exact calculation. But it is *quite* certainly *not* the case, as was demonstrated with helium, where we found any number of multiple periodic orbits'.[107]

What Born referred to as a 'catastrophe', and Heisenberg as both a 'catastrophe' and a 'misery', was conceived somewhat differently and not as quite as alarming by other physicists.[108] Van Vleck thought that the disagreement with experiments suggested that 'existing quantum conditions cannot be applied without some modification to atoms with more than one electron'.[109] Bohr's attitude is of particular interest. Well aware of the new calculations from Göttingen, he had commented on them even before the publication of the Born–Heisenberg paper, namely in a long paper on atomic structure that appeared in *Annalen der Physik*. In this paper he dealt at some length, if only qualitatively, with the helium atom, and in a footnote he referred to the forthcoming work of Born and Heisenberg. 'Not only is no explanation obtained of the non-occurrence of combinations between ortho- and parhelium terms, but not even an

Table 6.1. The experimentally known values of the Rydberg correction δ for ortho- and parhelium compared with the calculations of Born and Heisenberg for the excited states $k = 2$ to $k = 3$. The three models are those allowed by quantum theory. Model 1 refers to coplanar orbits in which the electrons move in the same direction, while in model 3 they move in opposite directions. Model 2 is the case of orbits inclined at nearly a right angle. Adapted from Born 1925, p. 341.

		$k = 2$	$k = 3$	$k = 4$
Experiment	ortho-helium	−0.060	−0.003	−0.001
	parhelium	+0.011	−0.002	−0.001
Theory	model 1	−0.063	−0.029	−0.017
	model 2	+0.014	+0.004	+0.002
	model 3	+0.078	+0.034	+0.019

approximate value of these terms can be obtained', he wrote. Moreover: 'This investigation may therefore be particularly well suited to provide evidence of the fundamental failure of the laws of mechanics to describe the finer details of the motion of systems with several electrons'.[110] At about the same time he wrote to Born:

> The result is certainly of great importance as it indicates the inadequacy of the present foundation of quantum theory in so far as it concerns systems with several electrons. However, I do not see this difficulty in quite the same light as you have suggested, for I believe . . . that it is possible to give a unified conception of the quantum theory in which the failure of mechanics for the stationary states fits naturally.[111]

Bohr agreed that the existing quantum theory had to be revised, but not that it was fundamentally wrong. As he saw it, the main problem was the assumption that classical mechanics was valid for several electrons moving in stationary orbits, but this was an assumption he had never considered fundamental to his research programme.[112] Whereas he was quite willing to abandon it, he was not yet willing to abandon the notion of electrons moving in definite orbits corresponding to the stationary states. At the time he thought that also a future quantum theory of atoms would retain three basic elements of the existing theory: the frequency condition ($\Delta E = h\nu$), the postulate of stationary states, and the correspondence principle.

Born was of the opinion that the helium catastrophe should have more drastic consequences, although he could not say exactly what consequences. In his contribution from the summer of 1923 to the Bohr issue of *Naturwissenschaften* he dealt with the helium problem and the failure of applying mechanics to the stationary states of this atom. It followed, he said, that atoms could be likened to planetary systems only in a very limited sense. 'However', he went on in the style of Bohr, 'the progress of quantum theory is not therefore blocked, for it will rather gain new strength by the existing contradictions. It becomes more and more likely that not only new assumptions in the usual sense of physical hypotheses will be necessary, but the entire system of concepts of physics must be rebuilt from the ground up'.[113]

6.6 THE PASCHEN–BACK ANOMALY AND OTHER DIFFICULTIES

In the years from about 1921 to 1925 the Bohr–Sommerfeld theory was challenged by several other problems distinct from the ones mentioned above, although the problems were not considered to be equally serious. By 1920 it was known that with a strongly increasing magnetic field strength, the complicated multiplets of the anomalous Zeeman effect gradually changed into the simpler patterns of the normal Zeeman effect (Figure 6.8). For atoms with more than one electron this so-called Paschen–Back effect could be roughly understood as caused by the angular momentum of the atomic

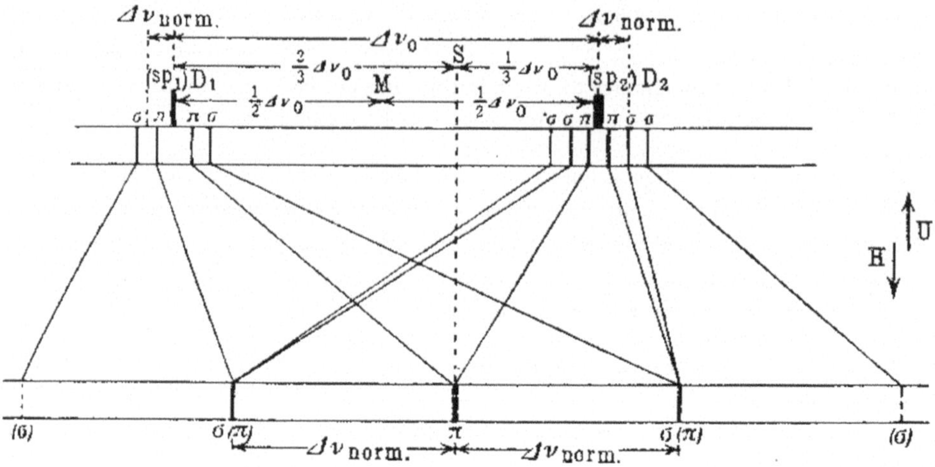

Fig. 6.8. The Paschen–Back effect. In a weak magnetic field the two components D_1 and D_2 of the yellow sodium line will exhibit an anomalous Zeeman effect (above); as the field increases in strength the pattern will change into a normal triplet (below).

Source: Sommerfeld 1924b p. 675.

core made up of all the electrons except the valence electron. This explanation implied that there should be no Paschen–Back effect in hydrogen (which has no core) and yet experiments indicated that the effect existed also in this case. As mentioned in Section 4.3, the possible existence of a Paschen–Back effect in one-electron systems such as H and He^+ was a source of worry to Sommerfeld, who in the first edition of *Atombau* suggested that it was inconsistent with his otherwise very successful model of the hydrogen atom.[114] However, the experimental evidence was not very clear and Sommerfeld chose to pay more attention to experiments that showed no Paschen–Back effect than to those which confirmed the effect in hydrogen.[115]

Among the first experiments made at Bohr's new institute in Copenhagen were spectroscopic measurements of the magnetic effect on the He^+ spectrum made in 1921 by Hans M. Hansen and Jens C. G. Jacobsen. Studying how the fine-structure components of the λ4686 line changed as the magnetic field increased up to 6,000 gauss, the two Danish physicists found that the separation pattern showed no Paschen–Back effect and thus agreed with the Bohr–Sommerfeld theory. Sommerfeld and other quantum physicists welcomed the Copenhagen measurements as a rescue from what might have been a challenge to the accepted atomic theory. They decided that this one affirmative experiment was more convincing than earlier results that contradicted the theory. In the third edition of *Atombau*, Sommerfeld spoke quite differently to the way he had done in the first edition:

> We are now convinced that these anomalies are caused by the effect of the electric fields present in the tubes. In fact, Hansen and Jacobsen have performed very careful measurements of the Zeeman effect of the He^+ line $\lambda = 4686$, and were able to establish that the

> Zeeman effect appears essentially normal and in any case shows no similarity with the Paschen–Back effect.[116]

Bohr agreed. In the second part of his treatise on the quantum theory of line spectra he had argued that 'the effect of a homogeneous magnetic field on the fine structure of the hydrogen lines will consist in the splitting up of every component in a normal Lorentz triplet', and it was with the purpose of testing this idea that the experiment of Hansen and Jacobsen was conducted. In early 1920 he wrote to Richardson:

> In the last months I have had a very busy time with the construction of the new laboratory, but I hope that the work spent on it will prove itself useful later. With some of the apparatus procured for the new laboratory . . . we have already started on some investigation of the effect of electric and magnetic fields on the components of the fine-structure of the helium spark lines, which as it is hoped may serve to test the considerations exposed in the 2nd Part of my paper in the Copenhagen Academy. Due to all this the publication of the 3rd Part has, as I am ashamed to say, again been postponed, but at present I am working hard at it and hope to have it finished in a few weeks.[117]

In the third part of the memoir, only published in 1922, Bohr stated triumphantly that the Copenhagen experiments 'prove that the effect of the field is of a type entirely different from the anomalous Zeeman effect on other spectra'.[118] But the anomaly had not disappeared, for in 1922–1923 new evidence turned up that confirmed a Paschen–Back effect in hydrogen of the same kind as the one known from the alkali metals. Careful experiments made by Otto Oldenburg in Munich and by Karl Försterling and Gerhard Hansen in Jena strongly suggested that the magnetic effect on the hydrogen lines could not be reconciled with the Bohr–Sommerfeld quantum theory.

Unable to criticize the new experiments, Sommerfeld was forced to return to his pessimistic view of 1919, that the Zeeman effect of hydrogen presented a real problem to his and Bohr's theory. This did not mean that he capitulated, for he believed that the experimental evidence could still be reconciled with a theory fortified with auxiliary hypotheses or perhaps be explained away as an effect of stray electric fields. However, none of these approaches worked and he thus ended up in the last refuge of theorists, with a call for more and better experimental investigations. Admitting that the theory might be incomplete (but not wrong), all what he could suggest was, 'At any rate, further experimental and theoretical clarifications are necessary'.[119]

Sommerfeld was clearly unwilling to accept a real contradiction between the quantum theory of atoms and the Paschen–Back effect in hydrogen, and he was not alone in this position. To mention but one example, Hans Falkenhagen, a physical chemist at the University of Cologne, tried to explain the experimental data by incorporating the motion of the atomic nucleus. His tedious calculations failed to give the wanted result, but without causing him to doubt the validity of the theory. Although he realized that a way out of the dilemma would be to admit that 'Sommerfeld's theory of the H atom is not quite in order', he dismissed the possibility because of 'the exemplary agreement that Sommerfeld's fine structure displays with the spark spectrum of He^{+}'.[120] The only possible alternative he could see was that one might have to use a different quantization procedure—he did not say which—for atomic systems that were not conditionally periodic.

In effect, the great success of the Bohr–Sommerfeld theory of one-electron atoms tended to make it immune to criticism and make physicists downplay the significance of the Paschen–Back anomaly. This effect is a manifestation of the electron's spin, a concept not known before Samuel Goudsmit and George Uhlenbeck introduced it from spectroscopic evidence in the summer of 1925. After the introduction of spin, Sommerfeld and his student Albrecht Unsöld succeeded in explaining the Paschen–Back effect in hydrogen. With the advantage of hindsight they wrote that the magnetic effect on the hydrogen lines 'constituted one of the most serious objections against the previous theory of the hydrogen atom'.[121] But this was not how physicists viewed the problem in the years 1922–1924, when the Paschen–Back anomaly was scarcely considered a sign of crisis in the existing quantum theory of atoms. As Heisenberg confirmed to Landé as late as February 1925, 'the hydrogen atom is in good shape'.[122]

There were other anomalies that played a similarly minor role in the crisis that eventually led to the fall of the Bohr–Sommerfeld theory. One of them was the Ramsauer effect, so named after the Heidelberg physicist Carl Ramsauer who, at a meeting of German scientists in Jena in September 1921, reported some startling results concerning the penetrability of slow electrons in an argon gas. A few earlier physicists had observed that slow cathode rays move more freely through a gas than fast ones, but it was only with Ramsauer's work that the effect became generally known and aroused widespread attention in the physics community. Franck, who had participated in the Jena meeting, reported to Bohr: 'In Jena I was particularly interested in a paper of Ramsauer that I am not able to believe, though I cannot show any mistake in the experiment. Ramsauer obtained the result that in argon the free path lengths are tremendously large at very low velocity of electrons . . . If this result is right, it seems to me fundamental'.[123] Bohr replied that he was very interested in the new result and wanted more information about it. He thought that the question was probably 'very closely connected with the general views of atomic structure'.[124]

Franck and Bohr were not the only physicists who found Ramsauer's experiments puzzling. In November Born wrote to Einstein about 'Ramsauer's quite crazy assertion (in Jena) that in argon the path length of the electrons tends to infinity with decreasing velocity (slow electrons pass freely through atoms!)'. He added that 'This we would like to refute!'[125] The initial scepticism with regard to the Ramsauer effect evaporated after it was confirmed by experiments carried out by, among others, Gustav Hertz in Eindhoven, Hans Mayer in Heidelberg, and Rudolph Minkowski and Hertha Sponer in Göttingen. Not only was the effect real, it also turned out that it was not limited to argon but appeared in the other noble gases as well and possibly was a general property of matter in the gaseous state.

The phenomenon defied theoretical explanation, whether in terms of classical theory or quantum theory. The first to take up the challenge was young Friedrich Hund, who at the time was a doctoral student under Born in Göttingen. Inspired by Franck, he developed a theory based on quantum conditions and the correspondence principle, from which followed that slow electrons would not be influenced by collisions with gas molecules.[126] Bohr was keenly interested in Hund's theory, which he knew in outline from his correspondence with Franck. Although the theory was somewhat unorthodox, he thought it agreed with the spirit of quantum theory. 'I see no other simple explanation of the Ramsauer experiment', he wrote to Franck, 'and am so sceptical about the established principles of physics that I do not feel justified in rejecting your [and Hund's] ideas as total nonsense'.[127] However, Hund's theory turned out to be untenable and it was not replaced by better theories. The Ramsauer effect thus remained unexplained, without the lack of explanation causing much concern at the time. Although physicists in Copenhagen and Göttingen were convinced that it was a quantum effect, other physicists thought of it in terms of classical gas theory or simply avoided attempts of explanation. At any rate, by far the most work on the Ramsauer effect was experimental, an autonomous line of research that was uninfluenced by quantum theory.

The Ramsauer effect was anomalous, but it was not obvious that the anomaly belonged to the domain of quantum theory. This may explain the limited role it played during the last years of the old quantum theory, when it was no more significant than the Paschen–Back effect. It is noteworthy that the Ramsauer anomaly did not appear in any of the editions of Sommerfeld's *Atombau* or in Born's *Atommechanik*. It only received a partial explanation in 1925, when Walter Elsasser used Louis de Broglie's new ideas of matter-waves to explain how slow electrons can penetrate almost freely through gases because of their very large wavelength (as given by the de Broglie formula $\lambda = h/mv$).[128] At about the same time Bohr returned to the effect and how to understand it in a broader physical context. He revealed some of his thoughts in a letter to Geiger of April 1925:

> Recently I have also felt that an explanation of collision phenomena, especially Ramsauer's results on the penetration of slow electrons through atoms, presents difficulties to our ordinary space-time description of nature similar in kind to those presented by the simultaneous understanding of interference phenomena and a coupling of changes of state of separated atoms by radiation. I believe that these difficulties exclude the retention of the ordinary space-time description of phenomena to such an extent that, in spite of the existence of coupling, conclusions about a possible corpuscular nature of radiation lack a sufficient basis.[129]

Related to the Ramsauer effect, at least as seen in retrospect, were experiments made at the Engineering Department of Western Electric (since 1925 the Bell Telephone Laboratories) by Clinton Davisson and his assistant Charles Kunsman. The two Americans bombarded metal surfaces with electrons at low speed in the hope of elucidating the inner structure of atoms. By analyzing the reflection patterns on the basis of a simplified atomic model devised by Davisson, they suggested that 'the nucleus is surrounded by a shell of electrons which does not completely compensate its charge'. In the case of magnesium (Z = 12) they inferred an inner shell with seven or eight electrons at a distance of about 1.5×10^{-9} cm from the nucleus. Although their result did not agree very well with Bohr's recent picture of the atom—according to which the two K electrons were surrounded by eight L electrons and two M electrons—they nonetheless concluded that it was 'reasonably in accord with the radius of the two quantum circular orbit of the magnesium atom as calculated on Bohr's theory'.[130]

The Davisson–Kunsman experiments attracted little interest in Europe, and none in Copenhagen. However, in Göttingen Minkowski and Sponer realized that some of the results with low-speed electrons obtained in America were anomalous and might be connected to the Ramsauer effect. In a review article of 1924 they attempted to explain both the Ramsauer effect and the Davisson–Kunsman experiments by assuming a restricted validity of classical electrodynamics in cases of elastic collisions of low-energy electrons with atoms.[131] By making use of the correspondence principle they argued that the results were in qualitative agreement with existing quantum theory rather than a problem for it. Indeed, the experiments of Davisson and Kunsman contributed to the crisis in quantum theory even less than the Ramsauer effect. Only in Elsasser's paper from the summer of 1925 were the American scattering experiments interpreted as a diffraction phenomenon inexplicable on the basis of ordinary atomic theory but understandable using the ideas of Einstein and de Broglie.

In the second part of his 1918 memoir on the quantum theory of line spectra Bohr argued that a perturbation of the electron in a hydrogen atom placed in both an electric and a magnetic field would not in general be multiply periodic. 'In such a case', he said, 'we cannot obtain a complete fixation of the stationary states, and we may conclude that the presence of external forces will not give rise to the splitting up of the hydrogen lines into a number of sharp components but to a diffusion of these lines'.[132] The crossed-field problem received considerable attention in the years 1922–1924, especially after Epstein had shown Bohr's intuition to be wrong. He and other physicists—including Otto Halpern, Wilhelm Lenz, and Oskar Klein—proved that in the case of crossed electric and magnetic fields there were multiply periodic solutions, which implied that also in this case the spectral lines of hydrogen would be sharp. These workers considered the problem an interesting exercise in perturbation theory rather than one that indicated a failure in the existing quantum theory.

Only at the end of 1924 did it become clear that the crossed-field case might constitute a problem for the Bohr–Sommerfeld theory of one-electron atoms. The difficulty perceived by Born, Pauli, and a few others was that one might begin with an allowed electron orbit and transform it by means of Ehrenfest's adiabatic principle into a forbidden one, such as a pendulum orbit coming arbitrarily close to the nucleus. 'An escape from this difficulty', said Pauli in his 1926 review of quantum theory, 'can be achieved only by a radical change in the foundation of the theory'.[133] He discussed the problem with Heisenberg at the time the latter developed his ideas of a fundamentally new interpretation of quantum theory. Shortly before the appearance of Heisenberg's

Umdeutung paper, Bohr gave an address to the sixth Scandinavian Mathematical Congress in Copenhagen in which he referred to the crossed-field problem:

> Difficulties of an especially grave nature, however, were brought to light by the interesting analysis by Klein and Lenz of the problem of a hydrogen atom in crossed eletric and magnetic fields. Here it was found impossible to satisfy Ehrenfest's condition, since suitable variation of the external forces would gradually transform orbits depicting stationary states which could not always be excluded from the manifold of these states, into orbits where the electron falls into the nucleus.[134]

It should be noted that Bohr spoke retrospectively. During the final year of the semiclassical Bohr–Sommerfeld theory the crossed-field problem was not generally seen as a serious one that necessitated a break with the theory.[135] Bohr also published his 1925 address in *Naturwissenschaften*, having first sent the German manuscript to Pauli, who had recently calculated the hydrogen atom according to the new quantum mechanics. Pauli pointed out that with his new theory, 'The difficulties with the crossed fields completely disappear'. The root of the difficulty, he now realized, was the arbitrary exclusion of pendulum orbits with $k = 0$ in Sommerfeld's relativistic fine-structure theory: 'This was the cause of all the evils and the original sin of all the considerations of the hydrogen atom resting on the quantization rules of periodic systems'.[136]

Superconductivity does not normally appear in works on the history of atomic and quantum theory, and that for good reasons since the phenomenon only received scant attention by the early generation of quantum physicists. Nonetheless, it may be argued that superconductivity was or should have been an anomaly, namely in so far that it was seen as a quantum effect. Following the discovery of superconductivity in 1911 there were several attempts to find the mechanism responsible for the sudden disappearance of resistance, most of them in the form of phenomenological theories of electrical conduction in metals. A few of these, such as a theory proposed by Wilhelm Wien in 1913, appealed to quantum theory, and in 1919 Fritz Haber invoked the Bohr–Sommerfeld quantization conditions in an attempt to understand the phenomenon.

Bohr did not refer to superconductivity in any of his publications from the period, but he was aware of it at an early date, directly related as it was to the electron theory of metals that he had thoroughly investigated in his doctoral dissertation. For example, in manuscript notes from the spring of 1914 he mentioned 'Kamerlingh Onnes's investigations of the electric conductivity at very low temperatures'. During a visit to the Leiden laboratory in April 1919 Kamerlingh Onnes demonstrated liquid helium to him, but only later, in the early 1930s, did Bohr become seriously involved in an unsuccessful attempt to explain superconductivity.[137]

During the 1921 Solvay conference, Kamerlingh Onnes suggested that superconductivity might be understood on the basis of the new atomic theory.[138] His suggestion was ignored by the experts in quantum physics except that Einstein, in a little known work of 1922, reviewed the problem of metallic conduction and its relation to superconductivity. Einstein, who had discussed the problem with Ehrenfest, assumed atomic electrons to move on quantum orbits with discretely defined momenta. If this were the case the velocity of the electrons would be fixed by the quantum conditions of Bohr's atomic model. However, nothing came out of either Einstein's or others' speculations. 'Given our ignorance of quantum mechanics of composite systems we are far away from being able to convert these vague ideas into a theory', Einstein pessimistically concluded.'[139] Incidentally, this may have been the first time the term 'quantum mechanics' appeared in a scientific text. It took more than thirty years until a satisfactory microscopic theory of superconducting metals was established, the celebrated BCS (Bardeen–Cooper–Schrieffer) theory of 1957. Superconductivity was an anomalous phenomenon in the sense that no theory of physics could account for it. But it was not specifically seen as a quantum anomaly and for this reason the lack of explanation was not considered a problem for the quantum theory of atoms. Physicists *could* have argued that the phenomenon was part of the domain of atomic quantum theory and therefore constituted a challenge to this theory. But they did not.

Notes for Chapter 6

1. Whittaker 1922. Bohr 1924, p. 35. Hendry 1983 has called attention to a 'magnetic tradition' in British atomic theory, including Thomson, McLaren, Allen, Whittaker, and Dirac. Adams 1923 included a careful account of Whittaker's theory, which he found to be 'of very great interest and importance'. On the other hand, he also admitted that the model proposed by Whittaker 'appears to be much more artificial than Bohr's model' (p. 104).

2. The first (1917) edition of the Berlin chemist Karl Hofmann's widely used textbook in inorganic chemistry did not mention Bohr's atomic theory, but the second edition of 1919 included a section on 'the structure of atoms according to Bohr and Debye' (Hofmann 1919, pp. 700–6).

3. Buchner 1915.

4. Untitled draft manuscript from the autumn of 1915 (Niels Bohr Archive, Bohr Scientific Manuscripts no. 101).

5. Lewis 1919, pp. 96–18.

6. See Kragh 1985b, on which part of the present section is based. Most historical accounts of the Bohr–Sommerfeld quantum atom ignore the chemical bond and the chemists' alternative models. Arabatzis 2006 is a welcome exception.

7. Kossel 1923, p. 599. Lepape 1922, p. 84.

8. Lecture of 2 December 1919, in Rud Nielsen 1976, p. 225. Bohr admitted that 'I shall be able to say very little about the questions that chemists may be most interested in, namely, the problems of the structure of molecules'.

9. Lecture of 13 February 1920, in Rud Nielsen 1976, p. 240.

10. Sommerfeld 1922a, p. 88.

11. Kossel 1920, p. 322. See also Kohler 1975.

12. Lodge 1924, p. 203, who was not the only one to compare Bohr with Newton. So did Hilbert, according to whom a reduction of chemistry to physics required 'a Newton of atomic theory, and this has been Niels Bohr, who on the basis of new physical ideas, namely, the quantum theory, made a deeper understanding of this area a possibility' (Sauer and Majer 2009, p. 509).

13. Born 1920, p. 382. Reprinted in Born 1922b, which the following year appeared in an English translation as *The Constitution of Matter*, Born's view was disseminated to a broad audience. The distinction between a higher mathematized physics and a lower empirical chemistry was by no means original. In his *Metaphysische Anfangsgründe* of 1786 Immanuel Kant famously argued that 'Chemistry can be nothing more than a systematic art or experimental doctrine, but never a proper science, because its principles are merely empirical, and allow of no a priori presentation in intuition'. The principles of chemistry, he went on, 'are not receptive to the application of mathematics'. Kant 2004, pp. 6–7.

14. Bjerrum 1949, p. 26, address of 8 March 1922 to the Royal Veterinary and Agricultural College of Copenhagen. Published in Danish in *Naturens Verden* **6** (1922), 145–54.

15. Kossel 1916. See also Stranges 1982, pp. 227–9.

16. Bohr to Kossel, 19 July 1921, in Rud Nielsen 1977, p. 707. The paper Bohr referred to was probably Bohr 1921c.

17. Lewis 1916, p. 773. Intraatomic forces different from the Coulomb force were not only considered by chemists, but also by some physicists. As we have seen in Section 3.7, Lewis was not convinced of Bohr's theory (see also Lewis 1923b, pp. 50–2). He suggested that the momentary position of an electron in its orbit had no physical effects, but that such effects belonged to the orbit as a whole. Moreover, he speculated that not only might electric and magnetic fields turn out to be discontinuous, the same might be the case with space and time (ibid., pp. 164–5).

18. Lewis 1923a, p. 163.

19. Lewis 1926a, p. 171.

20. Walker 1922, p. 743 and p. 744.

21. Born to Einstein, 29 November 1921, in Born 1971, p. 62. On the other hand, Born was less impressed by Lewis's understanding of physics and his criticism of Bohr's atomic model. He tried to convince him that atomic and molecular structure could only be understood on the basis of quantum theory. See Born to Lewis, 27 November 1920, as quoted in Greenspan 2005, p. 103.

22. Langmuir 1921a, p. 553. Langmuir 1919.

23. Andrade 1923, p. 239.

24. Langmuir 1921b, p. 293.

25. Ibid.

26. Bohr to Langmuir, 3 December 1920, in Rud Nielsen 1977, p. 725.

27. Bohr 1921c, p. 106. On this paper, a most important one, see Section 7.1. A few years later Bohr, while visiting the United States, met Langmuir in his laboratory in Schenectady: 'I was very much impressed by his personality and ingenuity and think that we understand each other far better now after having had opportunity to speak from the heart'. Bohr to Rutherford, 9 January 1924, in Stolzenburg 1984, p. 486.

28. Andrade 1923, p. 240.

29. Bohr 1922a, p. 75. Emphasis added.

30. Tolman 1922, p. 212.

31. Ibid., p. 226.

32. Millikan 1924, p. 1411 and p. 1414.

33. Langmuir 1920. Lewis 1923a, p. 56. See also Oxley 1921.

34. Campbell 1920.

35. Bohr 1921c, p. 104. See also Section 7.1.

36. Campbell 1923a. See also Campbell 1923b, pp. 141–4 for a clear discussion of the chemical and physical approaches to the problem of the covalent bond. For historical perspectives, see Kohler 1975, Kragh 1979, pp. 168–9, Kragh 1985b, and Stranges 1982, pp. 240–6.

37. Lewis 1923a, p. 56, and similarly in Lewis 1923b, where he he concluded that the Bohr atom and the chemical atom are 'now completely reconcilable'.

38. Loring 1921, p. 137.

39. Manuscript of 1923 on 'The Application of the Quantum Theory to Molecular Structure' (Archive for History of Quantum Physics, Bohr Scientific Manuscripts).

40. Fowler 1923, p. 461. This was a contribution to a meeting on 'The Electron Theory of Valency' held by the Faraday Society 13–14 July 1923 at Cambridge University. Other contributors included N. V. Sidgwick, J. J. Thomson, G. N. Lewis, T. M. Lowry, and W. Bragg. Lowry suggested in the general discussion that the orbit of an electron around two nuclei might not be elliptical but perhaps be a 'looped or figure-of-eight orbit'. His proposal was not taken seriously.

41. Main Smith 1923, p. 1073, who concluded that Bohr's theory 'must be held to be incapable of general application in chemistry'. See also Main Smith 1924b, p. 176. On his important contributions to the electron theory of the periodic system, see Section 7.6.

42. Main Smith 1924a, p. 323.

43. Born to Einstein, 7 April 1923, in Born 1971, p. 76.

44. On the origin and development of quantum chemistry, see Gavroglu and Simöes 2012.

45. Born and Heisenberg 1924.

46. Heisenberg to Pauli, 7 December 1923, in Hermann, von Meyenn, and Weisskopf 1979, p. 132.

47. The historical literature on this topic includes Fujisaki 1983, Assmus 1992a, and Assmus 1992b. See also Brand 1995, pp. 169–90, and Gearhart 2010, which pays particular attention to the connection between molecular spectroscopy and the problem of the specific heat of hydrogen in the old quantum theory.

48. Bohr 1923b, p. 41.

49. Bohr to Oseen, 20 December 1915, in Hoyer 1981, p. 568. On the works of Bjerrum and von Bahr, see Assmus 1992a and Fujisaki 1983.

50. Schwarzschild 1916. For this work, see also Section 4.3.

51. Rud Nielsen 1976, pp. 190–2.

52. Imes 1919. The gap also appeared in von Bahr's spectrum of 1913, but less conspicuously and without attracting much attention.

53. Reiche 1920, p. 293. Reiche 1921, pp. 155–9. On the half-quanta, see Mehra and Rechenberg 1999 and Gearhart 2010, p. 158.

54. Kratzer 1923.

55. Curtis 1923. On the mysterious half-quanta, see also Barker 1923.

56. Heurlinger 1920, dated 13 december 1919 (Reiche 1920 was dated ten days later).

57. Heurlinger 1919, which included no reference to Bohr's 1918a memoir. Heurlinger's paper was received by *Physikalische Zeitschrift* on 23 January 1919. Lenz 1919 referred to Bohr's 'analogy principle' and also to the Copenhagen memoir.

58. Birge to Kemble, 4 June 1923, quoted in Gearhart 2010, p. 157. See also Assmus 1992b, p. 20. On Heurlinger's work and troubled life, see Fujisaki 1991. Heurlinger was from 1920 plagued by nervous depressions and schizophrenia. For this reason he retired from research and was interned in a mental hospital in Uppsala. He died in 1927 from tuberculosis, at the age of 34.

59. Bohr to Kramers, 15 July 1922, original in Danish. In Rud Nielsen 1976, pp. 658–9.

60. Kramers and Pauli 1923, p. 360.

61. Kemble 1925, p. 1.

62. Einstein to Ehrenfest, ca. October 1913, in Klein et al. 1993, p. 564. See also Klein et al. 1995, pp. 270–3. The early discussion did not refer to atomic structure. For the history of the concept of zero-point energy, see Mehra and Rechenberg 1999, Gearhart 2010, and van Delft 2007, pp. 468–93.

63. Nernst 1921.

64. Mulliken 1924. The full report appeared in Mulliken 1925. For Heisenberg's theory of the Zeeman effect, see Section 8.1.

65. Pauli 1921. See also Enz 2004, pp. 61–3.

66. Zahn 1924, pp. 416–17.

67. Pauling 1926. Van Vleck 1932, p. 107. If Bohr were aware of the problematic nature of the case of the dielectric constant, he ignored it. It appears nowhere in his writings, whether published or not.

68. Kronig 1926. Mensing and Pauli 1926.

69. Eckert and Märker 2004, p. 48.

70. Nernst 1918, p. 153. Franck et al. 1919.

71. Planck 1919.

72. Rud Nielsen 1976, pp. 187–8. A few of Bohr's misspellings have been corrected. This is the first indication of the *Aufbau* or construction principle, which will be considered in Section 7.2.

73. Kunz 1918.

74. Soné 1920. Honda 1922.

75. Soné 1920. Pauli 1920, p. 204. The discrepancy between theory and experiment was also pointed out in McLennan 1923.

76. Lenz 1919. Nordheim 1923, p. 70, likened Lenz's model to a bicycle in which the two wheels spin in opposite directions.

77. Oxley 1920 and Oxley 1921. See also Andrade 1923, p. 279. Characteristic for the state of atomic theory in England, Oxley's qualitative and amateurish ideas of atomic structure were given space in the country's leading physics journals, including *Nature*, *Philosophical Transactions*, and *Proceedings of the Royal Society*. His ideas belonged to an entirely different tradition than the one cultivated by the quantum physicists in Munich, Göttingen, Copenhagen, and Leiden.

78. Born 1922a, p. 677. See also Mehra and Rechenberg 1982a, p. 422, and Born to Sommerfeld, 5 January 1923, in Eckert and Märker 2004, p. 136: 'It seems pretty certain that both of the H-atoms are coplanar (that is, Lenz's model), and perhaps one will succeed also in calculating the heat of dissociation'.

79. Nordheim 1923.

80. Allen 1923, who also constructed a static model of the triatomic hydrogen molecule as an alternative to Bohr's model of 1919.

81. Sommerfeld 1922a, pp. 93–6, 728–30. Reiche 1921, p. 154.

82. Bohr to Oseen, 13 February 1913 (in Danish). German translation in Hoyer 1973, p. 200.

83. Bohr to Rutherford, 1 July 1913, in Hoyer 1973, pp. 200–1.

84. Sommerfeld 1919, p. 508.

85. Pauli 1922. Niessen 1923. For accounts of the H_2^+ case in the old quantum theory, see Jensen 1984 and Mehra and Rechenberg 1982a, pp. 391–5.

86. Pauli 1922, p. 187. According to Darrigol 1992, p. 169, and Enz 2004, p. 65, Pauli's formulation foreshadowed the probabilistic and indeterministic features of the later quantum mechanics.

87. Bohr 1924, p. 12 (footnote).

88. The experiments and their interpretations are discussed in Jensen 1984.

89. Hogness and Lunn 1925, p. 55.

90. Born 1925, p. 281.

91. Rud Nielsen 1977, p. 125.

92. Kramers 1923a.

93. Kemble 1921, according to whom 'it seems probable that the application of the Bohr principle of selection [i.e., correspondence principle] to atoms returning to the normal state will have to be abandoned' (p. 133).

94. Epstein to Bohr, 23 December 1922, in Rud Nielsen 1977, p. 688. Epstein had moved to the United States in 1921, to become professor of theoretical physics at Caltech. He did not publish his calculations of the helium atom.

95. Heisenberg to Pauli, 6 March 1922, in, Hermann et al. 1979, p. 57. In Bohr 1923c, he sought to explain the diamagnetism by arguing that a noble gas atom 'can only orient itself with its axis of angular momentum at right angles to the direction of the field' (p. 278). An English translation of Bohr 1923c can be found in Rud Nielsen 1977, pp. 611–56. See also Sommerfeld to Bohr, 21 January 1923, where he criticizes Bohr's model and its '*awful* conception of magnetism' (Eckert and Märker 2004, p. 143)

96. Sommerfeld 1923.

97. Sommerfeld 1924b, p. 206.

98. Born 1971, pp. 69–70. The Born–Pauli collaboration resulted in an important paper published in the summer of 1922 (Born and Pauli 1922, see also Mehra and Rechenberg 1982a, pp. 410–21).

99. Born 1971, p. 71.

100. Heisenberg to Sommerfeld, 17 October 1922, in Eckert and Märker 2004, p. 123. Sommerfeld stayed at the time as a guest professor at the University of Wisconsin in Madison.

101. Heisenberg to Sommerfeld, 28 October 1922, in Eckert and Märker 2004, p. 127. See also Heisenberg to Pauli, 12 December 1922, in Hermann et al. 1979, p. 73. A lucid account of Heisenberg's work on the helium atom is presented in Cassidy 1992, pp. 145–9.

102. Heisenberg to Sommerfeld, 15 January 1923, in Eckert and Märker 2004, p. 140.

103. Heisenberg to Bohr, 2 February 1923, quoted in Cassidy 1992, p. 148.

104. Heisenberg to Pauli, 19 February 1923, in Hermann et al. 1979, p. 79. In addition to this 'second catastrophe', Heisenberg also mentioned another catastrophe related to the mechanical stability of the hydrogen molecule.

105. Born and Heisenberg 1923, p. 242. Born used the same phrase in his *Atommechanik*, which covered the helium anomalies both in the case of the normal and the excited atom (Born 1925, pp. 327–41).

106. Heisenberg to Pauli, 26 March 1923, in Hermann et al. 1979, p. 86. Heisenberg retrospectively considered the helium failure the beginning of the crisis in quantum theory that eventually led to quantum mechanics (Heisenberg 1929, p. 491).

107. Born to Einstein, 7 April 1923, p. 75 in Born 1971.

108. Born to Bohr, 4 March 1923 (Rud Nielsen 1977, p. 669), where Born reported the negative result of his and Heisenberg's calculations and asked about Bohr's opinion before publishing them. Heisenberg to Pauli, 19 February 1923, in Hermann et al. 1979, p. 80.

109. Van Vleck 1923, p. 373.

110. Bohr 1923c, p. 271. The paper was a contribution to a special issue of *Annalen*, dedicated to the spectroscopist Heinrich Kayser in honour of his seventieth birthday.

111. Bohr to Born, 2 May 1923, in Rud Nielsen 1977, p. 673.

112. See Darrigol 1992, pp. 176–8.

113. Born 1923, p. 542.

114. Sommerfeld 1919, p. 439. The theory of the hydrogen atom, he said, 'stands most likely in contradiction to experience' and is 'as yet unable to explain the complicated Zeeman types and their simplification into the Paschen–Back effect'.

115. On the Paschen–Back effect and its role in the old quantum theory, see Jensen 1984, Kragh 1985a, and Robotti 1992. Details and references to the primary literature can be found in these sources.

116. Sommerfeld 1922a, p. 604.

117. Bohr 1918a, p. 92 (Rud Nielsen 1976, p. 158). Bohr to Richardson, 26 February 1920 (Archive for History of Quantum Physics, Bohr Scientific Correspondence).

118. Bohr 1922c, p. 117 (Rud Nielsen 1976, p. 183).

119. Sommerfeld 1924b, p. 670. The fourth edition of *Atombau* covered the Paschen–Back effect over twenty pages.

120. Falkenhagen 1924, p. 10.

121. Sommerfeld and Unsöld 1926, p. 268. The first to give a satisfactory explanation of the Paschen–Back anomaly on the basis of the spin quantum number was John Slater at Harvard University.

122. Heisenberg to Landé, 18 February 1925, as quoted in Forman 1968, p. 173.

123. Franck to Bohr, 25 September 1921, in Thorsen 1987, p. 689. On the Ramsauer effect, see Mehra and Rechenberg 1982a, pp. 620–5. The experimental work carried out in Germany, England, and the United States is discussed in Soon Im 1995. The first to report the effect was Nils Åkesson, a Swedish physicist working in Heidelberg, who did so in a paper of 1914.

124. Bohr to Franck, 27 September, in Thorsen 1987, p. 691.

125. Born to Einstein, 29 November 1921, in Born 1971, p. 62. Born sketched some general ideas, based on the Maxwell–Boltzmann gas theory and not involving quantum theory, to explain the phenomenon.

126. Hund 1923. Hund justified his theory by arguing that it conformed with the recent views of Bohr, such as exposed in Bohr 1923a.

127. Bohr to Franck, 29 December 1922, in Thorsen 1987, p. 694. Hund had attended Bohr's Wolfkehl lectures in Göttingen in June 1921 and so Bohr had presumably met him. However, in his letter he referred to him as 'Haid', apparently a misreading of Franck's handwriting.

128. Elsasser 1925. See Medicus 1974.

129. Bohr to Geiger, 21 April 1925, in Stolzenburg 1984, p. 79, and similarly in a letter to Ralph Fowler of the same date (ibid., p. 81). For a published version of the speculation, see Bohr 1925b, which includes a first reference to de Broglie's ideas.

130. Davisson and Kunsman 1923, p. 253, who did not refer to Bohr's recent works on atomic structure and the periodic system. On the experiments and Davisson's later work, which led to the discovery of electron diffraction for which he was awarded the Nobel Prize in 1937, see Russo 1981.

131. Minkowski and Sponer 1924. In his correspondence with Pauli from June 1925, Heisenberg referred to the Ramsauer effect and the Davisson-Kunsman experiments, which he thought were in agreement with 'the Einstein theory' (he did not mention de Broglie). Heisenberg to Pauli, 24 June and 19 June 1925, in Hermann et al. 1979, pp. 226–9.

132. Bohr 1918b, p. 93.

133. Pauli 1926a, p. 164. See also Born 1925, p. 276.

134. Bohr 1925a, pp. 850–1.

135. See Mehra and Rechenberg 1982a, pp. 502–9 for a survey of the crossed-field problem and references to the literature. Their evaluation that it 'had been crucial in showing the failure of the Bohr–Sommerfeld theory of atomic structure' (1982b, p. 263) is undoubtedly an exaggeration.

136. Pauli to Bohr, 17 November 1925, in Hermann et al. 1979, pp. 257–8. When Bohr's paper appeared in *Naturwissenschaften* (**14**, 1926, pp. 1–10), it included a reference to Pauli's theory. In his quantum-mechanical theory of the hydrogen atom, Pauli concluded that 'special additional exclusion rules become superfluous in the new quantum mechanics, in which we have not conceived the stationary states as represented by particular electron orbits; hence the [crossed-fields] difficulties indicated above disappear automatically'. Pauli 1926b, p. 344.

137. On Bohr's ideas on superconductivity, which resulted in an unpublished paper on the subject ('Zur Frage der Supraleitung'), see Aaserud 2007, pp. 5–9. Van Delft 2007, p. 547, includes a picture showing Bohr in the Leiden laboratory together with Lorentz, Ehrenfest, and Kamerlingh Onnes.

138. See Dahl 1992 and Sauer 2007 for early theories of superconductivity. For Kamerlingh Onnes's suggestion, see Dahl 1992, p. 104, and for his continued interest in connecting the phenomenon to Bohr's atomic theory, Sauer 2007, p. 203.

139. Einstein's publication on superconductivity, entitled 'Theoretische Bemerkungen zur Supraleitung der Metalle', is analyzed in Sauer 2007. It is available in an English translation by Bjoern Schmekel (Einstein 1922). Bohr was aware of and 'enormously interested' in Einstein's ideas of superconductivity. See Ehrenfest to Einstein, 27 December 1921, in Buchwald et al. 2009, p. 396.

7

A Theory of the Chemical Elements

In the second part of his 1913 trilogy Bohr suggested electron configurations for many-electron ring atoms and sketched a theory of the periodic system, but this was an area to which he gave low priority in his subsequent research. He did not abandon it, however, and during the years 1920–1923 he returned to it, developing it into a comprehensive theory of the structure of all chemical elements.

This 'second atomic theory' differed in important respects; not only from Bohr's original conception of the atom but also from the conceptions of most other physicists. It was a detailed theory of the atom that relied on a mixture of empirical facts and principles of a general nature rather than calculations based on established quantum theory. The theory was received very positively, even enthusiastically, but also with some reluctance because of its unusual character, which to a large extent reflected Bohr's intuition and personal style of science. Substantial parts of Bohr's reasoning could not be reproduced or understood by other physicists because they expressed what Einstein called Bohr's 'musicality' and 'unique instinct and tact'. How could one build up atomic structures on the basis of the correspondence principle and vague references to the harmonic interplay of symmetrically moving electrons? Bohr did not reach his results solely by a divine use of such general principles, but only by combining them with an imaginative use of experimental data, primarily from spectroscopy and chemistry. In addition, he derived several results from the idea of penetrating elliptical orbits, which was a technique more in line with the traditional methods of the Bohr–Sommerfeld theory.

Bohr's theory of the periodic system scored several successes, especially in the area of X-ray spectroscopy, where it was spectacularly confirmed by the discovery of element 72 (hafnium) in late 1922. We shall return to the question of whether the discovery of hafnium was really based on a prediction from Bohr's new version of his atomic theory.

Conceptually the Bohr theory of the periodic system can be seen as the crowning achievement of the old atomic theory. In a sense it was conservative, the culmination of the semiclassical models, which during the first two decades of the century were developed with increasing confidence. But it was not, contrary to his theory of 1913, destined to revolutionize physics. The approach to atomic physics that in the course of a few years created the new quantum mechanics was based on ideas that in many ways contradicted essential parts of Bohr's second theory. The contradiction was clear already before the emergence of quantum mechanics, and can be seen in Pauli's new and very successful theory of the periodic system that both built on and transcended Bohr's. There was a world of difference between Bohr's atomic models of the elements, half pictorial and half symbolic, and Pauli's formal theory, which did not 'need the crutch of the idea of unambiguously defined electron orbits and mechanical model'.[1]

7.1 BOHR'S SECOND ATOMIC THEORY

Apart from preparing his new institute, in the autumn of 1920 Bohr was not only contemplating the structure of the helium atom but also the structure of more complicated atoms. He no longer thought that these atoms could be understood on the basis of ring models, as he had suggested in the second part of his trilogy of 1913. His return to this question may have been prompted by his correspondence with Ladenburg, who in June 1920 asked him about his view of Kossel's shell model and the associated idea of the noble gases being characterized by an outer shell with eight electrons. 'Do you still prefer your old view of two electrons in the outer shell of the noble gases?' Ladenburg wanted to know.[2] In his letter of reply Bohr used the occasion to stress that chemical stability must be considered from a dynamical rather than a geometrical point of view:

> The difficulties with the problem lie especially in the rational use of the different proposed electron configurations in the explanation of the chemical properties of the elements in question. This is, after all, not only related to the geometrical character of the configuration, but first of all to the conditions of stability of the configurations . . . [I]t also seems that an assumption of rings already has to be given up because of insufficient stability and that we are forced to expect much more complicated motions of the electrons in the atom.[3]

Bohr offered the first version of a revised theory of atomic structure in his December 1920 lecture to the Physical Society in Copenhagen, where he introduced the concept of penetrating orbits as providing a coupling mechanism between inner and outer electrons (Figure 7.1). He discussed in purely qualitative terms the electron configurations of the first nine elements in the periodic system, from hydrogen to fluorine, and distanced his own view of the carbon atom from that held by Landé and Born. The special properties of this element, such as its four tetrahedral valencies, could not be explained on the basis of cubical symmetry, he pointed out against the *Würfelatom* model, for if this were the case one would expect beryllium to have the same properties as carbon. No, 'the characteristic properties of carbon must be assumed to arise from the fact that in the neutral atom there are four electrons which are coupled to the two inner electrons'.[4]

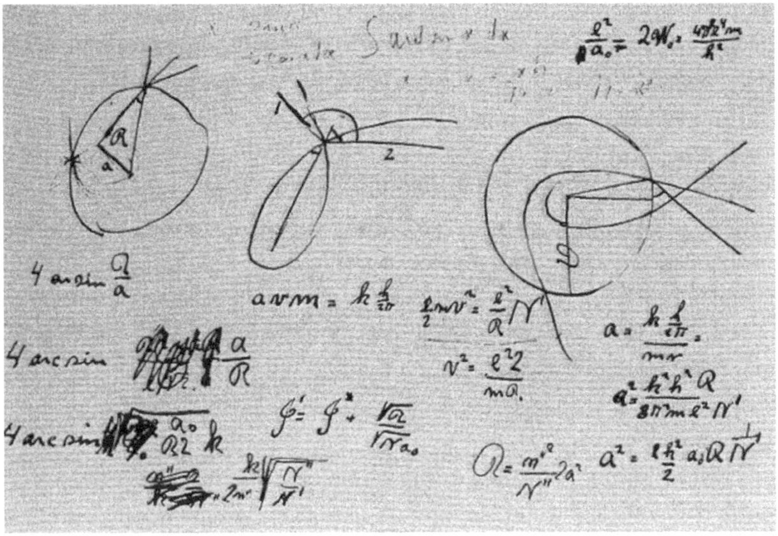

Fig. 7.1. Bohr's rough notes, probably from late 1920, for his new atomic theory, embodying penetrating electron orbits.
Source: Kragh 1985c, p. 53 (Niels Bohr Archive).

Four months after the Copenhagen lecture a fuller and more precise account of Bohr's new view of the constitution of atoms appeared in the letters section of *Nature*.[5] The occasion was a reply to Campbell's call for a reconcilation between the chemical–static and the physical–dynamic models of the atom, but this was a pretext only. In his *Nature* article Bohr not only criticized the chemical models, he also took issue with the physical theories of atoms with a polyhedral symmetry. This he had also done in his Copenhagen lecture, but he now based his arguments on the correspondence principle, which he claimed would 'remove the arbitrariness in the choice of the number and configurations of the electrons in the various groups, or shells, of the atom'. The same principle ruled out the symmetrical models of Sommerfeld, Landé, and Born, 'because the formation of such configurations would claim that all the electrons within each group should be originally bound by the atom at the same time'.[6] Why this consequence followed from the correspondence principle he did not explain, but then there were many claims in the article that appeared without a proper explanation and must have puzzled most readers. Nonetheless, the article aroused great interest, both among chemists and physicists. 'The announcement in your letter to *Nature* of the solution of the riddle of the periodic system has filled us chemists with particularly great expectations', Kasimir Fajans wrote in a letter to Bohr. 'When we once get oriented about the electron configurations in a large number of elements, I believe that many questions of chemistry will appear in a new light'.[7]

The letter to *Nature* contained the basic features of what can be termed Bohr's second atomic theory or his theory of the periodic system.[8] According to this theory, as presented in a rather provisional version in the *Nature* article, electrons make up various groups and subgroups corresponding to certain quantum numbers and approximate elliptic orbits. These groups were sometimes called 'shells', but they were not shells in the ordinary meaning of the term. During their orbital motions some of the electrons penetrate the region of internal electrons, causing a coupling of the revolving electrons, whose numbers and motions are governed by the correspondence principle. Although Bohr specified the orbits of the electrons by their principal quantum number n and their azimuthal quantum number k, he used only the former in his early classification of the electron configurations in the elements. In the later and more mature versions of the theory he used the nomenclature $(n_k)^x$ as indicating a group of x electrons with quantum numbers n and k.

Bohr avoided using the inner quantum number j that had recently been introduced by Sommerfeld as a means to understand the complex spectra. By subjecting it to the selection rule $\Delta j = \pm 1, 0$, Sommerfeld succeeded in making formal sense of the spectra, but the physical meaning of the new quantum number escaped him. The inner quantum number might perhaps correspond to a 'hidden rotation', he thought, but of its geometrical and physical significance he admitted ignorance. In a letter to Einstein he described the new quantum number as 'beautiful', yet also confessed that he was unable to 'imagine anything' about its nature.[9] Bohr fared no better. He could not find a way to associate the inner quantum number with a dynamical meaning in agreement with what he called the *Aufbau* or construction principle. It is not possible, he wrote in 1923 with regard to the helium atom, 'to furnish a rational classification of the stationary states of the two helium spectra with the help of the quantum numbers n, k, j on the current basis [of the quantum theory]'. He emphasized that 'the theory is unsettled with respect to the meaning and determination of the absolute value of the quantum number j'.[10] If the j quantum number was of no help in the case of helium, there was no reason to believe it would be helpful in the classification of the energy levels of the higher atoms.

In his letter to *Nature* of March 1921 Bohr denoted the number x of electrons in a particular group or shell characterized by the quantum number n as x_n. He suggested that the population numbers of the noble gases were as follows:

2 He	2_1	36 Kr	$2_1 8_2 18_3 8_2$
10 Ne	$2_1 8_2$	54 Xe	$2_1 8_2 18_3 18_3 8_2$
18 Ar	$2_1 8_2 8_2$	86 Rn	$2_1 8_2 18_3 32_4 18_3 8_2$.

Six months later, in another letter to *Nature*, he revised the classification, now concluding that the principal quantum number associated with each group of electrons increases by one unit as the atom is formed, implying that the value of n plays the role of a shell number. He did not reveal how he had come to the new and 'rigorous classification', except that it was based on the inevitable correspondence principle and 'a detailed examination of the parts of the orbital loops situated within the region of inner groups'.[11]

With the new classification scheme he wrote the configurations of argon and krypton as:

$$\mathrm{Ar} = 2_1 8_2 8_3 \text{ and } \mathrm{Kr} = 2_1 8_2 18_3 8_4$$

For xenon and radon he similarly revised the structures to:

$$\mathrm{Xe} = 2_1 8_2 18_3 18_4 8_5 \text{ and } \mathrm{Rn} = 2_1 8_2 18_3 32_4 18_5 8_6.$$

As shown by these examples, Bohr adopted the idea of intermediate shells (*Zwischenschalen*) that Ladenburg had introduced in 1920 but had been anticipated by Vegard and a few other scientists, Bohr and Sommerfeld included.[12] According to this idea, the number of electrons in the outer shell remains unchanged while an interior shell is completed with electrons. Bohr argued that the rare earth metals are characterized by a building up in the fourth shell ($n = 4$) from 18 electrons to 32 electrons, while the two outer shells remain unaffected by the building-up process. This implied that there had to be fourteen rare earth elements. In 1921 he did not state with which elements the rare earth group began and ended, a question he would take up the following year.

Shortly after his second letter to *Nature*, Bohr again lectured before the Physical Society in Copenhagen, this time at a meeting held jointly with the local Chemical Society. Greatly elaborating on his previous expositions he now dealt specifically with the periodic system and with the structure of all atoms. The published versions of the lecture—in Danish, German, English, French, and Russian—attracted much attention and helped his theory become widely known.[13] Ehrenfest, who at the end of 1921 stayed at Bohr's institute, wrote enthusiastically to Einstein about all the wonderful things that happened in Copenhagen. Bohr—whom he described as 'a tremendous physicist'—would soon publish his lecture: 'He presents how he has now interpreted the structure of *all* atoms. Both methodologically and as regards its results this is something quite enormous'.[14]

Even before the lecture appeared in the *Zeitschrift für Physik*, the Danish version was known to the physicists in Göttingen. 'I made a bungling translation and then Pauli and I tried to understand it', Franck wrote to Bohr. 'It is indeed wonderful what you have produced in the meanwhile, and the curiosity of the local physicists to get to know your methods mathematically as well is tremendous'.[15] It was generally believed at the time that Bohr's results rested on elaborate calculations, although this was not the case at all (see Section 7.5).

In addition to his scientific articles on the subject, Bohr covered it in two series of lectures, one given at Cambridge in March 1922 and the other being the Wolfkehl lectures at Göttingen three months later. He included a fairly detailed account of the theory in his Nobel lecture of December 1922 and it also appeared prominently in books written by Campbell (*The Structure of the Atom*), Sommerfeld (*Atombau*), Birtwistle (*The Quantum Theory of the Atom*), Andrade (*The Structure of the Atom*), Born (*Atommechanik*), Kramers and Holst (*The Atom and the Bohr Theory of its Structure*), and other authors. In France, scientists could get acquainted with the theory through the French translation of the Copenhagen lecture, which was known even earlier through a positive and insightful review by Léon Bloch. 'From a logical point of view, Bohr has reduced the question of the structure of the atom to a purely mathematical one', Bloch wrote, yet he was fully aware that in reality Bohr's theory was to a large extent based on empirical results.[16] In Germany, the Berlin chemist Fritz Paneth wrote an equally positive review in which he contrasted Bohr's theory—which 'cannot be rated highly enough'—with the 'conscious fictions' of people like Lewis and Langmuir. Bohr's theory, he said, 'claims to be the truth in as elevated a sense as one is justified at all to talk about an objectively real truth in any area of the sciences'.[17]

Bohr's theory rested heavily on general principles, and he frequently stressed this characteristic feature of the theory. He hoped to escape the Scylla of the unfruitful chemical models and the Charybdis of the hopelessly complex quantum-theoretical calculations by appealing to a qualitative understanding of the general principles of quantum physics. Of course, he often had to rely on empirical evidence, as he did in his original theory of 1913, but he did not think of his second theory as derived inductively from such evidence. Bohr rather thought that only by consistent use of general principles could one gain insight into the structure of atoms and thereby explain, for instance, the nature of Rydberg's numerological rule concerning the number of the elements in the various periods. However, his theory became vague or plainly unintelligible in the very parts in which he used general principles most extensively. Many readers of his works were fascinated by his qualitative arguments from correspondence and symmetry, but only few found them convincing. And understandably so, considering that he used the correspondence principle as a provisional tool in solving problems that were not yet ripe for theoretical clarification. Bohr tried to express what could not yet be expressed and ended up sounding cryptic. His works on atomic structure from the period 1920–1924 were very much permeated with his personal, almost idiosyncratic way of thinking. They could not possibly have been written by any other physicist.

In spite of his confidence in the general principles on which the theory rested, or ostensibly rested, Bohr nevertheless recognized that his extensive use of these principles was provisional and somewhat unsatisfactory. In Göttingen he spoke of symmetry considerations as 'very uncertain' at the present stage of the theory and acknowledged that 'it may be justified to ask how much weight to attach to these conclusions' drawn from them. 'A large amount of experimental material has contributed to the shaping of our view', he admitted. 'It is a matter of taste, I believe, whether to put the main emphasis on the general considerations or on the bare experimental facts'.[18] Bohr's taste clearly was for the general considerations, but it was a taste not shared by the majority of physicists.

By making eclectic use of general principles and empirical knowledge of the elements, Bohr was able to assign, for the first time, definite electron configurations to all the elements and in this way to reconstruct—not deduce—the periodic system. It was a reconstruction that in a not very clear way relied on different sources, some of them of a theoretical nature and others of an empirical nature. As expressed by Edward MacKinnon, in artistically blending atomic theory, general principles of quantum theory, optical and X-ray spectral data, and chemical knowledge, 'Bohr functioned more like a symphony director than a deductive logician'.[19]

The Copenhagen symphony director not only explained the change from one period to another, he also offered explanations of the transition groups and the rare earth group. The character of the theory led Bohr to believe that the principal problems of building up atoms were solved once he had established the general principles and tested them in the case of the lighter elements. He did not expect any new features to arise after the fourth period ending with krypton. The mechanical and therefore computational problems would of course increase with the number of electrons, but this was not a great obstacle for a semi-qualitative and non-mechanical theory of the kind that Bohr had devised. On the contrary, as he said in one of his Göttingen lectures: 'I hope that I have succeeded in showing that matters become simpler and simpler the farther we proceed in the periodic system. Formerly I believed that the difficulties would become increasingly greater the more electrons there are in the atoms. However, if we do not demand too much, it actually seems that we encounter simpler problems, and fewer new ones'. He cautiously added: 'I hardly need to emphasize how incomplete and uncertain everything still is'.[20]

In spite of his generally cautious attitude, Bohr did not hesitate to construct hypothetical transuranic elements, announcing to his audience in Göttingen that 'We might proceed further . . . and construct hundreds or thousands of elements; however, that is not the task of physics, which deals only with things that can be put to experimental test'. Not only did he write down the complete electron configuration for the uranium atom, the most complicated atom known at the time, he also predicted the configuration of the hypothetical element with atomic number $Z = 118$ (Figure 7.2). According to Bohr's theory, this element should be a noble gas with '8 electrons each in 5_1, 5_2, 5_3 and 5_4 orbits, 6 in each of 6_1, 6_2 and 6_3 orbits, and four each in 7_1 and 7_2 orbits'.[21] A few years later he asked Yoshio Nishina, a visiting physicist from Japan, to examine by means of X-ray spectroscopy whether there might be, as he suspected, elements of $Z = 93$, 94, or 96 homologous to uranium.[22]

In lectures in Göttingen, Cambridge, Copenhagen, and Stockholm, Bohr illustrated his theory with a new version of the periodic system that he had adapted from his fellow countryman, the chemist Julius Thomsen (Figure 7.3). Since his student days he had been acquainted with Thomsen's system of 1895, which was arranged in horizontal groups and vertical periods, in contrast to the standard version.[23] Thomsen, who

	1_1	$2_1 2_2$	$3_1 3_2 3_3$	$4_1 4_2 4_3 4_4$	$5_1 5_2 5_3 5_4 5_5$	$6_1 6_2 6_3 6_4 6_5 6_6$	$7_1 7_2$
1 H	1						
2 He	2						
3 Li	2	1					
4 Be	2	2					
5 B	2	2(1)					
– –	–	– –					
10 Ne	2	4 4					
11 Na	2	4 4	1				
12 Mg	2	4 4	2				
13 Al	2	4 4	2 1				
– –	–	– –	– –				
18 A	2	4 4	4 4				
19 K	2	4 4	4 4	1			
20 Ca	2	4 4	4 4	2			
21 Sc	2	4 4	4 4 1	(2)			
22 Ti	2	4 4	4 4 2	(2)			
– –	–	– –	– – –	–			
29 Cu	2	4 4	6 6 6	1			
30 Zn	2	4 4	6 6 6	2			
31 Ga	2	4 4	6 6 6	2 1			
– –	–	– –	– – –	– –			
36 Kr	2	4 4	6 6 6	4 4			
37 Rb	2	4 4	6 6 6	4 4	1		
38 Sr	2	4 4	6 6 6	4 4	2		
39 Y	2	4 4	6 6 6	4 4 1	(2)		
40 Zr	2	4 4	6 6 6	4 4 2	(2)		
– –	–	– –	– – –	– – –	–		
47 Ag	2	4 4	6 6 6	6 6 6	1		
48 Cd	2	4 4	6 6 6	6 6 6	2		
49 In	2	4 4	6 6 6	6 6 6	2 1		
– –	–	– –	– – –	– – –	– –		
54 X	2	4 4	6 6 6	6 6 6	4 4		
55 Cs	2	4 4	6 6 6	6 6 6	4 4	1	
56 Ba	2	4 4	6 6 6	6 6 6	4 4	2	
57 La	2	4 4	6 6 6	6 6 6	4 4 1	(2)	
58 Ce	2	4 4	6 6 6	6 6 6 1	4 4 1	(2)	
59 Pr	2	4 4	6 6 6	6 6 6 2	4 4 1	(2)	
– –	–	– –	– – –	– – – –	– – –	–	
71 Cp	2	4 4	6 6 6	8 8 8 8	4 4 1	(2)	
72 –	2	4 4	6 6 6	8 8 8 8	4 4 2	(2)	
– –	–	– –	– – –	– – – –	– – –	–	
79 Au	2	4 4	6 6 6	8 8 8 8	6 6 6	1	
80 Hg	2	4 4	6 6 6	8 8 8 8	6 6 6	2	
81 Tl	2	4 4	6 6 6	8 8 8 8	6 6 6	2 1	
– –	–	– –	– – –	– – – –	– – –	– –	
86 Em	2	4 4	6 6 6	8 8 8 8	6 6 6	4 4	
87 –	2	4 4	6 6 6	8 8 8 8	6 6 6	4 4	1
88 Ra	2	4 4	6 6 6	8 8 8 8	6 6 6	4 4	2
89 Ac	2	4 4	6 6 6	8 8 8 8	6 6 6	4 4 1	(2)
90 Th	2	4 4	6 6 6	8 8 8 8	6 6 6	4 4 2	(2)
– –	–	– –	– – –	– – – –	– – –	– – –	–
118 ?	2	4 4	6 6 6	8 8 8 8	8 8 8 8	6 6 6	4 4

Fig. 7.2. Electron configurations of the elements as Bohr presented them in his Nobel lecture of 1922. Notice the hypothetical element of atomic number 118 and also that element 72 has not yet become hafnium.

Source: Bohr 1923b, p. 40.

believed that the periodic system reflected the atomic structure of the elements, had formulated Rydberg's rule several years before Rydberg, namely in the form that the number of the elements in the periods can be written as $2n^2$ (to be precise, as $2n^2 - 1$, since the noble gases were not known at the time). Like Thomsen, Bohr found the system suggestive of atomic interpretations. 'Compared with usual representations of

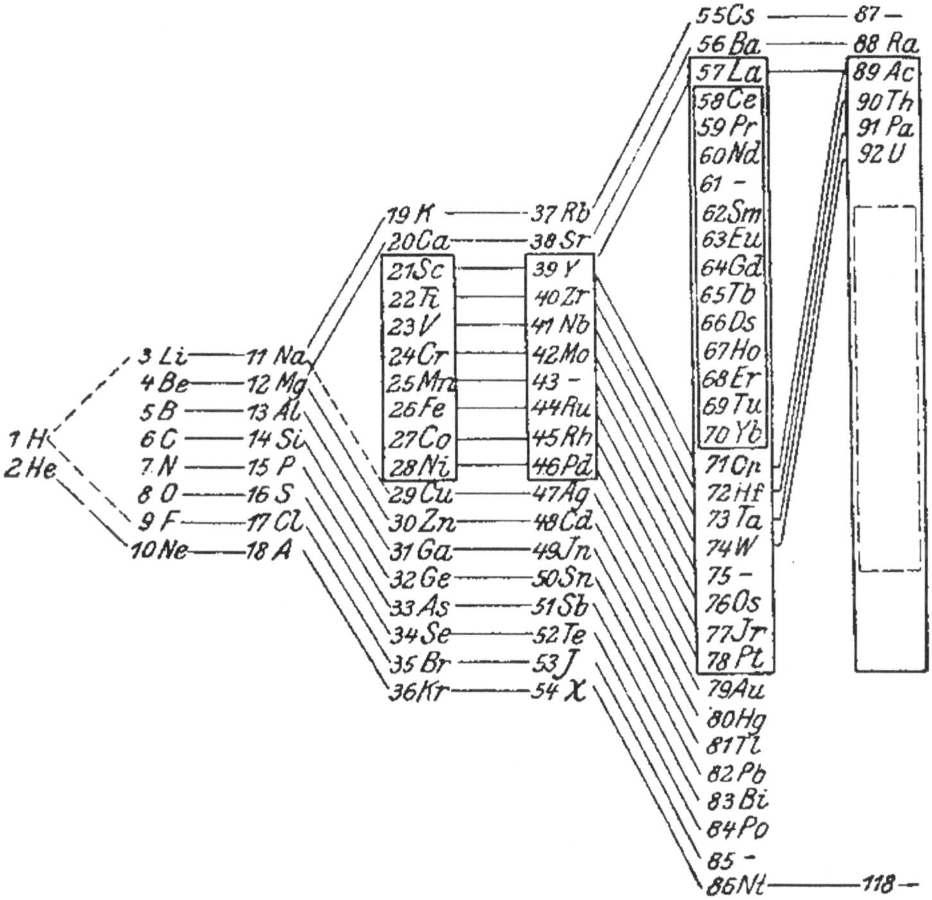

Fig. 7.3. The Thomsen–Bohr version of the periodic system. In this version element 72 is designated Hf = hafnium.

Source: Kramers and Holst 1925, p. 179.

the periodic system', he said, 'this method, proposed more than twenty years ago by Julius Thomsen, of indicating the periodic variations in the properties of the elements is more suited for comparison with theories of atomic constitution'.[24] The periodic system used by Bohr is sometimes known as the Thomsen–Bohr system.

In his lectures on atomic theory during the period in question, Bohr also made use of large plates that pictured the electron orbits in the atom of selected elements such as lithium, sodium, copper, and krypton. The orbits were drawn to scale and coloured in red and black. The plates, with their visual qualities and beautiful geometric forms, impressed the audience—in particular the picture of radium in which all 88 orbits were meticulously drawn (Figure 7.4). They gave the impression that this was really what atoms looked like, electrons moving about a nucleus in definite orbits much like planets or asteroids moving about the Sun. Although Bohr considered the pictures as symbolic

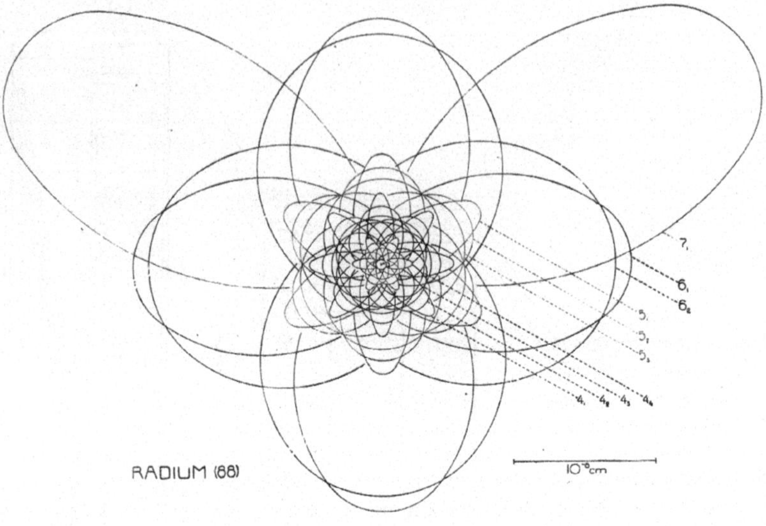

Fig. 7.4. The 88 electron orbits of radium according to Bohr's theory of 1921. In the original plates the orbits with odd principal quantum number were coloured in red, those with even quantum number in black. The elliptic orbits are shown closed for simplicity, but should really be slightly open, as the ellipses slowly precess.

Source: Kramers and Holst 1923.

rather than concrete representations, at the time he apparently had little doubt about the reality of the electron orbits or the fertility of the semimechanical model concept on which his theory relied.[25] A literal or at least semiliteral interpretation was certainly the message which his lectures and articles signaled to his colleagues in physics. His use of the plates only reinforced the message.

Bohr and his colleagues in Copenhagen were not alone in presenting visualizable models of the atom for pedagogical and educational purposes. Thus, as early as 1918 Sommerfeld made a sketch of the hydrogen atom, which a few years later, together with a model of Pauli's H_2^+ ion, was turned into a three-dimensional model displayed at the Deutsches Museum in Munich. Similar but more complicated models of the Bohr orbital atom, based on calculations made by Lawrence Bragg and Douglas Hartree, were shown at the British Empire Exhibition in London 1924–1925 and later at the Science Museum.[26]

To get a feeling of how Bohr reconstructed the periodic system, let us briefly follow his account of the structure of the lighter elements. While the two electrons in helium are in a 1_1 state, the binding of the next one will result in a 2_1 orbit, corresponding to lithium. Subsequent bindings will fill up the 2-quantum orbits until they are completed in neon, which ends the second period with the configuration Ne $= (1_1)^2(2_1)^4(2_2)^4$. This building-up scheme continues in the third period, where electrons are bound in 3-quantum orbits, ending with argon's eight electrons, namely $(3_1)^4(3_2)^4$. In potassium the electron does not go into a 3_3 state, but into the energetically favoured 4_1 state, and the same is the case with calcium of atomic number 20. From scandium to zinc the new electrons added go into 3_k orbits lying in the interior of the atom rather than into 4_1 orbits on its surface. In particular, from iron to copper they go into 3_1 and 3_2 orbits. In copper the 3_k orbits are completed with the structure $(3_1)^6(3_2)^6(3_3)^6$. In his Nobel lecture Bohr commented:

> Although we are not yet in a position to account in all details for the steps in the gradual development of the 3-quantum electron group, still we can say that with the help of the quantum theory we see at once why it is in the 4th period of the system of the elements that there occur for the first time successive elements with properties that resemble each other as much as the properties of the *iron group*; indeed, we can even understand why these elements show their well-known paramagnetic properties.[27]

As Fajans noted in his letter to Bohr, the new theory of atomic structure illuminated many chemical problems. However, it had disappointingly little to say of what interested chemists most, namely the nature of the chemical bond. Bohr vaguely suggested that a tetrahedral configuration of groups of 2-quantum orbits in carbon 'seems capable of furnishing a suitable foundation for explaining the structure of organic compounds', but then left the matter. As far as chemical combinations were concerned, he found it enough to deal with the easier case of ionic compounds. His argument was that 'the structures of combinations of atoms of the same element and of many organic compounds do not have the same significance for our purpose as those molecular structures in which the individual atoms occur as electrically charged ions'.[28] In other words, he chose to ignore the covalent bond because he judged it to be of little importance for atomic theory. Without saying so directly, he admitted that molecules were outside the reach of the theory. The electron orbits in molecules could probably not be classified as n_k states or understood in a similar simple way, and 'for just this reason it seems at present hardly possible to proceed, with the aid of the ideas of the quantum theory, far beyond the more qualitative considerations with which the chemist has explained the regularities of organic chemistry with so great success'.[29]

7.2 THE *AUFBAU* PRINCIPLE AND PENETRATING ORBITS

The point of departure of Bohr's theory was considerations of how an atom can be *formed* by the successive binding of electrons, this binding taking place in stages by transitions between stationary states according to quantum theory. When Ehrenfest presented Bohr's ideas at the third Solvay congress in Brussels, Lorentz found it 'very remarkable' that Bohr concerned himself primarily with the formation of the atom instead of with its structure—or rather, that he obtained insight into the structure of the atom by considering its formation.[30]

According to Bohr, in the building-up process a correspondence existed between the motion of the last captured electron and the occurrence of transitions between the stationary states corresponding to the various stages of the binding process. But how to choose between the numerous possibilities of the capture and binding of an electron? This is where the correspondence principle entered the picture: 'Among the processes which are conceivable and which according to the quantum theory might occur in the atom we shall reject those whose occurrence can not be regarded as consistent with a correspondence of the required nature'. He admitted that his atomic configurations could not 'in every detail be considered as the unambiguous result of applying the correspondence principle', but at the same time he maintained that these configurations were the only ones that agreed with both the physico-chemical properties of the elements and the general principles of quantum theory.[31]

What Bohr named the *Aufbauprinzip* (construction or building-up principle) was the general idea that the addition of electron number p to a partially completed atom with $p-1$ bound electrons will leave the quantum numbers of the $p-1$ electrons unchanged. When a new atom is imagined as being formed by capture processes, the principal quantum number will differ from that of the already bound electrons in the outer shell only if the atom being formed belongs to a new period in the periodic system. In 1923 he introduced what he called the 'postulate of invariance and permanence of quantum numbers' as follows: 'During a capture process the quantum numbers, characterizing the motions of each of the former bound electrons in the ground state of the building-up levels, will remain unchanged under the disturbing effect of the incoming electron'.[32] It followed by the construction principle that in each new period the principal quantum number increases by one unit, as he first pointed out in his second letter to *Nature*.

Bohr was well aware that the use of arguments based on the correspondence principle and the adiabatic principle presupposed the system in question to be either simply or multiply periodic, and that this was not the case for atoms with many electrons. He nonetheless applied his reasoning to such non-periodic systems, assuming that the principles mentioned would still be valid. In his article on the application of quantum theory to atomic structure from 1923 he argued as follows:

> But now, in order to give an account of the observed properties of the elements, we are forced to assume that the atoms, in the absence of external forces at any rate, always possess 'sharp' stationary states, although the general solution of the equations of motion for atoms with several electrons, even in the absence of external forces, exhibits no simple periodic properties of the kind mentioned... We shall try to show that, notwithstanding the uncertainty which the preceding considerations contain, it yet seems possible, even for atoms with several electrons, to characterise their motion in a rational manner by the introduction of quantum numbers. In the fixation of these quantum numbers, considerations which rest on the Adiabatic Principle, as well as on the Correspondence Principle... play an important role.[33]

This *demand* for the presence of stable and stationary states in many-electron atoms he called 'a general principle of the existence and permanence of the quantum numbers'.

Bohr's *Aufbau* principle was restricted to the electronic configurations in atoms, but a few years later it also came to play an important role in attempts to establish the distribution of electrons in the energy levels of simple molecules. Inspired by the similarity between atomic and molecular spectroscopy, Robert Mulliken, Friedrich Hund, and John E. Lennard-Jones suggested analogical uses of the principle in the molecular domain. In one of the early papers in the new science of quantum chemistry, Mulliken referred to 'Bohr's method of determining electron configurations in atoms by imagining all electrons removed, and then returning them one by one, each to the available orbit of lowest energy'.[34] Although Bohr's method was not directly applicable to molecules, indirectly it served as an important instrument in the early phase of molecular quantum physics.

In order to explain the finer details of the periodic system, and especially the transition groups and the group of the rare earths, Bohr made use of the hypothesis of penetrating orbits, which was essential to his entire line of argument. In his first letter to *Nature*, he stressed the significance of the idea: 'This coupling is the predominant feature of the whole picture, and is to be taken as a guide for the interpretation of all details as regards the formation of different groups and subgroups'.[35] The insight that an outer electron will sometimes perform elliptical orbits in such a way that it penetrates the inner shell was reached independently by Bohr and the soon-to-be famous Austrian physicist Erwin Schrödinger, then a newcomer to the quantum theory of atoms (Figure 7.5). At the time a professor at Breslau, Schrödinger had only recently taken up the Bohr–Sommerfeld theory, an area in which he had published a couple of papers. (Of course, a few years later his discoveries in wave mechanics would revolutionize quantum theory.) In a paper of early 1921 he investigated the emission spectrum of sodium on the basis of the Bohr–Sommerfeld theory, arguing that the valence electron in its elliptic motion would sometimes dip into the inner stable shell of eight

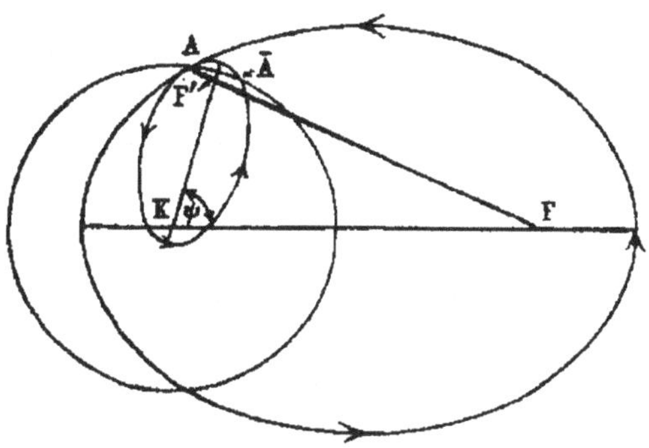

Fig. 7.5. Schrödinger's illustration of a valence electron penetrating into the inner electronic core. *Source*: Schrödinger 1921, p. 349.

electrons.[36] For reasons of simplicity he replaced the inner electrons by a thin shell of uniformly distributed negative electricity surrounding the nucleus. As a result of the penetration, the valence electron would go into a different orbit, what he called a *Tauchbahn*, and experience an effective nuclear charge closer to $Z^\star = 9$ than to $Z^\star = 1$.

Schrödinger sent a reprint of his paper to Bohr, expressing his hope that it 'will perhaps be of some little interest to you, as it seems to me to fit very well with the new and most interesting views, you have expressed in your letter to *Nature* on the necessary interaction between the electrons of the different "shells"'.[37] Bohr replied that the idea of penetrating orbits was not new to him, as he had exposed it already in December 1920, in his lecture to the Physical Society in Copenhagen. Indeed, in this lecture he discussed the idea at some length, if only qualitatively.[38] He gave a semi-quantitative and much more detailed account of it during the Göttingen lectures in June 1922. What his and Schrödinger's idea was about can be illustrated by the sodium atom and its spectrum. It was known from spectroscopy that the so-called sharp series of this element, which arises from a value of the azimuthal quantum number $k = 1$, can be represented by spectral terms given by the energies

$$E(n, k) = -\frac{Rh}{(n+\delta)^2} = -\frac{Rh}{n^{\star 2}},$$

where $n^\star = n + \delta$ is known as the effective quantum number and the Rydberg correction factor $\delta = \delta(k)$ is a measure of the deviation of the orbit of the valence electron from the corresponding hydrogenic orbit (cp. Section 6.5). In the case of sodium the first spectral term of the sharp series was found experimentally to be given by $n^\star = 1.63$. The value of n cannot be inferred directly from the observed value of $n^\star$, since the latter can be due to different combinations of n and δ, for example:

$$\begin{aligned} n^\star = 1.63 &= 1 + 0.63 \quad && (1_1\text{orbit}) \\ &= 2 - 0.37 && (2_1\text{orbit}) \\ &= 3 - 1.37 && (3_1\text{orbit}). \end{aligned}$$

While Sommerfeld had tentatively proposed the sharp series to arise from a 1_1 state, Schrödinger calculated on the assumption of a penetrating elliptic orbit that the state was 2_1. He obtained $\delta = -0.74$, which he thought was sufficiently close to $\delta = n^\star - n = 1.63 - 2 = -0.37$. In his first letter to *Nature*, Bohr arrived at a similar result, that the electrons added to a neon configuration had to be $n = 2$. However, in his more elaborate version of his theory, as presented in Göttingen, he realized that the penetration effect was more likely to result in a 3_1 state for the valence electron: 'We must assume that, because of the peculiar stability of the electron configuration already present, the eleventh electron will now be bound in a new type of orbit, namely, in a 3_1 orbit. Considerations of the firmness of the binding of this last electron lead us to the same conclusion'.[39]

In Bohr's theory, the penetrating orbits not only accounted for the large Rydberg corrections of some spectra, but above all they played the role of a coupling effect.[40] For some atoms the electrons move in undisturbed elliptical orbits ('of the first kind') due to the Coulomb field of the nucleus and the inner electrons. The penetrating orbits ('of the second kind') can be divided into two parts: outside the core of the atom the optical electrons move in approximate Keplerian ellipses exhibiting a perihelion precession, as, for example, the two valence electrons that move around an 18-electron core in calcium. When the electrons penetrate the region of the core, the field is changed and the internal orbit is no longer a simple continuation of the outer elliptical orbit. Instead, they perform an orbit much closer to the nucleus and therefore are more effectively bound. Assuming the configuration $(1_1)^2(2_1)^4(2_2)^4$ for neon, in the case of sodium the dimension of the core is determined by the major axis of the 2_1 ellipse. If the eleventh electron were in a 2_1 state as well, there would be no penetration, which would make it difficult to explain the value of the effective quantum number. Applying Sommerfeld's quantization conditions to the different parts of the orbit of the valence electron, Bohr found after a short calculation the result

$$n^\star \approx (n - m) + k,$$

where m denotes the maximum principal quantum number of the core electrons. Since k and m are constant for a given series, it follows that for the same series $n^\star$ increases by one unit, in agreement with experiments. The value of the Rydberg correction is given by

$$\delta = n^\star - n \approx -(m - k).$$

With the assumption $n = m + 1$ this gives an approximate value for the effective quantum number, $n^\star \approx 1 + k$, or $n^\star \approx 2$ and $\delta \approx -1$. This results in a principal quantum number for sodium of $1.63 + 1 = 2.63$, which Bohr considered to be in reasonable agreement with $n = 3$. He further argued that the difference between the internal and external orbits would decrease with increasing nuclear charge, and that this explained the chemical rule of increasing electropositivity through a group, for example from lithium to cesium. Using a refined version of Schrödinger's method, the Dutch physicist Theodor van Urk, at the University of Leiden, later confirmed Bohr's assignment of $n = 3$ to sodium and $n = 4$ to potassium.[41] Van Urk communicated his results in a letter to Bohr, and in response Bohr gave an account of how he had arrived at his new view of the structure of atoms:

> Concerning the relationship between the results given in my work in Zeitschrift für Physik and the considerations of Schrödinger, then I had originally made exactly the same deliberations and calculations as Schrödinger did; the results in my first letter to 'Nature' (24 March 1921) relied essentially on these, although of course I realized that such a shell calculation could only give a rough approximation. Shortly thereafter I saw that the question became much simpler when one took into consideration the real dynamical

constitution of the inner electron groups, and I could then derive in what I think is a rigorous way the results that were announced in my second letter to 'Nature'.[42]

The penetration effect played a role not only in explaining the alkali metals, but also in the explanation of the transition groups. Argon does not repeat the complete symmetry of neon, since the circular 3_3 orbits are completely outside the core and thus energetically less stable than the penetrating 3_2 orbits. Bohr explained the 4_1 electron in potassium in the same way as the 3_1 electron in sodium. But with increasing nuclear charge and a 3-quantum level that is not fully occupied, the situation is not quite analogous to the one of the second period:

Because of the decrease, with increasing nuclear charge, of the relative difference between the force fields inside and outside the region of the first 18 electrons, the dimensions of the parts of a 4_1 orbit lying outside this region will more and more approach the dimensions of a 4_1 orbit calculated with neglect of the interaction between the electrons.[43]

Such an orbit is more loosely bound than a 3_3 orbit, which consequently replaces the continuous building up of the 4-quantum electrons. With this sound, although qualitative argument, Bohr explained the presence of the first transition group. The building up of internal levels in the transition groups constituted a theoretical justification of Ladenburg's intermediate shells, except that there were no shells in Bohr's model of the atom. 'We do not assume a definite intermediate shell as something lying between the outermost and the inner orbits; rather, we have found a real reason for the peculiarity of the fourth period'.[44]

Bohr's new theory of atomic structure not only led to speculations of the possible existence of transuranic elements, such as a noble gas with atomic number $Z = 118$, it also led to speculations about a maximum value of the atomic number. Working at Bohr's institute in Copenhagen, the young Norwegian physicist Svein Rosseland discussed in 1923 whether radioactivity was a truly spontaneous phenomenon or caused by the external influence of the orbital electrons. As he pointed out, the quantum postulates alone offered no answer to the question. But according to the Bohr–Sommerfeld theory the shortest distance between a nucleus and an elliptically moving electron would be attained for electrons with $k = 1$ and be approximately given by

$$r = \frac{a_0}{2Z}(1 - \alpha^2 Z^2),$$

where a_0 is the Bohr radius and α the fine-structure constant. Since r will diminish with increasing Z, and the size of the nucleus will increase, Rosseland suggested that there would exist an upper limit for Z. 'It does not appear excluded', he said, 'that the presence of radioactivity among the heaviest known elements as well as the apparent absence of elements of higher atomic numbers may be connected with some sort of interaction between the nuclear and the external electrons'.[45] Rosseland's speculations were undoubtedly cleared with Bohr, who the same year stated without proof that an electron in an n_k orbit would fall into the nucleus if

$$\frac{Z}{k} \geq \frac{hc}{2\pi e^2} = 137.$$

He commented: 'For the heavier elements, the penetration of the series electron into the interior of the atom would lead to the result that orbits with $k = ½$ cannot exist. Even for $k = 1$, the electron in these elements comes to distances from the nucleus of the same order of magnitude as the value of the nuclear dimensions obtained from radioactive data. As remarked by Rosseland, this circumstance alone offers a hint toward an understanding of the limitation in the atomic number of existing elements'.[46] Bohr's remark was elaborated upon by Sommerfeld in the fourth edition of *Atombau*, where he proved that if $k < \alpha Z$ an electron would not perform an elliptic motion but instead spiral toward the nucleus, approaching it with almost the speed of light. For a $k = 1$ orbit, $Z = 1/\alpha = 137$ would therefore be the maximum atomic number.[47]

7.3 ATOMIC STRUCTURE AS REVEALED BY X-RAYS

Data from X-ray spectroscopy had, since the days of Moseley, played an important role in atomic theory, where they were used in particular to infer the population numbers in the inner rings of atoms (Section 4.6). However, this line of research, as cultivated by Sommerfeld, Debye, Vegard, and others, was based on the assumption of simple ring or shell atoms, an assumption which Bohr and most other physicists had abandoned by the early 1920s. Whereas optical spectroscopy served as a very important source for Bohr's second atomic theory, X-ray spectra played no significant role in its creation and early phase. In his letter to *Nature* of March 1921, Bohr mentioned the subject, but only casually, and it was also not prominent in his important lecture to the Physical Society of October the same year. Only at the end of the lecture did he deal with the topic, noting the 'possibility of obtaining information about the configurations of the electrons in the interior of the atoms from the X-ray spectra'.[48] In the English translation of May 1922 he provided more details and referred to new data that confirmed the predictions of the theory. During 1922–1923 X-ray spectroscopy became very important to the theory, providing what was perhaps its strongest confirmation.

Whereas in the decade 1913–1923 the theory of X rays was developed primarily by German physicists, and especially by Sommerfeld's group in Munich, towards the end of the period most of the important work in experimental X-ray spectroscopy was done in three laboratories outside Germany: Maurice de Broglie's in Paris, Manne Siegbahn's at Lund in Sweden, and William Duane's at Harvard. Of these, it was the two European laboratories that primarily contributed to the understanding of atomic structure. Maurice de Broglie had been an outstanding contributor to X-ray research since his measurements of X-ray absorption edges in 1914 and his development of the rotating crystal technique for obtaining good photographic plates.[49] Among his collaborators were his younger brother Louis de Broglie, René Ledoux-Lebard, and Alexandre Dauvillier. At Lund, Siegbahn worked with, among others, Elis Hjalmar, Václav Dolejšek, Ivar Malmer, and Dirk Coster. In 1924 he published *Spektroskopie der Röntgenstrahlen*, an authoritative account of X-ray spectroscopy and for a long time a standard reference work; the following year he was awarded the Nobel Prize in physics (for the year 1924).[50] The Paris and Lund laboratories competed with each other in their work on X-ray spectra and their interpretation, a rivalry that became obvious in the priority debate concerning element number 72.

Realizing the importance of X-ray spectroscopy for his new theory, Bohr established close ties with Siegbahn's laboratory, which he, together with Sommerfeld, had first visited in 1919. Siegbahn had even earlier expressed his desire to 'have the opportunity of talking to you [Bohr] about all sorts of questions, and especially on the subject of atomic model X-ray spectra'.[51] One of the most active and promising scientists in Lund was Coster, a Dutch physicist who had recently completed his doctoral dissertation under Ehrenfest in Leiden and now specialized in the investigations of X-ray lines in the rare earths and the heavier elements. In two long papers in the *Philosophical Magazine* from the summer of 1922, Coster applied his experimental results to Bohr's new theory, giving it, in fact, its first strong support. Summarizing his many new results, he concluded that they 'are in beautiful agreement with Bohr's theory as regards the successive development of the shells of electrons in the atom'. Coster elaborated:

> According to Bohr, at different stages of the periodic table we meet with atoms for which an inner shell of electrons is completed. Thus the M-shell is completed in the neighbourhood of the iron group, the inner N-shell is partly completed in the region of the Pd group and again definitely for the rare-earth metals . . . This conception is found to be in agreement with the experimental results.[52]

Working at Lund, Coster benefited not only from Siegbahn's expertise and excellent X-ray apparatus, but also from Bohr's proximity and interest in his work. Immediately after the publication of Coster's paper, Bohr invited him to the new institute in Copenhagen, where he stayed from September 1922 to September 1923. The first important result of his stay was a joint paper with Bohr on the significance of X-ray spectroscopy to the theory of atomic structure, which was received by *Zeitschrift für Physik* in early November 1922.

As Bohr had done on earlier occasions, in his joint paper with Coster he stressed the different natures of optical and X-ray spectra, pointing out that according to the theory a building up of internal orbits should be manifested in the X-ray lines:

> The typical periodicity of the chemical properties and of the optical spectra thus depends on the fact that the effective quantum numbers of the outermost electron orbits, in contrast to the principal quantum numbers, only change a little when one passes from one element to the homologous element in the next period of the system of the elements. The remarkable

> lack of periodicity in the essential features of the X-ray spectra, on the other hand, is due to the fact that it is here a matter of the phenomena of the innermost electrons in the atom, where we deal with electron groups that are already completely filled up and that are repeated in the same way by all following elements.[53]

Bohr and Coster based their work on the idea that even though the normal Moseley diagram does not show any discontinuities corresponding to the building up of the various electron groups, the building up should manifest itself in the first appearances of the various X-ray lines, which according to Bohr's theory were supposed to appear together with the electron group from which they originate. According to Coster, Bohr's levels of 1921 had to be classified not only according to their n and k quantum numbers, but to be further subdivided into $n(k_1, k_2)$, where k_1 is again the azimuthal quantum number ($\Delta k_1 = \Delta k = \pm 1$), and k_2 is a third quantum number obeying the selection rule $\Delta k_2 = 0, \pm 1$. The quantum number k_2 was taken to be the X-ray analogue of Sommerfeld's inner quantum number j. Thus, for example, whereas there were three $n = 3$ levels in Bohr's original theory, namely 3_1, 3_2 and 3_3, there were five levels under Coster's classification: 3(1, 1), 3(2, 1), 3(2, 2), 3(3, 2) and 3(3, 3). With these components, Bohr and Coster carried out their programme and concluded that 'it has been displayed that a level that, so far as the experimental values allow us to distinguish, belongs to certain values of n and k actually appears for the first time at a place where, according to the theory, there appears a level in the normal atom with similar values of n and k'.[54]

As visual evidence for the agreement between theory and experiment, Bohr and Coster presented a variant of the Moseley diagram showing the dependence of the term values $\sqrt{v/R}$ of the various elements on their atomic number (Figure 7.6). The diagram was based on the best X-ray data available at the time, especially those obtained in Lund and Harvard. (They avoided using data from Paris, which they suspected were inaccurate and unreliable.) Two aspects of this 'Bohr–Coster diagram' are of special interest. First, the diagram shows, although only crudely, where the building up of the electrons at the various levels begins. Second, the diagram enables one to almost visualize the parts of the periodic system in which the building up occurs at the intermediate, still incomplete levels. Both features were in excellent agreement with Bohr's theoretical results, even if the experimental material did not rigorously confirm the theory in detail.

Bohr and Coster paid much attention to the rare earths, which according to Bohr's theory were characterized by a gradual building up in the 4-quantum level, from 3×6 electrons to 4×8 electrons. Of the building up of the 4_4 sublevel, they wrote that it took place 'so to speak in competition with the valence electrons in the six-quantum orbits'. They further assumed that the process started in lanthanum ($Z = 57$) and ended in lutetium ($Z = 71$). However, the experimental data were not sufficiently exact in 1922 to infer with certainty the completion of the 4_4 sublevel and to decide unambiguously the electron structure of the unknown element of atomic number 72. The two physicists admitted that 'due to the incompleteness of the measurements there is some uncertainty as to the progress of the curves of the N levels for the elements in

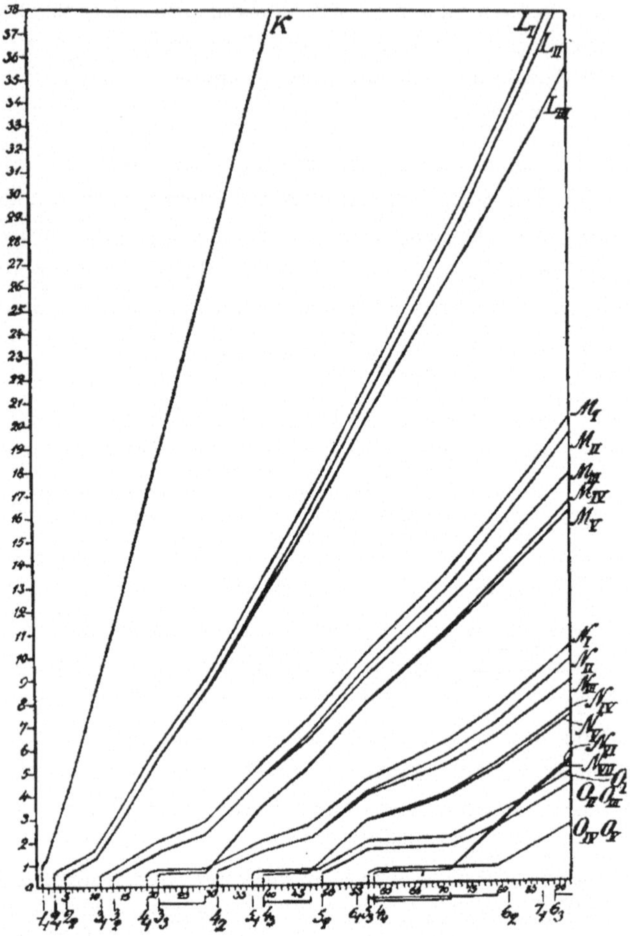

Fig. 7.6. The Bohr–Coster diagram, here in the version from the 1924 edition of Bohr's *Theory of Spectra and Atomic Constitution*. The numbers below the horizontal axis show the beginnings of the quantum groups according to Bohr's theory.

the neighborhood of 72, which marks the end of the family of the rare earths'.[55] The uncertainty lasted until the end of 1924, when the Japanese physicist Yoshio Nishina, working at Bohr's institute, provided the first experimental proof that the rare earth group ends with element 71 and comprises fourteen elements (Figure 7.7). Nishina's measurements showed that 'A plotting of the square roots of the level values as a function of the atomic number confirm in a striking way the general conclusions of Bohr and Coster'.[56]

Although the Bohr–Coster diagram and other X-ray evidence thus did not unambiguously and definitively prove Bohr's theory, they constituted an imaginative and convincing argument for the correctness of the basis of the theory. Through X-ray spectroscopy the theory gained respectability, in that its results were now derived from physical measurements. This made it more acceptable to the many scientists who were

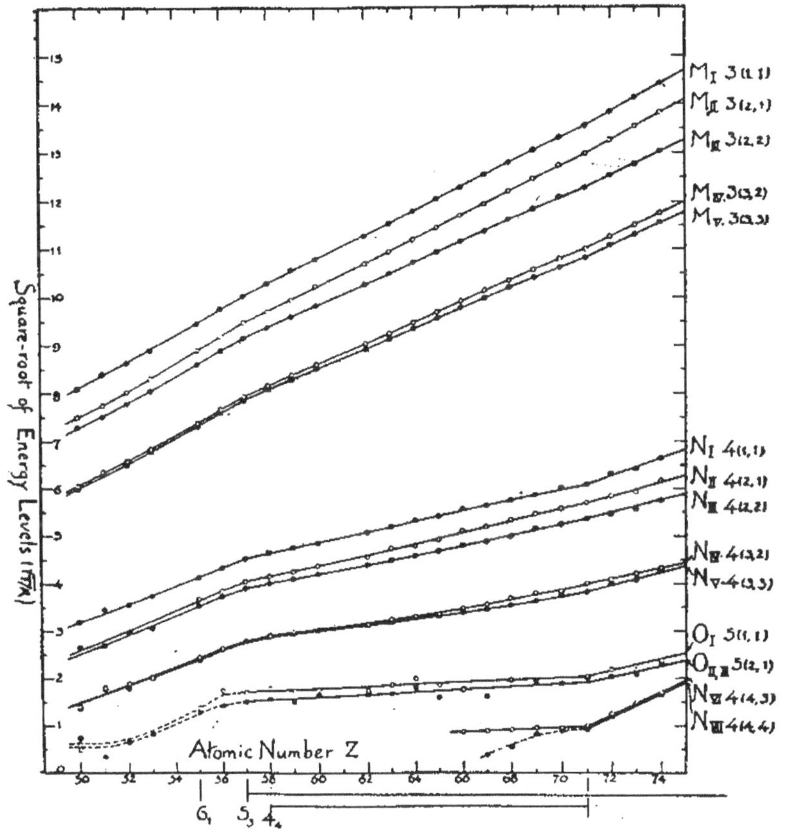

Fig. 7.7. X-ray spectroscopic measurements of the energy levels from Z = 50 to Z = 74 made by Nishina in 1924, while working in Copenhagen. The data provided the first proof that the rare earth level starts with Z = 58 and ends with Z = 71, in accordance with Bohr's theory.

Source: Nishina 1925, p. 534.

unimpressed by Bohr's vague talk about correspondence and symmetry. Siegbahn, generally recognized as a leader of X-ray spectroscopy, had little doubt that the theory was correct, at least approximately. In *Spektroskopie der Röntgenstrahlen* he expressed his confidence as follows: 'The atomic system thus built up by Bohr from entirely general principles has proved itself very satisfactory [and]... is in accord with the fundamental features of X-ray spectra'. Although experimental difficulties prevented verification of some of Bohr's predictions with regard to the beginning of new energy levels, 'In no case, however, has an energy level been found which is not represented in the Bohr table by the corresponding type of electron orbit'.[57]

In Paris, Dauvillier and Louis de Broglie followed an approach that to some extent anticipated the one used by Bohr and Coster, and which resulted in an X-ray diagram similar to theirs. Feeling that their work was not properly acknowledged, they claimed that priority in the diagram belonged to them and not to the Copenhageners.[58] The goal of the French physicists was the same as that of Bohr and Coster, and no less ambitious: to infer the atomic structure of all the elements in the periodic system by means of X-ray data. However, there were important differences between the methods and results of the two groups. As to methodology, whereas Bohr's system rested on general principles of quantum theory, the one of Dauvillier was derived inductively from experimental data. In fact, this was also how he and de Broglie conceived Bohr's theory of the periodic system. They valued his system, but had little respect for his general principles:

> Even if certain deductions by this scientist [Bohr] do not seem to be imposed with the desired rigor and even if it seems to be based on considerations of analogy, on symmetry or on experimental facts (like chemical valence and optical spectra) rather than on purely theoretical reasoning, it nevertheless constitutes a major work which should be considered as the logical development of his theory of spectra.[59]

As to results, from the slopes of the curves in their modified Moseley diagram, Dauvillier and de Broglie found quantum numbers for the X-ray levels that in some cases disagreed with those favoured in Copenhagen and Lund. For example, in 1922 Dauvillier derived from his data that the valence electron in sodium should be in a 1_1 state, whereas Bohr had used the penetration effect to argue that the correct state was 3_1.[60] The following year the two Paris physicists obtained population numbers for the L levels that were equally discordant with Bohr's numbers.[61] For the carbon atom they arrived at the configuration $(1_1)^2(2_1)^2(2_2)^2$, contrary to Bohr's $(1_1)^2(2_1)^4$ with its tetrahedral symmetry. Dauvillier and de Broglie's proposal for the L levels turned out to be closer to the truth than Bohr's. It was adopted by the British physicist Edmund Stoner in an important work of 1924 that marked a definite improvement over Bohr's system of electron configurations (Section 7.6).

7.4 HAFNIUM OR CELTIUM?

As described above, neither X-ray data nor theoretical arguments based on quantum theory provided an unequivocal answer to the beginning and end of the rare earth group. In his Copenhagen lecture of October 1921, Bohr considered the formation of atoms heavier than barium with $Z = 56$. 'With increasing nuclear charge', he wrote

> we shall have to expect not only that an electron in a 5_2 orbit will be bound more firmly than in a 6_1 orbit, but we must also expect that a moment will arrive when during the formation of the atom a 4_4 orbit will represent a firmer binding of the electron than an orbit of 5 or 6-quanta, in much the same way as in the elements of the fourth period a new stage in the development of the 3-quanta group was started when a point was reached where for the first time the 19th electron was bound in a 3_3 orbit instead of in a 4_1 orbit. We shall thus expect in the sixth period to meet with a new stage in the development of the group with 4-quanta orbits.[62]

In the spring of 1922 Bohr had convinced himself that his theory required completion of the 4_4 subgroup at element 71 and that the unknown element 72 therefore had to be a homologue to zirconium rather than a rare earth as assumed by many (but not all) chemists. 'If our ideas are correct', he said in Göttingen, 'the not yet discovered element with atomic number 72 must have chemical properties similar to those of zirconium and not to those of the rare earths'.[63] Denoting the completed levels of $n = 1$ to $n = 3$ as K, L, and M, respectively, he wrote the electron structure of the two elements as

$$\text{Zr} \quad \text{K L M} \quad (4_1)^4(4_2)^4(4_3)^2(5_1)^2$$

$$72 \quad \text{K L M}(4_1)^8(4_2)^8(4_3)^8(4_4)^8(5_1)^4(5_2)^4(5_3)^2(6_1)^2$$

Although Bohr was himself convinced, his arguments for the position of element 72, which relied on a postulated 'harmonic interaction' of the 4×8 N electrons, did not necessarily convince his audience. Indeed, this part of his theory relied to a considerable degree on his fine instinct, perhaps even more so than other parts. Shortly after his return from Göttingen he received the unwelcome information that element 72 was, apparently, a rare earth after all. According to Rutherford, who reported the news in a note in *Nature*, the missing element had recently been discovered by two French scientists who suggested the name celtium and the symbol Ct for the new element.[64] Their 'very definite conclusions' seemed to settle the question of the nature of element 72 and its place in the periodic system.

Celtium was not really a novelty, for its existence had been claimed as early as 1911, when the eminent French chemist Georges Urbain, in close competition with Carl Auer von Welsbach from Austria, announced that he had found the element in a fraction of ytterbium minerals. His method of identification was based on measurements of spectral lines and magnetic properties. Urbain's claim failed to win general acceptance, but with Moseley's successful application of X-ray spectroscopy for chemical purposes, he eyed a possibility to have celtium confirmed by the new and powerful method invented by the English physicist. Having examined Urbain's sample, Moseley concluded that it produced no X-ray lines corresponding to element 72, yet Urbain chose to interpret the result not as a disconfirmation but as a failed confirmation. Absence of evidence is not evidence of absence, he seems to have thought. As far as he was concerned, celtium was most likely a reality. To convince other scientists he needed positive confirmation from X-ray spectroscopy, for which reason he teamed up with Dauvillier in Paris. By measuring the L spectrum of a mixture of lutetium and ytterbium oxides, in the spring of 1922 Dauvillier detected two weak lines, which he identified as the $L\alpha_1$ and $L\beta_2$ lines of the missing element 72. Consequently, 'It is now unquestionable that the element of atomic number 72 is actually celtium'.[65]

Rutherford's note on and endorsement of celtium worried Bohr, who asked Coster about his opinion: 'The question is of paramount interest, since, as you know, the ideas of atomic structure seem to require that a substance with atomic number 72 must have essentially different properties than the ones that are shown by the rare earths'.[66] For a brief while he was inclined to accept the discovery claim and admit a failure in his own line of reasoning. 'The only thing I still know for sure about my lectures in Göttingen is that several of the results communicated are already wrong', he wrote to Franck. 'A first point is the constitution of element 72, which, as shown by Urbain and Dauvillier, contrary to expectation has turned out to be a rare earth after all'.[67] However, after having reconsidered the question he returned to his theoretically founded belief that element 72 could not possibly belong to the group of rare earths. It undoubtedly contributed to Bohr's change of mind that the X-ray specialists in Lund expressed strong doubts with respect to the accuracy of Dauvillier's measurements. Thus, Coster reported to Bohr that Siegbahn had recently been in Paris and there had the opportunity to examine Dauvillier's photographic plates with the celtium lines:

> [Siegbahn] had absolutely no confidence in Dauvillier's work on element 72. He could *not* see the lines $L\alpha_1$ and $L\beta_2$ from 72 (the only lines which D. claims to have found for this element) *at all*. When he pointed this out to D. (especially that he did not see any trace of these lines) Mr. Dauvillier said: Yes, that is quite possible, that is because we do not have clear weather today!![68]

Some time before August 1922 Bohr finally decided that element 72 had to be a chemically relative of zirconium and that the discovery claim of the two French scientists was probably based on a misinterpretation of the X-ray lines. After having convinced himself of the soundness of his theory, he was satisfied to let the matter of celtium rest.

The search for the true element of atomic number 72 was initiated by Hevesy, who stayed at Bohr's institute and had been encouraged by his friend Paneth to start looking for it in zirconium-rich minerals. Paneth already believed, in part for chemical reasons, that the element was a member of the fourth subgroup. 'It is to be recommended that one looks for element 72 in zirconium minerals', he wrote in the summer of 1922.[69] This was what happened, the search team in Copenhagen consisting of Hevesy and Coster. Within a couple of weeks they found two fairly strong lines of $Z = 72$ from zirconium minerals, and after chemical purification they concluded in early December that they had discovered a new element that could not possibly be celtium (Figure 7.8). Their discovery was made just in time for Bohr to announce it in his Nobel lecture in Stockholm on 11 December 1922: 'Dr Coster and Prof. Hevesy . . . have been able to establish the existence in the minerals investigated of appreciable quantities of an element with atomic number 72, the chemical properties of which show a great similarity to those of zirconium and a decided difference from those of the rare earths'.[70] This is undoubtedly the only time a major discovery has been announced in a Nobel lecture.

Continued investigations at Bohr's institute soon confirmed the existence of a considerable amount (2–8%) of element 72 in zirconium minerals. At the beginning of the new year Coster had measured six X-ray lines of wavelengths that agreed

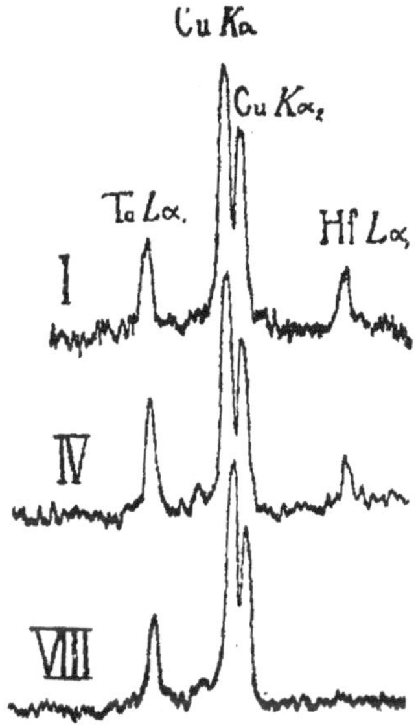

Fig. 7.8. Photometric curve of an early hafnium sample, probably from January 1922. The intensity of the X-rays is plotted against the wavelength. The three curves are from different stages of chemical precipitation with $Na_3(PO_4)_2$, with no content of hafnium in stage VIII. The tantalum line of constant intensity is a calibration standard from which the amount of hafnium can be estimated, and the two copper lines are due to the copper anticathode in the X-ray tube.

Source: Hevesy 1925, p. 27.

perfectly with those interpolated from the Moseley or Bohr–Coster diagram. On 2 January 1922 he and Hevesy officially announced their discovery in a note to *Nature*, stressing how well it agreed with Bohr's atomic theory.[71] They proposed naming the new element 'hafnium' after the Latin name for Copenhagen. During the following months the scientists at Bohr's institute were busy determining the chemical and physical properties of hafnium, including its atomic weight and optical spectrum. Within a year the first patents for the technical use of hafnium were taken out.[72] The discovery was widely seen as a triumph of Bohr's theory, which supposedly had guided Coster and Hevesy in their experimental search. Congratulating his former research student on the discovery, Rutherford found it to be 'an admirable example of the cooperation of theory and experiment. I am sure it is highly gratifying to Bohr, even if he tries to suppress it'. A year later Walther Nernst confided to Hevesy that had it not been for the discovery of hafnium, he would not have believed in Bohr's theory.[73]

The announcement of hafnium was, at the same time, a denunciation of celtium and the beginning of a long and bitter priority dispute between scientists in Copenhagen and Paris. Urbain and Dauvillier were grudgingly forced to admit that the original celtium claim of 1911 was a mistake, but even then they insisted that the observation of the two weak X-ray lines in May 1922 constituted a conclusive proof that they had discovered celtium, or what the Copenhageners wrongfully called hafnium. Unfortunately for their discovery claim, they were unable to produce good X-ray spectra of celtium or otherwise provide documentation for their claim, which solely rested on the two exceedingly faint lines reported in words only in 1922. (Bohr, Paneth, and Siegbahn believed for a time that the lines were due to self-delusion.) Nevertheless, during 1923 the controversy escalated, fueled by professional jealousy, personal disagreements, and the tense political and ideological climate in the aftermath of World War I.

By the end of 1924, the scientists in Copenhagen had prepared large amounts of pure hafnium, while celtium remained two elusive spectral lines and a name in French periodic tables. In spite of the missing evidence in favour of celtium, the controversy dragged on, more for political than scientific reasons. Only in 1930 did IUPAC, the International Union of Pure and Applied Chemistry, officially sanction hafnium as element number 72 in the periodic system.

Even though Bohr was not directly involved in the controversy, it disturbed him greatly. At the height of the priority debate he wrote to Rutherford about 'this terrible muddle about the new element, in which we quite innocently have dropped'. He continued: 'We had never dreamt of any competition with the chemists in the hunt for new elements, but wished only to prove the correctness of the theory. In the letter from Urbain which the editor of *Nature* kindly sent us for possible comments, he tries however to shift the whole matter, paying no regard to the important scientific discussion of the properties of the element 72, but tries only to claim a priority for announcing a detection of such an element'.[74]

Did the discovery of hafnium 'prove the correctness of the theory', as suggested by Bohr? Did his atomic theory actually *predict* the existence of an element with the properties of hafnium? According to the great philosopher Karl Popper it most certainly did:

> I still remember vividly the excitement of the discovery of the element 72 (Hafnium) in 1922, as a result of Niels Bohr's marvelous quantum theory of the periodic system of elements. It struck us then as the great moment when chemistry had been reduced to atomic theory . . . Bohr's theory led not only to the prediction of the chemical properties of elements, and thereby to the prediction of of the properties of the still unknown element 72 and thus to its discovery, but it also allowed the prediction of some of the properties of the chemical compounds. It was a great moment in the history of matter.[75]

Popper's view, equally triumphalist and reductionist, is undoubtedly exaggerated, although it was shared by some scientists at the time. If one understands 'prediction' in a strict and deductive sense, where a consequence is derived from a mathematically formulated theory, Bohr's atomic theory did not predict the place of element 72 in the periodic system. Bohr argued that the element had to be a homologue of zirconium, but his arguments, although in many ways compelling, were not deductions from quantum theory. They were based on a mixture of empirical evidence, partly of a chemical nature, and general principles related to the quantum theory of atoms. In so far as he presented a series of cogent arguments that element 72 should be a member of the fourth subgroup, Bohr did predict hafnium. The fact that his prediction did not flow deductively from a fundamental theory, and that calculations were conspicuously absent from his prediction, is in this respect of minor importance. Many predictions in physics, chemistry, and astronomy are of the same character.

From a historical point of view it is relevant to point out that Bohr's arguments were widely conceived of as a prediction, at least in a vague sense, and the discovery of hafnium was seen as a confirmation of his atomic theory. According to Kramers and Holst, writing from Copenhagen, 'The theory was able to predict that the element with atomic number 72 [must] . . . show a resemblance to the tetravalent elements zirconium (40) and thorium (90)'.[76] The same version was reported by Andrade in 1924: 'Bohr's theory . . . achieved a brilliant triumph when it predicted that the then unknown element of atomic number 72 should resemble zirconium and thorium, and not the trivalent rare earths. This prediction led to the work of Hevesy and Coster, who discovered the new element hafnium'.[77]

Bohr was not the first to suspect that the missing element did not belong to the rare earth group. For example, Paneth shared the belief and so did several other chemists. Bohr was well aware that he had no priority. In his Nobel lecture as well as on other occasions he referred to Julius Thomsen's old periodic system and to the recent work of Charles Bury, who in the spring of 1921 had stated that element 72 'would have the structure (2, 8, 18, 32, 8, 4), and would resemble zirconium'.[78] However, what interested Bohr was not so much the nature and properties of the missing element as the theoretical reasons for its position in the periodic system. He saw the element as a test case for his theory and gave arguments based on quantum theory why it must have a particular atomic constitution. These arguments were qualitative and could of course be criticized, but they were nonetheless rational and scientific arguments. Incidentally, Bohr never spoke of his arguments as a 'prediction', but typically said that his theory led him to 'expect' element 72 to have a certain constitution and place in the periodic system.[79]

7.5 THE RECEPTION OF BOHR'S THEORY

German physicists received the news of Bohr's theory of complex atoms with high expectations and a surprising lack of a critical attitude. Many were enthusiastic, for example Landé, who knew Bohr from his stay in Copenhagen in the fall of 1920 and who offered to arrange a German translation of his note in *Nature*. The note 'will be of the very greatest significance for the future atomic theory', he wrote to him. 'On the whole it seems to me that until your full and detailed exposition appears there is no sense in doing any more theoretical work in atomic theory'.[80] Having learned from

Landé about Bohr's new work, a galvanized Born wrote to Franck, who at the time was staying with Bohr in Copenhagen:

> Wouldn't that be grand! But the tragedy is that we understand not a word of how he does it. He only says that it is based on the correspondence principle ... Dear, good Franck, be a nice chap and write to us about it as well as you understand it, or ask Bohr or Kramers to write something clear. Otherwise, we will explode from curiosity. Or let me come to Copenhagen in eight days. At such an exciting event I would like being close by. Bohr is an astonishing man.[81]

Sommerfeld initially shared the enthusiasm of Landé and Born, as he told Landé in early March: 'Your remark that Bohr has broken like a bomb is also valid for Munich. From Bohr I got a copy of his *Nature* letter. We must relearn completely!' And in a postcard to Bohr a few days later, thanking him for the manuscript of the *Nature* letter: 'It evidently represents the greatest advance in atomic structure since 1913'.[82]

Although Sommerfeld's attitude to the quantum theory of atoms differed in important respects from Bohr's, his initial response to the new atomic theory was completely positive. Like several other German physicists, he thought that Bohr's atomic configurations were based on calculations and therefore a major step towards the mathematization of chemistry he aimed at. His only complaint was that what he thought was Bohr's mathematical chemistry rested on formal principles rather than the established methods of quantum theory: 'Although I am certainly convinced that your way is the right one; if, as it seems, you can mathematically reconstruct the numbers of the periods 2, 8, 18, ... then that is indeed the fulfillment of the most daring hopes in atomic physics. However, I am enough of a heretic to think that some day that will be possible in another, less formal and more uniform manner'.[83] In the third edition of *Atombau*, completed in the winter of 1921–1922, he praised 'Bohr's considerations based on calculations', while Gregor Wentzel at about the same time classified Bohr's theory as mathematical-deductive rather than inductive-empirical.[84]

Sommerfeld was not alone in reading mathematical calculations into Bohr's note, which in fact did not contain a single equation. In lectures of 1921–1923 given at Copenhagen, Hamburg, and Zurich, the great Göttingen mathematician David Hilbert dealt with the epistemological questions raised by the new physics of quanta and relativity. In his lectures in Hamburg in July 1923 he discussed Bohr's new atomic theory, which he suggested gave a full explanation of the periodic system, although not one that was derived from the truly fundamental physics represented by Hilbert's *Weltgleichungen* (world equations). Using an argument that closely followed Bohr's exposition in Göttingen the previous year he gave a fairly detailed account of the electron configurations of the elements. He thought it was possible 'to deduce the intimate properties of matter, even the particulars of the chemical elements, from the laws of field and motion alone, and to recognize them as necessary thought consequences [*denknotwendige Folgerungen*]'.[85] As he saw it, Bohr's theory was an important step on the road towards a complete mathematization of the sciences of matter, if only a first step.

Hilbert, the mathematical imperialist, paid no attention to the general and non-mathematical principles of quantum theory that according to Bohr provided his theory with a secure foundation.

The Finnish mathematical physicist Gunnar Nordström, professor at the University of Helsinki and a specialist in the theory of general relativity, shared the enthusiasm of most other physicists. Having studied Bohr's latest work, he wrote to him: 'The great and simple lines that you—in spite of the complex relationships—have found to govern the structure of atomic systems provide your new theory with a suggestive power which makes it appear convincing even in the absence of compelling proof from quantum theory. You have taken a giant step in understanding the world of atoms, and I heartily congratulate you with this new progress'.[86] Much of the attractiveness of the theory was indeed due to what Nordström called its 'suggestive power'.

While fascinated by Bohr's theory of the atomic structure of the chemical elements, some physicists expressed their puzzlement over Bohr's vague and enigmatic formulations. How were they to be understood? Rubinowicz, to whom Bohr had sent a reprint of his first note in *Nature*, responded as follows: 'Even if your letter to *Nature* does not let us see the finer details of the wonder-world which you now again has opened up to us, but only foreshadows them, I believe nevertheless to have gathered so much from it that by your work you once more lead us to a summit from which we now again can view a wide field as a clear and beautiful harmonic unity'.[87] It did not occur to Rubinowicz that if the finer details were missing, the reason might be that there were no finer details. As usual, Rutherford went more directly to the heart of matter. Having received for publication Bohr's second note to *Nature*, he wrote about his impression of the manuscript: 'Your letter is very interesting but of course it is very difficult for me to form an idea of how you arrive at your conclusions. Everybody is eager to know whether you can fix the 'rings of electrons' by the correspondence principle or whether you have recourse to the chemical facts to do so'.[88]

Ehrenfest, too, had received a copy of the manuscript before its publication. Like Rutherford, he wanted to know how Bohr had applied that magical wand of Copenhagen physics, the correspondence principle:

> I have read your *Nature* letter with eager interest (in Jena only Sommerfeld had it and he carefully kept his monopoly . . .). The picture you sketch there is really so much more beautiful than that of the quantum numbers decreasing again toward the outside. Of course it now interests me even more how you saw it all in considerations of correspondence. It interests me especially: what were the *difficulties* like that taught you to find the way from the conception of the first *Nature* letter to that of the second. I am VERY curious to know that.[89]

The chemists did not share the enthusiasm of the physicists, but they nevertheless expressed interest in and sometimes admiration for the theory. For example, Nevil Sidgwick argued that it was useful in elucidating the structure of coordination compounds, and at the British Association meeting in Liverpool in 1923 he lectured on the

Bohr atom and the periodic system. He presented the essence of the theory as follows: 'The arrangement of the orbits in an atom is arrived at by means of Bohr's correspondence principle, resting on a *mathematical analysis* of possible orbits, and is largely supported by the evidence of the optical spectra'. And in a footnote, referring to the Bohr–Coster paper: 'It is also entirely confirmed by the X-ray spectra'.[90] H. M. Hansen, who had talked with Langmuir during a stay in the United States, reported to Bohr that the American physical chemist 'was enthusiastic over your work on the periodic system and fully realized what a tremendous progress it meant for the things he dealt with'.[91]

If Bohr had developed a mathematical theory of the elements, his calculations would have been of three kinds. First, calculations of Fourier coefficients of various orbits and corresponding intensities of electron transitions associated with arguments of correspondence; second, calculations of binding energies of the external and internal parts of penetrating orbits due to a suitable field of force; and third, perturbation calculations associated with stability. Quite apart from the forbidden complexity of such imagined calculations and their absence from Bohr's papers and manuscripts, he most likely never considered calculating the properties of the elements. Kramers knew Bohr and his style of working better than most. In 1935 he gave an account of how Bohr had arrived at his theory, a contrast to how many of his colleagues in physics thought he had proceeded:

> It is interesting to recollect how many physicists abroad thought at the time of the appearance of Bohr's theory of the periodic system that it was extensively supported by unpublished calculations which dealt in detail with the structure of the individual atoms, whereas the truth was, in fact, that Bohr had created and elaborated with a divine glance a synthesis between results of a spectroscopic nature and of a chemical nature.[92]

By 1923 most physicists realized that Bohr's theory was not, contrary to what they had expected, based on calculations, but rather on empirical data combined with qualitative arguments based on symmetry and correspondence. As a result, their enthusiasm cooled. A year after Bohr's visit to Göttingen, Sommerfeld stated in a detailed treatment of the theory that 'Bohr's theory is, of course, not based on mathematics but is conceived intuitively'—a view quite different from his earlier conception of it.[93] In the fourth edition of *Atombau* he clearly expressed his scepticism about Bohr's intuitive use of symmetry and correspondence. Whereas Bohr meant to give a theoretical derivation of the completion of the electron groups in the noble gases from these concepts, Sommerfeld viewed this part of Bohr's theory as purely phenomenological. As he wrote: 'When we describe the octet shells of the noble gases as particularly stable, it is in no way a theoretical explanation but just an expression for the empirical matter'.[94] His former student, Pauli, may have agreed. In *Drei Aufsätze* Bohr had explained that one 'must expect' the eleventh electron in sodium to go into a 3-quantum orbit. Pauli, obviously annoyed with Bohr's statement, wrote in the margin of his copy of the book: 'No reason to expect anything; you concluded it from the spectra!!'[95]

When in 1924 Stoner proposed a new arrangement of the quantum numbers of the electrons of the elements, Sommerfeld quickly adopted the new system, to which he briefly referred in the preface of the fourth edition of *Atombau*. His comparisons of the two schemes shows how he then considered Bohr's theory, which he judged as generally 'sound' but no longer correct: 'Being based on incontestable experience with respect to number and order of X-ray levels and on the association of quantum numbers with these, Stoner's scheme is much more trustworthy than Bohr's. It has an arithmetic rather than geometric-mechanical character; without assuming any symmetry of orbits it exploits not just some but all available data of X-ray spectroscopy'.[96]

Many scientists outside Germany shared Sommerfeld's second thoughts about Bohr's second theory, namely that it was largely of the inductive-empirical type. As mentioned, this was how Dauvillier and Louis de Broglie looked at it, and it was also the reconsidered view of Rutherford. At the meeting of the British Association in Liverpool in September 1923 he praised Bohr's theory for its predictive power, which he thought was inductively rather than deductively rooted:

> The theory of Bohr, like all living theories, has not only correlated a multitude of isolated facts known about the atom, but has shown its power to predict new relations which can be verified by experiment. For example, the theory predicted the relations which must subsist between the Rydberg constants of the arc and spark spectra . . . a prediction so strikingly confirmed by Paschen's work of the spectrum of doubly ionised aluminium and [Alfred] Fowler's work on the spectrum of trebly ionised silicon. Finally, it predicted with such great confidence the chemical properties of the missing element, number 72, that it gave the necessary incentive for its recent discovery.[97]

The ever critical Pauli, who soon was to unravel the secrets of the periodic system, had no great faith in Bohr's theory. From Copenhagen he reported to Sommerfeld about the strengths and weaknesses of the theory. Mentioning the explanation of the rare earth group and the transition metals as important successes, he stressed that these were results due to the penetration effect. 'It is not a question of correspondence, but only a question of the strengths of the binding of the various n_k orbits'.[98] As to the weaknesses he referred to the theory's inability to provide a satisfactory explanation of the number of elements in the periods, that is, Rydberg's $2n^2$ rule. Pauli appreciated the correspondence principle when used in a restricted sense, but he did not believe it was of any value in the explanation of the periodic system. When it came to the question of the closing of the electron groups he sided with Sommerfeld, who in *Atombau* had expressed 'greater hope in the magic power of the quanta . . . than in correspondence and stability considerations'.[99]

In late 1924 Pauli wrote to Bohr about his sceptical attitude: 'I have already often said to you that I am of the opinion that the correspondence principle has in reality nothing to do with the problem of the closing of the groups in the atom . . . There is moreover no need whatever to talk of harmonic interplay'.[100] Pauli's criticism was not limited to the theory of the structure of atoms but concerned also the very basis of Bohr's theory, namely, the existence of electrons in stationary orbits. In early 1924 he was prepared to abandon the concept of electron orbits, which Bohr still considered a necessity.[101]

7.6 TOWARD A THEORY OF THE PERIODIC SYSTEM

Bohr was not the only scientist searching for an explanation of the periodic system in terms of electron configurations. Shortly after the appearance of his first note to *Nature*, the British chemist Charles Bury, at the University College of Wales, submitted a paper in which he argued that the electron shells do not fill in a sequential order, contrary to what Langmuir had claimed. 'Successive layers can contain 2, 8, 18, and 32 electrons', Bury wrote. 'Groups of 8 and 18 electrons in a layer are stable, even when that layer can contain a larger number of electrons. The maximum number of electrons in the outer layer is 8; more than 8 electrons can exist in a shell only when there is an accumulation of electrons in an outer layer.'[102] As mentioned above, he argued that the missing element 72 should be chemically related to zirconium, and he also speculated that the unknown element of atomic number 94 (the later plutonium) might be the first of a series of seven transition elements. At the end of his paper he noted that 'structures similar to those suggested by the author for the inert gases have been proposed by Bohr'.

Whereas Bury did not refer to either quantum theory or quantum numbers, three years later his compatriot and fellow chemist John Main Smith, at Birmingham University, did just that. In continuation of an earlier critique of the Bohr atom, he argued that Bohr's configurations needed to be revised and extended. For example, in the case of the L level ($n = 2$) Bohr had made use of only two subgroups (2_1, 2_2), each filled with four electrons, whereas Main Smith argued from chemical evidence as well as X-ray and optical spectroscopy that there had to be three subgroups populated by two, two, and four electrons, respectively. Referring to Dauvillier's recent determinations of X-ray absorption bands: 'It is certain that the 2 quanta group or L level consists of at least 3 levels, the 3 quanta group or M level of 5 levels'. More generally he concluded: 'Bohr's scheme of equal subgroups is improbable for the electron arrangement in any atom, and, in consequence, chemical theories based on Bohr's equal subgroup scheme . . . are based on an insecure foundation'.[103]

Main Smith's carefully argued electron configurations were superior to those proposed by Bohr, but his paper made no impact on the physicists' considerations of the structure of the elements. Bohr did not refer to it at the time.[104] Indeed, the physicists may not have been aware of his paper in the somewhat obscure *Journal of the Society of Chemical Industry*, which only became known some time after its main results had independently been developed by 24-year-old Edmund Stoner at Cambridge University.

Table 7.1. The Landé–Stoner classification of quantum states in the first three levels.

Level	K	L			M				
Sublevel		L_I	L_{II}	L_{III}	M_I	M_{II}	M_{III}	M_{IV}	M_V
n	1	2	2	2	3	3	3	3	3
k	1	1	2	2	1	2	2	3	3
j	1	1	1	2	1	1	2	2	3

Contrary to Bohr, Stoner was convinced that the classification of quantum levels in an atom had to incorporate the inner quantum number j, meaning that a particular state must be given by the three numbers n, k, and j. For this purpose he adopted a scheme proposed by Landé, according to which there were three L sublevels and five M sublevels, as independently argued by Dauvillier and Main Smith (Table 7.1). For example, the L sublevels were given by $n_{kj} = 2_{11}$, 2_{21}, and 2_{22}. Stoner recognized the 'remarkable feature' that the largest possible number N of electrons of the type n_{kj} that a single atom can accommodate is $N(n_{kj}) = 2j$. In his own words:

> The number of electrons in each completed level is equal to double the sum of the inner quantum numbers as assigned, there being in the K, L, M, N levels, when complete, 2, 8 (2 + 2 + 4), 18 (2 + 2 + 4 + 4 + 6), 32 . . . electrons. It is suggested that the number of electrons associated with each sub-level separately is also equal to double the inner quantum number.[105]

Using this insight he could account for the electron distributions of the elements in a more natural and fine-grained way than in Bohr's scheme, which 'is based on somewhat arbitrary arguments as to symmetry requirements'. Although Stoner's scheme differed from Bohr's in several respects, such as the distribution of electrons in the completed levels, he stressed that it was in harmony with the essential features of Bohr's theory and not an attempt to overthrow it.

As to the differences, they may be illustrated by comparing Bohr's structure of argon with the one suggested by Stoner. Bohr's structure as of 1922 was

$$\text{Ar}\,(1_1)^2(2_1)^4(2_2)^4(3_1)^4(3_2)^4,$$

while Stoner's structure, disregarding the subdivision due to the inner quantum number, can be written

$$\text{Ar}\,(1_1)^2(2_1)^2(2_2)^6(3_1)^2(3_2)^6.$$

More generally, in Stoner's system the maximum number of electrons in a given n_k subgroup came out as $2(2k - 1)$. It followed that the number of electrons in a group characterized by a given principal quantum number was

$$N(n) = \sum_{k=1}^{n} 2(2k-1) = 2n^2,$$

thus accounting for Rydberg's rule.

Stoner admitted that it was difficult to test his electron distributions, but thought they were supported by X-ray data and chemical evidence. As he pointed out, recent spectroscopic experiments by Alfred Fowler on ionized carbon had revealed a doublet structure that disagreed with Bohr's assignment of four 2_1 electrons in the neutral atom. On the other hand, Fowler's experiments agreed nicely with Stoner's classification—and also with the one of Dauvillier and de Broglie—according to which the electrons in the outer shell of the C^+ ion were two 2_1 electrons and one 2_2 electron. In Stoner's scheme the tetravalence of neutral carbon was not attributed to four equivalent orbits, as in Bohr's theory, but to the four *vacant* 2_2 states that were only filled up in the noble gas neon. Even before Stoner completed his paper Bohr had recognized that Fowler's investigations proved that the C^+ ion in its normal state possesses an electron moving in a 2_2 orbit.[106]

Bohr praised Stoner's classification for its 'formal beauty and simplicity', as he phrased it in a letter to Coster, but the beauty and simplicity did not make him adopt the new classification.[107] To his mind, electron orbits must be specified in the sense of the correspondence principle, and he was unable to see how this could be done with Stoner's scheme. What attracted Pauli's attention to Stoner's paper was not its aesthetical qualities but the connection it made between the electron distributions and the problem of the multiplet Zeeman spectra. 'Twice the inner quantum number does give the observed term multiplicity as revealed by the spectra in a weak magnetic field', Stoner wrote. 'The number of possible states of the (core + electron) system is equal to twice the inner quantum number'.[108] These were words that could not fail to galvanize Pauli, who had first become aware of the problem of the periodic system by listening to Bohr's lectures in Göttingen in June 1922. 'It made a strong impression on me', he recalled, 'that Bohr at that time and in later discussions was looking for a general explanation which should hold for the closing of every electron shell and in which the number 2 was considered to be as essential as 8 in contrast to Sommerfeld's approach'.[109]

More than two years later, in the fall of 1924, Pauli was investigating the difficult problem of the anomalous Zeeman effect, which he thought was somehow related to the problem of explaining the periodic system. In the course of this work he reached the conclusion that the then popular core model of atomic structure was incorrect (see also Section 8.2). According to this model or hypothesis, as formulated in different ways by Sommerfeld, Landé, and Heisenberg, one might consider an alkali atom as consisting of two parts: the valence electron and the core (*Rumpf*) made up of the nucleus and the inner electrons. The core was assumed to possess an angular momentum and a corresponding magnetic momentum. However, Pauli did not share Heisenberg's belief that the core was responsible for the complex Zeeman pattern, and he found a strong argument for his disbelief by calculating the relativistic mass variation of the K electrons in different elements.[110] The result of his calculations was that on the magnetic-core hypothesis the Zeeman splitting would depend on a correction factor

$$\gamma = 1 + \frac{W}{m_0 c^2},$$

involving the total energy W of the atom. Since W was given by Sommerfeld's fine-structure formula, the splitting would vary with the factor $\alpha^2 Z^2 / n^2$, where α is the fine-structure constant. After Landé had informed him that even in the case of thallium ($Z = 81$, $n = 1$) no such atomic-number effect existed, Pauli concluded that atomic angular momenta were due to the valence electrons alone.

In a letter to Landé of November 1924, Pauli formulated his conclusion in this way: 'In alkali atoms the optical electron [*Leuchtelektron*] is itself responsible for both the complex structure and the anomalous Zeeman effect. There is no question of a coupling with the noble-gas-like atomic core (even in other elements). The optical electron is able, in a mysterious unmechanical way, to run about in two states with the same k but with different angular momenta'. A few weeks later, in a letter to Bohr including his manuscript on the exclusion principle, Pauli dismissed the assumption of an atomic core with a magnetic moment twice that of the classical one. This assumption, he wrote, 'contradicts my physical feeling most vigorously. For it is completely incomprehensible and *isolated* in the quantum theory'. The mysterious behaviour was to be ascribed the valence electron, not the core: 'The same classically non-describable two-valuedness [*Zwedeutigkeit*], which causes the doublet structure, is also the reason for the anomalous Zeeman effect'. By that time Pauli was somewhat unsure about the validity of his new scheme of quantum numbers. 'I believe', he said to Bohr, 'that what I am doing here is no greater nonsense than the hitherto existing interpretation of the complex structure. My nonsense is conjugate to the hitherto customary one'.[111]

Stoner's paper served as an important stimulus for Pauli to develop his ideas into a theory of the periodic system. 'I am really very enthusiastic about Stoner's paper', he said in his letter to Bohr. 'The more I read it the more I like it. It was an eminently clever idea to connect the number of electrons in the closed subgroups with the number of Zeeman terms of the alkali spectra'.[112] In his classical paper that appeared in *Zeitschrift für Physik* in March 1925, and in which the exclusion principle was introduced, Pauli extended the Bohr–Coster classification system to four quantum numbers. While n and k_1 were the electron's principal and azimuthal quantum number, respectively, the number k_2, which he conceived as another part of the classically describable k quantum number, 'has a value for each of the two terms of a doublet, i.e. $(k_1 - 1)$ and k_1, and it varies in transitions by +1 or 0'.[113] Finally, the magnetic quantum number m_1, 'which determines the component of the angular momentum along an external field', could attain the 2 k_2 values ranging from $(-k_2 - ½)$ to $(k_2 - ½)$. As Pauli pointed out, 'The number of stationary states in a magnetic field for given k_1 and k_2 is $2j + 1$, and the number of these states for both doublet terms with given k_1 is altogether $2(2k_1 - 1)$'. The electron's peculiar *Zweideutigkeit* was absorbed into the new magnetic quantum number.

In spite of relying on Stoner's earlier classification, Pauli did not make explicit use of the inner quantum number, related as it was to the concept of the atomic core. Somewhat modestly he introduced the exclusion principle as a useful and 'very plausible' rule that, he said, could not be further justified:

> There can never be two or more equivalent electrons in an atom for which in strong fields the values of all quantum numbers n, k_1, k_2, m_1 . . . are the same. If an electron is present in the atom for which these quantum numbers (in an external field) have definite values, this state is 'occupied'.[114]

The quantum numbers were assigned to individual electrons, which was the reason why Pauli supposed the atom to be placed in a strong magnetic field, for in this case the coupling between the electrons would be negligible and they would be in individual stationary states. In order to ensure that the rule was valid also in the absence of a field, he made use of thermodynamic arguments based on adiabatic changes from strong to weak fields.

Pauli's new scheme was successful in so far as it resolved major problems, was more general and unified than earlier schemes, and retained the *Aufbau* principle and most previous rules for assigning the values of quantum numbers. 'We can retain completely on the basis on this classification the principle of permanence of quantum numbers (building-up principle) also for the complex structure of the spectra and the anomalous Zeeman effect'.[115] The closing of the shells, and then Rydberg's $2n^2$ rule, could now be seen as a consequence of a new rule or principle that prohibited any two electrons in an atom from having the same four quantum numbers. In less than a year Pauli's new magnetic quantum number would become the electron's spin, but in early 1925 neither Pauli nor others thought of spinning electrons. From a theoretical point of view the exclusion principle, or what Heisenberg called the *Pauli Verbot* (prohibition), was not entirely satisfactory, since it was a formal principle that lacked a foundation in fundamental physics. Pauli was well aware of this problem, but thought that it could be solved only when the basic principles of quantum theory had been improved.[116]

One problem that faced Pauli's theory of the periodic system was its relationship to the correspondence principle, which he mentioned at the end of his paper. What he had said privately to Bohr and Sommerfeld, he now stated publicly: the problem of the completion of electron groups in an atom has nothing to do with the correspondence principle. In his letter to Bohr of 12 December 1924 Pauli further commented on the correspondence principle and its role in the future development of quantum theory. He pointed out that although the correspondence principle was undoubtedly valid also for many-electron atoms in some form, there was no exact formulation of the principle in this case. As Pauli saw it, the key to an improved formulation of the quantum theory did not lie in an extension of the correspondence principle, but in a radical revision of the fundamental mechanical concepts on which the theory rested. In the spirit of positivism he wrote:

> I have avoided the term 'orbit' altogether in my paper... I think that the energy and [angular] momentum values of the stationary states are something much more real than the 'orbits'. The (not yet reached) aim must be to deduce these and all other physical and real observable properties of the stationary states from the (integral) quantum numbers and to deduce quantum-theoretical laws. But we should not tie the atoms in the chains of our prejudices (to which in my opinion belongs the existence of electron orbits in the sense of ordinary kinematics); on the contrary, we must adapt our concepts to experience.[117]

To the mind of Pauli, but not quite to that of Bohr, it was a virtue that no meaning in terms of a model could be given to the fourth degree of freedom of the electron. It made the exclusion principle more difficult to understand, yet also more attractive from his methodological point of view. Half a year later, when Uhlenbeck and Goudsmit introduced the spinning electron, the same attitude at first made Pauli resist the idea of spin because of its character of a mechanical model.

By 1920 it was known that some spectra of the heavy elements possessed a 'satellite' structure, a large number of lines with a separation of about one-thousandth that of the ordinary fine structure. Hantaro Nagaoka in Japan suggested in 1924 that these lines had their origin in the isotopic structure of the nucleus, and in the same year Pauli ascribed the 'hyperfine structure' to the nucleus having an intrinsic angular momentum that could only assume quantized values. Such a momentum would combine with the angular momentum of the electron shells to produce the observed satellite lines. Pauli's discovery turned out to be very important to the nascent field of nuclear physics, but at the time he thought it was rather 'crazy', as he said in a letter to Sommerfeld.[118]

Notes for Chapter 7

1. Pauli to Bohr, 31 December 1924, in Hermann et al. 1979, p. 197.
2. Ladenburg to Bohr, 18 June 1920, in Rud Nielsen 1977, p. 710. Ladenburg had just returned from a meeting in Halle, where he had given an address on atomic structure and the periodic system.
3. Bohr to Ladenburg, 16 July 1920, in Rud Nielsen 1977, p. 711. See also Bohr to Ladenburg, 29 September 1920 (ibid., p. 713).
4. The unpublished lecture manuscript on 'Some Considerations of Atomic Structure' is translated in Rud Nielsen 1977, pp. 43–69 (quotation on p. 68).
5. Bohr 1921c. The length of Bohr's letter was most unusual for a contribution to the letter section of *Nature*.
6. Bohr 1921c. See Darrigol 1997, pp. 156–7 for a possible interpretation of Bohr's claim that symmetrical models are in disagreement with the correspondence principle.

7. Fajans to Bohr, 25 June 1921, in Rud Nielsen 1977, p. 691.

8. On this theory, see Kragh 1979 and Kragh 1985c. Parts of the chapter rely on material from these two sources.

9. Sommerfeld 1922a, pp. 446–56. Sommerfeld to Einstein, 17 October 1921, in Hermann 1968, p. 93. For details about Sommerfeld's and Landé's use of the inner quantum number, see Forman 1970.

10. Bohr 1923c, p. 270 and p. 285. On the role of the inner quantum number in the building up of atoms, see Heilbron 1983.

11. Bohr 1921d.

12. Ladenburg 1920. The British chemist Charles Bury incorporated the intermediate shells more fully in a paper of 1921 in which he gave population numbers for most of the elements similar to those proposed by Bohr (Bury 1921). His paper was received on 28 April 1921, before Bohr's second letter to *Nature*.

13. Bohr 1922d, which also appeared in Bohr 1922e and in English in Bohr 1922a. The French translation, made by A. Corvisy, was Bohr, *Les Spectres et la Structure de l'Atome* (Paris: Hermann, 1923). A Russian translation of Bohr 1922e made by Sergei Vavilov, a physicist at Moscow's Institute of Physics and Biophysics, was published in 1923.

14. Ehrenfest to Einstein, 27 December 1921, in Buchwald et al. 2009, p. 396.

15. Franck to Bohr, 21 February 1922 (Bohr Scientific Correspondence, Niels Bohr Archive), quoted in Kragh 1979, p. 139. On Bohr's paper and his theory of the periodic system, see also El'yashevich 1970.

16. Bloch 1922, p. 113.

17. Paneth 1922, p. 397.

18. Göttingen lectures, 21 June 1922, in Rud Nielsen 1977, p. 397.

19. MacKinnon 1982, p. 178.

20. Göttingen lectures, 21 June 1922, in Rud Nielsen 1977, p. 406.

21. Ibid., p. 405. The electron configuration of element number 118 also appeared in Bohr's Nobel lecture (Bohr 1923b, p. 41).

22. According to Kim 2007, p. 26. Nishina stayed at Bohr's institute from 1923 to 1928, mainly working on X-ray spectroscopy.

23. Thomsen 1895, which Bohr evidently had studied. On Thomsen's speculations on the complexity of atoms and the periodic system, see Kragh 1982.

24. Bohr 1922a, pp. 69–70. See also Bohr 1932, p. 366.

25. Kramers and Holst (1923, p. 192) stressed that the pictures must be considered 'as largely symbolic', which was also the way Bohr saw them. Pauli recalled that at the time (1922–1923) he was impressed by 'the cautiousness with which these classical models were used by Bohr in contrast to other physicists' (Pauli 1945, p. 99). Some of the pictures from the Kramers-Holst book also appeared in Andrade 1924 (p. 53), McLennan 1923, and elsewhere in the popular literature.

26. For the models in Munich and London, see Schirrmacher 2009.

27. Bohr 1923b, p. 42.

28. Ibid., p. 93.

29. Ibid., p. 138.

30. See Rud Nielsen 1976, p. 392. During the discussion following Ehrenfest's exposition, Bohr's theory was discussed by Lorentz, Millikan, Langevin, Rutherford, and other participants.

31. Bohr 1922a, p. 85 and p. 116.

32. Bohr 1923c, p. 256.

33. Bohr 1924, pp. 15–16. Scerri 2007, p. 195, finds Bohr's attitude as expressed in the passage to be close to obscurantism.

34. Mulliken 1928, p. 190. The analogical use of Bohr's construction principle in molecular physics is described in Soon Park 2001.

35. Bohr 1921c.

36. Schrödinger 1921.

37. Schrödinger to Bohr, 7 February 1921, in Rud Nielsen 1977, p. 737. Bohr's note only appeared in the 24 March issue of *Nature*, so either Schrödinger must have known of it in advance or, what is more likely, the date of his letter is wrong. According to Mehra and Rechenberg (1982a, p. 352), Bohr had sent a copy of his paper to Ladenburg in Breslau, where Schrödinger stayed. But this seems to be based on a misreading of Schrödinger's letter to Bohr, which refers to Bohr's 1918 memoir from the Royal Danish Academy. At any rate, the recently published collection of Schrödinger's correspondence dates the letter May–June 1921 (Meyenn 2011, p. 75).

38. Bohr to Schrödinger, 15 June 1921, in Rud Nielsen 1977, p. 738. For the part of the Copenhagen lecture that includes Bohr's discussion of penetrating orbits, see pp. 60–2. Referring to the penetrating orbits Schrödinger admitted that Bohr had 'much more general reasons for the probablity of this conception' than himself. Schrödinger to Bohr, 28 June 1921, in Meyenn 2011, p. 79.

39. Rud Nielsen 1977, pp. 394.

40. The following summarizes Bohr's fifth Göttingen lecture (Rud Nielsen 1977, pp. 388–96). See also Kragh 1979, pp. 149–52.

41. Van Urk 1923. Born 1925, pp. 194–9.

42. Bohr to van Urk, 5 September 1922. Van Urk to Bohr, 2 September 1922. Bohr Scientific Correspondence, Archive for History of Quantum Physics.

43. Rud Nielsen 1977, p. 400.

44. Ibid., p. 402.

45. Rosseland 1923b. Rosseland worked at the time with the application of quantum theory to radioactive decay, arguing that radiationless transitions should occur not only in atoms but also in nuclei (Jensen 2000, pp. 99–102). For a historical survey of the idea of a maximum atomic number, see Kragh and Carazza 1995.

46. Bohr 1923c, p. 266 (footnote).

47. Sommerfeld 1924b, pp. 465–70.

48. Bohr 1922a, p. 120. This section relies on Kragh 1979.

49. For Maurice de Broglie's early work on X-ray spectroscopy, see Heilbron 1967.

50. Siegbahn had as close connections with Sommerfeld in Munich as he had with Bohr in Copenhagen. For the Siegbahn–Sommerfeld connection, see Kaiserfeld 1993 and the many letters reproduced in Eckert and Märker 2000–2004.

51. Siegbahn to Bohr, 21 September 1916, quoted in Kaiserfeld 1993, p. 314.

52. Coster 1922, p. 572. Coster acknowledged discussions with Siegbahn and Bohr.

53. Bohr and Coster 1923, p. 357. This was Bohr's first paper with a coauthor.

54. Ibid., p. 363.

55. Ibid., p. 370.

56. Nishina 1925, p. 552.

57. Siegbahn 1925, pp. 195–6.

58. De Broglie and Dauvillier 1925, which includes a complete diagram of the Bohr–Coster type, only with the axes reversed.

59. Dauvillier and de Broglie 1924, p. 1.

60. Dauvillier 1922. His configurations for the noble gases were the same as those Bohr had proposed in his second letter to *Nature*. Kragh 1979 gives further details about the work of Dauvillier and de Broglie in the area of X-ray spectroscopy.

61. Dauvillier and de Broglie 1924. Dauvillier 1924.

62. Bohr 1922a, pp. 109–10.

63. Rud Nielsen 1977, p. 155. Bohr first stated his belief in element 72 being a homologue to zirconium in an appendix to the German translation of his Copenhagen lecture (Bohr 1922d, p. 147, preface dated April 1922).

64. 'Very definite conclusions have been reached at to the identity of Ct with the mising element of no. 72 on the Moseley classification' (Rutherford 1922). The discovery of element 72 is well described in the literature. Apart from Kragh 1979 and Kragh 1980, see also Heimann 1967 and Scerri 1994.

65. Urbain 1922, p. 1349. The complex history of element 72 and its relation to the rare earth elements of atomic numbers 70 and 71 (ytterbium and lutetium) is examined in Kragh 1996 and Thyssen and Binnemans 2011.

66. Bohr to Coster, 3 July 1922 (Archive for History of Quantum Physics, Bohr Scientific Correspondence).

67. Bohr to Franck, 15 July 1922, in Rud Nielsen 1977, p. 696. In an appendix to *Drei Aufsätze über Spektren und Atombau* from the summer of 1922, Bohr publicly accepted Dauvillier's identification and its consequences (Bohr 1922d, p. 147).

68. Coster to Bohr, 16 July 1922, in Rud Nielsen 1977, p. 677. See also Coster to Bohr, 15 July 1922, as quoted in Kragh 1979, p. 183. In their joint paper on X-ray spectroscopy, Coster and Bohr added a critical note in which they expressed their lack of confidence in the discovery claim of Urbain and Dauvillier (Bohr and Coster 1922, p. 343).

69. Paneth 1922, p. 383.

70. Bohr 1923b, p. 44.

71. Coster and Hevesy 1923.

72. Hevesy 1925 is an informative account of early hafnium research.

73. Rutherford to Hevesy, 8 January 1922, reproduced in Hevesy 1963, p. 30. In 1923 Rutherford explicitly interpreted the discovery of hafnium as a result of Bohr's prediction (Section 7.5). For Nernst's comment, see Kragh 1980, p. 297. For Hevesy and hafnium, see also Levi 1985, pp. 50–9.

74. Bohr to Rutherford, 9 February 1923, quoted in Kragh 1980, p. 282. The reference is to Urbain and Dauvillier 1923. Details about the controversy over element 72, which included not only celtium but also 'oceanium', are given in Kragh 1980.

75. Popper 1982a, pp. 163–4. For a critical assessment of Popper's view, see Scerri 1998.

76. Kramers and Holst 1923, p. 204.

77. Andrade 1924, p. 52.

78. Bury 1921, p. 1608.

79. The question is discussed in Scerri 1994 and Scerri 2007, who concludes that Bohr's theory predicted hafnium in a weak sense, but that the prediction was mainly due to the empirical aspects of the theory (chemical and spectroscopic facts) and not to the aspects related to quantum theory. I consider this to be a not unreasonable conclusion.

80. Landé to Bohr, 21 February 1921, in Rud Nielsen 1977, p. 722. Bohr had sent him a copy of his manuscript and indicated that he was about to develop his ideas into a more detailed version (Bohr to Landé, 14 February 1921, ibid., p. 721).

81. Born to Franck, 22 February 1921, as quoted in Greenspan 2005, pp. 104–5.

82. Sommerfeld to Landé, 3 March 1921 (Archive for History of Quantum Physics, Landé correspondence). Sommerfeld to Bohr, 7 March 1921, in Rud Nielsen 1977, p. 740.

83. Sommerfeld to Bohr, 25 April 1921, in Rud Nielsen 1977, p. 741.

84. Sommerfeld 1922a, p. 134. Wentzel 1922.

85. Hilbert's manuscript of his 1923 lecture, as reproduced in Sauer and Majer 2009, on p. 414. In March 1921 Hilbert gave three lectures in Copenhagen on the occasion of receiving an honorary doctorate from the University of Copenhagen. During his stay in Copenhagen he met with Niels Bohr and his brother, the mathematician Harald Bohr. For Hilbert's interest in atomic theory and his appreciation of Bohr's work, see further Hilbert to Bohr, 23 August 1923 (Archive for History of Quantum Physics, Bohr Scientific Correspondence). As mentioned in Section 4.2, also the Swedish mathematician Ivar Fredholm took an interest in Bohr's atomic theory.

86. Nordström to Bohr, 2 February 1922 (Archive for History of Quantum Physics, Bohr Scientific Correspondence).

87. Rubinowicz to Bohr, 18 April 1921, in Rud Nielsen 1977, p. 733.

88. Rutherford to Bohr, 26 September 1921 (Archive for History of Quantum Physics, Bohr Scientific Correspondence).

89. Ehrenfest to Bohr, 27 September 1921, quoted in Kragh 1979, p. 137.

90. Sidgwick 1923, p. 726, emphasis added.

91. Hansen to Bohr, 28 October 1922 (Archive for History of Quantum Physics, Bohr Scientific Correspondence).

92. Kramers 1935, p. 90. For his contemporary account of Bohr's theory, see Kramers 1924c.

93. Sommerfeld 1924b, p. vi.

94. Ibid., p. 192. Whereas this edition of *Atombau* described Bohr's theory extensively, in the third edition he just mentioned (on p. 134) Bohr's noble gas configurations as communicated in his first letter to *Nature*.

95. See Weisskopf 1970, p. 18.

96. Sommerfeld 1925, p. 70.

97. Rutherford 1923, pp. 14–15. The Liverpool meeting had strong participation from Copenhagen. Bohr lectured on the correspondence principle, Hevesy on hafnium, and Coster on X-ray spectrocopy. Bohr's new theory of the periodic system also appeared prominently in the address of McLennan (1923), president of Section A (mathematics and physics) of the British Association.

98. Pauli to Sommerfeld, 6 June 1923, in Hermann et al. 1979, p. 95. See also Pauli to Landé, 23 September 1923 (ibid., p. 122).

99. Sommerfeld 1924b, p. 192. Pauli to Sommerfeld, 6 December 1924, in Eckert and Märker 2004, p. 178.

100. Pauli to Bohr, 12 December 1923, in Hermann et al. 1979, pp. 187–8.

101. Pauli to Bohr, 21 February 1924, in ibid., p. 148.

102. Bury 1921, p. 1602. See Davies 1986, who argues that priority in understanding the periodic system belongs to the chemist Bury rather than the physicist Bohr. Davies finds Bohr 1921c to be nothing but a 'verbose effort'.

103. Main Smith 1924a, p. 549. An extended version of his ideas of electron configurations appeared in Main Smith 1924b. For his earlier critique of Bohr's atom, see Main Smith 1923.

104. In Bohr 1925a he cited Main Smith's contribution. See also Bohr 1932, p. 367.

105. Stoner 1924, p. 726. For a detailed analysis of Stoner's work and its significance for Pauli's explanation of the periodic system, see Heilbron 1983. His background and early career is described in Cantor 1994.

106. Fowler 1924. Bohr, p. 138 in the second edition of 1924 of *The Theory of Spectra and Atomic Constitution*.

107. Bohr to Coster, 10 December 1924, in Rud Nielsen 1977, p. 681. Coster to Bohr, 7 December 1924 (ibid., p. 679), found the justification of Stoner's classification to be 'rather doubtful', but nonetheless viewed it as a significant progress.

108. Stoner 1924, p. 725.

109. Pauli 1946, p. 213, an address given at the Institute for Advanced Study on the occasion of Pauli's award of the Nobel Prize for that year.

110. Pauli 1925a. The history of the origin and early development of Pauli's exclusion principle is described in detail in Heilbron 1983 and Massimi 2005. See also van der Waerden 1960 and Mehra and Rechenberg 1982a, pp. 666–84. The road from the exclusion principle to spin is critically reviewed in Robotti 1990.

111. Pauli to Landé, 24 November 1924, in Hermann et al. 1979, p. 177. Pauli to Bohr, 12 December 1924, in ibid., p. 187.

112. Pauli to Bohr, 12 December 1924, in ibid., p. 188.

113. Pauli 1925b, p. 766, with English translation in ter Haar 1967, pp. 184–203. The complicated question of the meaning of the quantum numbers in Pauli's system is considered in Seth 2009. The numbers can be translated into modern notation by $k_1 = l + 1$, $k_2 = j + ½$, and $m_1 = m_j$, where $j = l \pm ½$.

114. Pauli 1925b, p. 776. The 'equivalent electrons' in the quotation refer to electrons with the same binding energy or principal quantum number.

115. Pauli 1925b, p. 768.

116. Massimi 2005 deals in detail with the changing status of Pauli's prescription, arguing that it started as a 'rule' and only became a proper 'principle' in 1940, when Pauli proved that it follows from the structure of relativistic wave equations.

117. Pauli to Bohr, 12 December 1924, in Hermann et al. 1979, p. 188–9 (translation in Darrigol 1992, p. 208). Bohr responded on 22 December, suggesting that Pauli's belief that the correspondence principle could not explain the completion of the groups was premature.

118. Pauli 1924. Pauli to Sommerfeld, 29 September 1924, in Hermann et al. 1979, p. 161. On Pauli's note, see Belloni 1982.

8

Crisis: The End of the Bohr Model

At the time when Bohr's atomic model would have been celebrating its tenth anniversary, problems began to accumulate. Some of these were experimental anomalies, while others were of a logical and conceptual nature; some of them were old, others new. To the list of old problems belonged the anomalous Zeeman effect, which turned out to be intimately connected to the complex spectra and the assignment of quantum numbers to many-electron atoms. The physicists could account for the spectroscopic details of the Zeeman effect, but only by proceeding phenomenologically and abandoning any explanation of the effect based on established atomic theory. The core models introduced by Landé and Heisenberg did not fit easily with Bohr's principles, among other reasons because they operated with half-integral quantum numbers and seemed to contradict the construction principle on which Bohr had based his theory of the periodic system. Generally speaking, the years after 1922 saw a drastic shift in the status and nature of atomic models, which became increasingly abstract, symbolic, and formal, soon leading to proposals of abandoning the model concept at all.

In the summer of 1925, while completing a major review paper on atomic and quantum theory, Van Vleck commented on the most recent developments: 'Modern physics certainly is passing through contortions in its attempt to explain the simultaneous appearance of quantum and classical phenomena', he said; 'but it is not surprising that paradoxical theories are required to explain paradoxical phenomena'.[1] This was an evaluation with which Bohr fully agreed.

The feeling of a crisis in atomic physics gave rise to several proposals of a radical nature, perhaps the most radical one being the idea that energy and momentum are not conserved in individual atomic processes. This idea, which featured prominently in a theory offered by Bohr, Kramers, and Slater (BKS) in 1924, was an attempt to understand radiation from atoms without making use of the concept of light quanta, which were particles *non grata* in Copenhagen. The BKS atom with its orchestra of 'virtual oscillators' was as original as it was controversial. Rejected by a majority of physicists, the theory was short-lived but nonetheless influential. Its influence was particularly strong in the dispersion theories that attracted much attention in the final phase of the old quantum theory and were greatly significant in the revolutionary change to a new 'quantum mechanics', a term that Born had introduced about a year before it became a reality (but, as mentioned in Section 6.6, the term was used by Einstein as early as 1922). In the process that led to the new mechanics and at the same time made the Bohr–Sommerfeld orbital atomic theory obsolete, the centres of theoretical physics in Göttingen and Copenhagen played a dominant role. The key players in the development included the leaders of the two institutions, Born and Bohr, but no less more important were the three young physicists Kramers, Pauli, and Heisenberg.

The crisis that culminated in the spring and summer of 1925 was the result of accumulated empirical anomalies and a growing dissatisfaction with the conceptual foundation of atomic physics. It led to the abandonment of electron orbits and attempts to base atomic theory on discrete rather than continuous quantities, and it was influenced by the positivistic doctrine that a satisfactory theory of the atomic world should include only such symbols as referred to measurable quantities. The latter doctrine, known as the observability principle, featured prominently in Heisenberg's epoch-defining paper of the late summer of 1925, but the actual role it played in the foundation of quantum mechanics was more limited. Although the emergence of the Göttingen quantum mechanics was indeed a break with the past, the break with Bohr's old atomic theory was not complete. The basic postulates, meaning the notion of stationary states and the frequency condition $\Delta E = h\nu$, survived. Perhaps the strongest link of continuity was provided by the correspondence principle, which in a mathematically sophisticated form permeated much of Heisenberg's new theory.

8.1 THE RIDDLE OF THE ZEEMAN EFFECT

'A colleague who met me strolling rather aimlessly in the beautiful streets of Copenhagen said to me in a friendly manner, "You look very unhappy"; whereupon I answered fiercely, "How can one look happy when he is thinking about the anomalous Zeeman effect?".'[2] This is what Pauli recalled from his stay in Copenhagen, perhaps in the spring of 1923. He was not the only physicist to worry about this spectroscopic effect, which was seen as a problem for atomic theory no less serious than the helium anomaly, an indication that something was quite wrong with the Bohr–Sommerfeld theory of atomic structure. The same year, in his contribution to the Bohr issue of *Die Naturwissenschaften*, Born suggested that 'the entire complex of questions connected with the anomalous Zeeman effect' was reason enough to reconsider the existing atomic theory and possibly to reconstruct it entirely.[3] The complex of questions he referred to was not only the anomalous Zeeman effect, but also the related magneto-optical Paschen–Back effect, the higher multiplet spectra discovered in 1922, and the relationship between multiplets in optical spectra and X-ray spectra. For the sake of simplicity, in this section I shall consider only the anomalous Zeeman effect.

Physicists had been studying the complex or anomalous Zeeman effect ever since it was discovered in 1897 by the Irishman Thomas Preston. Their efforts resulted in various empirical rules, for example that the frequencies of the components were always rational fractions of those of the normal Zeeman effect. However, contrary to this effect, which Lorentz had explained immediately after its discovery, the anomalous case defied explanation in terms of both electron and quantum theory. The rules established by Preston, Runge, Sommerfeld, and other physicists were useful, but they were basically of a numerological kind. For the normal Zeeman effect, as analyzed by Debye and Sommerfeld in their papers of 1916, the magnetic contribution to the energy of a spectral term was given by $mh\nu_L$, where $\nu_L = eH/4\pi m_e c$ is the classical Larmor precession frequency, with m_e denoting the mass of the electron. The quantum number m was interpreted as the component of the azimuthal quantum number in the direction of the magnetic field H.

The first substantial progress in accounting for the anomalous Zeeman effect was made by Landé, who in February 1921 told Bohr about his ongoing research. Presenting it as empirically rather than theoretically based, he wrote: 'With regard to the complicated types of the Zeeman effect, I have found a few empirical rules which go considerably beyond Sommerfeld's compilation in the *Ann. d. Phys.* and permit one to make predictions regarding the neon spectrum. But what these rules signify is entirely incomprehensible to me'.[4] Believing at the time that Bohr might be a rival in the quest to understand the anomalous Zeeman effect, he was 'in a hurry with this [paper] because Bohr obviously is thinking about these things; and why should a foreign country forestall us in this'. Sommerfeld informed him that Bohr was too busy with other matters and, at any rate, 'I don't look upon him as a foreigner'.[5] In fact, Bohr was at the time working hard on his new theory of the periodic system, not to mention his new institute of theoretical physics, and he had little time to think about the regularities of complex spectra.

In a paper that appeared in June 1921 Landé expressed the energy of a spectral Zeeman term by the generalized formula

$$W = W_0 + gmh\nu_L,$$

where W_0 is the energy of the unperturbed state. The proportionality constant g was introduced empirically, the only justification for it and its assigned values being that the observed components could then be reproduced. The constant or 'splitting factor', as Landé called it, was assigned the value $g = 1$ for singlets (normal Zeeman effect), but for doublets $g = 2j/(2k - 1)$. For triplets the dependence of g on j and k was more complicated, but in all cases it could be expressed as rational fractions similar to the one of the doublet case. Assuming as a working hypothesis that Sommerfeld's inner quantum number j was a measure of the total angular momentum and m determined its space quantization in a magnetic field, Landé argued that for singlets and triplets m attained $2j + 1$ values, namely $m = 0, \pm1, \pm2, \ldots, \pm j$. However, in the case of doublets, where $j = k$ and $j = k - 1$, he had to assign half-integral values to the magnetic or 'equatorial' quantum number:

$$m = \pm 1/2, \pm 3/2, \ldots, \pm (j - \tfrac{1}{2}).$$

Since m can attain $2j$ values, the total number of magnetic states for a doublet pair is $2k + 2(k - 1) = 2(2k - 1)$. Not only did half-integral values of m (and then also of j) lack any theoretical basis, they also seemed to contradict Bohr's atomic theory, which allowed only integral quantum numbers. In 1921 Landé did not propose a physical interpretation of his theory, except that he suggested that the g factors might be interpreted as 'apparently anomalous values of e/m' (where m stands for the electron's mass, not the magnetic quantum number).[6] Sommerfeld, who knew of Landé's work in advance of its publication, was critical of his approach but nonetheless responded with enthusiasm: 'Bravo, you are a magician! Your construction of the doublet Zeeman types is *very* beautiful'.[7] Without referring to Landé by name, he told Einstein about the recent progress in understanding the anomalous Zeeman effect on the basis of the inner quantum number: 'Light, or better, dawn really is coming to spectroscopy'.[8] Perhaps so, but the dawn did not extend to the underlying quantum theory, which was still shrouded in fog.

In another important work on the Zeeman effect, published in May 1923, Landé developed his phenomenological theory on the basis of the core model of the atom expressed in terms of angular momentum vectors. Apart from the total angular momentum J, these vectors were the core angular momentum R and the angular momentum K of the valence electron, the three vectors being connected by the equation $J = R + K$. Using this model he obtained an expression for the g factor:

$$g = 1 + \frac{(J^2 - 1/4) + R^2 - K^2}{2(J^2 - 1/4)} = \frac{3}{2} + \frac{R^2 - K^2}{2(J^2 - 1/4)}.$$

Landé's new formula agreed impressively with experiments and thus brought a much needed order to the area. However, being an empirical formula summarizing experimental data, rather than a deduction from a theory or model, it still lacked a proper theoretical explanation. And there were other problems: it postulated half-integral quantum numbers, it required that the core contributed twice to the magnetic energy of the atom, and it seemed to violate Bohr's generally accepted construction principle. In view of the controversial half-quanta, Landé cautiously referred to the quantities appearing in his formula as '"apparent" angular momenta, from which the real momenta may perhaps deviate'.[9] In spite of the obvious weaknesses in his work it aroused much attention in the German physics community and was widely seen as an important step towards the final understanding of the anomalous Zeeman effect. Heisenberg found Landé's work to be 'very beautiful' and Pauli congratulated him with his 'wonderful scheme of the multiplets of the Zeeman effect'.[10]

The complex structure of spectral lines and their behaviour in magnetic fields was not among Bohr's primary concerns, but of course he was well aware of the work that Sommerfeld and others had done in this area. He referred to it in his Berlin lecture of 1920, where he suggested that the problems might be understood on the basis of the correspondence principle. As he saw it at the time, the anomalous Zeeman effect did not indicate a failure of ordinary electrodynamics, but was rather to be seen as 'connected with an effect of the magnetic field on that intimate interaction between the motion of the inner and outer electrons which is responsible for the occurrence of the doublets'.[11] About two years later, in his Göttingen lectures of June 1922, Bohr dealt briefly with the new theories of the anomalous Zeeman effect proposed by Landé and Heisenberg, but apart from noting that the effect proved the classical theory to be inadequate he did not go into details. He considered Heisenberg's recent theory to be interesting, but also to rest on assumptions that were unacceptable. His main objection to the works of Landé and Heisenberg was the half-integral quantum numbers, which he found contradicted his theory of the electron configurations of the elements. 'My viewpoint is this', he complained to Landé, 'that the entire manner of quantization (half integer quantum numbers etc.) does not appear reconcilable with the basic principles of the quantum theory, especially not in the form in which these principles are used in my work on atomic structure'.[12]

The intense discussions Bohr had with Pauli during the latter's stay in Copenhagen in 1922–1923 made him pay greater attention to the magneto-optical problems, which he took up in his survey of line spectra and atomic structure published in 1923 in *Annalen der Physik*. Referring to Landé's first paper on the anomalous Zeeman effect, he interpreted the results by assuming that 'the change in the motion of the atom as a whole consists in a superposed precession around an axis through the nucleus, and parallel to the field, with a frequency of precession independent of the orientation of the atom relative to the field'.[13] This interpretation he claimed to rest 'on the basis of the correspondence principle', as usual without elaborating or clarifying the comment. Although he noted that Landé's assigment of half-integral quantum numbers disagreed with a 'direct analogy with the quantum theory of periodic systems', he offered an explanation of Landé's findings.

Bohr's hypothesis was qualitative and quite speculative, its essence being a postulate of a non-mechanical coupling between the core's inner orbits and the outer orbit of the valence electron. In order to explain Landé's result that an atom placed in a weak magnetic field can take on $2(2k - 1)$ states, he suggested the following view:

> Because of stability properties of the atom which cannot be described mechanically, the coupling of the series electron to the atomic core is subject to a constraint [*Zwang*] which is not analogous to the effect of an external field, but which forces the atomic core to assume two different positions in the atom, instead of the single orientation possible in a constant external field, while at the same time, as a result of the same constraint, the outer electron, instead of $2k$ possible orientations in an external field, can only assume $2k - 1$ orientations in the atomic assemblage.[14]

In a later work Bohr, translating *Zwang* as 'strain', spoke of 'a mechanically undescribable "strain" in the interaction of the electrons which prevents a unique assignment of quantum indices on the basis of mechanical pictures'.[15]

By means of this hypothesis of a constraint or *Zwang*, blatantly ad hoc as it was, Bohr could maintain the construction principle, get the right number of anomalous Zeeman states, namely $2(2k - 1)$, and even account for the stability of the helium atom. The cost was that he had to accept half-integral quantum numbers. This he did reluctantly and only for the quantities j and m, which he considered to be of a more formal nature with no clear model interpretation. The numbers that really mattered, and which characterized the electron configurations of the elements, were n and k, and these he insisted had to be integers. The assumption of a half-integral azimuthal quantum number not only had consequences that 'seem to contradict our experience about spectra', it also 'must be regarded as a departure from this theory [the quantum theory of periodic systems] which can hardly be substantiated'.[16] In an unpublished manuscript on the applications of quantum theory, a planned sequel to his 1923 paper in *Zeitschrift für Physik*, Bohr returned to the problem of the anomalous Zeeman effect and the idea of a peculiar intra-atomic constraint responsible for the failure of Larmor's theorem:

> In the electronic assemblage in an atom we have to do with a coupling mechanism which does not permit a direct application of the quantum theory of mechanical periodicity systems... We are led to the view that the interplay between series electron and atomic core, at least as far as the relative orientation of the orbit of the series electron and the electronic orbits in the core is concerned, conceals a 'constraint' [*Zwang*] that cannot be described by our mechanical concepts and has the effect that the stationary states of the atom, in essential respects, cannot be compared with those of a mechanical periodicity system... According to our view, it is just this constraint that finds its expression in the regularities of the anomalous Zeeman effect and, in particular, is responsible for the failure of the Larmor theorem.[17]

Bohr's discussions with Pauli resulted in a joint manuscript on the anomalous Zeeman effect, in which the effect was explained on the basis of integral quantum numbers only. Enclosing a copy of the manuscript in a letter to Landé, Bohr wrote: 'As you will see from our note, we are always so conservative here in Copenhagen that we stick sincerely to the integral quantum numbers'.[18] Bohr and Pauli submitted the manuscript to *Zeitschrift für Physik* in early 1923, but decided to withdraw it because they came to doubt its validity.

In another letter to Landé, Bohr wrote that his and Pauli's manuscript was 'a desperate attempt [*Verzweiflungsversuch*] to remain true to the integral quantum numbers; we hoped to find in the very paradoxes an indication of the path along which one should search for a solution of the anomalous Zeeman effect. However, in the meantime the doubts about the necessity of those paradoxes have continued to grow'.[19] Latest by March, the 'conservative' physicists Pauli and Bohr had come to accept half-integral inner quantum numbers. Bohr, said Pauli in a letter to Landé, 'keeps to the integral quantization of *k*', whereas he believed that 'in the question of the quantization of *j* and *m* one has to do with novel phenomena that defy the current foundation of quantum theory'.[20]

8.2 HEISENBERG'S CORE MODELS

Frustrated over his failed paper with Bohr, Pauli nonetheless continued investigating the anomalous Zeeman effect, and in late April his efforts resulted in a paper of a formal rather than model-based nature. Instead of starting with the complex spectral terms of the anomalous effect, Pauli started with the simpler strong-field terms and worked his way backward by means of the Paschen–Back effect. To do so he made use of a new rule, which he introduced without attempting to justify it: 'The sum of the energy values in all those stationary states belonging to given values of *m* and *k*, remains a linear function of the field strength during an entire transition from weak to strong fields'.[21] In this way he was able to derive Landé's *g* factors in the case of weak fields and without relying on specific model interpretations.

In spite of this result Pauli was far from satisfied. 'I have for very long vexed myself with the anomalous Zeeman effect and often lost my way', he confided to Sommerfeld. 'For a time I was quite desperate. On Bohr's insistence I have finally send a small paper to the Zeitschrift für Physik . . . I have written all of this with a tear in the corner of my eyes and am anything but delighted'.[22] At the end of his paper he expressed the same kind of dissatisfaction, pointing out that his derivation was of a formal nature only. A model interpretation 'on the basis of the currently known principles of quantum theory is hardly possible', he concluded. One problem was the failure of Larmor's theorem, which required $g = 1$; another was that 'the appearance of half-integral values for *m* and *j* already implies a fundamental break with the framework of the quantum theory of multiply periodic systems'. This was also what he wrote to Landé: 'The question of the normalization of the quantum numbers is presently of secondary importance to me. For I am convinced that in the anomalous Zeeman effect there is no conditionally periodic model and that something essentially new must be done'.[23]

Although Pauli presented his theory as phenomenological and model-independent, he was not yet ready to consider this a virtue rather than a deficiency. In his letter to Landé of 23 May he wrote that his theory of the anomalous Zeeman effect was 'restricted to the phenomenological level and leaves out all model considerations', yet he also said that '[I am] very unhappy that I have not yet succeeded in finding a satisfactory model interpretation for these so remarkably simple law-like regularities'. In fact, his theory was not entirely model-independent, as he confessed to Sommerfeld: 'I would never have reached the given representation of the spectral terms in strong fields if I had not been guided by model representations'.[24] Sommerfeld and Pauli were at the time both moving from a model-based understanding of atomic structure to one based on empirical rules, only was Pauli moving faster. He described his vector model—at the time sometimes referred to as an *Ersatz* (substitution) model—in letters to Landé and Sommerfeld, and he published a detailed version of it in a paper of early 1924. As he pointed out, for weak fields his theory led to the expression

$$g = \frac{3}{2} + \frac{R^2 - K^2}{2J^2},$$

which he could not bring into agreement with Landé's empirical formula. In his letter to Sommerfeld of 19 July he made the proposal that the classical differential expression might have to be replaced by a difference expression:

> The structure of the expressions is rather similar but one *cannot* make the difference disappear by changing the normalization of the indefinite additive constants involved in R, K and J. . . . One may also say that the two expressions are related to one another like the differential quotient $\frac{d}{dJ}\frac{1}{J}$ to the difference quotient $\left(\frac{1}{J} - \frac{1}{J-1}\right)$, which seems to indicate something non-mechanical.

Much of the discussion about atomic models and the anomalous Zeeman effect in the years 1922–1925 relied on the contributions of the young Werner Heisenberg, who in his very first scientific paper from the beginning of 1922 presented an atomic model to account for the regularities that Landé had established phenomenologically. However, it was a model that differed radically from the ordinary kind of model associated with the Bohr–Sommerfeld theory. Heisenberg developed a slightly earlier semiclassical model of Sommerfeld's into a theory which accomodated all empirical data and reproduced Landé's formula for doublets and triplets as well as the transition from weak to strong fields as given by the Paschen–Back effect.[25] To do so, he made use of half-integral quantum numbers for k and m. He justified his results by inventing a core model in which, in the case of alkali atoms, the valence electron was assumed to share half a unit of angular momentum with the core, leaving the electron with $(k - ½)$ $h/2\pi$ and the core with ½ $h/2\pi$. The doublet terms would arise from the two orientations of the angular momenta of the core and the valence electron being either parallel or antiparallel:

$$j = (k - ½) + ½ = k \text{ and } j = (k - ½) - ½ = k - 1.$$

The electron's angular momentum vector was assumed to precess around the momentum vector of the core and thereby create an inner magnetic field at the site of the core, which would orient itself in, or opposite to, the direction of the total field made up of the inner field and the external field.

The assumptions of Heisenberg's model were arbitrary and tailored to give the results known empirically, but they did yield an explanation of the magneto-optical anomalies, at least of a sort, and for this reason the model was widely seen as a significant advance. However, the conceptual price to be paid for the instrumental success was high, for the assumptions violated several of the most cherished theoretical principles of quantum theory. Among these were the Bohr–Sommerfeld quantum conditions, the selection rules, and the restriction of the total angular momentum to a set of discrete values. In Heisenberg's system no such restriction could be justified. Moreover, Bohr's construction principle was incompatible with the sharing of angular momentum between the valence electron and the core. For example, an argon atom had a closed shell or core with zero momentum, but when an electron was added, turning it into a sodium structure, the core would, according to Heisenberg, suddenly gain an angular momentum of $\frac{1}{2}\, h/2\pi$. This contradicted Bohr's construction principle, which was based on the permanence of quantum numbers during the building up of atoms. It was no wonder that Bohr responded negatively to the new theory from Munich.

Heisenberg's suggestive theory was not only problematical for methodological reasons, but also because of its empirical consequences. For example, in the case of alkali metals it led to values of the internal magnetic fields that were much larger than ordinarily assumed. McLennan also pointed out that according to Heisenberg electrons could be in a state characterized by an azimuthal quantum number $k = \frac{1}{2}$, which seemed impossible for very heavy atoms. If the electronic orbit closest to a uranium nucleus was of this type, it would imply a shortest distance of approach equal to 4×10^{-12} cm, whereas the radius of the uranium nucleus was known to be 6.5×10^{-12} cm. 'For reasons of this character', argued McLennan in his address to the 1923 meeting of the British Association, 'we are practically precluded from assigning to k, the azimuthal quantum number, a value less than 1 in defining the electronic orbits in atoms'.[26] As mentioned in Section 7.2, Bohr arrived at the same conclusion, namely that an electron with quantum number $k = \frac{1}{2}$ could only exist in atoms with $Z < 68$.

The various problems did not shake Heisenberg's confidence in his core model, which was based solely on the fact that it worked. Having an opportunistic attitude, he believed that 'success justifies the means' (*Der Erfolg heiligt die Mittel*).[27] Contrary to Bohr, who placed the highest value on consistency and clarity, Heisenberg was at the time willing to work with a model inconsistent with established physics if it resulted in answers that agreed with experiments. This might be an unphilosophical attitude, as Pauli later complained to Bohr,[28] but from Heisenberg's perspective this was an objection of no significance. While Bohr, Landé, and Pauli were critical, each in their own way, Sommerfeld had a greater appreciation for the work of his prodigy. To Einstein he wrote of Heisenberg's forthcoming paper:

> I have in the meantime uncovered wonderful numerical laws for line combinations in connection with the Paschen measurements and presented them in the third edition of my book. A pupil of mine (Heisenberg, 3rd semester!) has even explained these laws and those of the anomalous Zeeman effect with a model... Everything works out, but yet in the deepest sense remains unclear. I can only advance the craft of the quantum, you must make its philosophy... Set yourself to it![29]

But Einstein had no wish to engage fully in either quantum theory or quantum philosophy.

By the summer of 1923, Pauli had reached the conclusion that 'there is no [satisfactory] model for the anomalous Zeeman effect and that we must create something fundamentally new'. He was at that time convinced that the problem of atoms with several electrons was of a physical rather than mathematical nature. In so far that one could still speak of individual electron orbits, he suggested that they 'behave more as a system of oscillators in which the frequencies are associated not with the motion but with the transitions'.[30]

A few months later Heisenberg came up with something new, yet his new conception was still formulated within the framework of the core model, which Pauli distrusted. Heisenberg outlined his theory of a modification of the formal rules of quantum theory in a letter to Pauli of 9 October 1923, in which he emphasized that it was a model only in a symbolic and heuristic sense. He did not subscribe to atomic models in a physical sense, he said, but wanted to extract from them such information as could be used in constructing a new theory of a symbolic kind. This theory of the future—hopefully a near future—he envisaged as being consistently discrete. According to Bohr's second quantum postulate of 1913, the frequency of a transition was obtained by taking the difference between the energies of two stationary states:

$$h\nu = E_2 - E_1 = \Delta E.$$

Heisenberg pointed out, as Bohr had recognized from the very beginning, that this was a strange mixture of classical and quantum concepts. To remove the strangeness he suggested replacing the classical (and hence continuous) energies with quantum energies based on a difference equation of the form $E = \Delta F$, where $F = F(r, j, k, m)$ was a function of the quantum numbers. What he called 'the new Göttingen theory of the anomalous Zeeman effect' included the programmatic statement that 'model representations have in principle only a formal sense, they are the classical analogues of the "discrete" quantum theory'.[31] In late 1923 Heisenberg wrote to Sommerfeld about his new and still unpublished theory:

> In the classical theory one would have $\partial F/\partial j = H$, $\partial H/\partial J = \nu$. One now replaces the differential quotient $\partial F/\partial j$ with the corresponding *difference* quotient, and then everything comes out right, in particular the construction principle, the summation rule, Rydberg's displacement law, etc. When one then contemplates what has really been done, one sees clearly that model conceptions have no real meaning. The orbits are not real, neither with respect to frequency nor energy. Obviously, only the coordinates J have a real meaning, and that only at discrete points.[32]

Relying on Pauli's *Ersatzmodell* Heisenberg found it necessary to introduce a new quantum rule, namely, that 'A determinate value of the coupling energy between the electron and the atomic core is not associated, as has been so far assumed, to one value of the inner quantum number j, but rather with two values'.[33] In addition, he formulated a 'new quantum principle' which amounted to a correspondence between the quantum-theoretical Hamiltonian of a system and the classical Hamiltonian. From Heisenberg's symbolic core model followed Landé's *g* factors and all other results of the complex spectra. To the list of negative aspects belonged the fact that 'now one does not understand quantum theory at all', as he phrased it in his letter to Pauli of 9 October. But this he did not necessarily see as a disadvantage:

> This, however, I find rather congenial. Now the real goal must be to arrive from the symbols *in a unique way* at the discrete states; whether the formulae thus obtained make comprehensible sense, is doubtful to me. Born summarizes our task for the time to come in the word 'discretization of atomic physics'. Now I wonder how much of it turns out to be correct when I continue to think about it; and I am anxious to see how you digest it.[34]

Heisenberg's new theory of the anomalous Zeeman effect worked even better than his first one, and it had the methodological advantage that it was more consistent with established principles of quantum physics. In particular, it no longer violated the *Aufbau* principle, which helped make it acceptable in Copenhagen. In his published paper Heisenberg emphasized as a virtue how his theory provided a natural interpretation of Bohr's principle, as he also explained privately to Bohr: 'I can hardly imagine another interpretation of the difficulties described by you in the Kayser issue of the *Annalen*'.[35] To his father, August Heisenberg, he wrote about Bohr's influence: 'I realize ever more that Bohr is the only person who, in the philosophical sense, understands something of physics'.[36] While Bohr eventually approved of Heisenberg's theory, Pauli remained intransigent. He considered it an 'ugly' theory, which did not even 'yield an explanation of the half-integral quantum numbers and of the failure of Larmor's theorem'.[37] Pauli came to the conclusion that the core model was fundamentally wrong, even in the symbolic representation offered by Heisenberg. With his theory of the exclusion principle he thought he had buried the model.

In a letter to Bohr of February 1924, Pauli revealed how he looked upon the contemporary situation in the quantum theory of atoms. There were, he said, two groups of atomic physicists in Germany: one group that had began with integral quantum numbers and another that instead started with half-integral quantum numbers:

> Both groups, however, have the characteristics in common that there is no a priori argument to be had from their theories that tells which quantum numbers and which atoms should be calculated with half-integral values of the quantum numbers and which should be calculated with integral values. They can decide this only a posteriori by comparison with experience. I myself have no taste for this kind of theoretical physics... I am far more radical than the 'half-integral-number' atomic physicists.[38]

Referring to the problem of interpreting the failure of Larmor's theorem, he added: 'Unfortunately, with respect to this main point Heisenberg's considerations do not lead us beyond what we already know'. As to the anomalous Zeeman effect, Heisenberg had achieved a formal description of it, but not *explained* it on a basis consistent with quantum theory. Pauli had no better alternative to offer, and thus his critique remained on the negative level.

There was however one important point in Heisenberg's work with which the critical Pauli fully agreed, and this he spelled out in his letter to Bohr of 21 February: 'According to my point of view, Heisenberg hits the truth precisely when he doubts that it is possible to speak of determinate trajectories. Doubts of this kind Kramers has never considered as reasonable. I must nevertheless insist upon this, because the point appears to me to be very important'. Pauli expressed the same kind of doubt in a letter to Eddington, in which he emphasized that 'the quantum theory in no way calls for only a modification of the theory of light, but generally calls for a new definition of the concept of the electromagnetic field for non-static processes'.[39] He believed that some of the contradictions in physics, such as the one between the wave phenomena of light and Einstein's light quanta, were rooted in 'the fact that we give up the *laws* of classical theory, but still continue to work with the *concepts* of that theory'. Among the concepts that had no operational meaning and therefore had to be abandoned he counted the concept of electrons moving in fixed orbits.

Heisenberg's references to the new Göttingen theory and Born's discretization programme in atomic physics reflected ideas that were discussed by the Göttingen physicists in the years 1923–1924, when Born gave an autumn term course on 'Perturbation Theory Applied to Atom Mechanics'. Although these ideas were never developed into a proper theory, they may nevertheless have played a role in the development of quantum theory. In relation to his own and Heisenberg's failed attempt to calculate the helium atom, Born remarked in his *Atommechanik* that calculations between two or more electrons were based on classical differential equations; on the other hand, in the interaction between atoms and radiation a difference equation, namely Bohr's frequency condition $h\nu = \Delta E$, replaced the differential equations. With respect to this asymmetry, he declared that 'The systematical transformation of the classical mechanics into a discontinuous atomic mechanics is the goal towards which quantum theory strives'.[40]

Born's interest in a discretization of physics was not new. In 1919 young Pauli had offered a critical analysis of Hermann Weyl's new unified theory of gravitation and electromagnetism in which he objected that it was physically meaningless to speak of the field strength in the interior of an electron because the field would be unobservable even in principle.[41] For this reason he suggested that the field concept needed to be modified. Inspired by Pauli's comments, and probably also by his own work on the theory of crystal lattices, Born related them to the problems of quantum theory. In a letter to Pauli of late 1919, he wrote:

> The way out of all quantum difficulties must be sought by starting from entirely fundamental points of view. One is not allowed to carry over the concept of space-time as a four-dimensional continuum from the macroscopic world of experience into the atomistic

> world; the latter evidently demands another type of number-manifold to give an adequate picture. However, I have no idea of how to do that.[42]

The Göttingen discretization programme reflected Born's belief that the quantum riddles could probably be solved by introducing new mathematical techniques, such as difference equations. Bohr, on the other hand, was convinced that a new mathematical formulation had to come after a physical understanding of the problems, not before it. According to Heisenberg's recollection: 'Bohr would always say, "Well, first we have to understand how physics works. Only when we have completely understood what it is all about can we then hope to represent it by mathematical schemes". But Born would argue the other way, and would say, "Well, perhaps some new mathematical tool is a decisive help in understanding physics"'.[43] Heisenberg was as much influenced by Born as by Bohr, and when it came to the question of the role of mathematics he tended to side with the professor in Göttingen.

While a fundamental discretization was part of the Göttingen research programme, other physicists argued that quantum phenomena did not necessitate a change from differential to difference equations. On the contrary, Einstein thought that quantum discontinuities might well be generated as solutions of continuous differential equations. Convinced that quantum theory lacked a proper logical foundation, he believed that the quantum conditions should be deduced from a set of classical differential equations, as he told Born in 1920:

> I myself do not believe that the solution to the quanta has to be found by giving up the continuum . . . Pauli's objection is directed not only against Weyl's theory, but also against anyone else's continuum theory . . . I believe now, as before, that one has to look for redundancy in determination by using differential equations so that the *solutions* themselves no longer have the character of a continuum. But how?[44]

At any rate, the hypothesis of a quantum physics based on a discrete rather than continuous picture of space-time, or of replacing all differential equations with difference equations, did not lead to useful results. Yet it was not forgotten and was later developed in various directions. For example, in 1930 Heisenberg proposed in a letter to Bohr that if space were discrete, structured like a three-dimensional lattice with a smallest length of $h/Mc \cong 10^{-15}$ m, where M is the mass of the proton, some of the problems of quantum mechanics might disappear. He developed a theory on this basis, but never published it.[45]

8.3 LIGHT QUANTA AND VIRTUAL OSCILLATORS

At the time when Heisenberg developed his second core theory of the anomalous Zeeman effect, the focus of leading quantum physicists had to some extent shifted from atomic structure to radiation theory. The unresolved question of the nature of light and its interaction with matter was not new, of course, and, as indicated by Pauli's letter of 1923 to Eddington, it continued to occupy the minds of the physicists. As mentioned in Section 5.2, Bohr was convinced that Einstein's light quanta caused insuperable difficulties and could not possibly be reconciled with the wave nature of light so solidly demonstrated by interference and diffraction phenomena. However, it was only at the end of 1923 that the problem began to play a major role in Bohr's thoughts about progress in atomic theory. In the letter to Bohr of December 1923, in which Heisenberg outlined his second core model and its application to the anomalous Zeeman effect, he suggested that his new quantum principle was connected with the interaction between atoms and radiation fields, a suggestion which undoubtedly captured Bohr's attention with regard to the proposed theory.[46] At that time Heisenberg was not yet aware that Bohr, in collaboration with Kramers and Slater, had recently arrived at some new and quite radical ideas concerning radiation and atoms.

In his Nobel lecture in Stockholm on 11 December 1922, Bohr briefly discussed Einstein's explanation of the photoelectric effect in terms of light quanta. He admitted it to be of 'heuristic value', but nonetheless dismissed it because it 'is quite irreconcilable with so-called interference phenomena [and] not able to throw light on the nature of radiation'.[47] He was at the time unaware that Arthur Compton ten days earlier had presented a paper to the American Physical Society that gave strong support to Einstein's light quanta. Compton, at the time a professor at Washington University, St. Louis, showed in experiments from 1922 that when monochromatic X-rays of wavelength λ—he used the K_α line of molybdenum—were scattered by graphite, the scattered radiation included a component of longer wavelength λ'. The shift in wavelength $\Delta\lambda$ was found to depend on the scattering angle θ, but to be independent of the scattering material. Energy conservation (for $v << c$) yields $hc/\lambda - hc/\lambda' = \frac{1}{2}mv^2$, where m is the mass of the electron, and momentum conservation results in two equations involving θ. As Compton showed, a simple calculation gives for the 'Compton effect' the formula

$$\lambda' - \lambda = \Delta\lambda = \frac{h}{mc}(1 - \cos\theta).$$

The constant quantity $\lambda_C = h/mc = 0.0243$ Å soon became known as the Compton wavelength. It is about 20 times smaller than the Bohr radius, the relation between the two lengths being $\lambda_C/a_0 = 2\pi\alpha$. In a classic paper of May 1923 Compton argued that his result was incomprehensible on the basis of the classical theory of radiation but could be fully explained on the assumption of radiation quanta carrying energy and momentum given by

$$E = \frac{hc}{\lambda} \quad \text{and} \quad p = \frac{h}{\lambda}.$$

'This remarkable agreement between our formulas and the experiments can leave but little doubt that the scattering of X rays is a quantum phenomenon', he concluded.[48] Formulae identical to Compton's were independently derived by Debye in Zurich, for which reason the effect was sometimes referred to as either the Compton–Debye effect or the Debye–Compton effect, but Compton's work was earlier and the only one which included precise experiments as support of the radiation quantum hypothesis.[49] Compton's conclusions were questioned on experimental grounds by William Duane and other American physicists, and it took nearly a year until they were corroborated and generally accepted as proof of the corpuscular nature of X-rays. By the end of 1923 consensus had not yet been achieved, although a majority of physicists already agreed that the Compton–Debye interpretation was probably correct and that light quanta were real.[50]

Bohr did not belong to the majority, and neither did his close collaborator Kramers. According to the Dutch physicist, the explanatory power of Einstein's hypothesis did not compensate for the problems it created. 'The theory of light quanta may thus be compared with medicine which will cause the disease to vanish but kill the patient', he opined.[51] Bohr knew about Comptons's work, but saw no reason to accept his conclusions with regard to the light quantum. In the late autumn of 1923 he visited the United States and Canada, giving lectures at the University of Toronto and several American universities. At Yale he was invited to give the Silliman lectures, a series of six lectures between 6 and 15 November, which attracted considerable attention.[52] During his stay in the United States, he had conversations with Duane, Michelson, and other physicists concerning the Compton effect and related matters. On one occasion they discussed the possibility of testing the nature of individually scattered X-ray quanta and thereby decide whether or not energy and momentum were absolutely conserved. The possibility of such a test, which was later realized in the Compton–Simon experiment mentioned below, was suggested by the British-American physicist William Swann in a conversation he had with Bohr and Compton in November 1923.[53] Swann was at the time entertaining the idea that the role of the electromagnetic field in radiation processes might be to guide light quanta.

During their discussions in the United States, Duane confirmed Bohr in his belief that the Compton effect could be explained without abandoning the wave theory. 'I have no doubt', he wrote to Bohr in March 1924, 'that the shift in the wave-length that A. H. Compton has been writing about should be ascribed to this tertiary radiation [produced by photoelectrons]'. In a letter of reply Bohr informed Duane about 'the new view as regards the connection between radiation and transition processes described in a recent paper in Philosophical Magazine by Dr Slater, Dr Kramers and myself'.[54] This paper, often known as the BKS paper, had its origin in an idea of John Slater that was, however, transformed into a theory which included as a crucial element the radical hypothesis that energy and momentum are not conserved in individual atomic processes. Speculations of energy nonconservation predated Slater's idea and coloured the way Bohr understood it.

These speculations were in part inspired by ideas due to the Cambridge physicist Charles Darwin, Bohr's old colleague from his time in Manchester, who in July 1919 prepared a manuscript on 'Critique of the Foundations in Physics'. In this unpublished work, which he enclosed in a letter to Bohr, he expressed in a general way his frustration about the conceptual basis of physics. 'It may be', he said, 'that it will prove necessary to make fundamental changes in our ideas of space and time or to abandon the conservation of matter and electricity or even as a last forlorn hope to endow electrons with free will'. The conflict between quantum theory and the classical wave theory rested on the assumption of strict energy conservation, and 'I therefore claim that the possibilities to be deduced from denying the exact conservation should be thoroughly exhausted before further modifications are made'.[55] Finding Darwin's considerations greatly interesting, Bohr offered his own considerations in a letter, which was not, however, actually sent:

> As regards the wave theory of light I feel inclined to take the often proposed view that . . . all difficulties are concentrated on the interaction between the electromagnetic forces and matter. Here I feel on the other hand inclined to take the most radical or rather mystical views imaginable. On the quantum theory conservation of energy seems to be quite out of question and the frequency of the incident light would just seem to be the key to the lock which controls the starting of the interatomic process.[56]

In 1922–1923 Darwin developed his vague thoughts into a dispersion theory in which energy conservation was only satisfied statistically, in the same sense as in Boltzmann's probabilistic theory of the second law of thermodynamics. According to Darwin's theory, when an atom was hit by a wave it would acquire a certain probability of emitting a secondary spherical wave that interfered with the incident one to produce the dispersed wave. He suggested that the emission of a spherical wave with a particular energy would be triggered by some favourable configuration of the electron orbits. As to the photoelectric effect, he found it to be 'an impossibility in conjunction with the wave theory if energy is exactly conserved, but if only a statistical balance is required, then it becomes nothing more than one unexplained problem among others'.[57]

The idea of statistical energy conservation was at the time in the air and not a novelty to Bohr, who had contemplated it some years earlier (Section 5.2). Nor was it a novelty to Sommerfeld, who in the third edition of *Atombau* suggested that the 'mildest cure' for reconciling wave theory and quantum theory might be to abandon energy conservation for individual radiation processes.[58] Having studied Darwin's dispersion theory, Bohr pointed out some serious problems with it, among which he did *not* count its questioning of energy conservation (which Bohr considered to be 'of a most formal nature') and of course also not its dismissal of light quanta. What he did not like was that the theory challenged the real existence of excited stationary states, which according to Darwin might be just an epiphenomenon caused by changing electric fields.[59] Bohr insisted that the stationary states were real and beyond discussion. More concretely he pointed out that it followed from Darwin's theory that for low-intensity incident light the probability of interference should be small, which contradicted experiments showing that even very weak light produced the same interference pattern as strong light.[60]

The 23-year-old Harvard physicist John Slater conceived his idea of a new radiation theory reconciling light quanta with classical wave theory while staying with Ralph Fowler in Cambridge, England, in the autumn of 1923. From Cambridge he moved on to Copenhagen, where he worked at Bohr's institute from December 1923 until the following June. At the time he was 'profoundly dissatisfied with Bohr's hypothesis of instantaneous jumps from one stationary state to another', not so much because of the jumps, but because they supposedly occurred instantaneously.[61] Not having problems with light quanta he imagined that they were guided in their paths by classical wave fields generated by vibrating 'corresponding charges' (not orbiting electrons) in the stationary states of the atoms. Emission of radiation was not restricted to abrupt transitions from one stationary state to another, as in Bohr's original model, but occurred continuously and with all possible transition frequencies while the atom was in a stationary state—which according to Bohr's theory was inactive.

However, there was no real contradiction between Slater's mechanism and the Bohr theory, for Slater's guiding field was not an ordinary electromagnetic field but a kind of ghost field that carried neither energy nor momentum. Still, in a purely formal sense it could be described as the electromagnetic field containing all the possible Bohr transition frequencies. These frequencies were imagined to be those of a set of virtual harmonic oscillators, not the frequencies belonging to the actual motion of electrons. Thus while an atom in a stationary state emits a virtual field, it is still stationary in Bohr's sense since it does not emit a real electromagnetic wave. When, according to Slater, a real light quantum is emitted, the atom changes discontinuously to a new stationary state and begins emitting virtual waves with frequencies corresponding to this state. In the case of a collection of many atoms, each one contributes with its own virtual field to the fields produced by all the others. The atoms 'communicate' with one another. Bohr convinced himself that Slater's picture was 'far more harmonious from the point of view of the correspondence principle' than the traditional one, where spontaneous radiation was connected only with transitions between stationary states.[62]

In early December 1923 Slater described his hypothesis in a letter to Kramers: 'It seems possible to suppose that there is an electromagnetic field, produced not by the actual motion of the electrons, but with motions with the frequency of possible emission lines . . . and amplitudes determined by the correspondence principle, the function of this field being to determine the motion of the quanta'.[63] The amplitudes of the wave field determined the probability of a quantum being emitted with a particular frequency, but Slater did not think of the probabilistic feature as very important. 'It should be noted', he wrote in one of his manuscripts, 'that the only place where chance and discontinuity comes into the theory is in emission; once a quantum is emitted, the rest of the process is prescribed exactly as in the classical theory'. He hoped that 'when the dynamics of the inside of atoms are better known, chance may be eliminated there also'.[64]

Slater's idea of light quanta guided by a system of virtual ghost-like waves had some similarity to Louis de Broglie's contemporary ideas associated with his new theory of matter waves. It has been suggested that Slater was inspired by de Broglie's theory, parts of which were probably known to him, but according to Slater this was not the case. 'De Broglie's work was not known in Copenhagen during the time, the first half of 1924, when I was there', he recalled.[65] Although de Broglie's theory of matter waves was probably unknown to Bohr and his associates at the time, his work on atomic structure and X-ray spectroscopy were well known (if not appreciated, see Section 7.4). In the late development of the Bohr–Sommerfeld theory, de Broglie's theory of matter waves played almost no role at all. Bohr did refer to the famous doctoral thesis of the French physicist, but only in July 1925.

When Slater arrived in Copenhagen, his ideas were immediately taken up, discussed and substantially revised—almost beyond recognition—by Bohr and Kramers. The note he submitted to *Nature* in late February 1924 was in important ways influenced by the views of the two Copenhagen physicists, one result being that the light quanta no longer appeared in a significant way. In fact, Slater mentioned that Kramers' objections to his original idea of a radiation field guiding discrete quanta had forced him to abandon the idea. At the time he did not seem to have considered it a defeat or great loss. As he explained in a note from the summer of 1925, shortly after the Bohr–Kramers–Slater theory had been refuted by experiments, 'I became persuaded that the simplicity of the mechanism obtained by rejecting a corpuscular theory more than made up for the loss involved in discarding conservation of energy and rational causation'.[66] On the other hand, he later recalled differently: 'To my consternation I found that they [Bohr and Kramers] completely refused to admit the real existence of the photons . . . This conflict, in which I acquiesced to their point of view but by no means was convinced by any arguments they tried to bring up, led to a great coolness between me and Bohr, which was never completely removed'.[67]

The mechanism that Slater described in early 1924 was in terms of 'virtual fields' and 'virtual oscillators', terms that were coined by Bohr and Kramers, but, as far as the concept of virtual oscillators are concerned, can be found in an earlier paper by Ladenburg and Reiche. According to Slater:

> Any atom may, in fact, be supposed to communicate with other atoms all the time it is in a stationary state, by means of a virtual field of radiation, originating from the oscillators having the frequencies of possible quantum transitions, and the function of which is to provide for statistical conservation conservation of energy and momentum by determining the probabilities of quantum transitions . . . The part of the field originating from the given atom itself is supposed to induce a probability that that atom loses energy spontaneously.[68]

Slater's ideas were quickly transformed into the more elaborate and significantly different Bohr–Kramers–Slater (BKS) theory, which was, however, the product of

Bohr and Kramers rather than Slater.[69] In spite of its break with the radiation mechanism of Bohr's atomic model, Bohr found Slater's idea of emission of (virtual) radiation during a stationary state appealing. He had recently introduced a 'coupling principle', which included the notion of 'latent' forces that controlled transition probabilities in atoms. Slater's virtual fields were somewhat similar to Bohr's latent forces, a similarity Bohr alluded to in a letter of early 1925, when the BKS theory was still alive and well:

> It was just the completion which your suggestion of radiative activity of higher quantum states apparently lent to the general views on the quantum theory with which I had been struggling for years which made me welcome your suggestion so heartily. Especially I felt it was far more harmonious from the point of view of the correspondence principle to connect the spontaneous radiation with the stationary states themselves and not with the transitions.[70]

The intimate connection between the BKS theory and the correspondence principle also appears in a letter Bohr wrote to Michelson in February 1924. Albert Michelson, America's first Nobel laureate in physics, had established his entire career on the wave theory of light, and Bohr assured him that 'it appears to be possible for a believer in the essential reality of the quantum theory to take a view which may harmonize with the essential reality of the wave theory conception'. The reason for Bohr's assertion that quanta and waves could be harmonized was this:

> In fact on the basis of the correspondence principle it seems possible to connect the discontinuous processes occurring in atoms with the continuous character of the radiation field in a somewhat more adequate way than hitherto perceived... I hope soon to send you a paper about these problems written in cooperation with Drs. Kramers and Slater.[71]

He said much the same in the published paper, in which he expressed his belief that the new theory would allow them 'to arrive at a consistent description of optical phenomena by connecting the discontinuous effects occurring in atoms with the continuous radiation field in a somewhat different manner from what is usually done'.[72]

The BKS paper on 'The Quantum Theory of Radiation' appeared in the May issue of *Philosophical Magazine* and at the same time a German translation was published in *Zeitschrift für Physik*. It was a paper very much in Bohr's style—verbose, discursive, repetitive, and qualitative—and was presented as a further development of Bohr's research programme based on the correspondence principle. 'The present paper', it said in the introduction, 'may in various respects be considered as a supplement to the first part of a recent treatise by Bohr, dealing with the principles of quantum theory'.[73] The lengthy paper included only a single equation, which was Bohr's frequency condition $h\nu = E_1 - E_2$. Among the issues discussed in the BKS paper was the time of duration of the stationary states and the corresponding sharpness of the states, an issue which Bohr had previously dealt with in his memoir on the application of quantum theory to atomic structure. Bohr now argued that there is a 'limit of definition of the motion and of the energy in the stationary states' which manifests itself in a finite line width. The postulate of the stationary states, he said, 'imposes an a priori limit to the accuracy with which the motion in these states can be described by means of classical electrodynamics'.[74] In other words, he suggested an inverse relationship between the uncertainty in energy ΔE and the lifetime Δt, a relationship of roughly the same form as the one that later appeared in Heisenberg's famous energy–time uncertainty relation based on quantum mechanics:

$$\Delta E \Delta t \geq \frac{h}{2\pi}.$$

The new Copenhagen theory was basically an attempt to reconcile the continuous electromagnetic field with the physical picture of the discontinuous quantum transitions in atoms (but not, as in Slater's original idea, with the light quanta). To harmonize continuity with discontinuity, or at least to correlate the two concepts, strict causality had to be abandoned, and from this it followed that energy and momentum conservation changed from being absolutely valid to having only statistical validity. As mentioned above, Bohr had discussed this possibility in a conversation with Compton and Swann in November 1923, before he knew about Slater's ideas.

Following Slater, the basic physical assumption of the BKS theory was to associate each atom with an unspecified number of virtual oscillators that produced a virtual radiation field through which the atoms entered in a mutual communication. It remained unclear what the enigmatic virtual oscillators were, except that they were abstract and unphysical quantities that could not be directly observed. While the virtual field was not endowed with measurable properties, it transmitted probabilities for transitions in other atoms. In this way, the field established a communication between distant atoms, but one which merely changed the probability that a transition would occur. Bohr and Kramers (and, nominally, Slater) expressed the basis of their theory as follows:

> We will assume that a given atom in a certain stationary state will communicate continually with other atoms through a time-spatial mechanism which is virtually equivalent with the field of radiation which on the classical theory would originate from the virtual harmonic oscillators corresponding with the various possible transitions to other stationary states . . . As regards the occurrence of transitions, which is the essential feature of the quantum theory, we abandon on the other hand any attempt at a causal connexion between the transitions in distant atoms, and especially a direct application of the principles of conversation of energy and momentum so characteristic for the classical theories.[75]

According to the BKS view, the experimentally established validity of the conservation laws was nothing but the result of a statistical average over a great number of individual events. For the individual events the laws would not be exactly satisfied. Considering the Compton effect, Bohr and his coauthors argued that Compton's light-quantum interpretation was not the only one and that the effect could easily be explained on the basis of the BKS theory: 'In contrast to this [Compton's] picture, the scattering of the radiation by the electrons is, in our view, considered as a continuous phenomenon to which each of the illuminated electrons contributes through the emission of coherent secondary wavelets'.[76] In the BKS explanation of the Compton effect, the shift in wavelength was attributed to a Doppler shift. However, this explanation indicated what the three authors called 'a feature strikingly unfamiliar to the classical conceptions', namely an incompatibility between the kinematical and the wave-theoretical description. Kramers and Heisenberg later called attention to 'the curious fact that the centre of these spherical waves moves relative to the excited atom'.[77]

While hoping that the new Copenhagen theory would turn out to be correct, Kramers and Bohr tended to emphasize its unfinished nature. According to Kramers, writing at a time when experiment had not yet settled the matter: 'The new conceptions of the postulates, which require the independence of the atom processes, is in no way a completed theory; it is only an attempt to throw a little light in the great darkness of our ignorance about the course of the atomic processes, and it should, for the time being, be conceived as essentially a working-programme for the theorists'. This is how he wrote in a chapter added to the German version of his and Holst's book on the Bohr atom, in which he gave a pedagogical and fairly detailed account of the BKS theory, or what he consistently called 'Bohr's new view'. As to the theory's element of acausality, he preferred to consider it 'rather a matter of taste', although there is little doubt that his own taste (and Bohr's as well) was acausal. To Kramers, the principle of causality was a fact of experience rather than a logical necessity, and 'one could easily imagine that it breaks down for atomic processes'. Similarly, he thought that one should keep an open mind with respect to a violation of the law of energy conservation. Interestingly, Kramers suggested that large-scale energy nonconservation processes might occur in the depths of space. There are indications, he said, that 'in the hot stars, ... the principle of energy conservation cannot be used just like that, but that in these bodies there occurs, so to speak, a spontaneous creation of energy which contributes to maintain the enormous radiation of energy of the stars into space'.[78]

The BKS theory aroused immediate and great attention, if little enthusiasm. The chemist Fritz Haber, who visited Bohr in 1924, was exposed to a one-and-a-half-hour lecture on the BKS radiation theory of which he understood little. And what he thought to have understood was a misunderstanding based on Bohr's rejection of the light quantum. Bohr, he reported to Einstein, 'strives with all fibres back to the classical world'.[79] As to any enthusiasm outside Copenhagen, it was largely limited to Born and Schrödinger. In the United States, Swann also found the theory attractive. In a letter to Bohr of May 1924 Schrödinger declared himself 'extremely sympathetic' to the main parts of the theory and in particular to its renunciation of strict causality and energy conservation, which he, as a student of the Viennese physicist Franz Exner, had long been fond of. 'Your new point of view means a far-reaching return to classical theory, as far as radiation is concerned', he wrote, wondering about the difference between 'virtual' radiation and the 'real' radiation caused by transitions between stationary orbits.[80]

Schrödinger misunderstood Bohr's intentions, which he interpreted in accordance with his own preferences. Contrary to what Haber and Schrödinger thought, Bohr had no wish to 'return to classical theory'. In a paper of September 1924 published in *Die Naturwissenschaften*, Schrödinger publicly endorsed the BKS theory, which he interpreted in accordance with his own holistic approach to physics: 'A certain stability of the world order *sub specie aeternitatis* can only exist through the interrelationship of each individual system with the rest of the entire world'.[81] Moreover, Schrödinger spoke of the 'Exner–Bohr conception of the energy principle', which he related to speculations of the statistical nature of thermodynamics in a manner foreign to Bohr's thinking.

Most other physicists, including Sommerfeld, Heisenberg, Compton, Ehrenfest, Franck, and Pauli, were either sceptical or directly opposed to the BKS theory.[82] In response to a letter from Bohr, including a draft version of the BKS theory, Pauli declared that he felt mystified: 'I laughed a little . . . about your warm recommendations of the words "communicate" and "virtual" . . . On the basis of my knowledge of these two words . . . I have tried to guess what your paper is all about. But I have not succeeded'.[83] Of particular interest is the response of Einstein, who had earlier expressed his worries over causality in relation to radiation theory. 'Can the quantum absorption and emission of light ever be understood in the sense of the complete causality requirement, or would a statistical residue remain?' he asked Born in 1920. Four years later, after having learned about the BKS theory, he spoke out forcefully against it:

> Bohr's opinion about radiation is of great interest. But I should not want to be forced into abandoning strict causality without defending it more strongly than I have so far. I find the idea quite intolerable that an electron exposed to radiation should choose *of its own free will*, not only its moment to jump off, but also its direction. In that case, I would rather be a cobbler, or even an employee in a gaming-house, than a physicist.[84]

As we learn from a letter to Ehrenfest, Einstein gave a colloquium on the BKS theory in May 1924, in which he gave his reasons for rejecting the theory. 'This idea is an old acquaintance of mine, whom, however, I do not regard as a respectable fellow [*einen reellen Kerl*']', he told Ehrenfest.[85] Born was more inclined to consider the BKS theory a respectable fellow. Seriously interested in the theory, he thought that the valuable part of it was the idea of emission of radiation while an atom stays in a stationary state. In a reply to a letter from Born on the subject, Bohr stressed that 'the assumption of a coupling between the changes of state in distant atoms by means of [virtual] radiation precludes the possibility of a simple description of the occurrences in terms of visualizable pictures'.[86]

Bohr and Kramers realized that their theory, if it were to be seriously considered in the physics community, had to make testable predictions. Applying it to the Compton effect they concluded that the direction of a recoil electron after scattering by a monochromatic X-ray would not be uniquely determined, as required by the conservation laws, but display a wide statistical distribution. As early as June 1924, Walther

Bothe and Hans Geiger in Berlin proposed an experiment to test the theory by simultaneously measuring the scattered X-rays and the recoil electrons. This was one of the first experiments using electronic coincidence devices, and it was not until April 1925 that they had their final result ready; it was 'incompatible with Bohr's interpretation of the Compton effect'.[87] Using a cloud chamber to determine the direction of electrons recoiling from X-ray scattering, Compton and Alfred Simon reached the same conclusion a little later (Figure 8.1). In what they described as 'a crucial test' they provided 'a direct and conclusive proof that . . . energy and momentum are conserved during the interaction between radiation and individual electrons'.[88]

Einstein was pleased, if by no means surprised. In a short manuscript written during a visit to Rio de Janeiro, he referred to the still unpublished experiment of Bothe and Geiger. 'At the time of my departure from Europe', he ended the manuscript, 'the experiment was not yet completed. However, according to the results attained so far, such statistical dependence appears to exist. If this is confirmed, then there is a new important argument for the reality of light quanta'.[89]

Bohr had for a year defended the BKS theory and taken it very seriously; as seriously as his coauthor Kramers and much more so than his other coauthor Slater. The

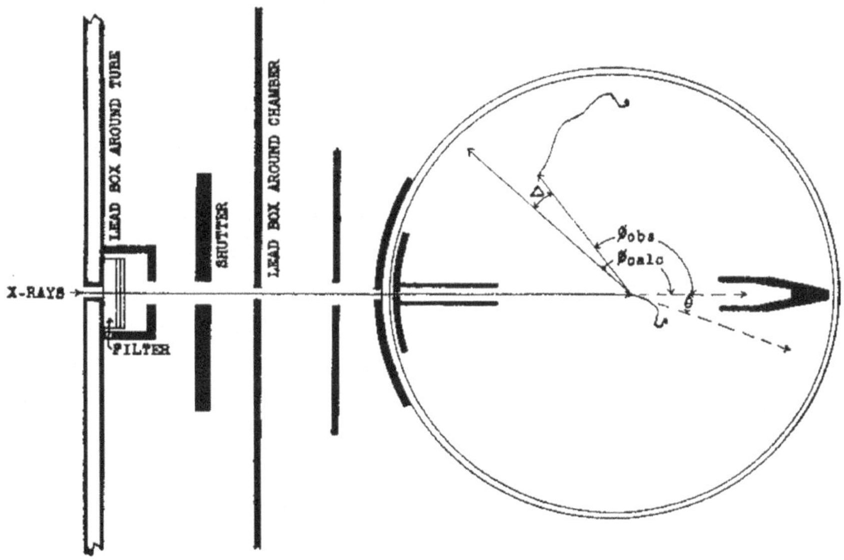

Fig. 8.1. The Compton–Simon experiment of 1925 was designed to test the effect of X-ray quanta scattered by electrons. On the radiation quantum hypothesis, if a recoil electron is ejected an angle θ, the scattered X-ray quantum proceeds in a definite direction ϕ. On the BKS theory, the scattered X-rays are spherical waves and there should be no correlation between the directions of recoil electrons and the directions in which the effects of scattered X-rays are observed. Compton and Simon used a cloud chamber in which each recoil electron produced a visible track. From a series of stereoscopic photographs they concluded that there was a strong correlation, hence that the BKS theory was wrong.

Source: Compton and Simon 1925, p. 291.

experimental verdict was a disappointment, yet it was inescapable and generally accepted as a crucial refutation of the theory.[90] In a mood of resignation he wrote to Fowler that 'there is nothing else to do than to give our revolutionary efforts as honourable a funeral as possible'.[91] While Bohr accepted that the BKS theory was wrong, he did not yet accept the light quantum or (as it would be coined the following year) the photon.[92] Having received the news from Geiger, he maintained that, 'in spite of the existence of a coupling [between quantum processes], conclusions about a possible corpuscular nature of radiation lack a sufficient basis'.[93] Kramers was equally convinced that the refutation of the BKS theory did not provide a final answer to the problem of the nature of light. In a letter to the American physicist Harold Urey, who had stayed in Copenhagen 1923–1924, he wrote: 'Still the wave description of optical phenomena must be preserved in some way, and we think that Slater's original hypothesis contains a good deal of truth'.[94]

As to Slater, his response to the failure of the BKS theory was different from those of Bohr and Kramers: he merely returned to his original view of a virtual field guiding the radiation quanta.[95] After having returned to Harvard, but before the final Geiger–Bothe experiment, Slater developed his ideas into a theory of quantum optics that appeared in the spring of 1925. The theory made use of some of the key ideas of the BKS paper, including that energy and momentum were not precisely conserved for individual atomic processes. It followed from Slater's theory (as it did from the BKS theory) that, as a result of the finite lifetime of the stationary states, spectral lines must have a minimum width, an inference that he later saw as an anticipation of the energy-time uncertainty relation.[96] By adopting the BKS idea of frequencies emitted for all possible transitions, and not only for the ones turning up spectroscopically, he thought he had solved the old conundrum of the 'free will' of electrons. For then 'the atom is under no necessity of knowing what transitions it is going to make ahead of time'.[97]

Bohr publicly abandoned the BKS theory in an addendum to a paper of July 1925 in which he had adopted a BKS-inspired approach to investigate collision and capture processes. While arguing in the paper that some kinds of atomic collision violated strict conservation of energy and momentum, in the addendum he admitted this to be a mistake. 'If this way out is now closed', he suggested, 'we are probably forced to recognize in the capture phenomenon a new feature of the supramechanical stability of the stationary states that cannot be described in space-time pictures'.[98]

According to Pauli, the theory of Bohr, Kramers, and Slater was a 'reactionary *Kopenhagener Putsch*', and he considered it 'a magnificent stroke of luck' that the revolt had been crushed so quickly by the experiments in Berlin and St. Louis. As he stressed in a letter to Kramers, it was the kinematic concepts that needed modification, not the concept of energy:

> It is true that, in cases where interference phenomena are present, we cannot define definite 'trajectories' for the electrons in an atom; and just as little as it would be justified to doubt the existence of electrons for this reason, would it be justified to doubt the existence of light

> quanta because of interference phenomena. For probably every unprejudiced physicist it can now be regarded as proved that light quanta are just as much (and just as little) physically real as electrons. But to neither of them should the classical kinematic concepts in general be applied.[99]

What Pauli called a reactionary Copenhagen revolt was a failed revolution, but the failure was not without fruitful consequences. In spite of its short lifetime, the BKS theory was important in paving the way for a greater understanding that methods and concepts of classical physics could not be carried over to a future quantum mechanics. In particular, the theory appeared prominently in Kramers' theory of dispersion of 1924 and its further development into the Kramers–Heisenberg dispersion theory of 1925, the final step before Heisenberg's formulation of quantum or matrix mechanics. In 1929, looking back on the development of quantum theory, Heisenberg singled out the BKS theory as the high point in the crisis that led to the new quantum mechanics. In his view, the theory 'contributed more than any other work at that time to a clarification of the situation in quantum theory'.[100] However, the historical significance of the BKS theory should be understood in the proper and restrictive sense that *one element* in it—the idea of virtual oscillators—played a fruitful role.[101] The important dispersion theory of Kramers was formulated in the language of the BKS theory, but it did not actually rely on it (see the next section). Likewise, it was the virtual oscillators, and not the BKS theory as a whole, that played an essential role in the reasoning that led Heisenberg to quantum mechanics.

8.4 DISPERSION THEORIES

The dispersion of light in gases and other refractive media was successfully explained by classical electron theory by assuming the atom to contain a number of elastically bound electrons with characteristic frequencies ν_0.[102] When hit by an electromagnetic wave of frequency ν the electrons would vibrate, the collective result being a characteristic dependence of the refraction index n (or dielectric constant ϵ, related to the refraction index as $n \cong \sqrt{\epsilon}$) on the frequencies ν and ν_0. In terms of the polarizability α, which for weak fields is the ratio P/E between the dipole moment and the electric field, the classical dispersion formula could be written in the form

$$\alpha = \frac{e^2}{4\pi^2 m} \sum_i \frac{f_i}{\nu_{0i}^2 - \nu^2}.$$

The term f_i was interpreted as either the 'strength' of the oscillating charge or the number of dispersion electrons in an atom. With the advent of Bohr's atomic model, the picture of elastically bound electrons lost its validity and had to be replaced by one with electrons moving on stationary orbits around a nucleus. But how, then, could dispersion be explained? This was an old question, going back to the very beginning of Bohr's theory. At the meeting of the British Association in September 1913 a sceptical Lorentz had asked the young Dane how his model of the atom could possibly account for the well-established classical dispersion formula (Section 3.1). Bohr could offer no answer.

The first theories of dispersion relating to Bohr's model, as proposed by Debye and Sommerfeld in 1915 and by Davisson in 1916, were based on particular atomic and molecular models and assumed that the perturbations of the electrons in their stationary orbits could be calculated by classical electrodynamics. Understandably, Bohr found theories of this kind to be untenable because they conflicted with his fundamental distinction between orbital frequencies and the radiation frequencies caused by quantum jumps (see Section 3.9). In his withdrawn 1916 paper to *Philosophical Magazine*, he severely criticized the Debye–Sommerfeld theory, pointing out that it involved 'considerable formal difficulties due to the contrast between the assumption of stationary states and ordinary electrodynamics'. We must assume, he said, that

> [T]he dispersion in sodium and potassium vapour depends essentially on the same mechanism as the transition between different stationary states, and that it cannot be calculated by application of ordinary electrodynamics from the configuration and motions of the electrons in these states . . . If the above view is correct and the dispersion depends on the mechanism of transition between different stationary states, we must, on the other hand, assume that this mechanism shows a close analogy to an ordinary electrodynamic vibrator.[103]

The 'close analogy' would soon become known as the correspondence principle.

A more satisfactory quantum dispersion theory was first proposed by Rudolf Ladenburg, who in 1921 discussed 'the probability of different spontaneous transitions and its connection to the Bohr theory'. His basic insight was to interpret the nominator in the dispersion formula, the f_i factor, in terms of the transition probabilities given by the *A* and *B* coefficients introduced by Einstein in 1916. He thus replaced the picture of atomic electrons vibrating as a result of an incident wave by a picture in which the atom makes an induced transition from a state *i* to a higher state *k* and then spontaneously falls back to the original state: 'The absorption is produced by a transition of the molecules from a state *i* to a state *k*, and the strength of the absorption is determined by the probability of such transitions $i \rightarrow k$'. The result follows from Einstein's theory, which 'leads to an important relation between this probability factor and the probability of the spontaneous (reverse) transition from state *k* to state *i*'.[104] Although Ladenburg's paper did not at first attract much attention, it was received favourably by Bohr, who in an unpublished manuscript of 1921 made some comments on the 'unsolved problem [of] how a detailed theory of dispersion can be developed on the basis of the quantum theory'. He thought that a 'promising beginning' had been made with 'the interesting considerations about this phenomenon, recently published by Ladenburg'.[105]

In his 1923 paper on the fundamental postulates of quantum theory, Bohr stated that 'the phenomena of dispersion must thus be so conceived that the reaction of the atom on being subjected to radiation is closely connected with the unknown mechanism which is answerable for the emission of the radiation on the transition between stationary states'. He praised the work of Ladenburg, who 'has tried in a very interesting and promising manner, to set up a direct connection between the quantities which are important for a quantitative description of the phenomena of dispersion according to the classical theory and the coefficients of probability appearing in the deduction of the law of temperature radiation by Einstein'.[106] Together with Fritz Reiche, his colleague at the University of Breslau, Ladenburg offered a more detailed version of his dispersion theory in the 1923 issue of *Naturwissenschaften* marking the tenth anniversary of Bohr's atomic theory.[107] In this paper, Ladenburg and Reiche made use of the correspondence principle to argue that although the detailed mechanism behind dispersion was unknown, the classical theory could be carried over to the domain of quantum physics.

A major improvement of Ladenburg's dispersion theory occurred with two notes published by Kramers in 1924. They were both to some extent based on the BKS theory, but Kramers' results did not rely crucially on this theory. 'You perhaps noticed his [Kramers's] letter to *Nature* on dispersion', Slater wrote to Van Vleck in late July 1924; 'the formulas & that he had before I came, although he didn't see the exact application'.[108] What Kramers used in his published derivation was only the idea of the atom as an assemblage of virtual oscillators, but he did not rely on either statistical energy conservation or the lack of coupling between atomic processes. Moreover, he emphasized in his second note that the virtual oscillators should be understood 'only as a terminology suitable to characterise certain main features of the connexion between the description of optical phenomena recognized and the theoretical interpretation of spectra'.[109] Likewise, Van Vleck thought that the virtual oscillators were 'in some ways very artificial', yet he found the idea valuable because it seemed to be the only way to avoid the 'almost insuperable difficulty that it is the spectroscopic rather than the orbital frequencies . . . which figure in dispersion'.[110]

A recognized master in the application of the correspondence principle, Kramers argued that Bohr's principle required an extension of Ladenburg's result, which was only valid for transitions to the ground state. For this reason it had nothing to say about transitions in the limit of large quantum numbers. To satisfy the requirements of the correspondence principle Kramers needed a dispersion formula consisting of two parts: one corresponding to processes where an atom decays spontaneously from a higher to a lower state, and another in which the atom returns to the excited state by a transition induced by the incident light. While the first part corresponded to absorption frequencies ν_a, the second related to emission frequencies ν_e. The resulting formula can be written as

$$\alpha = \frac{e^2}{4\pi^2 m}\left(\sum_a \frac{f_a}{\nu_a^2 - \nu^2} - \sum_e \frac{f_e}{\nu_e^2 - \nu^2}\right).$$

Kramers referred to the second part, which formally corresponds to emission oscillators of negative mass, or negative value of e^2/m, as 'negative dispersion'. For atoms in the ground state this term vanishes and the formula reduces to Ladenburg's result. The appearance of a negative dispersion term lacked experimental evidence, but Kramers argued that it was needed to establish connection by means of the correspondence principle to the classical dispersion formula. It could, he suggested, be understood as an instance of the stimulated emission predicted by Einstein in his radiation theory of 1916. The negative term was criticized by the American physicist Gregory Breit, at the University of Minnesota, whose critical note gave rise to Kramers' second letter to *Nature*.[111] Experimental support for the negative dispersion was first reported in 1928 by Ladenburg and his collaborators.

Two other features of Kramers' dispersion theory need to be mentioned. First, he emphasized as an advantage of his formula that 'it contains only such quantities as allow of a direct physical interpretation on the basis of the fundamental postulates of the quantum theory of spectra and atomic constitution, and exhibits no further reminiscence of the mathematical theory of multiple periodic systems'.[112] Second, to come from the classical frequency of motion to the radiation frequency in the limit of large quantum numbers, he replaced the differential quotients of the classical formula by difference quotients. This would soon turn out to be an important step. Kramers's dispersion programme culminated in a joint paper with Heisenberg, but even before this paper his work was recognized as an important advance. Ladenburg told him 'how much it pleases me that you have managed to give a correspondence derivation of the relation between dispersion and transition probabilities'. He thought that, 'In this way a solid basis has now been created'.[113]

Kramers' work was also appreciated by Van Vleck, who regarded it 'a distinct advance in the problem of reconciling dispersion with quantum phenomena'.[114] Van Vleck was himself a main contributor to the theory of dispersion, in which field he competed with the European physicists and obtained results of no less significance than they did. Thus, he arrived independently at Kramers' dispersion formula, including the negative dispersion term, and he did so without relying of the BKS theory. Having learned from Slater about Kramers' work, he wrote to his colleague and rival in Copenhagen: 'The concept and introduction of the virtual-oscillator formula is entirely yours, and I always refer to the "Kramers dispersion formula", but I had developed the perturbation theory method for absorption etc. prior to learning of any of your work'.[115]

Although Bohr was strongly interested in dispersion theory and closely followed Kramers' work on the subject, he did not participate actively in it. Nonetheless, the theory was sometimes—rather unfairly to Kramers, it would seem—considered the result of a collaboration between the two Copenhagen physicists. According to Heisenberg, writing to Landé, 'The beautiful thing about the new Bohr and Kramers dispersion theory is precisely that one now knows (or suspects)... *how* the quantum mechanics will look'.[116]

An important paper by Born published in August 1924 took its starting point in the 'considerable progress' made with the BKS theory and Kramers' use of it to explain the dispersion of light. Born's paper, entitled 'Über Quantenmechanik', merits attention not only because it introduced 'quantum mechanics' as the name for the quantum theory of the future, but also because it was a serious attempt to formulate the general structure of this still unborn theory. Although Born made use of virtual oscillators as 'the real primary thing', he stressed that his theory was independent of the 'still disputed conceptual framework of that theory [BKS], such as the statistical interpretation of energy and momentum transfer'.[117] The abstract theory included Kramers's dispersion theory, which it connected to Heisenberg's theory of multiplets and the anomalous Zeeman effect. From a formal point of view its most important result was that it offered a 'formal passage from classical mechanics to a "quantum mechanics" ... in the sense of a transition from differential to difference equations'.[118] According to Born, if a transition from a stationary state of quantum number n to another state of quantum number $n - \tau$ was characterized by some classical function $F(n)$, the function could be translated into its quantum-theoretical analogue by a replacement of the kind

$$\frac{\partial F(n)}{\partial n} \to \frac{1}{\tau}\left[F(n) - F(n - \tau)\right].$$

What has been called 'Born's correspondence rule' was known earlier to Kramers, who referred to it in his second note on dispersion theory, but Born arrived at it independently.[119] He emphasized, again in agreement with Kramers, that whereas the classical dispersion formula as derived from perturbation theory contained action variables and other reminiscences of mechanical orbits, in its quantum-theoretical translation only observable quantities appeared. However, the appeal to observable quantities was of a purely formal nature: Born's theory did not lead to calculations that could be compared to spectroscopic or other measurements.

In a programmatic passage in his *Atommechanik*, Born summarized what he called the fundamental ideas of a future quantum mechanics, a theory that would be based solely on observable properties such as the frequency and intensity of light emitted by an atomic system. As to the orbital frequencies and distances of electrons, as they appeared in the existing theory, 'these quantities are, as a matter of principle, not accessible to observation'. For this reason, 'our procedure is just a formal computational scheme which, for certain cases, allows us to replace the still unknown quantum laws by computations on a classical basis'. Concerning the true laws of quantum physics, he wrote:

> Of these true laws we would have to require that they only contain relations between observable quantities, that is, energy, light frequencies, intensities, and phases. As long as these laws are still unknown, we must always face the possibility that our provisional quantum rules will fail; one of our main tasks will be to delimit the validity of these rules by comparison with experience.[120]

Kramers' two notes of 1924 were rather sketchy and left out many details. It was only when he joined forces with Heisenberg, who spent the period from September 1924 to April 1925 at Bohr's institute, that his theory reached completion. The result of their collaboration, a detailed article published only in March 1925, was mostly written by Kramers and was to a large extent an elaboration of his earlier ideas. Based on the correspondence principle, the two physicists gave a full and systematic explanation of dispersion, extending Kramers' theory to cover also the kind of incoherent or anomalous dispersion predicted by the Austrian physicist Adolf Smekal. According to Smekal, who based his arguments on the light-quantum hypothesis, the scattered radiation should include terms with a frequency smaller and larger than the frequency ν of the incident radiation: if $\Delta\nu$ is the frequency difference between two stationary states, the incoherent radiation would have the frequencies $|\nu \pm \Delta\nu|$. In agreement with the Copenhagen attitude to light quanta, Kramers and Heisenberg reproduced the Smekal effect by means of the correspondence principle and without making use of Einstein's hypothesis of the corpuscular nature of light.[121]

As in Kramers's earlier theory, the Kramers–Heisenberg dispersion theory was detached from any intuitive model of the atom and visualizable concepts such as the orbits and velocities of revolving electrons. Its formulae contained 'only the frequencies and amplitudes which are characteristic for the transitions, while all those symbols which refer to the mathematical theory of periodic systems will have disappeared'. Again, some of its results were achieved by following the discretization procedure, namely 'by interpreting the differential quotients . . . as differences between two quantities, in analogy to Bohr's procedure in the case of frequencies'.[122] Kramers and Heisenberg acknowledged the close connection of their theory to the recent BKS theory, which was however more important to the senior author than the junior author. While Kramers tended to see the work as a triumph of the BKS approach, to Heisenberg the connection was of little significance. What mattered to him was the formal structure of the theory, the formulae themselves and the fact that they agreed with the observability principle by depending only on observable quantities.

In his Faraday Lecture of 1930, Bohr described Heisenberg's symbolic quantum mechanics as 'a most ingenious completion of the trend of ideas characterised by Kramers' adaption of Lorentz' classical theory of the optical dispersion phenomena to the quantum theory of spectra'. He likewise singled out Kramers' general dispersion theory in a memorial address of 1952. This work, he said, 'actually proved a stepping stone for Heisenberg, who shortly afterwards in a most ingenious way accomplished a rational formulation of quantum mechanics in which all direct reference to ordinary pictures of mechanical motion was finally abandoned'.[123] Heisenberg agreed that Kramers' work had a significant impact on the process that led to quantum mechanics: 'One felt that one had now come a step further in getting into the spirit of the new mechanics. Everybody knew that there must be some new kind of mechanics behind it. Nobody had a clear idea about it, but still one felt that the dispersion formula was a good step in the right direction'.[124]

While recognized as an important theoretical advance, from an empirical point of view there was little to recommend the Kramers–Heisenberg theory. In fact, there was no direct experimental evidence for its results—as little as there was for Kramers's earlier version or for Born's generalized theory. The explanation of the Smekal effect related to experiment, but this phenomenon was only demonstrated in 1928, when the Indian physicist Chandrasekhara Raman, in collaboration with his student Kariamanikam Krishnan, discovered the Raman effect or what at the time was sometimes referred to as the Smekal–Raman effect.[125] The lack of empirical confirmation did not count heavily in how physicists received the Kramers–Heisenberg theory. Experts in atomic and quantum theory found the work important and stimulating for theoretical and conceptual reasons, not because it was able to account for known experimental phenomena relating to dispersion and scattering or because it predicted new phenomena. To Heisenberg, it was a fruitful 'failure'. Its importance was methodological, or so he later recalled: 'This attempt was a failure. It got me into an impenetrable thicket of complicated mathematical formulae, from which I found no way out. But the attempt confirmed me in the attitude that one should not ask about electron orbits inside the atom at all.'[126]

If there were no electron orbits, then there was no Bohr model of the atom in the classical sense either. Indeed, by the summer of 1924 the visualizable Bohr or Bohr–Sommerfeld model of the atom was fading and was no longer considered a candidate for the real structure of atoms. The constitution of the atom in terms of a tiny positive nucleus surrounded at great distances by a system of electrons was left untouched, and so was the postulate of stationary states (not orbits), but few leading physicists believed in the planetary analogy—that the electrons actually moved in definite orbits whose geometry was characterized by quantum numbers. Objections to the orbital model had been around for some time, raised in particular by Pauli. By Christmas time 1923 he and Heisenberg agreed that 'model conceptions have no real meaning' and that 'the orbits are not real', as Heisenberg expressed it in his letter to Sommerfeld of 8 December.[127] One year later, Pauli praised his former professor for having left out atomic models in his exposition of complex spectra in the fourth edition of *Atombau*:

> The model concepts are now in a severe, fundamental crisis, which, I believe, will finally end with a further radical sharpening of the contrast between classical and quantum theory . . . [T]he conception of definite and unambiguously determined electron orbits in the atom can hardly be sustained. One now has the strong impression with all models, that we speak there a language that is not sufficiently adequate for the simplicity and beauty of

the quantum world. For this reason I found it so beautiful that your presentation of the complex structure is completely free of all model-prejudices.[128]

The shift in attitude to orbital models may be illustrated by the 6 July 1923 issue of *Die Naturwissenschaften* celebrating the tenth anniversary of the Bohr atom. The issue included articles by Coster and Kramers on Bohr's 'second atomic theory' accompanied by two-colour figures of the orbital structure of atoms from hydrogen to radium, taken from the plates Bohr used in his lectures.[129] These pictorial models still had a certain credibility, but not for long. In the same issue Born explicitly criticized the planetary atom analogy routinely used in more popular expositions of the theory. 'The similarity of atoms to planetary systems has only limited validity', he pointed out. 'It becomes increasingly probable that not only new assumptions will be needed in the ordinary sense of physical hypotheses, but that the entire system of concepts in physics will have to be restructured in its foundations'.[130]

Even more than Born and Heisenberg, Pauli campaigned against the kind of orbital pictorial models that he seems to have associated with Kramers rather than Bohr. It was as if such models emotionally offended him. In his letter to Bohr of 12 December 1924, in which he enclosed his manuscript on the exclusion principle, Pauli proudly mentioned that he had avoided the term 'orbit' in his work altogether. 'I believe that the energy and [angular?] momentum values of the stationary states are something much more real than "orbits"', he said. In a footnote he poked fun at 'our good friend Kramers and his colourful picture books', a reference to the Kramers–Holst book and its pictorial atomic models. In his letter to Kramers of 27 July 1925, quoted above, Pauli once again stressed his lack of belief in orbits, adding that in a recent report on superconductivity from the Leiden laboratory, 'I have again noticed with horror traces of your spirit in the form of figures!'[131] (Figure 8.2). As Pauli saw it, in so far as one could speak of atomic models at all, it had to be a mathematical and not a pictorial model.

8.5 FROM ANOMALIES TO CRISIS

Among many physicists in the period 1923–1925 there was a mounting feeling that the existing quantum theory of the atom had come to a dead end and that further search for substantial progress within the framework of the Bohr–Sommerfeld theory was futile. The old framework had to be replaced by a new one. What is often referred to as a deep-seated attitude of *crisis* included the expectation that a way out of the morass of anomalies and inconsistencies necessitated a radically new foundation of atomic physics, probably one where there was no place for visualizable models and where mathematical symbols referring to observables would replace mechanical terms. The general feeling of crisis was not new to German physicists, but it acquired a new and more specific dimension with the experimental and conceptual problems that plagued quantum theory.[132]

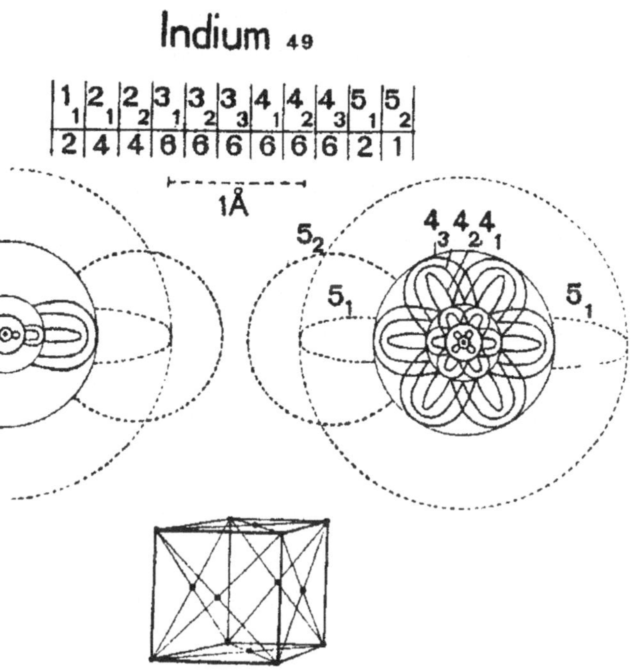

Fig. 8.2. The electronic and lattice structure of indium according to Bohr's atomic theory and as used by Kamerlingh Onnes in a paper of 1924 on superconductivity. Kamerlingh Onnes was in contact with Kramers in Copenhagen, who provided him with this and other pictures of metal atoms.

Source: Kamerlingh Onnes 2004.

As demonstrated by statements from Pauli, Born, Heisenberg, and Bohr, by the beginning of 1925 the quantum crisis was real. With the recognition that the simple Bohr–Sommerfeld orbital theory was wrong, earlier successes became puzzles. This is what Heisenberg referred to in a letter to Pauli of 21 June 1925 in which he reported what he felt was his painfully slow progress in 'fabricating quantum mechanics'. Heisenberg expressed surprise that Pauli in an earlier letter had wondered about the failure of mechanics, because: 'If anything like mechanics were true, one would never understand the existence of atoms. Evidently, there exists another, a "quantum mechanics", and one must only wonder about the fact that the hydrogen atom agrees, as far as the energy constant is concerned, with something computed classically'.[133] Shortly after the quantum revolution, Bohr considered the same issue, the surprising empirical success of the semi-classical model of the hydrogen atom. 'The atomic theory has passed through a serious crisis in these last years', he wrote to Oseen in Sweden. Looking back on the development, he reflected:

> At the present stage of the development of the quantum theory we can hardly say whether it was good or bad luck that the properties of the Kepler motion could be brought into such

> simple connection with the hydrogen spectrum, as was believed possible at one time. If this connection had merely had the asymptotic character which one might expect from the correspondence principle, then we should not have been tempted to apply mechanics so crudely as we believed possible for some time. On the other hand, it was just these mechanical considerations that were helpful in building up the analysis of the optical phenomena which gradually led to quantum mechanics.[134]

Although the perception of a serious crisis was real, it was far from shared by all physicists. It was restricted to the relatively few who were actively engaged in foundational research related to atomic structure, perhaps a dozen or so. Even Sommerfeld, the leader of the important Munich school in atomic physics, did not identify the situation leading up to quantum mechanics as a crisis preceding a revolution. 'The new development', he wrote in 1929, 'does not signify a revolution, but a joyful advancement [*eine erfreuliche Weiterbildung*] of what was already in existence, with many fundamental clarifications and sharpenings'.[135] Another physicist who did not consider the turn to quantum mechanics revolutionary was Einstein. To him, the shift from the old to the new quantum theory was an evolution rather than a revolution. He did not consider quantum mechanics a fundamental theory but of the same phenomenological nature as the Bohr–Sommerfeld theory.

The Bohr orbital model was not abandoned because it was challenged by a new and better theory of the atom; witness that the Bohr model lost its credibility about a year before the emergence of quantum mechanics. The reason why physicists lost faith in the model and looked forward to a new theory to replace it, was a combination of empirical problems and theoretical shortcomings.[136] For one thing, persistent discrepancies between the Bohr–Sommerfeld theory and experimental results called for more and more radical revisions of the basic assumptions of the theory. Heisenberg's second core model had little in common with Bohr's orbital theory of the chemical elements of 1922. Many of the experimental challenges came from attempts to account for spectroscopic observations in terms of quantum numbers naturally connected to the theory. The anomalous Zeeman effect and other observations suggested that $l = k - 1$ rather than k was of fundamental significance, yet the l quantum number had no natural interpretation within Bohr's theory. Band spectra and complex spectra indicated, indeed required, half-integral quantum numbers, and sometimes the inner quantum number j needed to be represented by $\sqrt{j(j+1)}$. Such oddities had no place in the Bohr model and could only be grafted onto it.

It should not be forgotten that Bohr's atomic model, plagued by difficulties as it came to be, was eminently fruitful in a vast domain and within a short span of time scored a large number of empirical successes. Table 8.1 lists some of these successes; that is, explanations or predictions which at the time were generally considered as confirmations or support of the theory.[137] Most of them have been discussed in earlier chapters. Some of the empirical successes were evidently regarded as more important than others. While the explanation of the hydrogen atom was an impressive and unqualified

Table 8.1. Partial list of experimental successes and failures in the history of the Bohr atom 1913–1925. There is no correlation between the left and right columns.

Successes	Failures
One-electron atoms (H, He^+)	Helium
Spectrum, ionization energy etc.	Spectrum, ionization energy etc.
One-electron atoms (H, He^+) fine structure	Anomalous Zeeman effect
Franck–Hertz experiments	Many-electron atoms, complex spectra
Stern–Gerlach experiments	Ramsauer effect
Space quantization	(Davisson–Kunsman experiment)
Stark effect	Paschen–Back effect in hydrogen
Normal Zeeman effect	Hydrogen molecule ion (H_2^+)
Periodic system, explanation of	Zero-point energy (molecular spectra)
Prediction of element $Z = 72$ (Hf)	Behaviour of electrons in crossed electric and magnetic fields
X-ray spectra	Molecules; covalent bond
Optical dispersion	Superconductivity

success, one which was almost too good to be true, Bohr's later theory of the periodic system was only temporarily successful. It was soon replaced by better theories (Stoner's and Pauli's) which were within the general framework of Bohr's atomic theory but still differed from it in various respects. To repeat from Section 4.4, Sommerfeld's explanation of the fine structure of one-electron atoms was considered a great triumph of the Bohr–Sommerfeld theory, yet it turned out that the triumph was illusory and the nearly perfect agreement between theory and data fortuitous. As Bertrand Russell moralized:

> A theory which explains all the known relevant facts down to the minutest particular may nevertheless be wrong. There may be other theories, which no one has yet thought of, which account equally well for all that is known. We cannot accept a theory with any confidence merely because it explains what is known.[138]

Sommerfeld's fine-structure formula was correct for the wrong reasons, but in the present context this is less relevant since it became recognized only after the emergence of quantum mechanics. Yet it is an instructive case, both from a historical and a philosophical point of view.[139]

The experiments on space quantization in a magnetic field that Otto Stern and Walther Gerlach performed in Frankfurt am Main in 1921–1922 have a certain similarity to Paschen's experiments on the fine structure of the spectra of hydrogen and ionized helium. They both produced experimental evidence in support of a wrong theory. According to Sommerfeld's theory of 1916, atoms with one valence electron should possess a magnetic moment of one Bohr magneton, given by $M = eh/4\pi mc$; if placed in a magnetic field they should assume only two orientations, either aligned with the direction of the field or opposite to it. Stern and Gerlach reasoned that if space quantization were a reality, a beam of silver atoms passing a non-uniform magnetic field should split up into two separate beams and not, as would be expected from the classical Larmor theory, merely be broadened. This effect, wrote Stern in a preliminary paper, 'unequivocally decides between the quantum-theoretical and classical interpretation'.[140] In their experiment of 1922 they found a distinct separation into two beams. The splitting was only 0.2 mm, but it was enough to prove the space quantization expected from quantum theory and to determine the magnetic moment of the silver atom to one Bohr magneton.[141]

Of interest here is merely that the Stern–Gerlach experiment was considered a triumph of the Bohr–Sommerfeld quantum theory because it confirmed that an electron orbit can only assume discrete orientations in space. The results were not only a confirmation of Sommerfeld's theory, they also agreed with the ideas of Bohr, who had theoretical reasons to believe that 'the magnetic axis of the silver atom is always directed parallel to the field and never perpendicular to it'. When Gerlach informed Bohr of the results in a postcard of 8 February 1922 (Figure 8.3), he congratulated him 'on the confirmation of your theory'.[142] Paschen even claimed that the experiment 'proves for the first time the reality of Bohr's stationary states'.[143]

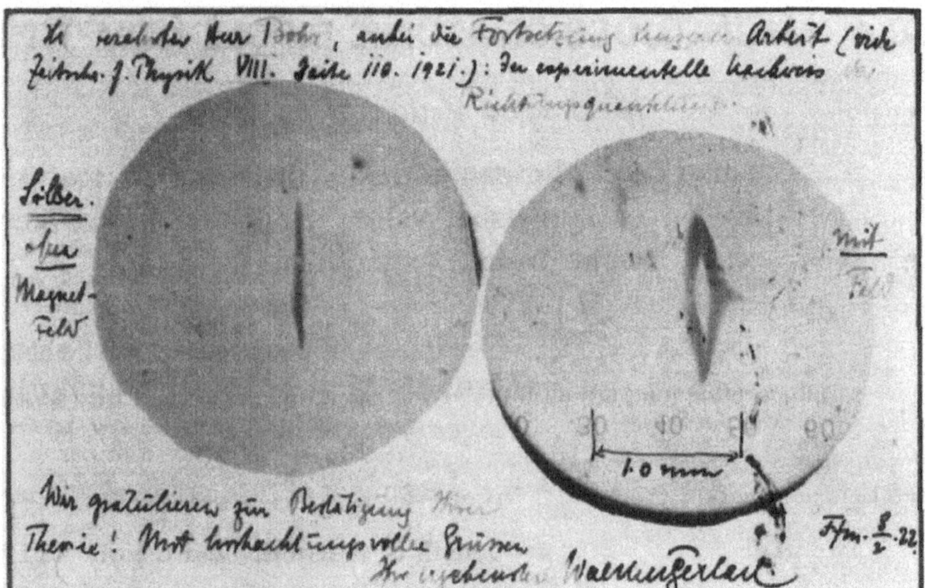

Fig. 8.3. Postcard from Gerlach to Bohr of 8 February 1922, informing him of the magnetic splitting of a beam of silver atoms.

Source: Friedrich and Herschbach 2003, p. 56 (Niels Bohr Archive).

However, the situation was somewhat ambiguous, for a closer consideration of the Stern–Gerlach effect indicated that it presented difficulties for the ordinary understanding of the quantum theory, in particular the unrestricted validity of the adiabatic hypothesis. In a critical analysis of the experiment, Einstein and Ehrenfest discussed the difficulties and referred to 'Bohr's view—that in complicated fields there will be no sharp quantization at all'.[144] Although known, the difficulties were not seen as posing a threat to the fundamental assumptions of the theory, and it was only after the emergence of spin quantum mechanics that it became clear that what Stern and Gerlach had observed did not really support the Bohr–Sommerfeld quantum theory. In retrospect they had discovered the two spin states of the electron.

Basically successful as Bohr's theory was, its empirical success was limited and countered by a growing number of anomalies. The most important of the experimental problems were undoubtedly the helium atom and the anomalous Zeeman effect, but there were several other phenomena that the theory could not account for satisfactorily (Table 8.1). Anomalies are of course bad news for a theory, but they come in different kinds and are often evaluated in widely different ways by the scientific community. Some are regarded as crucially important—the kind of stuff that may cause a wholesale refutation of a theory—while others are hardly noticed and have almost no effect at all on how the theory is judged. In the case of the Bohr atom, physicists paid great attention to the anomalies related to the helium atom and the Zeeman effect, which, as we have seen, were characterized as nothing less than catastrophes. On the other hand, there were also anomalies that indicated flaws in the theory but nonetheless were ignored for all practical purposes or at least assigned very little weight.

One example was the evidence of a Paschen–Back effect in hydrogen, where according to the Bohr–Sommerfeld theory there should be none. The inconsistency between theory and experiment was recognized by leading physicists, but without considering it a problem that threatened the theory. The contradiction did not really disturb the physicists, who convinced themselves that it should not count as a genuine anomaly. We have a somewhat similar case in the failed attempt to understand the hydrogen molecule ion H_2^+. This anomaly—and it clearly was one—was not much noticed at the time and did not contribute to the rise of the crisis that began in 1923. As a third and rather remarkable example of a quantum anomaly that was not taken seriously and scarcely recognized as a proper anomaly, one may mention the covalent bond that ties atoms together in molecules. As early as 1915 it was evident that Bohr's theory could not account for the H_2 molecule, yet the sustained impotence of the theory in the area of molecular constitution was of no importance in the crisis of the old quantum theory. It was of importance to the reputation of the theory in chemical circles, but this was something few physicists cared about.

The case of zero-point energy, or the corresponding notion of half-integral quantum numbers, was an anomaly only in the sense that it did not fit naturally with the Bohr–Sommerfeld quantum theory. As we have seen, half-quanta turned up in both theories of the anomalous Zeeman effect and in molecular spectroscopy. The zero-point energy that Mulliken inferred in 1925, and which had been suggested before him, could not be theoretically justified by existing quantum theory, and this constituted a problem. However, it is worth pointing out that Mulliken's 'discovery' of the zero-point energy was not much noticed in the community of atomic and quantum physicists, possibly because it was made in a chemical rather than physical context. In fact, the physicists in Copenhagen and Munich seem to have ignored his work or, more likely, been unaware of it. It did not contribute to the quantum crisis any more than the hydrogen molecule failure did. At any rate, Bohr much disliked the half-quanta, arguing that they were irreconcilable with the basic principles of quantum theory, while most physicists learned to live with them, if not love them. According to Van Vleck, they were necessary, yet 'in many ways decidedly illogical'.[145] It was only with Heisenberg's quantum mechanics from the summer of 1925 that the problem of zero-point energy disappeared. As Heisenberg showed, it followed from his new theory that the energy of a harmonic oscillator is given by

$$W = (n + \tfrac{1}{2})h\nu, \ n = 0, 1, 2, \ldots,$$

and not, as in the older theory, by $W = nh\nu$. This was not the end of the zero-point energy problem, but only the beginning of a new chapter, where it came to be seen as the energy density of vacuum. It was eventually interpreted in terms of the cosmological constant introduced by Einstein in 1917. But this new chapter, fascinating as it is, does not belong to the history of the Bohr atomic model.

In addition to recognized failures or anomalies, there were also a few consequences of the quantum theory that disagreed with experiments but were recognized only after the crisis had culminated and the old quantum theory been replaced by quantum mechanics. One may speak of a *post hoc* anomaly, although of course it does not count as an anomaly in the real, historical sense. Consider the fine structure, where Hansen's identification of a IIc component in hydrogen (Section 4.4) was anomalous but only clearly recognized as such at the end of 1925. On the basis of the Sommerfeld–Bohr–Kramers theory there should exist in the He^+ spectrum a IIIb component, which did not in fact exist (Section 5.3), and this anomaly or inconsistency *could* have been known as early as 1919. However, it went unnoticed until Goudsmit and Uhlenbeck pointed it out in their paper in *Physica* in which they introduced their new classification scheme. Since this paper only appeared in September 1925, their observation was of no consequence in the crisis of the old quantum theory.

Another interesting and little known case is provided by the theory of dielectric constants (Section 6.3), where Pauling in 1927 concluded that because of the space quantization, 'the old quantum theory definitely requires that the application of a strong magnetic field to a gas such as hydrogen chloride produce a very large change in the dielectric constant of the gas'. Even a magnetic field of moderate strength should have a drastic effect on the dielectric constant, a prediction that was flatly contradicted by experiments. Therefore, 'it provides an instance of an apparently unescapable and yet definitely incorrect prediction of the old quantum theory'.[146] The new quantum mechanics fortunately led to the result that a magnetic field does not influence the dielectric constant, in agreement with experiment. Pauling's result could presumably have been derived several years earlier, and would then have constituted an anomaly. But this is not what happened.

After quantum mechanics had stabilized, a few physicists looked back to consider the problems that had faced the Bohr–Sommerfeld theory and made it a theory of the past. In April 1928, at a symposium of the American Chemical Society, Van Vleck presented a list of the inadequacies of the old quantum theory in order to illustrate to the chemists the superiority of the new quantum mechanics. As to the successes of the old theory, he emphasized that all of these were reproduced by the Heisenberg–Schrödinger–Dirac theory. He called attention to the parts of the old theory that survived in, and could be better understood using, the new theory:

> Although we cannot attach as much reality and vividness to electron orbits in stationary states as previously, nevertheless the new theory has the vital feature of a discrete succession of energy levels, with spectroscopic frequencies proportional to their differences. Thus we can still keep our beloved energy level diagrams, selection principles, and the like.[147]

According to Van Vleck, the Bohr–Sommerfeld theory had failed on a number of counts, which may be compared with those listed in Table 8.1. The failures mentioned by Van Vleck were the following:

(1) *The fine details of the hydrogen and ionized helium spectrum*, to which he not only counted the Paschen–Back effect and the appearance of the forbidden IIc line, but also the arbitrary exclusion of pendulum orbits ($k = 0$) 'on the ground that in them the electron would collide with the nucleus'.

(2) *The spectrum of neutral helium.*

(3) *The intensity of spectral lines.* While admitting that in some cases intensities could be estimated by means of the correspondence principle, Van Vleck found that the method 'lacked adequate logic' and did not result in unambiguous quantitative results. The old quantum theory was never able to treat intensity in the same rigorous way as frequency, in the way quantum mechanics could.

(4) *The anomalous Zeeman effect.*

(5) *Half-quanta.* Whereas the half-integral quantum numbers were introduced phenomenologically in the old quantum theory, and never provided with a rigorous justification on the basis of the theory, in the new quantum mechanics 'half quantum numbers [are] no longer a bugaboo'.

(6) *Dielectric constants and magnetic susceptibilities.* On this inadequacy, see p. 350 and in Section 6.3.

(7) *Dispersion and the emission and absorption of radiation.*

(8) *Collisions of electrons with atoms*, including the Ramsauer effect.

Another retrospective list of problems in the old quantum theory was made by Owen Richardson, who, like Van Vleck, used it to confront the now abandoned Bohr–Sommerfeld theory with the new quantum mechanics.[148] Several of the problems in Richardson's list also appeared in Van Vleck's, including the helium atom, the anomalous Zeeman effect, half-integral quantum numbers, and the Ramsauer effect. In addition he mentioned experiments on soft X-rays and the reflection of electrons on metals studied by Davisson and Kunsman.

8.6. OBSERVABILITY AND *UMDEUTUNG*

There could be no question of rejecting the Bohr–Sommerfeld theory just because it faced some insurmountable empirical problems such as the anomalous Zeeman effect and the spectrum of helium. In the absence of an alternative theory it would have left the physicists with a wealth of unexplained phenomena, a drastic loss of knowledge instead of a gain in knowledge. By 1924 it was realized that the Bohr–Sommerfeld theory was inadequate, perhaps even fundamentally wrong, and that its basic architecture was a hopeless mess, a patchy combination of incomparable elements. But it was also realized that the theory after all held a strong empirical record and that any theory of the future would have to be at least as strong. According to Bohr, this future theory needed to contain the existing theory as a limiting case and also, like the existing theory, to keep the connections to classical theory. 'It is necessary for us', he wrote to Born, 'to regard the usual formulation of the quantum theory, as it is elaborated for atomic structure, as a limiting case of a more general theory'.[149]

The experimental problems did not themselves show the way to the more general and superior theory the physicists dreamed of. To find the way it was necessary to identify the methodological and conceptual flaws in the existing theory, flaws that were supposedly reminiscences of its semi-mechanical origin. A small group of physicists consequently focused on the methodological and logical structure of the theory, asking themselves about its very foundation. In this process fundamental notions were questioned, not only about the existence of electron orbits but also of the use of differential equations in the basic theory and the possibility of a space-time description of the interior of the atom. Such radical ideas were not foreign to Bohr, who in a letter to the Danish philosopher Harald Høffding from September 1922 expressed his view of the situation in atomic physics as follows:

> Here we are in the peculiar situation that we have acquired certain information about the structure of the atom which can be considered as being as certain as any one of the facts in the natural sciences. On the other hand we encounter difficulties of such a deeply-rooted nature that we do not even faintly see the road to their solution; in my personal opinion these difficulties are of such a nature that they hardly allow us to hope that we shall be able, in the world of the atom, to carry through a description in space and time of a kind which

> corresponds to our ordinary sensory images. Under these circumstances one must naturally constantly bear in mind that one is operating with analogies, and this step, in which the application of these analogies is delimited in each case, is of decisive significance for progress.[150]

Among the methodological issues that played a role in the declining phase of the old quantum theory was the doctrine or desideratum that physical theories should build solely on concepts which refer to quantities that can, at least in principle, be observed. As mentioned, this so-called observability principle had been advocated since 1919 by Pauli and Born in particular, and in a mild version it also entered the dispersion theories of Kramers, Heisenberg, and Born. On the other hand, Bohr seems not to have appreciated the observability principle as a heuristic tool, although he advocated a limited use of it as early as 1913. In his *Atommechanik* completed in November 1924, Born stated about the 'unknown and true quantum laws' that 'we must require that they involve only observable quantities such as energies, frequencies of light, intensities and phases'.[151] About half a year later he wrote a paper together with 22-year-old Pascual Jordan, his new assistant, in which the observability principle was presented as 'A fundamental axiom of large range and fruitfulness [which] states that the true laws of nature involve only such quantities as can be observed and determined in principle'.[152] Although the operationalist observability criterion was thus assigned an important role in parts of the German physics community, it was not closely connected to the hostile attitude to electron orbits that Pauli expressed in particular.

Neither did the criterion motivate Heisenberg to eliminate the unobservable orbits, as he did in his seminal paper from the late summer of 1925 which marks the birth of quantum mechanics. In his earlier papers he did not refer to the observability principle, which he first mentioned in a letter to Pauli of late June in which he adopted the doctrine stated by Born and Jordan. In Heisenberg's formulation the *Grundsatz* (basic axiom) read: 'In calculating any quantities, like energy, frequency, etc., only relations between those quantities should occur, which can be controlled in principle'.[153] Less than two weeks later, he confirmed his nearly complete agreement with Pauli's views, not only with regard to observability but also with regard to orbital or other mechanical models. Referring to Bohr's model of the hydrogen atom, Heisenberg now wrote: 'It is really my conviction that an interpretation of the Rydberg formula in terms of circular and elliptical orbits . . . does not have the slightest physical significance. And all my wretched efforts are devoted to killing totally the concept of an orbit—which one cannot observe anyway—and replace it by a more suitable one'.[154]

Heisenberg's observability criterion is sometimes, and not without reason, seen as a version of the operationalist philosophy developed by the American physicist-philosopher Percy Bridgman, a Nobel laureate of 1946 for his works in high-pressure physics. Yet, although Bridgman recognized the similarity of the quantum observability principle to operationalism, he refrained from endorsing Heisenberg's use of it or considering it a vindication of the true operationalist method. Like some modern historians and philosophers of physics, he did not consider Heisenberg's method to be a constructive principle, but suspected that it was instead a philosophical rationale added after the fact. 'I have always wondered', he said in 1936, 'whether perhaps this requirement of Heisenberg was not formulated after the event as a sort of philosophical justification for its success, rather than having played an indispensable part in the formulation of the theory'.[155]

The result of Heisenberg's hard thinking and intense intellectual interaction with Pauli was a paper that appeared in the 18 September 1925 issue of *Zeitschrift für Physik* with the title 'Über Quantentheoretische Umdeutung Kinematischer und Mechanischer Beziehungen'; that is, on the quantum-theoretical reinterpretation of kinematic and mechanical relations. Historians have pointed out that in an earlier paper on the anomalous Zeeman effect, Sommerfeld used the magical word *Umdeutung* in the title, and that the similarity between the two papers is not limited to the titles but extends to the applied methodologies. Whether this coincidence implies a real connection, an inspiration from Sommerfeld in 1922 to Heisenberg in 1925, seems more doubtful. At any rate, other German physicists spoke of *Umdeutung* in a quantum context as well. Thus, in the summer of 1925 the Breslau physicists Fritz Reiche and Willy Thomas wrote a paper that both terminologically and methodologically has a great deal of similarity to Heisenberg's. 'We try to arrive at a general relation', they wrote, referring to dispersion, 'by maintaining the reinterpretation [*Umdeutung*] of classical quantities into quantum-theoretical ones for all quantum numbers'.[156]

Umdeutung apart, Heisenberg introduced his paper by referring to some of the well-known problems that faced existing quantum theory, such as the treatment of many-electron atoms and the case of a hydrogen atom placed in crossed electric and magnetic fields. Having reminded his readers that 'the Einstein–Bohr frequency condition' $\Delta E = h\nu$ already represented a complete break with classical mechanics, he appealed to observability as a guiding principle:

> In this situation it seems sensible to discard all hope of observing hitherto unobservable quantities, such as the position and period of the electron, and to concede that the partial agreement of the quantum rules with experience is more or less fortuitous. Instead it seems more reasonable to try to establish a theoretical quantum mechanics, analogous to classical mechanics, but in which only relations between observable quantities occur.[157]

In his symbolic translation or reinterpretation of classical mechanics into a new formalism of quantum mechanics, Heisenberg relied heavily on Kramers' dispersion theories. No less heavily did he rely on Bohr's correspondence principle, which played an important if mostly implicit role in his reasoning. 'Heisenberg's quantum mechanics', writes Olivier Darrigol, 'may in fact be seen as the ultimate stage of the evolution of the correspondence principle'.[158] Heisenberg later spoke of his *Umdeutung* paper in words that support this view:

> For me [it] represented in a certain sense the quintessence of our discussions in Copenhagen – a mathematical formulation of Bohr's correspondence principle. I hoped that, by means of a

> mathematical method which for me was still new and very strange, I had found a way to the remarkable relations, which had already been glimpsed from time to time during discussions with Bohr and Kramers.[159]

The development of quantum mechanics proper is however outside the scope of the present work, for which reason I shall not deal further with Heisenberg's remarkable paper.

Bohr was not actively involved in the creation of Heisenberg's new mechanics in the summer of 1925, a period in which there was actually no interaction between the two physicists. Yet Bohr should be given the last word. On 24 August and again on 31 August Bohr gave talks on the recent developments in quantum theory, the first time in Oslo and the second time in Copenhagen to the Sixth Scandinavian Mathematical Congress. On neither of these occasions did he mention Heisenberg's theory, of which he was most likely unaware. A few days after his address in Copenhagen he received a letter from Heisenberg, then vacationing in Austria: 'I committed the crime of writing a paper on quantum mechanics, about which I should like to hear your opinion. It will presumably appear in the next issue of the Zeitschrift'.[160]

A revised version of Bohr's Copenhagen address was later published in *Nature*, and here Bohr did comment on Heisenberg's work and its further development by Born and Jordan into a theory of matrix mechanics. In a manuscript for this publication he described Heisenberg's theory as 'probably of extraordinary scope' and 'a brilliant realization of the efforts which hitherto have served as a guideline in the development of the atomic theory'. He further noted that, 'In the nature of matter, there is no longer question of an independent correspondence principle in the above-mentioned scheme [Heisenberg's], but rather the whole formulation of quantum mechanics may well be regarded as a sharpening of the content of this principle'.[161]

As far as Bohr was concerned, the quantum revolution of 1925 was not a proper revolution. He tended to agree with Sommerfeld's view that it was rather 'a joyful advancement of what was already in existence'.

Notes for Chapter 8

1. Van Vleck 1926, p. 287.
2. Pauli 1946, p. 214. On the history of the anomalous Zeeman effect, see Forman 1970, Massimi 2005, pp. 47–52, and Mehra and Rechenberg 1982a, pp. 445–84.
3. Born 1923, p. 541. As early as 1920 Sommerfeld had called the effect one of the two unsettled questions of atomic physics', but at that time the question was not considered a burning one. Although he thought it called for 'entirely new things', Sommerfeld did not consider it more serious than another unsettled question, the one of the polyhedral structure of the atom. See Sommerfeld 1920a and also Sommerfeld 1920b, where he characterized the anomalous Zeeman effect as a *Zahlenmysterium* (number mystery).
4. Landé to Bohr, 4 February 1921, reproduced in Forman 1970, p. 238.

5. Landé to Sommerfeld, 17 March 1921. Sommerfeld to Landé, late March 1921. Both letters are reproduced in Forman 1970, pp. 260–1.

6. Landé 1921, p. 239.

7. Sommerfeld to Landé, 25 February 1921, in Mehra and Rechenberg 1982a, p. 470.

8. Sommerfeld to Einstein, 17 October 1921, in Hermann 1968, p. 94.

9. Landé 1923, p. 198.

10. Heisenberg to Pauli, 21 February 1923, and Pauli to Landé, 10 March 1923, in Hermann et al. 1979, p. 81 and p. 83.

11. Bohr 1922a, p. 59. Bohr was kept informed about the recent works on the anomalous Zeeman effect by letters from Landé, who in February 1921 reported his latest results (see Mehra and Rechenberg 1982a, p. 469).

12. Bohr to Landé, 15 May 1922 (Archive for History of Quantum Physics, Bohr Scientific Correspondence).

13. Bohr 1923c, in Rud Nielsen 1977, p. 629.

14. Ibid., p. 646. Cassidy (1978 and 1979, p. 222) suggests that Bohr was stimulated by a non-mechanical 'mystery force' that Van Vleck had previously introduced in his attempt to understand the helium atom. According to Van Vleck, such a hypothetical force, as proposed by Langmuir and a few others, would be negligible except at atomic distances and 'not have a mechanism based on the Maxwell field-equations' (Van Vleck 1922, p. 851). However, he did not advocate the idea and I know of no evidence that it stimulated Bohr's thinking. For Langmuir's mystery force, see Section 6.1.

15. Bohr 1925a, p. 850.

16. Bohr 1923c, in Rud Nielsen 1977, p. 647.

17. Rud Nielsen 1976, p. 558. The manuscript is in German.

18. Bohr to Landé, 14 February 1923 (Archive for History of Quantum Physics, Bohr Scientific Correspondence).

19. Bohr to Landé, 3 March 1923, quoted in Richter 1979, p. 34. See also Heisenberg to Sommerfeld, 4 January 1923, in Eckert and Märker 2004, p. 133.

20. Pauli to Landé, 23 May 1923, in Hermann et al. 1979, p. 90. Pauli also reported about Bohr's views in his letter to Sommerfeld of 6 June 1923.

21. Pauli 1923, p. 162.

22. Pauli to Sommerfeld, 6 June 1923, in Hermann et al. 1979, p. 97.

23. Pauli to Landé, 17 August 1923, in ibid., p. 110.

24. Pauli to Sommerfeld, 19 July 1923, in ibid., p. 105.

25. Heisenberg 1922. On this paper, see Cassidy 1978 and, for a more detailed and scholarly version, Cassidy 1979.

26. McLennan 1923, p. 58, who relied on Rosseland 1923b. For $k = 1$, the minimum distance to the uranium nucleus was found to be 16×10^{-12} cm, hence permissible.

27. Heisenberg to Pauli, 19 November 1921, in Hermann et al. 1979, p. 38. Stressing its opportunistic character, Cassidy 1978 calls Heisenberg's first core model 'Machiavellian'.

28. 'Heisenberg... is very unphilosophical, he does not care about a clear elaboration of the fundamental assumptions and of their relation to the previous theories'. Pauli hoped that Heisenberg, after having stayed with Bohr in Copenhagen, would 'return home with a philosophical orientation to his thinking'. Pauli to Bohr, 11 February 1924, in Hermann et al. 1979, p. 143.

29. Sommerfeld to Einstein, 11 January 1922, in Hermann 1968, p. 96. On Sommerfeld as a quantum craftsman, see Seth 2010. In the third edition of *Atombau*, Sommerfeld added an appendix on 'the riddle of the anomalous Zeeman effect' in which he expounded his own and Heisenberg's ideas on the subject (Sommerfeld 1922a, pp. 497–504).

30. Pauli to Sommerfeld, 6 June 1923, in Hermann et al. 1979, p. 95, and similarly in his letter to Landé of 17 August. On Pauli's negative attitude to the core model, see Serwer 1977.

31. Heisenberg to Pauli, 9 October 1923, in Hermann et al. 1979, p. 125. Heisenberg also gave an account of his new theory in a letter to Bohr of 22 December 1923 (Archive for History of Quantum Physics, Bohr Scientific Correspondence). In a letter of 7 December 1923, Heisenberg sent his manuscript on *Zeemangemüse mit Quantensosse* (Zeeman vegetables with quantum sauce) to Pauli, informing him that he was hoping for Bohr's approval—'papal blessing'. He did receive the blessing, and after some modifications the paper was published in August 1924.

32. Heisenberg to Sommerfeld, 8 December 1923, in Eckert and Märker 2004, p. 157. The symbol J denotes an action variable.

33. Heisenberg 1924, p. 292. See Darrigol 1992, pp. 196–200.

34. In Hermann et al. 1979, pp. 126–7.

35. Heisenberg to Bohr, 22 December 1923, as quoted in Cassidy 1992, p. 171. The reference is to Bohr 1923c, which appeared in a special issue of *Annalen der Physik* in honour of the seventieth birthday of the spectroscopist Heinrich Kayser. On Bohr's invitation, Heisenberg spent some time in Copenhagen in March 1924, where they discussed the theory and its formulation, including its relation to radiation theory.

36. Letter of 29 November 1923, quoted in Cassidy 1992, p. 171.

37. Pauli to Landé, 14 December 1923, in Hermann et al. 1979, p. 134. On Pauli's evaluation of Heisenberg's theory, see also Pauli 1926a, pp. 237–9.

38. Pauli to Bohr, 21 February 1924, in Hermann et al. 1979, pp. 147–8.

39. Pauli to Eddington, 20 September 1923, in Hermann et al. 1979, p. 116.

40. Born 1925, p. 34. Born's remarks in *Atommechanik* echoed the introduction to his important paper on 'quantum mechanics' from the summer of 1924 (Born 1924, see also Section 8.4).

41. Weyl to Pauli, 9 December 1919, and Pauli to Eddington, 20 September 1923, in Hermann et al. 1979, p. 6 and p. 116. Pauli 1919, pp. 749–50. In his *Enzyklopädie* article on relativity theory of 1921, Pauli expressed the same criticism, that the field strength at a given point in the interior of an electron is 'unobservable, by definition, and thus fictitious and without physical meaning' (Pauli 1958, p. 206). For Pauli's operationalism and his early views concerning unified field theories, see Hendry 1984.

42. Born to Pauli, 23 December 1919, in Hermann et al. 1979, p. 10. See also Darrigol 1992, p. 196.

43. Interview with Heisenberg by T. S. Kuhn on 19 February 1963. American Institute of Physics, Niels Bohr Library & Archives. http://www.aip.org/history/ohilist/4661_6.html.

44. Einstein to Born, 27 January 1920, in Born 1971, p. 21. Einstein published his idea of reconciling quanta and fields, or rather deriving quanta from fields, in Einstein 1923. His question of how to produce discontinuous solutions from differential equations would be answered with Schrödinger's wave mechanics, if not quite to Einstein's satisfaction. Later in life, Einstein considered on various occasions the possibility of replacing the continuous field concept with an 'algebraic physics' based on discrete space-time, but nothing came out of these speculations. See Stachel 1993.

45. Heisenberg to Bohr, 26 February 1930 (Archive for History of Quantum Physics, Bohr Scientific Correspondence), translated and analysed in Carazza and Kragh 1995. Some modern theories of quantum gravity, in particular the theory of loop quantum gravity developed since about 1990, assumes space to be built up of smallest cells.

46. Heisenberg to Bohr, 22 December 1923 (Archive for History of Quantum Physics, Bohr Scientific Correspondence). The connection also appeared in Heisenberg's published paper (Heisenberg 1924, p. 299 and p. 307).

47. Bohr 1923b, p. 33.

48. Compton 1923, p. 501. Remarkably, he did not explicitly connect his X-ray quanta with Einstein's theory of light quanta and in fact did not mention Einstein's name in his paper. For details about Compton's experiments and theoretical ideas, see Stuewer 1975.

49. Debye 1923. On the relationship between the works of Compton and Debye, see Stuewer 1975, pp. 234–7.

50. Stuewer 1975. Hendry 1981 disagrees that Compton's discovery had a decisive effect on physicists' view on the nature of electromagnetic radiation.

51. Kramers and Holst 1923, p. 175. The two authors stressed that Einstein's theory of the light quantum 'in no way has sprung from the Bohr theory, to say nothing of its being a necessary consequence of it'.

52. Bohr's notes for the Silliman lectures are included in Rud Nielsen 1976, pp. 581–601. The lectures were covered by *The New York Times*, which gave summaries of all of them. 'Likened to solar system, he pictures the atom with nucleus corresponding to sun, and electrons to planets', it informed its readers on 7 November 1923. See also the summary account in *Science* **58** (1923), 459–60. Earlier Silliman lecturers included J. J. Thomson (1905) and Rutherford (1906).

53. Swann's suggestion is mentioned in a footnote on p. 290 in Compton and Simon 1925. See also Stuewer 1975, p. 301.

54. Duane to Bohr, 5 March 1924, and Bohr to Duane, 17 May 1924, in Stolzenburg 1984, p. 323 and p. 326. Duane believed at the time that the Compton shift was caused by photoelectrons emitted from the K and L levels by the primary X-rays. See also Bohr to Rutherford, 9 January 1924, in ibid., p. 487.

55. Quoted from the extracts in Stolzenburg 1984, pp. 13–14. Darwin's unpublished manuscript exists in at least one other version (Archive for History of Quantum Physics), which is quoted and discussed in Jammer 1966, pp. 171–2.

56. Bohr to Darwin, July 1919 (draft), in Stolzenburg 1984, p. 16.

57. Darwin 1923, p. 771 and Darwin 1922. On Darwin and his works in quantum physics, see Navarro 2009. See also Konno 2000 for Darwin's dispersion theory.

58. Sommerfeld 1922a, p. 311. The passage did not appear in the first edition of 1919.

59. Bohr to Darwin, 21 December 1922, in Stolzenburg 1984, p. 316. Darwin, in a letter to Bohr of 23 November 1922, had 'left it open whether the upper states really are stationary, but in private [I] may say that I personally do not believe they are' (ibid., p. 17).

60. Bohr 1924, p. 40, who referred to experiments reported in Gans and Miguez 1917.

61. Slater 1967, p. 6.

62. Bohr to Slater, 10 January 1925, in Stolzenburg 1984, p. 67.

63. Slater to Kramers, 8 December 1923, in Stolzenburg 1984, p. 492. See also Hendry 1981. Slater further explained that his ideas might explain the helium spectrum, if only 'by making a whole lot of guesses'.

64. Manuscript of 4 November 1923, quoted in Petruccioli 1993, p. 113. On the development of Slater's views, see Konno 1983.

65. Slater 1967, p. 6.

66. Slater 1925b.

67. Slater 1975, pp. 12–14.

68. Slater 1923. Ladenburg and Reiche 1923, p. 590, who introduced classical *Ersatzoszillatoren* (substitute oscillators) in dispersion theory.

69. Bohr et al. 1924. The theory has attracted considerable historical attention. Apart from Hendry 1981 and Petruccioli 1993, see also Stolzenburg 1984, Duncan and Janssen 2007, pp. 597–617, and Dresden 1987, pp. 41–78, 159–215, who deals in particular with Kramers' contribution to and view of the theory.

70. Bohr to Slater, 10 January 1925, quoted in Stuewer 1975, p. 293. On Bohr's coupling principle and latent forces, see Bohr 1924, p. 36.

71. Bohr to Michelson, 7 February 1924, in Stolzenburg 1984, p. 405. See also Livingston 1973, pp. 301–2, according to whom (p. 292) Michelson had met Bohr at the 1921 Solvay congress, where he, 'feeling somewhat lost', listened to Bohr lecturing on atomic theory. However, Bohr did not attend the Solvay conference this year and he only met with Michelson during his trip to the United States in the autumn of 1923.

72. Bohr et al. 1924, p. 786.

73. Ibid. The treatise referred to was Bohr 1924. The BKS paper is reprinted in van der Waerden 1967, pp. 159–76.

74. Bohr et al. 1924, p. 795. Bohr 1924, pp. 29–30. The emphasis on the limited definability of motion and energy would later play an important role in Bohr's complementarity philosophy. For this connection, see Tanona 2004.

75. Bohr et al. 1924, pp. 790–1. The BKS theory maintained statistical energy conservation. In 1929 Bohr went further, proposing that energy and momentum might not be conserved even on a macroscopic level. This much more radical hypothesis he thought justified in order to explain the continuous beta spectrum and also stellar energy production. See Jensen 2000, pp. 146–56.

76. Bohr et al. 1924, p. 799.

77. Kramers and Heisenberg 1925, p. 685.

78. Kramers and Holst 1925, pp. 123–40. The preface of the book was dated March 1925, the new chapter on the interaction of light and matter, including the BKS theory, being the work of Kramers. The chapter also appeared separately, as a paper in a Danish physics journal (Kramers 1925). On Kramers' view of causality in relation to the BKS theory, see Radder 1983 and Dresden 1987, pp. 191–5.

79. Haber to Einstein, 1924, quoted in Stolzenburg 1984, p. 26.

80. Schrödinger to Bohr, 24 May 1924, in Stolzenburg 1984, p. 490. On Exner's ideas of indeterminism and acausality, and the way they influenced Schrödinger's thinking and response to the BKS theory, see Hanle 1979.

81. Schrödinger 1924, p. 724.

82. According to Paul Forman, it is 'only by reference to the widespread acausal sentiment that one can understand the immediate and widespread assent which the [BKS] theory received in Germany' (Forman 1971, p. 99). However, there was no such widespread assent among German physicists.

83. Pauli to Bohr, 21 February 1924, in Hermann et al. 1979, p. 147.

84. Einstein to Born, 27 January 1920, and Einstein to Hedwig and Max Born, 29 April 1924, in Born 1971, p. 23 and p. 82. Einstein also objected to the theory by means of more technical arguments based on thermodynamics. See Stolzenburg 1984, pp. 23–38, where the reactions of Einstein and other physicists are recounted.

85. Einstein to Ehrenfest, 31 May 1924, quoted in Stolzenburg 1984, p. 27.

86. Bohr to Born, 1 May 1925, a reply to Born's letter of 24 April, both in Stolzenburg 1984, pp. 308–11.

87. Bothe and Geiger 1925.

88. Compton and Simon 1925, p. 299.

89. Einstein's manuscript, dated 7 May 1925, is reproduced in Tolmasquim and Moreira 2002, pp. 240–2. The final results of the Bothe–Geiger experiments were published in the 15 May issue of *Naturwissenschaften* with a detailed report following in June in *Zeitschrift für Physik*.

90. On this question and the later philosophical interest in it, see the appendix (Chapter 9).

91. Bohr to Fowler, 21 April 1925, in Stolzenburg 1984, p. 82.

92. The name 'photon' was proposed by G. N. Lewis in December 1926, but in a sense widely different from Einstein's light quantum. Lewis's photon was a 'hypothetical new atom, which is not light but plays an essential part in every process of radiation'; also contrary to the light quanta, his photons were conserved quantities, 'uncreatable and indestructable' (Lewis 1926b; see also Stuewer 1975, pp. 323–6). In spite of the speculative nature of Lewis's ideas of radiation, Bohr found them interesting. See Bohr to Richardson, 16 February 1926, Archive for History of Quantum Physics.

93. Bohr to Geiger, 21 April 1925, in Stolzenburg 1984, p. 79. Much later, in 1936, Paul Dirac advocated a reintroduction of a BKS-like theory in order to explain experiments by Robert Shankland that seemed to imply energy nonconservation in individual atomic processes. Interestingly, in this case Bohr argued against Dirac's proposal, while Einstein was in favour of it. Referring to the old BKS theory, Bohr pointed out that the situation was now 'quite different' from what it had been in the days of the old quantum theory (Bohr 1936). For this episode, see Kragh 1990, pp. 169–73.

94. Kramers to Urey, 16 July 1925, in Stolzenburg 1984, p. 86.

95. Slater 1925b.

96. Slater 1925a. Slater 1967, p. 8.

97. Slater 1925a, p. 398. On Bohr's response to Slater's paper, see Bohr to Slater, 10 January 1925, in Stolzenburg 1984, pp. 66–8.

98. Bohr 1925b, p. 156. See also Darrigol 1992, pp. 249–51, and Bohr's correspondence with Fowler, Born, and Franck concerning collision and capture processes, as quoted in Stolzenburg 1984, pp. 71–4.

99. Pauli to Kramers, 27 July 1925, where he speaks of 'the Scylla of the number-mystical Munich school and the Charybdis of the reactionary Copenhagen Putsch [coup or revolt]' (Stolzenburg 1984, pp. 443–4). Pauli was at the time aware of Heisenberg's manuscript that marked the beginning of a new quantum mechanics and which he praised in his letter to Kramers.

100. Heisenberg 1929, p. 492. According to Jammer 1966, p. 187, 'It is hard to find in the history of physics a theory which was so soon disproved after its proposal and yet was so important for the future development of physical thought' as the BKS theory. On the other hand, Duncan and Janssen (2007) argue that while the idea of virtual oscillators was important in the

development that led to quantum mechanics, the BKS theory played only a minor role. Significantly, in his 1951 historical review of the development of quantum theory, Sommerfeld did not even mention the BKS theory (Sommerfeld and Bopp 1951).

101. Dresden 1987, p. 244. Duncan and Janssen 2007, pp. 613–16.

102. For electromagnetic dispersion theories at the end of the nineteenth century and in the early twentieth century, see Buchwald 1985 and Loria 1914.

103. Bohr, 'On the Application of the Quantum Theory to Periodic Systems' (1916), in Hoyer 1981, p. 449. Bohr's views of dispersion theories are surveyed in Konno 2000.

104. Ladenburg 1921, p. 451, translated in van der Waerden 1967, pp. 139–57. Duncan and Janssen (2007, p. 584) wrongly assert that the correspondence principle is not mentioned anywhere in Ladenburg's paper. It appears in fact on p. 459.

105. Bohr, 'Application of the Quantum Theory to Atomic Problems in General' (1921), in Rud Nielsen 1976, p. 414. For Bohr's appreciation of Ladenburg's work, see also Bohr 1925a, p. 851.

106. Bohr 1924, p. 39.

107. Ladenburg and Reiche 1923. Ladenburg communicated his and Reiche's results to Bohr in a letter of 14 June 1923 (Stolzenburg 1984, p. 400). Ladenburg's two papers are analyzed in Konno 1993. See also interview with Reiche of 30 March 1962, by T. S. Kuhn and G. Uhlenbeck: http://www.aip.org/history/ohilist/4841_1.html. American Institute of Physics, Niels Bohr Library & Archives.

108. Slater to Van Vleck, 27 July 1924, quoted in Duncan and Janssen 2007, p. 588. See also Slater 1967, p. 7. The details of Kramers' original reasoning, probably dating from November 1923, are not known, but Darrigol (1992, pp. 225–8) offers a reconstruction of it. See also Radder 1982a and Dresden 1987, pp. 150–9. In a detailed analysis of Kramers' papers, Konno (1993) suggests that the BKS theory was in fact instrumental in his derivation of the dispersion formula.

109. Kramers 1924b, p. 674. The first paper, submitted on 25 March, was Kramers 1924a.

110. Van Vleck 1924, p. 344.

111. Breit 1924.

112. Kramers 1924b, p. 311.

113. Ladenburg to Kramers, 2 April 1924, in Duncan and Janssen 2007, p. 589.

114. Van Vleck 1924, the first part of which is reproduced in van der Waerden 1967, pp. 203–22 (quotation on p. 219). On Van Vleck's contributions to radiation theory and their relation to Kramers' theory of dispersion, see Duncan and Janssen 2007 and Konno 1993.

115. Van Vleck to Kramers, 22 September 1924, in Konno 1993, p. 146.

116. Heisenberg to Landé, 6 July 1924, quoted in Cassidy 1992, p. 180. Hendry (1984, p. 46) calls Kramers's theory of 1924 'the Bohr–Kramers dispersion theory'.

117. Born 1924, pp. 386–7, with English translation in van der Waerden 1967, pp. 181–98.

118. Born 1924, p. 380.

119. Kramers 1924b. Dresden 1987, p. 222. Jammer 1966, p. 193, calls the translation formula 'Born's correspondence rule'.

120. Born 1925, pp. 113–14.

121. Kramers and Heisenberg 1925, with English translation in van der Waerden 1967, pp. 223–52. Smekal 1923. The Kramers–Heisenberg dispersion theory is discussed in Dresden 1987, pp. 215–22, and in Mehra and Rechenberg 1982b, pp. 170–90.

122. Kramers and Heisenberg 1925 (van der Waerden 1967, p. 234).

123. Bohr 1932, p. 369. The memorial lecture, presented in the Kamerlingh Onnes Laboratory in Leiden, is reprinted in Aaserud 2007, pp. 355–60 (quotation on p. 358).

124. Conversation with Jagdish Mehra, quoted in Mehra and Rechenberg 1982b, p. 189. See also Born's appreciation of Kramers' (and Heisenberg's) work on dispersion theory, according to whom 'It was the first step from the bright realm of classical mechanics into the still dark and unexplored underworld of the new quantum mechanics' (Born 1978, p. 216).

125. Raman 1928, who made his discovery while searching for an optical analogue of the Compton effect. For details about Raman and his work, see Singh 2002. Bohr was aware of the work done by Raman and his collaborators at the University of Calcutta. On Raman's instigation, in 1924–1925 his student Bidhubnusan B. Ray did postdoctoral work in X-ray spectroscopy in Copenhagen. See Raman to Bohr, 21 March 1923, and Bohr to Raman, 18 May 1923 (Archive for History of Quantum Physics, Bohr Scientific Correspondence).

126. Heisenberg 1969, p. 87.

127. Eckert and Märker 2004, p. 157.

128. Pauli to Sommerfeld, 6 December 1924, in Eckert and Märker 2004, p. 177.

129. Coster 1923. Kramers 1923b. The figures were reproduced from Kramers and Holst 1923. See Section 7.1.

130. Born 1923, p. 542.

131. Pauli to Bohr, 12 December 1924, and Pauli to Kramers, 27 July 1925, in Stolzenburg 1984, p. 427 and p. 443. Kamerlingh Onnes 1924 contained pictures of atomic structures provided by Kramers. For Kamerlingh Onnes's interest in atomic models, see Sauer 2007, p. 203. For Pauli's view on Kramers as a believer in electron orbits, see also Pauli to Bohr, 21 February 1924, as quoted in Section 8.2.

132. For two different evaluations of the crisis in the old quantum theory, see Seth 2007 and Hentschel 2009. According to a classical study by Paul Forman from 1971, the very possibility of the crisis in the old quantum theory depended on the particular *Zeitgeist* that permeated German culture in the early 1920s and predisposed physicists toward a symbolic and acausal theory of the atom (Forman 1971). In Kragh 1999, pp. 153–4, I have listed some arguments against the strong version of Forman's thesis in so far as it concerns the crisis in atomic theory in the years 1923–1925. As mentioned above, the reception of the BKS theory in Germany does not support the Forman thesis. On perspectives on Forman's thesis, see Carson et al. 2011.

133. Heisenberg to Pauli, 21 June 1925, in Hermann et al. 1979, p. 219. In 1929, looking back at the development, Heisenberg remarked that the quantitative understanding of the hydrogen atom 'appeared more and more as an accidental and incomprehensible result of the theory' (Heisenberg 1929, p. 491).

134. Bohr to Oseen, 29 January 1926, in Stolzenburg 1984, pp. 238–9.

135. Sommerfeld 1929, p. v, who referred to both Schrödinger's wave mechanics and Heisenberg's quantum mechanics. For Sommerfeld's relaxed attitude to the problems in atomic theory, see Sommerfeld 1924a. And for his emphasis on the need to base atomic theory on empirical regularities rather than models, Seth 2009.

136. For an early analysis of the decline of Bohr's atomic theory, see Margenau and Wightman 1944.

137. The list is based on Kragh 2002b.

138. Russell 1927 (originally published 1923), p. 88, in connection with Sommerfeld's relativistic theory of the hydrogen atom.

139. Kragh 1985a. Vickers 2011 provides a philosophical perspective on Sommerfeld's strikingly successful but nonetheless wrong theory.

140. Stern 1921, p. 252. English translation in *Zeitschrift für Physik D* **10** (1998), 114–16.

141. Gerlach and Stern 1922. For the experiments and their interpretations, see Mehra and Rechenberg 1982a, pp. 433–45, and Friedrich and Herschbach 1998 and 2003. Weinert 1994 analyzes the experiments from a philosophical perspective, suggesting that Stern and Gerlach 'had done the right experiment, but on the wrong theory'.

142. Bohr to Gerlach, 18 February 1922 (Archive for History of Quantum Physics, Bohr Scientific Correspondence). Gerlach to Bohr, 8 February 1922, reproduced in Friedrich and Herschbach 2003. For Bohr's view on the Stern–Gerlach effect, see Bohr 1924, p. 27.

143. Paschen to Gerlach 1922, quoted in Mehra and Rechenberg 1982a, p. 443. Karl Darrow concurred. Space quantization, he said in 1925, is 'the most spectacular of all the remarkable consequences of Bohr's interpretation of the stationary states, [and] also the only one which has ever been directly verified' (Darrow 1925–1926, p. 684).

144. Einstein and Ehrenfest 1922, p. 34.

145. Van Vleck 1926, p. 230.

146. Pauling 1927, p. 145. See also Van Vleck 1932, pp. 113–15.

147. Van Vleck 1929, p. 490.

148. Richardson 1927. Yet another list of shortcomings in the old quantum theory appeared in Lindsay and Margenau 1936 (pp. 392–3), who mentioned many-electron atoms, dispersion, anomalous Zeeman effect, half-quanta, and the Ramsauer effect.

149. Bohr to Born, 9 December 1924, in Stolzenburg 1984, p. 73.

150. Bohr to Høffding 22 September 1922. The letter is reproduced in full in Faye 1991, p. 108.

151. Born 1925, p. 114. The role of the observability principle in the birth of quantum mechanics is discussed in, for example, Mehra and Rechenberg 1982b, pp. 273–90 and Camilleri 2009, pp. 17–31.

152. Born and Jordan 1925, p. 493.

153. Heisenberg to Pauli, 24 June 1925, in Hermann et al. 1979, p. 227. Mehra and Rechenberg 1982b, Camilleri 2009, Darrigol 1992, and MacKinnon 1977 all agree that the observability criterion did not play the crucial role in Heisenberg's thinking that it is often ascribed and which Heisenberg himself expressed at some occasions.

154. Heisenberg to Pauli, 9 July 1925, in Hermann et al. 1979, p. 231.

155. Bridgman 1936, p. 65. Bridgman's sceptical attitude with regard to admitting quantum mechanics as an operationalist theory is discussed in Walter 1990, pp. 203–7.

156. Reiche and Thomas 1925, p. 512, as pointed out in Duncan and Janssen 2007. Sommerfeld 1922b. For the Sommerfeld–Heisenberg similarity, see Cassidy 1979 and Seth 2009.

157. Heisenberg 1925, p. 880, translated in van der Waerden 1967, pp. 261–76. There are several good analyses of Heisenberg's work, such as MacKinnon 1977, Darrigol 1992, pp. 263–76, and Mehra and Rechenberg 1982b, pp. 290–305. For those aspiring to understand the mathematical details of the paper, Aitchison et al. 2004 is a helpful source.

158. Darrigol 1997, p. 558.

159. Heisenberg 1967, p. 100.

160. Heisenberg to Bohr, 31 August 1925, in Stolzenburg 1984, p. 366.

161. Bohr 1925a. Manuscript on 'Atomic Theory and Mechanics', in Stolzenburg 1984, pp. 264–5.

9

Appendix: The Philosophers' Atom

Bohr's theory of atomic structure attracted almost no philosophical interest during its lifetime, although it did attract conceptual criticism of a kind that may perhaps be called semi-philosophical (Section 4.5). One of the very few philosophers who referred to the theory, Harold C. Brown at Stanford University found it 'revolutionary' that the very existence of electrons during quantum jumps was brought into question, 'for it deprives matter of its eternal existence, an essential attribute under the old conception'.[1] After the theory lost its scientific actuality with the emergence of the new quantum mechanics there seemed little reason to deal with it in a philosophical context. Only a few authors, including Karl Popper, Susan Stebbing, and Henry Margenau, found it worthwhile to comment on it. The situation changed with the historical turn in philosophy of science in the 1960s, and since then various aspects of the theory have been studied by a growing number of philosophers. Of course, compared with the massive interest in the philosophy of quantum mechanics, the philosophically oriented literature dealing with Bohr's old model is negligibly small. Nonetheless, it is a literature that historians of modern atomic theory should be acquainted with and make use of in their studies.

While some philosophers have examined the dynamical aspects of the theory or research programme—that is, followed it from its birth in 1913 to its death in 1925—others have focused on the original theory and offered critical reconstructions of Bohr's reasoning. How did he really proceed? Or perhaps: how ought he to have proceeded? To this latter category belong the insightful studies by Peter Achinstein and John Norton.[2] Others again have examined the Bohr atom in order to discuss and exemplify general problems of philosophy of science such as realism, explanations, domains, models, and consistency.

For example, although Bohr stressed that his theory of 1913 did not provide an *explanation* of why light is emitted from atoms (Section 2.4), he did account for—or explain—the lines of the hydrogen spectrum on the basis of his theory. What kind of explanation was it? How does it relate to other explanations in science? Such questions were asked by some contemporary physicists, and they have been reconsidered by a few modern philosophers.[3] Others have discussed whether the explanatory and predictive power of Bohr's theory resides in the model itself or in the laws of nature that were part of the model.

Some philosophers have looked at particular elements of Bohr's theory, mostly the correspondence principle, which has also served as inspiration for proposals of extending Bohr's principle into rules of a methodological and epistemological nature that relate new scientific theories to earlier theories. Such a proposal may first have been enunciated by Ernest Hutten, who in 1956 chose to understand the correspondence principle as the prescription that a new and better theory must contain the old one as a special case.[4] According to the version argued by Heinz Post and others, the 'general correspondence principle' states that any acceptable new theory should explain the well-confirmed part of its predecessor by producing the results of that theory under the conditions it has been confirmed.[5] The idea behind such proposals is of obvious philosophical (and also historiographical) interest, but it has little to do with what Bohr, Kramers, and Heisenberg understood by the correspondence principle in the days of the old quantum theory. Bohr himself restricted the principle to the case of the transition from classical theory to quantum theory. 'The correspondence principle,' he said in 1925, 'expresses the tendency to utilise in the systematic development of the quantum theory every feature of the classical theories in a rational transcription appropriate to the fundamental contrast between the postulates and the classical theories'.[6]

Another relevant issue of a general kind is the one of the *domain* of a theory; that is, the total body of information for which a theory is expected to account. Obviously, the adequacy or inadequacy of a theory can be judged only in relation to its domain—it makes no sense to test quantum theory by recording the exact orbit of Venus around the Sun. One kind of inadequacy is incompleteness, where a theory is able to account only for parts of its domain.[7] Was Bohr's theory incomplete? How incomplete can a theory be and still be judged acceptable? We have seen that Bohr's theory was indeed incomplete, among other reasons because its domain included the world of molecules, a world it largely failed to account for. But we have also seen that although some areas, such as molecules and superconductivity, belonged to the domain of the theory, its failure in these areas was not a matter of much concern. The domain of a theory is not a fixed entity, but one which is negotiable and subject to change. In principle the domain of Bohr's theory was all of atomic and molecular physics (and chemistry and astrophysics as well), but in reality it was much more restricted.

'In the early 1920s we knew that Bohr's theory of the atom could not possibly be quite correct', Heisenberg recalled in one of his last essays. 'But we guessed that it pointed in the right direction, and we hoped that we would be able some day to avoid the *inconsistencies* and to replace Bohr's theory by a more satisfactory picture.'[8] While Heisenberg's comment related to the later phase of the Bohr atom, where one can indeed find inconsistencies, most of the philosophical discussion has been concerned with the original theory as presented in 1913. Inconsistencies are bad, but there are different kinds of inconsistencies, and they are not all equally bad. It is common to distinguish between external and internal inconsistencies, where the first class concerns the theory in question and its relation to other well established theories relevant to it.

Thus, Bohr's theory was inconsistent *with* classical electrodynamics, which was considered a serious problem by many physicists in the early phase. The universal validity of electrodynamics had a nearly paradigmatic status, for which reason it was difficult to accept a theory which by fiat made it invalid inside the atom. However, this kind of inconsistency need not be very problematic, as it is intertheoretical and does not relate to the structure of the theory itself. It may even be argued that all new theories of a fundamental or revolutionary kind are by their very nature 'inconsistent' in this respect (Copernicus' heliocentric theory was in conflict with Aristotelian physics, the special theory of relativity with classical mechanics, etc.)

It is a more serious matter if a theory is internally or logically inconsistent, meaning that there are elements in it which contradict other of the elements of which it is made up. That Bohr's theory was inconsistent in this deep sense is what some philosophers have argued and a few physicists and chemists, including G. N. Lewis and J. Stark, argued in the period from 1913 to about 1920. For example, the notion of quantum jumps might seem to be inconsistent with the postulated restriction of electrons to stationary orbits. After all, if an electron moves from one stationary state to another, how can it avoid being in the 'forbidden zone' between the two states? Does the atom exist at all during the transition?

In an important essay of 1970 the Hungarian-British philosopher Imre Lakatos claimed that Maxwellian electrodynamics was an integral part of Bohr's theory, which for this reason was inconsistent. He described the theory as 'a research programme progressing on inconsistent foundations'.[9] According to electrodynamics, an orbiting electron must radiate energy, which contradicts Bohr's concept of stationary states, including the concept of a ground state. From a classical point of view there can be no such thing as the ground state of an atom of the Rutherford type. Bohr recognized of course the conflict between his theory and classical electrodynamics, but he did not acknowledge the latter as an essential part of his theory. Although he readily admitted the validity of 'the admirably coherent group of conceptions which have been rightly termed the classical theory of electrodynamics',[10] he nonetheless postulated that this admirable theory was invalid inside the atom, that is, he restricted its domain. Several modern philosophers have argued along similar lines and in detail examined the structure of Bohr's original theory, which they acquit from the charge of internal inconsistency.[11]

With regard to this issue, the exchange of ideas between Heisenberg and Thomas Kuhn in 1963 is of considerable interest. Kuhn suggested that while Bohr was always eager to get rid of inconsistencies there was a sort of 'higher inconsistency' which he was willing to accept. In response, Heisenberg pointed out that one should distinguish between 'paradox' and 'inconsistency', and that Kuhn's 'higher inconsistency' might belong to the first class. Bohr, he said, felt strongly that paradoxes would remain even in the final version of quantum theory:

> Because he was so much impressed by these paradoxes which were apparently unavoidable, he counted always on the possibility, 'Well, these paradoxes may even, in the long run, mean some kind of inconsistency which cannot be avoided'. On the other hand, in the course of the development, Bohr realized that an inconsistency is something still much worse than a paradox because inconsistency means that you just talk nonsense – that you do not know what you are talking about. A paradox may be very disagreeable but still you can make it work. An inconsistency can never be made to work.[12]

The model of atomic structure developed in the period of the old quantum theory involved several theoretical innovations, most notably the atomic theories of Rutherford, Bohr, and Sommerfeld. The relationship between these three theories can be seen as a case of increasing 'truth approximation', which leads to a corresponding increase in explanatory power and success. Whereas Bohr's theory can be characterized as a specialization of Rutherford's, Sommerfeld's atomic theory is better described as a concretization of Bohr's theory.[13]

Bohr's atomic theory, considered as a research programme evolving over time, was an important theme in the debate over the dynamics of science that started in the 1960s as a reaction against the ahistoric tendencies of positivism and Popperian critical realism. The three leading philosophers of this important turn—Kuhn, Lakatos, and Feyerabend—all dealt with the case, arguing that it fitted with their respective (and conflicting) views of the nature and development of science. Although their versions were different, they had in common that experiments were assigned a very limited role in the process that led from the early quantum atom to the new quantum mechanics. In neither of the versions does one get a sense of Bohr's intense preoccupation with experimental facts and his continual endeavours to relate his theory to experiments or extract testable predictions from it.

In his influential essay of 1970 Lakatos certainly downplayed the role of experiments in early quantum theory, stating that experimental tests 'may well be a waste of time'. He gave a fairly detailed account of the development of Bohr's research programme, which he saw as the unfolding of a kind of masterplan that Bohr had had from the very beginning. Having solved the apparent anomalies of the hydrogen spectrum, such as the Pickering–Fowler lines, Bohr 'thought of taking the possible spin of the electron into account', we are told. Of course, Bohr never actually did so, but this is Lakatos's 'rational reconstruction' of the history. While the Bohr atom was at first empirically progressive, eventually it entered a phase of degeneration: 'The programme lagged behind the discovery of "facts". Undigested anomalies swamped the field. With ever more sterile inconsistencies and ever more ad hoc hypotheses, the degenerating phase of the research programme set in.'[14]

As critics have pointed out, Lakatos's account of the later phase of Bohr's theory is unsatisfactory, selective, and highly incomplete.[15] According to Lakatos, the background problem was 'the riddle of how Rutherford atoms . . . can remain stable; for, according to the well-corroborated Maxwell–Lorentz theory of electromagnetism they should collapse'. As we have seen, this riddle played no role at all for Bohr in the creation of his theory. For another thing, Lakatos's claim that 'Wave mechanics soon caught up with, vanquished and replaced Bohr's programme', has no basis in historical fact either. But then it comes from a philosopher who admitted that his historical statements should 'be taken not with a grain, but with tons, of salt'.[16] It is a claim that Lakatos may have felt forced to make in order to maintain his thesis that a scientific theory can only be falsified in the presence of a better rival theory. In any case, wave mechanics never competed with the Bohr–Sommerfeld quantum theory. What happened was that the theory was replaced (not challenged) by the new Göttingen version of quantum mechanics, which was not a rival to it but rather grew out of what little was left of it—its ruins.

Moreover, Lakatos's claim that in this later phase Bohr's theory was empirically sterile is an exaggeration. It can be and has been argued that the strongly modified version of the Bohr model that existed by 1924 was in fact empirically progressive in so far as it led to predictions of novel facts, as did the BKS theory and Kramers' theory of dispersion. To Popper, the BKS theory—'an intellectual achievement [which] might perhaps even rank with Bohr's theory of the hydrogen atom of 1913'—was refuted by experimental facts.[17] Not believing in crucial experiments, Lakatos considered Popper's view an example of naïve falsificationism. As he saw it, Popper was wrong, for the experiments of Bothe, Geiger, Compton, and Simon were insufficient to refute the BKS theory, the fall of which was only brought about with the emergence of wave mechanics.[18] This is an interpretation that does not agree with the course of historical events.

Lakatos was not the only philosopher of his generation who misrepresented Bohr's motivations for introducing his atomic model of 1913. So did Popper, who in a later work repeated the mistake. There was a problem in atomic theory, he said, 'thought by Bohr to be *the* fundamental one: the problem of *atomic stability*, or of the stationary ground state of non-radiating electrons in the atom'.[19]

An early critic of Lakatos's account, Kuhn pointed out that Bohr did not take his starting point in the riddle of the radiative instability of Rutherford's atom, but in an 'entirely normal puzzle' connected with the absorption of charged particles in matter. In the perspective of Kuhn, Bohr's initial research programme was normal science. The case 'illustrates with particular clarity the revolutionary efficacy of normal research puzzles'. What for Lakatos was a degenerating phase of a research programme was for Kuhn 'a case book example of crisis' in excellent agreement with his general philosophy of scientific revolutions: 'History of science, to my knowledge, offers no equally clear, detailed, and cogent example of the creative functions of normal science and crisis.'[20] Kuhn did not speak of Bohr's theory as a paradigm, nor did he characterize the transition to quantum mechanics as a paradigm shift. And it was not, at least not in the strong sense originally proposed by Kuhn, where two different paradigms were considered to be incommensurable. Given that the creators of the new Göttingen mechanics had been actively involved in developing (and destroying) Bohr's atomic theory, it would be hard to argue that the two theories were either incommensurable or incommunicable.

The third of the trio of philosophers, Paul Feyerabend, did not agree with Kuhn's assertion that Bohr's atomic theory had its origin in puzzles of normal science. In a long paper primarily aimed at defending the later Bohr against Popper's charges of dogmatism, Feyerabend maintained that Bohr's works were from the very beginning 'non-normal', being guided throughout by philosophical principles.[21] The way he portrayed Bohr was as a philosopher, not a physicist, and how can a philosopher be engaged in something as mundane as normal science? Even when Bohr occupied himself with puzzle-solving research, it was from a philosophical perspective, namely in order to turn even minor problems into difficulties of a fundamental nature. The disagreement between Feyerabend and Kuhn seems to reflect different understandings of what constitutes normal science. At any rate, it is generally agreed that none of the general schemes of science development proposed by Kuhn, Lakatos, Feyerabend, or other philosophers adequately capture the complex history of Bohr's atomic theory and its transition into Heisenberg's quantum mechanics.

Notes for Chapter 9

1. Brown 1925, p. 203, based on an address given to the American Philosophical Association on 28 November 1924. Brown, who was 'fearful of getting beyond my depths in the intricacies of modern physics', relied on the popular exposition of Bohr's theory in Kramers and Holst 1923. Although written by a distinguished philosopher, Russell 1927 (first edition 1923) did not comment on the philosophical aspects of the Bohr atom.

2. Achinstein 1993. Norton 2000. See also Petruccioli 1993.

3. Achinstein 1983, pp. 119–22.

4. Hutten 1956, pp. 165–70.

5. Radder 1991. French and Kamminga 1993.

6. Bohr 1925a, p. 849.

7. Shapere 1977.

8. Heisenberg 1973, p. 6. Emphasis added.

9. Lakatos 1970, p. 140. According to Brown 1992, Bohr's characteristic combination of classical and non-classical principles was not only 'a logically risky game', but the principles were also 'inconsistent with each other' (p. 399).

10. Copenhagen lecture of 20 December, 1913. Bohr 1922a, p. 19.

11. Bartelborth 1989. Vickers 2008. Further references to the philosophical literature can be found in these sources. See also Norton 2000, p. 88.

12. Interview by T. S. Kuhn of 25 February 1963. American Institute of Physics, http://www.aip.org/history/ohilist/4661_8.html.

13. Hettema and Kuipers 1995.

14. Lakatos 1970, p. 146 and p. 154.

15. See Radder 1982b and Hettema 1995. A logical reconstruction of Bohr's atomic theory in the style of Lakatos, but much more detailed and much more abstract (and hence historically less relevant), is offered in Zoubek and Lauth 1992.

16. Lakatos 1970, p. 141, p. 154 and p. 140.

17. Popper 1963, pp. 242–3.

18. Lakatos 1970, pp. 168–9.

19. Popper 1982b, p. 47.

20. Kuhn 1970, pp. 256–9. The pioneering study of Heilbron and Kuhn (1969) did not comment on the relation to Kuhn's philosophy of science but kept strictly to the historical record. The same was the case with Kuhn's detailed historical analysis of the origin and early development of quantum theory (Kuhn 1978).

21. Feyerabend 1969.

BIBLIOGRAPHY

Aaserud, Finn (2001). 'I know how little I have deserved this . . . ', pp. 272–312 in Henry Nielsen and Keld Nielsen, eds, *Neighbouring Nobel: The History of Thirteen Danish Nobel Prizes*. Aarhus: Aarhus University Press.

Aaserud, Finn, ed. (2007). *Niels Bohr. Collected Works*, vol. 12. Amsterdam: Elsevier.

Abegg, Richard (1904). 'Die Valenz und das periodische System: Versuch einer Theorie der Molekülarverbindungen', *Zeitschrift für anorganische Chemie* **39**, 330–80.

Achinstein, Peter (1983). *The Nature of Explanation*. New York: Oxford University Press.

Achinstein, Peter (1993). 'How to defend a theory without testing it: Niels Bohr and the "logic of pursuit" ', *Midwest Studies in Philosophy* **18**, 90–120.

Adams, Edwin P. (1923). *The Quantum Theory*. Washington, D.C.: National Research Council.

Aitchison, Ian J. R., David A. MacManus, and Thomas M. Snyder (2004). 'Understanding Heisenberg's "magical" paper of July 1925: A new look at the calculational details', *American Journal of Physics* **72**, 1370–9.

Allen, H. Stanley (1915a). 'The magnetic field of an atom in relation to theories of spectral series', *Philosophical Magazine* **29**, 40–9.

Allen, H. Stanley (1915b). 'An atomic model with a magnetic core', *Philosophical Magazine* **29**, 714–24.

Allen, H. Stanley (1923). 'A static model of the hydrogen molecule', *Proceedings of the Royal Society of Edinburgh* **43**, 180–96.

Andrade, Edward N. da C. (1923). *The Structure of the Atom*. London: G. Bell and Sons.

Andrade, Edward N. da C. (1924). 'The structure of the atom,' pp. 43–55 in Edward F. Armstrong, ed., *Chemistry in the Twentieth Century*. London: Ernest Benn Ltd.

Andrade, Edward N. da C. (1958). 'The birth of the nuclear atom', *Proceedings of the Royal Society A* **244**, 437–55.

Angelino, Roberto (2010). 'On the cultural relationship between Niels Bohr and Harald Høffding', *Nuncius* **25**, 317–256.

Arabatzis, Theodore (2006). *Representing Electrons: A Biographical Approach to Theoretical Entities*. Chicago: University of Chicago Press.

Assmus, Alexi (1992a). 'The molecular tradition in early quantum theory', *Historical Studies in the Physical and Biological Sciences* **22**, 209–31.

Assmus, Alexi (1992b). 'The Americanization of molecular physics', *Historical Studies in the Physical and Biological Sciences* **23**, 1–34.

Aston, Francis W. (1913). 'A new elementary constituent of the atmosphere', Report, British Association for the Advancement of Science **83**, 403.

Badash, Lawrence (1969). *Rutherford and Boltwood: Letters on Radioactivity*. New Haven: Yale University Press.

Banet, Leo (1970). 'Balmer's manuscripts and the construction of his series', *American Journal of Physics* **38**, 821–8.

Barker, E. F. (1923). 'Molecular spectra and half-quanta', *Astrophysical Journal* **58**, 201–7.

Bartelborth, Thomas (1989). 'Kann es rational sein, eine inkonsistente Theorie zu akzeptieren?', *Philosophia Naturalis* **26**, 91–120.

Beller, Mara (1999). *Quantum Dialogue: The Making of a Revolution*. Chicago: University of Chicago Press.

Belloni, Lanfranco (1982). 'Pauli's 1924 note on hyperfine structure', *American Journal of Physics* **50**, 461–4.

Benz, Ulrich (1975). *Arnold Sommerfeld: Lehrer und Forscher and der Schwelle zum Atomzeitalter*. Stuttgart: Wissenschaftlicher Verlagsgesellschaft.

Bergia, Silvio and Luis Navarro (2000). 'On the early history of Einstein's quantization rule of 1917', *Archives Internationales d'Historie des Sciences* **50**, 321–73.

Birge, Raymond T. (1921). 'The Balmer series of hydrogen, and the quantum theory of line spectra', *Physical Review* **17**, 589–609.

Birge, Raymond T. (1956). 'Physics and physicists of the past fifty years', *Physics Today* **9** (September) 20–8.

Bjerrum, Niels (1914). 'Über die ultraroten Spektren der Gase. III. Die Konfiguration der Kohlendioxydmoleküls und die Gesetze der intramolekularen Kräfte', *Verhandlungen der Deutschen Physikalischen Gesellschaft* **16**, 737–53.

Bjerrum, Niels (1949). *Niels Bjerrum. Selected Papers*. Copenhagen: Munksgaard.

Bloch, Léon (1918). 'Quelques récents progrès de la physique (1914–1918)', *Revue Générale des Sciences Pures et Appliquées* **29**, 166–75, 198–208.

Bloch, Léon (1922).'La structure des atomes d'après les travaux récents de N. Bohr', *Journal de Physique et le Radium* **5**, 110–24.

Bohr, Niels (1913a). 'On the theory of the decrease of velocity of moving electrified particles on passing through matter', *Philosophical Magazine* **25**, 10–31.

Bohr, Niels (1913b). 'On the constitution of atoms and molecules, part I', *Philosophical Magazine* **26**, 1–25.

Bohr, Niels (1913c). 'On the constitution of atoms and molecules, part II', *Philosophical Magazine* **26**, 476–502.

Bohr, Niels (1913d). 'On the constitution of atoms and molecules, part III', *Philosophical Magazine* **26**, 857–75.

Bohr, Niels (1913e). 'The spectra of helium and hydrogen', *Nature* **92**, 231–2.

Bohr, Niels (1914a). 'On the effect of electric and magnetic fields on spectral lines', *Philosophical Magazine* **27**, 506–24.

Bohr, Niels (1914b). 'Atomic models and X-ray spectra', *Nature* **92**, 553–4.

Bohr, Niels (1915a). 'The spectra of hydrogen and helium', *Nature* **95**, 6–7.

Bohr, Niels (1915b). 'On the quantum theory of radiation and the structure of the atom', *Philosophical Magazine* **30**, 394–415.

Bohr, Niels (1915c). 'On the series spectrum of hydrogen and the structure of the atom', *Philosophical Magazine* **29**, 332–5.

Bohr, Niels (1918a). 'On the quantum theory of line-spectra, part I: On the general theory', *Kgl. Danske Videnskabernes Skrifter, Naturvidenskabelig og Mathematisk Afdeling*, 8. Række IV.1, 1–36.

Bohr, Niels (1918b). 'On the quantum theory of line-spectra, part II: On the hydrogen spectrum', *Kgl. Danske Videnskabernes Skrifter, Naturvidenskabelig og Mathematisk Afdeling*, 8. Række IV.1, 37–100.

Bohr, Niels (1919). 'On the model of a triatomic hydrogen molecule', *Meddelanden från K. Vetenskapsakademiens Nobelinstitut* 5, no. 28.

Bohr, Niels (1920). 'Über die Serienspektra der Elemente', *Zeitschrift für Physik* **2**, 423–69.

Bohr, Niels (1921a). 'Zur Frage der Polarization der Strahlung in der Quantentheorie', *Zeitschrift für Physik* **6**, 1–9.

Bohr, Niels (1921b). *Abhandlungen über den Atombau aus den Jahren 1913–1916*. Braunschweig: Vieweg & Sohn.

Bohr, Niels (1921c). 'Atomic structure', *Nature* **107**, 104–7.
Bohr, Niels (1921d). 'Atomic structure', *Nature* **108**, 208–9.
Bohr, Niels (1922a). *The Theory of Spectra and Atomic Constitution*, trans. A. D. Udden. Cambridge: Cambridge University Press.
Bohr, Niels (1922b). 'On the selection principle of the quantum theory', *Philosophical Magazine* **43**, 1112–16.
Bohr, Niels (1922c). 'On the quantum theory of line-spectra, part III: On the spectra of elements of higher atomic number', *Kgl. Danske Videnskabernes Skrifter, Naturvidenskabelig og Mathematisk Afdeling*, 8. Række IV.1, 101–8.
Bohr, Niels (1922d). 'Der Bau der Atome und die physikalischen und chemischen Eigenschaften der Elemente', *Zeitschrift für Physik* **9**, 1–67.
Bohr, Niels (1922e). *Drei Aufsätze über Spektren und Atombau*. Braunschweig: Vieweg & Sohn.
Bohr, Niels (1923a). 'Über die Anwendung der Quantentheorie auf den Atombau. I: Die Grundpostulate der Quantentheorie', *Zeitschrift für Physik* **13**, 117–65.
Bohr, Niels (1923b). 'The structure of the atom', *Nature* **112**, 29–44.
Bohr, Niels (1923c). 'Linienspektren und Atombau', *Annalen der Physik* **71**, 228–88.
Bohr, Niels (1923d). 'The effect of electric and magnetic fields on spectral lines', *Proceedings of the Physical Society* **25**, 275–302.
Bohr, Niels (1924). 'On the application of the quantum theory to atomic structure. Part I: The fundamental postulates', *Proceedings of the Cambridge Philosophical Society* (Supplement) **22**, 1–44.
Bohr, Niels (1925a). 'Atomic theory and mechanics', *Nature* **116**, 845–52.
Bohr, Niels (1925b). 'Über die Wirkung von Atomen bei Stössen', *Zeitschrift für Physik* **34**, 142–57.
Bohr, Niels (1928). 'The quantum postulate and the recent development of atomic theory', pp. 565–88 in *Atti del Congresso Internazionale dei Fisici*, vol. 2. Bologna: Zanichelli.
Bohr, Niels (1932). 'Chemistry and the quantum theory of atomic constitution', *Journal of the Chemical Society*, **131**, 349–84.
Bohr, Niels (1936). 'Conservation laws in quantum theory', *Nature* **138**, 215–16.
Bohr, Niels (1955). 'Rydberg's discovery of the spectral laws', *Lunds Universitets Årsskrift*, Avd. 2, **50**, no. 21, 15–21.
Bohr, Niels (1961). 'Reminiscences of the founder of nuclear science and of some developments based on his work', *Proceedings of the Physical Society* **78**, 1083–115.
Bohr, Niels and Dirk Coster (1923). 'Röntgenspektren und periodisches System der Elemente', *Zeitschrift für Physik* **12**, 342–74.
Bohr, Niels, Hendrik A. Kramers, and John C. Slater (1924). 'The quantum theory of radiation', *Philosophical Magazine* **47**, 785–802.
Bokulich, Alisa (2010). 'Bohr's correspondence principle', *Stanford Encyclopedia of Philosophy* (http://plato.stanford.edu/entries/bohr-correspondence/).
Born, Max (1909). 'Über das Thomsonsche Atommodell', *Physikalische Zeitschrift* **10**, 1031–4.
Born, Max (1920). 'Die Brücke zwischen Chemie und Physik', *Naturwissenschaften* **8**, 373–82.
Born, Max (1922a). 'Über das Modell der Wasserstoffmolekel', *Naturwissenschaften* **10**, 677–8.
Born, Max (1922b). *Der Aufbau der Materie: Drei Aufsätze über Moderne Atomistik und Elektronentheorie*. Berlin: Springer.
Born, Max (1923). 'Quantentheorie und Störungsrechnung', *Naturwissenschaften* **11**, 537–42.
Born, Max (1924). 'Über Quantenmechanik', *Zeitschrift für Physik* **26**, 379–95.
Born, Max (1925). *Vorlesungen über Atommechanik*. Berlin: Julius Springer.
Born, Max (1971). *The Born-Einstein Letters*. London: Macmillan.
Born, Max (1978). *My Life: Recollections of a Nobel Laureate*. New York: Charles Scribner's Sons.
Born, Max and Pascual Jordan (1925). 'Zur Quantentheorie aperiodischer Vorgänge', *Zeitschrift für Physik* **33**, 479–505.

Born, Max and Werner Heisenberg (1923). 'Die Elektronenbahnen im angeregten Heliumatom', *Zeitschrift für Physik* **16**, 229–43.

Born, Max and Werner Heisenberg (1924). 'Zur Quantentheorie der Molekeln', *Annalen der Physik* **74**, 1–31.

Born, Max, Werner Heisenberg, and Pascual Jordan (1926). 'Zur Quantenmechanik II', *Zeitschrift für Physik* **35**, 557–615.

Born, Max and Wolfgang Pauli (1922). 'Über die Quantelung gestörter mechanischer Systeme', *Zeitschrift für Physik* **10**, 137–58.

Borrelli, Arianna (2009). 'The emergence of selection rules and their encounter with group theory, 1913–1927', *Studies in History and Philosophy of Modern Physics* **40**, 327–37.

Bothe, Walther and Hans Geiger (1925). 'Experimentelles zur Theorie von Bohr, Kramers und Slater', *Naturwissenschaften* **13**, 440–1.

Bourget, Henri, Charles Fabry, and Henri Buisson (1914). 'Sur le poids atomique du nébulium et la température de la nébuleuse d'Orion', *Comptes Rendus* **158**, 1017–19.

Brackett, Frederick S. (1922). 'Visible and infra-red radiation of hydrogen', *Astrophysical Journal* **56**, 154–61.

Brand, John C. D. (1995). *Lines of Light: The Sources of Dispersive Spectroscopy, 1800–1930*. New York: Gordon and Breach.

Breit, Gregory (1924). 'The quantum theory of dispersion', *Nature* **114**, 310.

Brickwedde, Ferdinand G. (1982). 'Harold Urey and the discovery of deuterium', *Physics Today* **35** (September), 34–9.

Bridgman, Percy W. (1936). *The Nature of Physical Theory*. New York: Dover Publications.

Brillouin, Léon (1922). *La Théorie des Quanta et l'Atome de Bohr*. Paris: Librairie Scientifique Blanchard.

Brillouin, Léon (1960). *Wave Propagation and Group Velocities*. New York: Academic Press.

Brillouin, Marcel (1919). 'Actions mécaniques à hérédité discontinue par propagation; essai de théorie dynamique de l'atome à quanta', *Comptes Rendus* **168**, 1318–20.

Brock, William H. (1985). *From Protyle to Proton: William Prout and the Nature of Matter 1785–1985*. Bristol: Adam Hilger.

Brown, Bryson (1992). 'Old quantum theory: A paraconsistent approach', *Philosophy of Science Association* **2**, 397–411.

Brown, Harold C. (1925). 'The material world—snark or boojum?' *Journal of Philosophy* **12**, 197–214.

Bruzzaniti, G. and Nadia Robotti (1989). 'The affirmation of the concept of isotopy and the birth of mass spectrography', *Archives Internationales d'Histoire des Sciences* **39**, 309–34.

Buchner, Ernst H. (1915). 'The analogy between radicles and elements', *Nature* **95**, 701.

Buchwald, Diana K., Robert Schulmann, József Illy, Daniel J. Kennefick, and Tilman Sauer, eds (2004). *Collected Papers of Albert Einstein*, vol. 9. Princeton: Princeton University Press.

Buchwald, Diana K., Tilman Sauer, Ze'ev Rosenkranz, József Illy, and Virginia I. Holmes, eds (2006). *Collected Papers of Albert Einstein*, vol. 10. Princeton: Princeton University Press.

Buchwald, Diana K., Ze'ev Rosenkranz, Tilman Sauer, József Illy, and Virginia I. Holmes, eds (2009). *Collected Papers of Albert Einstein*, vol. 12. Princeton: Princeton University Press.

Buchwald, Eberhard (1923). *Das Korrespondenzprinzip*. Braunschweig: Vieweg & Sohn.

Buchwald, Jed Z. (1985). *From Maxwell to Microphysics: Aspects of Electromagnetic Theory in the Last Quarter of the Nineteenth Century*. Chicago: University of Chicago Press.

Bury, Charles R. (1921). 'Langmuir's theory of the arrangement of electrons in atoms and molecules', *Journal of the American Chemical Society* **43**, 1602–9.

Camilleri, Kristian (2009). *Heisenberg and the Interpretation of Quantum Mechanics: The Physicist as Philosopher*. Cambridge: Cambridge University Press.

Campbell, Norman R. (1913). *Modern Electrical Theory*. Cambridge: Cambridge University Press.

Campbell, Norman R. (1914). 'The structure of the atom', *Nature* **92**, 586–7.

Campbell, Norman R. (1920). 'Atomic structure', *Nature* **106**, 408–9.

Campbell, Norman R. (1923a). 'A static or dynamic atom?', *Nature* **111**, 659.

Campbell, Norman R. (1923b). *The Structure of the Atom*. Cambridge: Cambridge University Press.

Cantor, Geoffrey (1994). 'The making of a british theoretical physicist—E. C. Stoner's early career', *British Journal for the History of Science* **27**, 277–90.

Carazza, Bruno and Helge Kragh (1995). 'Heisenberg's lattice world: The 1930 theory sketch', *American Journal of Physics* **63**, 595–605.

Carazza, Bruno and Nadia Robotti (2002). 'Explaining atomic spectra within classical physics: 1897–1913', *Annals of Science* **59**, 299–320.

Caroe, G. M. (1978). *William Henry Bragg 1862–1942: Man and Scientist*. Cambridge: Cambridge University Press.

Carson, Cathryn, Alexei Kojevnikov, and Helmuth Trischler, eds (2011). *Weimar Culture and Quantum Mechanics: Selected Papers by Paul Forman and Contemporary Perspectives on the Forman Thesis*. New York: World Scientific Publishing Company.

Cassidy, David C. (1978). 'Heisenberg's first paper', *Physics Today* **31** (July), 23–8.

Cassidy, David C. (1979). 'Heisenberg's first core model of the atom: The formation of a professional style', *Historical Studies in the Physical Sciences* **10**, 187–224.

Cassidy, David C. (1992). *Uncertainty: The Life and Science of Werner Heisenberg*. New York: W. H. Freeman & Company.

Christiansen, Christian (1910). *Lærebog i Fysik*. Copenhagen: Gyldendal.

Christiansen, Christian and Johannes J. C. Müller (1921). *Elemente der Theoretischen Physik*. Leipzig: Ambrosius Barth.

Cole, Alfred D. (1915). 'Recent evidence for the existence of the nucleus atom', *Science* **41**, 73–81.

Compton, Arthur H. (1917). 'The intensity of X-ray reflection, and the distribution of the electrons in atoms', *Physical Review* **9**, 29–57.

Compton, Arthur H. (1923). 'A quantum theory of the scattering of X-rays by light elements', *Physical Review* **21**, 483–502.

Compton, Arthur H. and Alfred W. Simon (1925). 'Directed quanta of scattered X-rays', *Physical Review* **26**, 289–99.

Conn, G. K. T. and H. D. Turner (1965). *The Evolution of the Nuclear Atom*. London: Iliffe Books.

Conway, Arthur W. (1913). 'An electromagnetic hypothesis as to the origin of series spectra', *Philosophical Magazine* **26**, 1010–17.

Coster, Dirk (1922). 'On the spectra of X-rays and the theory of atomic structure', *Philosophical Magazine* **43**, 1070–107; **44**, 546–73.

Coster, Dirk (1923). 'Röntgenspektren und Bohrsche Atomtheorie', *Naturwissenschaften* **11**, 567–77.

Coster, Dirk and George Hevesy (1923). 'On the missing element of atomic number 72', *Nature* **111**, 79.

Courant, Richard (1981). 'Reminiscences from Hilbert's Göttingen', *Mathematical Intelligencer* **3**, 154–64.

Crehore, Albert C. (1913). 'On the formation of the molecules of the elements and their compounds, with atoms as constituted on the corpuscular-ring theory', *Philosophical Magazine* **26**, 25–84.

Crehore, Albert C. (1915). 'The gyroscopic theory of atoms and molecules', *Philosophical Magazine* **29**, 310–32.

Crehore, Albert C. (1921). 'An atomic model based on electromagnetic theory,' *Philosophical Magazine* **42**, 569–93.

Crookes, William (1886). 'Presidential address to Section B', Report, British Association for the Advancement of Science, 558–76.

Curtis, William F. (1914). 'Wavelengths of hydrogen lines and determination of the series constant', *Proceedings of the Royal Society* A **90**, 605–20.

Curtis, William F. (1923). 'Half quanta', *Nature* **109**, 713.

Cuthbertson, Clive and Maude Cuthbertson (1910). 'On the refraction and dispersion of air, oxygen, nitrogen, and hydrogen, and their relations', *Proceedings of the Royal Society* A **83**, 151–71.

Dahl, Per F. (1972). *Ludvig Colding and the Conservation of Energy Principle*. New York: Johnson Reprint Corporation.

Dahl, Per F. (1992). *Superconductivity: Its Historical Roots and Development from Mercury to the Ceramic Oxides*. New York: AIP Press.

Darrigol, Olivier (1992). *From c-Numbers to q-Numbers: The Classical Analogy in the History of Quantum Theory*. Berkeley: University of California Press.

Darrigol, Olivier (1993). 'Strangeness and soundness in Louis de Broglie's early works', *Physis* **30**, 303–72.

Darrigol, Olivier (1997). 'Classical concepts in Bohr's atomic theory (1913–1925)', *Physis* **34**, 545–67.

Darrow, Karl K. (1925–1926). 'Some contemporary advances in physics. The atom model', *Bell System Technical Journal* **4**, 407–58, 642–84; **5**, 96–142.

Darwin, Charles G. (1912). 'A theory of the absorption and scattering of the α rays', *Philosophical Magazine* **23**, 901–20.

Darwin, Charles G. (1922). 'A quantum theory of optical dispersion', *Nature* **110**, 841–2.

Darwin, Charles G. (1923). 'The wave theory and the quantum theory', *Nature* **111**, 771–3.

Darwin, Charles G. (1955). 'The discovery of atomic number', pp. 1–11 in Wolfgang Pauli, ed., *Niels Bohr and the Development of Physics*. London: Pergamon Press.

Dauvillier, Alexandre (1922). 'Analyse de la structure électronique des éléments', *Journal de Physique et le Radium* **3**, 154–77, 221–51.

Dauvillier, Alexandre (1924). 'Sur la distribution des électrons entre les niveaux L des éléments', *Comptes Rendus* **178**, 476–9.

Dauvillier, Alexandre and Louis de Broglie (1924). 'Le systéme spectral des rayons Röntgen et structure de l'atome', *Journal de Physique et le Radium* **6**, 1–19.

Davies, Mansel (1986). 'Charles Rugeley Bury and his contributions to physical chemistry', *Archive for History of Exact Sciences* **36**, 76–90.

Davis, Bergen and F. S. Goucher (1917). 'Ionization and excitation of radiation by electron impact in mercury vapor and hydrogen', *Physical Review* **10**, 101–15.

Davis, E. A. and Isobel J. Falconer (1997). *J. J. Thomson and the Discovery of the Electron*. London: Taylor & Francis.

Davisson, Clinton J. (1916). 'The dispersion of hydrogen and helium on Bohr's theory', *Physical Review* **8**, 20–7.

Davisson, Clinton J. and Charles H. Kunsman (1923). 'The scattering of low speed electrons by platinum and magnesium', *Physical Review* **22**, 242–58.

Dawson, H. M. (1913). [Abstract of Bohr 1913d], *Journal of the Chemical Society, Abstracts* **104**, ii 1045–6.

De Broglie, Louis (1924). *Recherches sur la Théorie des Quanta*. Paris: Masson et Cie.

De Broglie, Louis (1938). 'Le principe de correspondance et les interactions entre la matière et le rayonnement', *Actualités Scientifiques et Industrielles*, no. 704. Paris: Hermann et Cie.

De Broglie, Louis and Alexandre Dauvillier (1925). 'Semi-optical lines in the X-ray spectra', *Philosophical Magazine* **49**, 752–3.

Debye, Peter (1915a). 'Die Konstitution des Wasserstoff-Moleküls', *Sitzungsberichte der Bayerischen Akademie der Wissenschaften München*, 1–26.

Debye, Peter (1915b). 'Zerstreuung von Röntgenstrahlen', *Annalen der Physik* **46**, 809–23.

Debye, Peter (1916). 'Quantenhypothese und Zeeman-Effekte', *Physikalische Zeitschrift* **17**, 507–12.

Debye, Peter (1917). 'Der erste Elektronenring der Atome', *Physikalische Zeitschrift* **18**, 276–84.

Debye, Peter (1923). 'Zerstreuung von Röntgenstrahlen und quantentheorie', *Physikalische Zeitschrift* **24**, 161–6.

Debye, Peter and Paul Scherrer (1916). 'Interferenzen an regellos orientierten Teilchen im Röntgenlicht, I', *Physikalische Zeitschrift* **17**, 277–83.

DeVorkin, David H. and Ralph Kenat (1983a). 'Quantum physics and the stars (I): The establishment of a stellar temperature scale', *Journal for the History of Astronomy* **14**, 102–32.

DeVorkin, David H. and Ralph Kenat (1983b). 'Quantum physics and the stars (II): Henry Norris Russell and the abundances of the elements in the atmospheres of the sun and stars', *Journal for the History of Astronomy* **14**, 180–222.

Dirac, Paul A. M. (1925). 'The fundamental equations of quantum mechanics', *Proceedings of the Royal Society* A **110**, 561–79.

Dresden, Max (1987). *H. A. Kramers: Between Tradition and Revolution*. Berlin: Springer.

Duane, William and Gerald Wendt (1917). 'A reactive modification of hydrogen produced by alpha radiation', *Physical Review* **10**, 116–28.

Duhem, Pierre (1913). *Notices sur les Titres et Travaux Scientifique de Pierre Duhem*. Bordeaux: Gounouilhou.

Duncan, Anthony and Michel Janssen (2007). 'On the verge of *Umdeutung* in Minnesota: Van Vleck and the correspondence principle', *Archive for History of Exact Sciences* **61**, 553–671.

Eckert, Michael and Karl Märker, eds (2000). *Arnold Sommerfeld. Wissenschaftlicher Briefwechsel*, vol. 1. Berlin: Verlag für Geschichte der Naturwissenschaften und der Technik.

Eckert, Michael and Karl Märker, eds (2004). *Arnold Sommerfeld. Wissenschaftlicher Briefwechsel*, vol. 2. Berlin: Verlag für Geschichte der Naturwissenschaften und der Technik.

Ehrenfest, Paul (1913). 'Bemerkung betreffs der spezifischen Wärme zweiatomiger Gase', *Verhandlungen der Deutschen Physikalischen Gesellschaft* **15**, 451–7.

Ehrenfest, Paul (1917). 'Adiabatic invariants and the theory of quanta', *Philosophical Magazine* **33**, 500–13.

Ehrenfest, Paul (1923). 'Adiabatische Transformationen in der Quantentheorie und ihre Behandlung durch Niels Bohr', *Naturwissenschaften* **11**, 543–50.

Einstein, Albert (1916). 'Strahlungs-Emission und –Absorption nach der Quantentheorie', *Verhandlungen der Deutschen Physikalischen Gesellschaft* **18**, 318–23.

Einstein, Albert (1917a). 'Zum Quantensatz von Sommerfeld und Epstein', *Verhandlungen der Deutschen Physikalischen Gesellschaft* **19**, 82–92.

Einstein, Albert (1917b). 'Quantentheorie der Strahlung', *Physikalische Zeitschrift* **18**, 121–8.

Einstein, Albert (1922). 'Theoretical remark on the superconductivity of metals'. http://www.arxiv.org/physics/0510251.

Einstein, Albert (1923). 'Bietet die feldtheorie Möglichkeiten für die Lösung der Quantenproblems?', *Sitzungsberichte der Preussischen Akademie der Wissenschaften Berlin, Math.-Phys. Klasse*, 359–65.

Einstein, Albert (1949). 'Autobiographical notes', pp. 1–95 in Paul A. Schilpp, ed., *Albert Einstein: Philosopher-Scientist*. New York: Library of Living Philosophers.

Einstein, Albert and Paul Ehrenfest (1922). 'Quantentheoretische Bemerkungen zum Experiment von Stern und Gerlach', *Zeitschrift für Physik* **11**, 31–4.

Elsasser, Walter (1925). 'Bemerkungen zur Quantenmechanik freier Elektronen', *Naturwissenschaften* **13**, 711.

El'yashevich, M. A. (1970). 'The Mendeleev periodic law, atomic spectra, and atomic structure', *Soviet Physics Uspekhi* **12**, 1–23.

El'yashevich, M. A. (1986). 'Niels Bohr's development of the quantum theory of the atom and the correspondence principle (his 1912–1923 work in atomic physics and its significance)', *Soviet Physics Uspekhi* **28**, 879–909.

Enz, Charles P. (2004). *No Time to Be Brief: A Scientific Biography of Wolfgang Pauli*. Oxford: Oxford University Press.

Epstein, Paul S. (1916a). 'Zur Theorie des Starkeffektes', *Annalen der Physik* **50**, 489–521.

Epstein, Paul S. (1916b). 'Zur Theorie des Starkeffekte', *Physikalische Zeitschrift* **17**, 148–50.

Epstein, Paul S. (1918). 'Anwendungen der Quantenlehre in der Theorie der Serienspektren', *Naturwissenschaften* **17**, 230–53.

Evans, Evan J. (1913). 'The spectra of helium and hydrogen', *Nature* **92**, 5.

Evans, Evan J. (1915). 'The spectra of helium and hydrogen', *Philosophical Magazine* **29**, 284–97.

Eve, Arthur S. (1914). 'Modern views on the constitution of the atom', *Science* **40**, 115–21.

Eve, Arthur S. (1939). *Rutherford. Being the Life and Letters of the Rt Hon. Lord Rutherford, O.M.* Cambridge: Cambridge University Press.

Ewald, Paul P. (1913). 'Bericht über die Tagung der British Association in Birmingham', *Physikalische Zeitschrift* **14**, 1298–307.

Fadner, W. L. (1985). 'Theoretical support for the generalized correspondence principle', *American Journal of Physics* **53**, 829–38.

Fajans, Kasimir (1913). 'Die radioaktiven Umwandlungen und die Valenzfrage vom Standpunkte der Struktur der Atome', *Verhandlungen der Deutschen Physikalischen Gesellschaft* **15**, 240–59.

Falconer, Isobel (1988). 'J. J. Thomson's work on positive rays, 1906–1914', *Historical Studies in the Physical and Biological Sciences* **18**, 265–310.

Falkenhagen, Hans (1924). 'Zum Paschen-Back Effekt des H-Atoms', *Zeitschrift für Physik* **28**, 1–10.

Farrar, W. V. (1965). 'Nineteenth-century speculations on the complexity of the chemical elements', *British Journal for the History of Science* **2**, 297–323.

Farrar, W. V. (1969). 'Richard Laming and the coal-gas industry, with his views on the structure of matter', *Annals of Science* **25**, 243–53.

Favrholdt, David (1992). *Niels Bohr's Philosophical Background*. Royal Danish Academy of Sciences and Letters, Historisk-Filosofiske Meddelelser **63**. Copenhagen: Munksgaard.

Favrholdt, David, ed. (1999). *Niels Bohr. Collected Works*, vol. 10. Amsterdam: North-Holland.

Faye, Jan (1991). *Niels Bohr: His Heritage and Legacy*. Dordrecht: Kluwer Academic.

Fechner, Gustav T. (1828). 'Über die Anwendung der Gravitationsgesetzes auf die Atomlehre', *Kastners Archiv für die Gesamte Naturlehre* **15**, 257–91.

Fechner, Gustav T. (1864). *Über die Physikalische und Philosophische Atomenlehre*. Leipzig: Hermann Mendelsohn.

Fedak, William A. and Jeffrey J. Prentis (2002). 'Quantum jumps and classical harmonics', *American Journal of Physics* **70**, 332–44.

Feuer, Lewis S. (1974). *Einstein and the Generations of Science*. New York: Basic Books.

Feyerabend, Paul K. (1969). 'On a recent critique of complementarity, part II', *Philosophy of Science* **36**, 82–105.

Föppl, Ludwig (1912). 'Stabile Anordnungen von Elektronen im Atom', *Journal für Reine und Angewandte Mathematik* **141**, 251–302.

Föppl, Ludwig (1915). 'Über die Stabilität des Bohrschen Atommodelles', *Physikalische Zeitschrift* **15**, 707–12.

Foote, Paul D., William F. Meggers, and Fred L. Mohler (1922). 'A significant exception to the principle of selection', *Philosophical Magazine* **43**, 657–61.

Forman, Paul (1968). 'The doublet riddle and atomic physics circa 1924', *Isis* **59**, 156–74.

Forman, Paul (1970). 'Alfred Landé and the anomalous Zeeman effect, 1919–1921', *Historical Studies in the Physical Sciences* **2**, 153–261.

Forman, Paul (1971). 'Weimar culture, causality, and quantum theory, 1918–1927: Adaption of German physicists and mathematicians to a hostile intellectual environment', *Historical Studies in the Physical Sciences* **3**, 1–115.

Foster, J. Stuart (1924). 'Observation of the Stark effect in hydrogen and helium', *Physical Review* **23**, 667–84.

Fowler, Alfred (1912). 'Observations of the principal and other series in the spectrum of hydrogen', *Monthly Notices of the Royal Astronomical Society* **73**, 62–71.

Fowler, Alfred (1913). 'The spectra of helium and hydrogen', *Nature* **92**, 95–6.

Fowler, Alfred (1914). 'Series lines in spark spectra', *Philosophical Transactions of the Royal Society* A **214**, 225–66.

Fowler, Alfred (1924). 'The series spectrum of ionized carbon (CII)', *Proceedings of the Royal Society* A **105**, 299–310.

Fowler, Alfred and John W. Nicholson (1914). 'Series lines in spectra', *Monthly Notices of the Royal Astronomical Society* **74**, 354–61.

Fowler, Ralph H. (1923). 'Bohr's atom in relation to the problem of covalency', *Transactions of the Faraday Society* **19**, 459–68.

Franck, James and Gustav Hertz (1913). 'Messungen der Ionisierungsspannung in verschiedenen Gasen', *Berichte der Deutschen Physikalischen Gesellschaft* **15**, 34–44.

Franck, James and Gustav Hertz (1914). 'Über die Erregung der Quecksilberresonanzlinie 253,6 $\mu\mu$ durch Elektronenstösse', *Berichte der Deutschen Physikalischen Gesellschaft* **16**, 512–17.

Franck, James and Gustav Hertz (1916). 'Über Kinetik von Elektronen und Ionen in Gasen', *Physikalische Zeitschrift* **17**, 409–16, 430–40.

Franck, James and Gustav Hertz (1919). 'Die Bestätigung der Bohrschen Atomtheorie im opticsher Spektrum durch Untersuchungen der unelastichen Zusammenstösse langsamer Elektronen mit Gasmolekülen', *Physikalische Zeitschrift* **20**, 132–43.

Franck, James, Paul Knipping, and Thea Krüger (1919). 'Über einen Zusammenhang zwischen Stossionisation und der Dissoziationsarbeit neutraler Moleküle', *Verhandlungen der Deutschen Physikalischen Gesellschaft* **21**, 728–32.

Franck, James and Fritz Reiche (1920). 'Über Helium und Parhelium', *Zeitschrift für Physik* **1**, 154–61.

French, S. and H. Kamminga, eds (1993). *Correspondence, Invariance and Heuristics: Essays in Honour of Heinz Post*. Dordrecht: Kluwer Academic.

Friedrich, Bretislav and Dudley Herschbach (1998). 'Space quantization: Otto Stern's lucky star', *Dædalus* **127**, 165–91.

Friedrich, Bretislav and Dudley Herschbach (2003). 'Stern and Gerlach: How a bad cigar helped reorient atomic physics', *Physics Today* **56** (December), 53–9.

Fujisaki, Chiyoko (1982). 'P. Drude's theory of dispersion of light and atom model (1900–1913)', *Historia Scientiarum* no. 22, 19–68.

Fujisaki, Chiyoko (1983). 'From Deslandres to Kratzer: Development of the understanding of the origin of infrared band spectra', *Historia Scientiarum* no. 24, 53–86; no. 25, 57–86.

Fujisaki, Chiyoko (1991). 'The role of T. Heurlinger in the development of modern visible molecular spectroscopy', *Historia Scientiarum* **1**, 132–61.

Fulcher, Gordon S. (1915). 'The Stark effect and atomic structure: A critique of certain recent work', *Astrophysical Journal* **41**, 359–72.

Gans, Richard and A. P. Miguez (1917). 'Über die Brechbarkeit von Licht sehr geringer Intensität', *Annalen der Physik* **52**, 291–307.

Garbasso, Antonio (1914). 'Theoretisches über das elektrische Analogon des Zeemanphänomens', *Physikalische Zeitschrift* **15**, 123.

Gavroglu, Kostas and Ana Simões (2012). *Neither Physics nor Chemistry: A History of Quantum Chemistry*. Cambridge, Mass.: MIT Press.

Gearhart, Clayton A. (2010). '"Astonishing successes" and "bitter disappointment": The specific heat of hydrogen in quantum theory', *Archive for History of Exact Sciences* **64**, 113–202.

Gehrcke, Ernst (1914a). 'Über ein Modell zur Erklärung der Lichtemission', *Physikalische Zeitschrift* **15**, 123–6, 198–201, 344–8, 838–9.

Gehrcke, Ernst (1914b). 'Zur Theorie der elektrischen Aufspaltung der Wasserstoffserie', *Verhandlungen der Deutschen Physikalischen Gesellschaft* **16**, 431–5.

Gehrcke, Ernst (1920). 'Über ein Modell zur Erklärung der Lichtemission, V', *Physikalische Zeitschrift* **21**, 172–5.

Gehrcke, Ernst (1923). 'Die Spektren des Wasserstoffs und die neuere Atomtheorie', *Zeitschrift für Technische Physik* **5**, 194–9.

Geiger, Hans and Ernest Marsden (1913). 'The laws of deflexion of α particles through large angles', *Philosophical Magazine* **25**, 604–23.

Gerlach, Walther and Otto Stern (1922). 'Der experimentelle Nachweis der Richtungsquantelung', *Zeitschrift für Physik* **9**, 349–52.

Goldschmidt, Robert B., Maurice de Broglie, and Frederick A. Lindemann, eds (1921). *La Structure de la Matière. Rapports et Discussions du Conseil de Physique Tenu à Bruxelles du 27 au 31 Octobre 1913 Sous les Auspices de l'Institut International de Physique Solvay*. Paris: Solvay Institute.

Goudsmit, Samuel and George E. Uhlenbeck (1925). 'Opmerking over de spectra van waterstof en helium', *Physica* **5**, 266–70.

Graetz, Leo (1918). *Die Atomtheorie in ihrer Neuesten Entwickelung*. Stuttgart: J. Engelhorns Nachf.

Grandin, Karl (1999). *En Slags Modernism i Vetenskapen: Teoretisk Fysik i Sverige under 1920-Talet*. Uppsala: Uppsala University.

Grassmann, Robert (1862). *Körperlehre oder Atomistik*. Stettin: Stettin Buchhandlung.

Greenspan, Nancy T. (2005). *The End of the Certain World: The Life and Science of Max Born*. New York: Basic Books.

Grotrian, Walter (1920). 'Gestezmässigkeiten im Spektrum des Neons', *Physikalische Zeitschrift* **21**, 638–43.

Grotrian, Walter (1928). *Graphische Darstellung der Spektren von Atomen und Ionen mit Ein, Zwei und Drei Valenzelektronen*, 2 vols. Berlin: Springer.

Haas, Arthur E. (1910a). 'Über eine neue theoretische Bestimmung des elektrischen Elementarquantums und des Halbmessers des Wasserstoffsatom,' *Physikalisches Zeitschrift* **11**, 537–8.

Haas, Arthur E. (1910b). 'Der Zusammenhang der Planckschen elementaren Wirkungsquantums mit den Grundgrössen der Elektronentheorie', *Jahrbuch der Radioaktivität und Elektronik* **7**, 261–8.

Haas, Arthur E. (1912). 'Über Gleichgewichtlagen von Elektronengruppen in einer äquivalenten Kugel von homogener positiver Elektrizität', *Sitzungsberichte d. Kaiserliche Akademie d. Wissenschaften (Wien)* IIa, **120**, 1111–71.

Hahn, Dorothy A. and Mary E. Holmes (1915). 'The valence theory of J. Stark from a chemical standpoint', *Journal of the American Chemical Society* **37**, 2611–26.

Hanle, Paul A. (1979). 'Indeterminacy before Heisenberg: The case of Franz Exner and Erwin Schrödinger', *Historical Studies in the Physical Sciences* **10**, 225–69.

Hansen, Gerhard (1925). 'Die Feinstruktur der Wasserstofflinien', *Physikalische Zeitschrift* **26**, 678–80.

Hansen, Hans M. (1914). 'Über die Begleiterscheinungen des inversen Zeemaneffektes (die longitudinale zirkulare und transversale lineare Doppelbrechung) zwischen die Komponenten einer Zerlegten Absorptionslinie', *Annalen der Physik* **43**, 169–248.

Harkins, William D. and Ernest D. Wilson (1915). 'Recent work on the structure of the atom', *Journal of the American Chemical Society* **37**, 1396–421.

Heilbron, John L. (1964). *A History of the Problem of Atomic Structure from the Discovery of the Electron to the Beginning of Quantum Mechanics*. Ph.D. dissertation, University of California. University Microfilms no. 65–3004.

Heilbron, John L. (1966). 'The work of H. G. J. Moseley', *Isis* **57**, 336–64.

Heilbron, John L. (1967). 'The Kossel-Sommerfeld theory and the ring atom', *Isis* **58**, 450–585.

Heilbron, John L. (1968). 'The scattering of α and β particles and Rutherford's atom', *Archive for History of Exact Sciences* **4**, 247–307.

Heilbron, John L. (1974). *H. G. J. Moseley: The Life and Letters of an English Physicist, 1887–1915*. Berkeley: University of California Press.

Heilbron, John L. (1977a). 'J. J. Thomson and the Bohr atom', *Physics Today* **30**, 23–30.

Heilbron, John L. (1977b). 'Lectures on the history of atomic physics 1900–1922', pp. 40–108 in Charles Weiner, ed., *History of Twentieth Century Physics*. New York: Academic Press.

Heilbron, John L. (1981). 'Rutherford-Bohr atom', *American Journal of Physics* **49**, 222–31.

Heilbron, John L. (1983). 'The origins of the exclusion principle', *Historical Studies in the Physical Sciences* **13**, 261–310.

Heilbron, John L. and Thomas S. Kuhn (1969). 'The genesis of the Bohr atom', *Historical Studies in the Physical Sciences* **1**, 211–90.

Heisenberg, Werner (1922). 'Zur Quantentheorie der Linienstruktur und der anomalen Zeemaneffekte', *Zeitschrift für Physik* **8**, 273–97.

Heisenberg, Werner (1924). 'Über eine Abänderung der formalen Regeln der Quantentheorie beim Problem der anomalen Zeemaneffekte', *Zeitschrift für Physik* **26**, 291–307.

Heisenberg, Werner (1925). 'Über quantentheoretische Umdeutung kinematischer und mechanischer Beziehungen', *Zeitschrift für Physik* **33**, 879–93.

Heisenberg, Werner (1929). 'Die Entwicklung der Quantentheorie 1918–1928', *Naturwissenschaften* **17**, 490–6.

Heisenberg, Werner (1930). *The Physical Principles of the Quantum Theory*. New York: Dover Publications.

Heisenberg, Werner (1967). 'Quantum theory and its interpretation', pp. 94–108 in Stefan Rozental, ed., *Niels Bohr: His Life and Work as Seen by His Friends and Colleagues*. New York: Interscience Publishers.

Heisenberg, Werner (1969). *Der Teil und das Ganze: Gespräche im Umkreis der Atomphysik*. Munich: Piper & Co. Verlag.

Heisenberg, Werner (1973). 'Tradition in science', *Bulletin of the Atomic Scientists* **29**, 4–10.

Hendry, John (1981). 'Bohr-Kramers-Slater: A virtual theory of virtual oscillators and its role in the history of quantum mechanics', *Centaurus* **25**, 189–221.

Hendry, John (1983). 'Monopoles before Dirac', *Studies in History and Philosophy of Science* **14**, 81–7.

Hendry, John (1984). *The Creation of Quantum Mechanics and the Bohr-Pauli Dialogue*. Dordrecht: Reidel.

Hentschel, Klaus (2009). 'Quantum theory, crisis period 1923—early 1925', pp. 613–17 in Daniel Greenberger, Klaus Hentschel, and Friedel Weinert, eds, *Compendium of Quantum Physics: Concepts, Experiments, History and Philosophy*. Berlin: Springer.

Hermann, Armin, ed. (1965a). *Arthur Erich Haas. Der erste Quantenansatz für das Atom*. Stuttgart: Battenberg.

Hermann, Armin, ed. (1965b). *J. Stark, P. Epstein: Der Stark-Effekt*. Stuttgart: Battenberg.

Hermann, Armin, ed. (1967). *James Franck, Gustav Hertz. Die Elektronenstossversuche*. Stuttgart: Battenberg.

Hermann, Armin, ed. (1968). *Albert Einstein, Arnold Sommerfeld: Briefwechsel*. Basel: Schwabe.

Hermann, Armin, Karl von Meyenn, and Victor F. Weisskopf, eds (1979). *Wolfgang Pauli: Wissenschaftlicher Briefwechsel mit Bohr, Einstein, Heisenberg u.a.*, vol. 1. Berlin: Springer.

Herzfeld, Karl F. (1914). 'Der Zeemaneffekt in den Quantentheorien der Serienspektren', *Physikalische Zeitschrift* **15**, 193–198.

Hettema, Hinne (1995). 'Bohr's theory of the atom 1913–1923: A case study in the progress of scientific research programmes', *Studies in History and Philosophy of Modern Physics* **26**, 307–23.

Hettema, Hinne and Theo A. F. Kuipers (1995). 'Sommerfeld's Atombau: A case study in potential truth approximation', pp. 273–97 in T. A. F. Kuipers and Anne R. Mackor, eds, *Cognitive Patterns in Science and Common Sense*. Amsterdam: Rodopi.

Heurlinger, Torsten (1919). 'Zur Theorie der Bandspektren', *Physikalische Zeitschrift* **20**, 188–90.

Heurlinger, Torsten (1920). 'Über Atomschwingungen und Molekülspektren', *Zeitschrift für Physik* **1**, 82–91.

Hevesy, George (1913). 'Die Valenz der Radioelemente', *Physikalische Zeitschrift* **14**, 49–62.

Hevesy, George (1914). 'The diffusion and valency of the radio-elements', *Philosophical Magazine* **27**, 586–601.

Hevesy, George (1925). 'The discovery and properties of hafnium', *Chemical Reviews* **2**, 1–41.

Hevesy, George (1963). 'Gamle dage', *Fysisk Tidsskrift* **60**, 26–30.

Hicks, William M. (1914). [Contribution to] 'Discussion on the structure of atoms and molecules', Report, British Association for the Advancement of Science, 296–9.

Hirosige, Tetu and Sigeo Nisio (1964). 'Formation of Bohr's theory of atomic constitution', *Japanese Studies in History of Science* no. 3, 6–28.

Hoddeson, Lillian H. and Baym, G. (1980). 'The development of the quantum mechanical electron theory of metals: 1900–28', *Proceedings of the Royal Society* A **371**, 8–23.

Hoddeson, Lillian H., Ernest Braun, Jürgen Teichmann, and Spencer Weart, eds (1992). *Out of the Crystal Maze: Chapters from the History of Solid-State Physics*. New York: Oxford University Press.

Hofmann, Karl A. (1919). *Lehrbuch der Anorganischen Chemie*. Braunschweig: Vieweg & Sohn.

Hogness, T. R. and E. G. Lunn (1925). 'The ionization of hydrogen by electron impact as interpreted by positive ray analysis', *Physical Review* **26**, 44–55.

Holton, Gerald (1961). 'On the recent past of physics', *American Journal of Physics* **29**, 805–11.

Holton, Gerald (1970). 'The roots of complementarity', *Dædalus* **99**, 1015–55.

Holton, Gerald (1988). 'On the hesitant rise of quantum physics research in the United States', pp. 147–87 in G. Holton, *Thematic Origins of Scientific Thought: Kepler to Einstein*. Cambridge, Mass.: Harvard University Press.

Hon, Giora (1989). 'Franck and Hertz versus Townsend: A study of two types of experimental error', *Historical Studies in the Physical and Biological Sciences* **20**, 79–106.

Honda, Kotaro (1922). 'Bohr's model of the hydrogen molecules and their magnetic susceptibility', *Nature* **110**, 664–5.

Hoyer, Ulrich (1973). 'Über die Rolle der Stabilitätsbetrachtungen in der Entwicklung der Bohrschen Atomtheorie', *Archive for History of Exact Sciences* **10**, 177–206.

Hoyer, Ulrich (1974). *Die Geschichte der Bohrschen Atomtheorie*. Weinheim: Physik Verlag.

Hoyer, Ulrich, ed. (1981). *Niels Bohr. Collected Works*, vol. 2. Amsterdam: North-Holland.

Hoyt, Frank C. (1923). 'The relative intensity of X-ray lines', *Philosophical Magazine* **46**, 135–45.

Hoyt, Frank C. (1924). 'Relative probabilities of the transitions involved in the Balmer lines of hydrogen', *Philosophical Magazine* **47**, 826–31.

Hoyt, Frank C. (1925). 'Application of the correspondence principle to relative intensities in series spectra', *Physical Review* **26**, 749–60.

Hughes, Jeff (2003). 'Occultism and the atom: The curious story of isotopes', *Physics World* **16** (September), 31–6.

Hull, Albert W. (1917). 'The crystal structure of iron', *Physical Review* **9**, 84–7.

Hund, Friedrich (1923). 'Theoretische Betrachtungen über die Ablenkung von freien langsamen Elektronen in Atomen', *Zeitschrift für Physik* **13**, 241–63.

Hund, Friedrich (1967). *Geschichte der Quantentheorie*. Mannheim: Bibliographisches Institut.

Hutten, Ernest H. (1956). *The Language of Modern Physics*. London: George Allen & Unwin.

Ishiwara, Jun (1915). 'Die universelle Bedeutung des Wirkungsquantums', *Proceedings of the Tokyo Mathematical-Physical Society* **8**, 106–16.

Imes, Elmer S. (1919). 'Measurement in the near infrared absorption of some diatomic gases', *Astrophysical Journal* **50**, 251–76.

Jammer, Max (1966). *The Conceptual Development of Quantum Mechanics*. New York: McGraw-Hill.

Jeans, James (1901). 'The mechanism of radiation', *Philosophical Magazine* **2**, 421–55.

Jeans, James (1913). 'Discussion on radiation', Report, British Association for the Advancement of Science, 376–86.

Jeans, James (1914). *Report on Radiation and the Quantum-Theory*. London: The Electrician.

Jeans, James (1919). 'The quantum theory and new theories of atomic structure', *Journal of the Chemical Society, Transactions* **115**, 865–71.

Jenkin, John (2008). *William and Lawrence Bragg, Father and Son: The Most Extraordinary Collaboration in Science*. Oxford: Oxford University Press.

Jensen, Carsten (1984). 'Two one-electron anomalies in the old quantum theory', *Historical Studies in the Physical Sciences* **15**, 81–106.

Jensen, Carsten (2000). *Controversy and Consensus: Nuclear Beta Decay 1911–1934*. Basel: Birkhäuser Verlag.

Jørgensen, Sophus M. (1908). *The Fundamental Conceptions of Chemistry*. London: Society for the Promotion of Christian Knowledge.

Kaiser, Walter (2001). 'Electron gas theory of metals: Free electrons in bulk matter', pp. 255–303 in Jed Z. Buchwald and Andrew Warwick, eds, *Histories of the Electron: The Birth of Microphysics*. Cambridge, Mass.: MIT Press.

Kaiserfeld, Thomas (1993). 'When theory addresses experiment: The Siegbahn-Sommerfeld correspondence, 1917–1940', pp. 306–26 in Svante Lindqvist, ed., *Center on the Periphery: Historical Aspects of 20th-Century Swedish Physics*. Canton, Mass.: Science History Publications.

Kalckar, Jørgen (1985). *Niels Bohr. Collected Works*, vol. 6. Amsterdam: North-Holland.

Kamerlingh Onnes, Heike (1924). 'Rapports sur de nouvelles expériences avec les supraconducteurs', *Communications from the Physical Laboratory of the University of Leiden*, Supplement 50a.

Kant, Immanuel (2004). *Metaphysical Foundations of Natural Science*, ed. by Michael Friedmann. Cambridge: Cambridge University Press.

Kaufmann, Walter (1901). 'Die Entwicklung des Elektronenbegriffs', *Physikalische Zeitschrift* **3**, 9–15.

Kaye, George W. C. (1914). *X rays: An Introduction to the Study of Röntgen Rays*. London: Longmans, Green and Co.

Keller, Alex (1983). *The Infancy of Atomic Physics: Hercules in His Cradle*. Oxford: Clarendon Press.

Kemble, Edwin (1921). 'The probable normal state of the helium atom', *Philosophical Magazine* **42**, 123–33.

Kemble, Edwin (1925). 'The application of the correspondence principle to degenerate systems and the relative intensities of band lines', *Physical Review* **25**, 1–22.

Kenat, Ralph and David H. DeVorkin (1990). 'Quantum physics and the stars (III): Henry Norris Russell and the search for a rational theory of stellar spectra', *Journal for the History of Astronomy* **21**, 157–86.

Kim, Dong-Won (2007). *Yoshio Nishina: Father of Modern Physics in Japan*. New York: Taylor & Francis.

Kirchberger, Paul (1922). *Die Entwicklung der Atomtheorie, Gemeinverständlich Dargestellt*. Karlsruhe: C. F. Müllersche Hofbuchhandlung.

Klein, Martin J. (1970). *Paul Ehrenfest. Vol. 1: The Making of a Theoretical Physicist*. Amsterdam: North-Holland.

Klein, Martin J., A. J. Kox, and Robert Schulmann, eds (1993). *Collected Papers of Albert Einstein*, vol. 4. Princeton: Princeton University Press.

Klein, Martin J., A. J. Kox, and Robert Schulmann, eds (1995). *Collected Papers of Albert Einstein*, vol. 5. Princeton: Princeton University Press.

Klein, Oskar (1927). 'Elektrodynamik und Wellenmechanik vom Standpunkt des Korrespondenzprinzips', *Zeitschrift für Physik* **41**, 407–42.

Kleinert, Andreas and Charlotte Schönbeck (1978). 'Lenard und Einstein: Ihr Briefwechsel und ihr Verhältnis vor der Nauheimer Diskussion von 1920', *Gesnerus* **35**, 318–33.

Knudsen, Ole (2001). 'O. W. Richardson and the electron theory of matter, 1901–1916', pp. 227–54 in Jed Z. Buchwald and Andrew Warwick, eds, *Histories of the Electron: The Birth of Microphysics*. Cambridge, Mass.: MIT Press.

Kohler, Robert E. (1971). 'The origin of G. N. Lewis's theory of the shared pair bond', *Historical Studies in the Physical Sciences* **3**, 343–76.

Kohler, Robert E. (1974). 'Irving Langmuir and the "octet" theory of valence', *Historical Studies in the Physical Sciences* **4**, 39–87.

Kohler, Robert E. (1975). 'The Lewis-Langmuir theory of valence and the chemical community, 1920–1928', *Historical Studies in the Physical Sciences* **6**, 431–68.

Kojima, Chieko (2004). 'La physique Française avant Louis de Broglie', *Annales de la Fondation Louis de Broglie* **29**, 767–83.

Konno, Hiroyuki (1983). 'Slater's evidence for the Genesis of the Bohr-Kramers-Slater theory', *Historia Scientiarum* no. 25, 39–52.

Konno, Hiroyuki (1993). 'Kramers' negative dispersion, the virtual oscillator model, and the correspondence principle', *Centaurus* **36**, 117–66.

Konno, Hiroyuki (2000). 'Bohr's search for the quantum theory of dispersion: The number of dispersion electrons, absorption and emission of light and the oscillator model', *Historia Scientiarum* **10**, 163–76.

Konno, Hiroyuki (2002). 'Ritz's discovery of the Lyman series before 1913 and Lyman's indifference to the Bohr theory', *Centaurus* **44**, 127–39.

Kossel, Walther (1914a). 'Bemerkung zur Absorption homogener Röntgenstrahlen', *Verhandlungen der Deutschen Physikalischen Gesellschaft* **16**, 898–909.

Kossel, Walther (1914b). 'Bemerkung zur Absorption homogener Röntgenstrahlen, II', *Verhandlungen der Deutschen Physikalischen Gesellschaft* **16**, 953–63.

Kossel, Walther (1916). 'Über Molekülbildung als Folge des Atombaues', *Annalen der Physik* **49**, 229–362.

Kossel, Walther (1920). 'Die Valenzkräfte im Lichte der neueren physikalischen Forschung', *Zeitschrift für Elektrochemie* **26**, 314–23.

Kossel, Walther (1923). 'Die Beziehungen der Bohrschen Atomtheorie zur Deutung chemischer Vorgänge', *Naturwissenschaften* **27**, 598–604.

Kossel, Walther and Arnold Sommerfeld (1919). 'Auswahlprinzip und Verschiebungssatz bei Serienspektren', *Verhandlungen der Deutschen Physikalischen Gesellschaft* **21**, 240–59.

Kox, A. J., Martin J. Klein, and Robert Schulmann, eds (1996). *Collected Papers of Albert Einstein*, vol. 6. Princeton: Princeton University Press.

Kragh, Helge (1977). 'Chemical aspects of Bohr's 1913 theory', *Journal of Chemical Education* **54**, 208–10.

Kragh, Helge (1979). 'Niels Bohr's second atomic theory', *Historical Studies in the Physical Sciences* **10**, 123–86.

Kragh, Helge (1980). 'Anatomy of a priority conflict: The case of element 72', *Centaurus* **23**, 275–301.

Kragh, Helge (1982). 'Julius Thomsen and 19th-century speculations on the complexity of atoms', *Annals of Science* **39**, 37–60.

Kragh, Helge (1985a). 'The fine structure of hydrogen and the gross structure of the physics community, 1916–26', *Historical Studies in the Physical Sciences* **16**, 67–125.

Kragh, Helge (1985b). 'Bohr's atomic theory and the chemists, 1913–1925', *Rivista di Storia della Scienza* **2**, 463–86.

Kragh, Helge (1985c). 'The theory of the periodic system', pp. 50–67 in Anthony P. French and P. J. Kennedy, eds, *Niels Bohr. A Centenary Volume*. Cambridge, Mass.: Harvard University Press.

Kragh, Helge (1989a). 'The aether in late nineteenth century chemistry', *Ambix* **36**, 49–65.

Kragh, Helge (1989b). 'Concept and controversy: Jean Becquerel and the positive electron', *Centaurus* **32**, 204–40.

Kragh, Helge (1990). *Dirac: A Scientific Biography*. Cambridge: Cambridge University Press.

Kragh, Helge (1992). 'Relativistic collisions: The work of Christian Møller in the early 1930s', *Archive for History of Exact Sciences* **43**, 299–328.

Kragh, Helge (1996). 'Elements no. 70, 71 and 72: Discoveries and controversies', pp. 67–89 in C. H. Evans, ed., *Episodes from the History of the Rare Earth Elements*. Dordrecht: Kluwer Academic.

Kragh, Helge (1997a). 'J. J. Thomson, the electron, and atomic architecture', *The Physics Teacher* **35**, 328–32.

Kragh, Helge (1997b). 'The origin of radioactivity: from solvable problem to unsolved non-problem', *Archive for History of Exact Sciences* **50**, 331–58.

Kragh, Helge (1999). *Quantum Generations: A History of Physics in the Twentieth Century*. Princeton: Princeton University Press.

Kragh, Helge (2000). 'Conceptual changes in chemistry: The notion of a chemical element, ca. 1900–1925', *Studies in History and Philosophy of Modern Physics* **31**, 435–50.

Kragh, Helge (2002a). 'The vortex atom: A Victorian theory of everything', *Centaurus* **44**, 32–114.

Kragh, Helge (2002b). 'Atomic structure and the crisis in quantum theory, 1910–1924', pp. 19–36 in E. Beltrametti, G. Giuliani, A. Rimini, and N. Robotti, eds, *One Hundred Years of h. Conference Proceedings vol. 79*. Bologna: Italian Physical Society.

Kragh, Helge (2003a). 'The electron, the protyle, and the unity of matter', pp. 195–226 in Jed Z. Buchwald and Andrew Warwick, eds, *Histories of the Electron: The Birth of Microphysics*. Cambridge, Mass.: MIT Press.

Kragh, Helge (2003b). 'Magic number: A partial history of the fine-structure constant', *Archive for History of Exact Sciences* **57**, 395–431.

Kragh, Helge (2009). 'The solar element: A reconsideration of helium's early history', *Annals of Science* **66**, 157–82.

Kragh, Helge (2010a). 'Before Bohr: Theories of atomic structure, 1850–1913', *RePoss: Research Publications on Science Studies* **10**. Department of Science Studies, Aarhus University. http://www.ivs.au.dk/reposs.

Kragh, Helge (2010b). 'The childhood of H_3 and H_3^{+}', *Astronomy & Geophysics* **51**, 6.1–6.3.

Kragh, Helge (2011a). 'Resisting the Bohr atom: The early British opposition', *Physics in Perspective* **13**, 4–35.

Kragh, Helge (2011b). 'A controversial molecule: The early history of triatomic hydrogen', *Centaurus* **53**, 257–79.

Kragh, Helge (2011c). 'Conceptual objections to the Bohr atomic theory – do electrons have a "free will"?', *European Physical Journal H* **36**, 327–52.

Kragh, Helge and Bruno Carazza (1995). 'A historical note on the maximum atomic number of chemical elements', *Annales de la Fondation Louis de Broglie* **20**, 207–15.

Kramers, Hendrik A. (1919). 'Intensities of spectral lines', *Kgl. Danske Videnskabernes Skrifter, Naturvidenskabelig og Mathematisk Afdeling*, 8. Række, III.3, 1–100.

Kramers, Hendrik A. (1920). 'Über den Einfluss eines elektrischen Feldes auf die Feinstruktur der Wasserstofflinien', *Zeitschrift für Physik* **3**, 199–223.

Kramers, Hendrik A. (1923a). 'Über das Modell des Heliumatoms', *Zeitschrift für Physik* **13**, 312–41.

Kramers, Hendrik A. (1923b). 'Das Korrespondenzprinzip und der Schalenbau des Atoms', *Naturwissenschaften* **11**, 550–9.

Kramers, Hendrik A. (1924a). 'The law of dispersion and Bohr's theory of spectra', *Nature* **113**, 673–4.

Kramers, Hendrik A. (1924b). 'The quantum theory of dispersion', *Nature* **114**, 310–11.

Kramers, Hendrik A. (1924c). 'Die chemischen Eigenschaften der Atome nach der Bohrschen Theorie', *Naturwissenschaften* **12**, 1050–4.

Kramers, Hendrik A. (1925). 'Om vekselvirkningen mellem lys og stof', *Fysisk Tidsskrift* **23**, 26–40.

Kramers, Hendrik A. (1935). 'Atom- og kvanteteoriens udvikling i årene 1913–1925', *Fysisk Tidsskrift* **33**, 82–96.

Kramers, Hendrik A. and Helge Holst (1923). *The Atom and the Bohr Theory of its Structure: An Elementary Presentation*. London: Gyldendal.

Kramers, Hendrik A. and Helge Holst (1925). *Das Atom und die Bohrsche Theorie seines Baues*. Berlin: Springer.

Kramers, Hendrik A. and Werner Heisenberg (1925). 'Über die Streuung von Strahlung durch Atome', *Zeitschrift für Physik* **31**, 681–707.

Kramers, Hendrik A. and Wolfgang Pauli (1923). 'Zur Theorie der Bandenspektren', *Zeitschrift für Physik* **13**, 351–67.

Kratzer, Adolf (1923). 'Die Feinstruktur einer Klasse von Bandenspektren', *Annalen der Physik* **71**, 72–103.

Kronig, Ralph D. (1926). 'The dielectric constant of diatomic dipole gases on the new quantum mechanics', *Proceedings of the National Academy of Sciences* **12**, 488–93.

Kroo, Jan (1918). 'Der erste und zweite Elektronenring der Atome', *Physikalische Zeitschrift* **19**, 307–11.

Kuhn, Thomas S. (1970). 'Reflections on my critics', pp. 231–78 in Imre Lakatos and Alan Musgrave, eds, *Criticism and the Growth of Knowledge*. Cambridge: Cambridge University Press.

Kuhn, Thomas S. (1978). *Black-Body Theory and the Quantum Discontinuity, 1894–1912*. Oxford: Clarendon Press.

Kuntze, Friedrich (1909). 'Die Elektronentheorie der Brüder Hermann und Robert Grassmann', *Vierteljahrschrift für Wissenschaftliche Philosophie und Soziologie* **13**, 273–98.

Kunz, Jacob (1918). 'On Bohr's atom and magnetism', *Physical Review* **12**, 59–69.

Ladenburg, Rudolf (1920). 'Die Elektronenordnung in den Elementen der grossen Perioden', *Naturwissenschaften* **8**, 5–11.

Ladenburg, Rudolf (1921). 'Die quantentheoretische Deutung der Zahl der Dispersionselektronen', *Zeitschrift für Physik* **4**, 451–68.

Ladenburg, Rudolf and Fritz Reiche (1923). 'Absorption, Zerstreuung und Dispersion in der Bohrschen Atomtheorie', *Naturwissenschaften* **11**, 584–98.

Lakatos, Imre (1970). 'Falsification and the methodology of scientific research programmes', pp. 91–196 in Imre Lakatos and Alan Musgrave, eds, *Criticism and the Growth of Knowledge*. Cambridge: Cambridge University Press.

Laming, Richard (1845). 'Observations on a paper by Prof. Faraday concerning electric conduction and the nature of matter', *Philosophical Magazine* **27**, 420–3.

Landé, Alfred (1919a). 'Dynamik der räumlichen Atomstruktur', *Verhandlungen der Deutschen Physikalischen Gesellschaft* **20**, 230–9.

Landé, Alfred (1919b). 'Das Serienspektrum des Heliums', *Physikalische Zeitschrift* **20**, 228–34.

Landé, Alfred (1920). 'Würfelatome, periodisches System und Molekülbildung', *Zeitschrift für Physik* **2**, 380–404.

Landé, Alfred (1921). 'Über den anomalen Zeemaneffekt (Teil I)', *Zeitschrift für Physik* **5**, 231–41.

Landé, Alfred (1923). 'Termstruktur und Zeemaneffekt der Multipletts', *Zeitschrift für Physik* **15**, 189–205.

Langevin, Paul and Maurice de Broglie, eds (1912). *La Théorie du Rayonnement et les Quanta: Rapports et Discussions de la Réunion Tenue à Bruxelles, du 30 Octobre au 3 Novembre 1911*. Paris: Gauthier-Villars.

Langmuir, Irving (1912). 'The dissociation of hydrogen into atoms', *Journal of the American Chemical Society* **34**, 860–77.

Langmuir, Irving (1914). 'Note on the heat of formation of hydrogen from hydrogen atoms', *Philosophical Magazine* **27**, 188–9.

Langmuir, Irving (1915). 'The dissociation of hydrogen into atoms. Part II: Calculation of the degree of dissociation and the heat of formation', *Journal of the American Chemical Society* **37**, 417–58.

Langmuir, Irving (1919). 'The arrangement of electrons in atoms and molecules', *Journal of the American Chemical Society* **41**, 868–934.

Langmuir, Irving (1920). 'Theories of atomic structure', *Nature* **105**, 261.

Langmuir, Irving (1921a). 'Future developments of theoretical chemistry', *Chemical and Metallurgical Engineering* **24**, 553–7.

Langmuir, Irving (1921b). 'The structure of the static atom', *Science* **53**, 290–3.

Langmuir, Irving (1921c). 'The structure of the helium atom', *Physical Review* **17**, 339–53.

Larmor, Joseph (1895). 'A dynamical theory of the electric and luminiferous medium. Part II. Theory of electrons', *Philosophical Transactions of the Royal Society* **186**, 695–743.

Larmor, Joseph (1900). 'Presidential address to Section A', Report, British Association for the Advancement of Science, 613–28.

Larmor, Joseph (1929). *Mathematical and Physical Papers*, vol. 2. Cambridge: Cambridge University Press.

Larmor, Joseph (1931). 'The scientific environment of Clerk Maxwell', pp. 74–90 in J. J. Thomson et al., *James Clerk Maxwell: A Commemoration Volume 1831–1931*. Cambridge: Cambridge University Press.

LeBon, Gustave (1905). *The Evolution of Matter*. New York: Charles Scribner's Sons.

Lenard, Philipp (1903). 'Über die Absorption von Kathodenstrahlen verschiedener Geschwindigkeit', *Annalen der Physik* **12**, 714–44.

Lenz, Wilhelm (1919). 'Zur Theorie der Bandenspektren', *Verhandlungen der Deutschen Physikalischen Gesellschaft* **21**, 632–43.

Leone, Matteo, Alessandro Paoletti, and Nadia Robotti (2004). 'A simultaneous discovery: The case of Johannes Stark and Antonino Lo Surdo', *Physics in Perspective* **6**, 271–94.

Leone, Matteo and Nadia Robotti (2006). 'The entrance of quantum mechanics in Italy: From Garbasso to Fermi', pp. 225–37 in Claudio Garola, Arcangelo Rossi, and Sandro Sozzo, eds, *The Foundations of Quantum Mechanics*. New Jersey: World Scientific.

Lepape, Alexandre (1922). 'La discontinuité et l'unité de la matière', *Bulletin de la Société Chimique de France* **31**, 1–94.

Levi, Hilde (1985). *George de Hevesy. Life and Work*. Copenhagen: Rhodos.

Lewis, Gilbert N. (1916). 'The atom and the molecule', *Journal of the American Chemical Society* **38**, 762–85.

Lewis, Gilbert N. (1917). 'The static atom', *Science* **46**, 297–302.

Lewis, Gilbert N. (1923a). *Valence and the Structure of Atoms and Molecules*. New York: Chemical Catalog Company.

Lewis, Gilbert N. (1923b). 'Introductory address: valence and the electron', *Transactions of the Faraday Society* **19**, 452–8.

Lewis, Gilbert N. (1926a). *The Anatomy of Science*. New Haven: Yale University Press.

Lewis, Gilbert N. (1926b). 'The conservation of photons', *Nature* **118**, 874–5.

Lewis, William C. (1919). *A System of Physical Chemistry*, vol. 3. London: Longmans, Green and Co.

Lindemann, Frederick A. (1914a). 'Atomic models and X-ray spectra', *Nature* **92**, 500–1.

Lindemann, Frederick A. (1914b). 'Atomic models and X-ray spectra', *Nature* **92**, 631.

Lindemann, Frederick A. (1914c). 'Über die Grundlagen der Atommodelle', *Verhandlungen der Deutschen Physikalischen Gesellschaft* **16**, 281–94.

Lindsay, Robert B. (1927). '"Pendulum" orbits in atomic models', *Proceedings of the National Academy of Sciences* **13**, 413–19.

Lindsay, Robert B. and Henry Margenau (1936). *Foundations of Physics*. New York: John Wiley & Sons.

Livingston, Dorothy M. (1973). *The Master of Light: a Biography of Albert A. Michelson*. New York: Charles Scribner.

Lodge, Oliver (1903). 'Modern views of matter', *Annual Report of the Smithsonian Institution, 1903*, 215–28.

Lodge, Oliver (1906). *Electrons, or the Nature and Properties of Negative Electricity*. London: Bell and Sons.

Lodge, Oliver (1913). 'Continuity', Report, British Association for the Advancement of Science, 1–42.

Lodge, Oliver (1924). *Atoms and Rays: An Introduction to Modern Views on Atomic Structure and Radiation*. New York: George H. Doran Co.

Lorentz, Hendrik A. (1927). *Problems of Modern Physics*. Boston: Ginn and Company.

Lorentz, Hendrik A. (1952). *The Theory of Electrons and its Applications to the Phenomena of Light and Radiant Heat*. New York: Dover Publications.

Loria, Stanislaw (1914). *Die Lichtbrechung in Gasen als Physikalisches und Chemisches Problem*. Braunschweig: Vieweg & Sohn.

Loring, F. H. (1921). *Atomic Theories*. London: Methuen & Co.

Lyman, Theodore (1914). 'An extension of the spectrum in the extreme ultra-violet', *Nature* **93**, 241.

Lyman, Theodore (1915). 'The extension of the spectrum beyond the Schumann region', *Proceedings of the National Academy of Sciences* **1**, 368–71.

Lyman, Theodore (1916). 'The extension of the spectrum beyond the Schumann region', *Astrophysical Journal* **43**, 89–102.

MacKinnon, Edward M. (1977). 'Heisenberg, models, and the rise of matrix mechanics', *Historical Studies in the Physical Sciences* **8**, 137–88.

MacKinnon, Edward M. (1982). *Scientific Explanation and Atomic Physics*. Chicago: University of Chicago Press.

Maier, Clifford L. (1964). *The Role of Spectroscopy in the Acceptance of an Internally Structured Atom 1860–1920*. Ph.D. dissertation, University of Wisconsin. University Microfilms no. 64–10, 272.

Main Smith, John D. (1923). 'The Bohr atom', *Chemistry and Industry* **42**, 1073–8.

Main Smith, John D. (1924a). 'Atomic structure', *Chemistry and Industry* **43**, 323–5, 437, 548–9.

Main Smith, John D. (1924b). *Chemistry and Atomic Structure*. London: Ernest Benn.

Margenau, Henry (1950). *The Nature of Physical Reality*. New York: McGraw-Hill.

Margenau, Henry and Arthur Wightman (1944). 'Atomic and molecular theory since Bohr: historical survey', *American Journal of Physics* **12**, 119–30.

Massimi, Michela (2005). *Pauli's Exclusion Principle: The Origin and Validation of a Scientific Principle*. Cambridge: Cambridge University Press.

Maxwell, James C. (1965). *The Scientific Papers of James Clerk Maxwell*, ed. W. D. Niven. New York: Dover Publications.

McCormmach, Russell (1966). 'The atomic theory of John William Nicholson', *Archive for History of Exact Sciences* **3**, 160–84.

McCormmach, Russell (1982). *Night Thoughts of a Classical Physicist*. Cambridge, Mass.: Harvard University Press.

McLaren, Samuel B. (1913a). 'The theory of radiation', *Nature* **92**, 165.

McLaren, Samuel B. (1913b). 'The magneton and Planck's universal constant', *Philosophical Magazine* **26**, 800.

McLaren, Samuel B. (1913c). 'A theory of magnets', Report, British Association for the Advancement of Science, 391–2.

McLaren, Samuel B. (1913d). 'A theory of gravity', *Philosophical Magazine* **26**, 636–73.

McLennan, John C. (1923). 'On the origin of spectra', Report, British Association for the Advancement of Science, 25–58.

McLennan, John C. and David A. Keys (1916). 'On the ionisation of metallic vapours in flames', *Proceedings of the Royal Society* A **92**, 591–608.

Medicus, Heinrich A. (1974). 'Fifty years of matter waves', *Physics Today* **27**, 38–46.

Mehra, Jagdish (1975). *The Solvay Conferences on Physics: Aspects of the Development of Physics since 1911*. Dordrecht: Reidel.

Mehra, Jagdish and Helmut Rechenberg (1982a). *The Historical Development of Quantum Theory*, vol. 1. New York: Springer.

Mehra, Jagdish and Helmut Rechenberg (1982b). *The Historical Development of Quantum Theory*, vol. 2. New York: Springer.

Mehra, Jagdish and Helmut Rechenberg (1999). 'Planck's half-quanta: a history of the concept of zero-point energy', *Foundations of Physics* **29**, 91–132.

Meitner, Lise (1964). 'Looking back', *Bulletin of the Atomic Scientists* **20**, 2–7.

Mensing, Lucie and Wolfgang Pauli (1926). 'Über die Dielektrizitätskonstante von Dipolgasen nach der Quantenmechanik', *Physikalische Zeitschrift* **27**, 509–12.

Merton, Thomas R. (1915). 'On the origin of the "4686" series', *Proceedings of the Royal Society* A **91**, 382–7.

Merz, John T. (1965). *A History of European Thought in the Nineteenth Century*, 4 vols. New York: Dover Publications.

Meyenn, Karl von, ed. (2011). *Eine Entdeckung von Ganz Ausserordentlicher Tragweite: Schrödinger's Briefwechsel zur Wellenmechanik und zum Katzenparadoxon*. Heidelberg: Springer.

Meyer-Abich, Klaus (1965). *Korrespondenz, Individualität und Komplementarität: Eine Studie zur Geistesgeschichte der Quanthetheorie in den Beiträgen Niels Bohrs*. Wiesbaden: Franz Steiner Verlag.

Millikan, Robert A. (1917a). 'Radiation and atomic structure', *Science* **45**, 321–30.

Millikan, Robert A. (1917b). 'Theoretical considerations relating to the single-lined and the many-lined spectrum of mercury', *Physical Review* **9**, 378–82.

Millikan, Robert A. (1924). 'Atomism in modern physics', *Journal of the Chemical Society, Transactions*, **125**, 1405–17.

Millikan, Robert A. (1951). *The Autobiography of Robert A. Millikan*. London: MacDonald.

Minkowski, Rudolph and Hertha Sponer (1924). 'Über den Durchgang von Elektronen durch Atome', *Ergebnisse der Exakten Naturwissenschaften* **3**, 67–85.

Møller, Christian (1931). 'Über den Stoss zweier Teilchen unter Berücksichtigung der Retardation der Kräfte', *Zeitschrift für Physik* **70**, 786–95.

Moore, Ruth (1966). *Niels Bohr: The Man, His Science, and the World They Changed*. New York: Alfred A. Knopf.

Moseley, Henry G. J. (1913). 'The high-frequency spectra of the elements', *Philosophical Magazine* **26**, 1024–34.

Moseley, Henry G. J. (1914a). 'Atomic models and X-ray spectra', *Nature* **92**, 554.

Moseley, Henry G. J. (1914b). 'The high-frequency spectra of the elements, part II', *Philosophical Magazine* **27**, 703–13.

Mulliken, Robert S. (1924). 'The band spectrum of boron monoxide', *Nature* **114**, 349–50.

Mulliken, Robert S. (1925). 'The isotope effect in band spectra, II: The spectrum of boron monoxide', *Physical Review* **25**, 259–94.

Mulliken, Robert S. (1928). 'The assigment of quantum numbers for electrons in molecules, I', *Physical Review* **32**, 186–222.

Nägeli, Carl W. (1884). *Mechanisch-Physiologischen Theorie der Abstammungslehre*. Munich: R. Oldenbourg.

Nagaoka, Hantaro (1904). 'Kinetics of a system of particles illustrating the line and the band spectrum and the phenomena of radioactivity', *Philosophical Magazine* **8**, 445–55.

Navarro, Jaume (2009). '"A dedicated missionary": Charles Galton Darwin and the new quantum mechanics in Britain', *Studies in History and Philosophy of Modern Physics* **40**, 316–26.

Navarro, Luis and Enric Pérez (2006). 'Paul Ehrenfest: The genesis of the adiabatic hypothesis, 1911–1914', *Archive for History of Exact Sciences* **60**, 209–67.

Needell, Allan A. (1980). *Irreversibility and the Failure of Classical Dynamics: Max Planck's Work on the Quantum Theory 1900–1915*. Ann Arbor, Michigan: University Microfilms, No. 3058.

Nernst, Walther (1907). *Theoretische Chemie*. Stuttgart: Ferdinand Enke.

Nernst, Walther (1916). 'Über einen Versuch von quantetheoretischen Betrachtungen zur Annahme stetiger Energieänderungen zurückzukehren', *Verhandlungen der Deutschen Physikalischen Gesellschaft* **18**, 83–116.

Nernst, Walther (1918). *Die Theoretischen und Experimentellen Grundlagen des Neuen Wärmesatzes*. Halle: Wilhelm Knapp.

Nernst, Walther (1921). *Das Weltgebäude im Lichte der Neueren Forschung*. Berlin: Julius Springer.

Nernst, Walther (1922). 'Zum Gültigkeitsbereich der Naturgesetze', *Naturwissenschaften* **10**, 489–95.

Nersessian, Nancy J. and H. Floris Cohen, eds (1987). *Selected Works of H. A. Lorentz*, Vol. 5. Nieuwekerk: Palm Publications.

Nicholson, John W. (1911a). 'A structural theory of chemical elements', *Philosophical Magazine* **22**, 864–89.

Nicholson, John W. (1911b). 'The spectrum of nebulium', *Monthly Notices of the Royal Astronomical Society* **72**, 49–64.

Nicholson, John W. (1912a). 'On the new nebular line at λ4353', *Monthly Notices of the Royal Astronomical Society* **72**, 693.

Nicholson, John W. (1912b). 'The constitution of the solar corona, III', *Monthly Notices of the Royal Astronomical Society* **72**, 729–39.

Nicholson, John W. (1913a). 'The theory of radiation', *Nature* **92**, 199.

Nicholson, John W. (1913b). 'A possible extension of the spectrum of hydrogen', *Monthly Notices of the Royal Astronomical Society* **73**, 382–5.

Nicholson, John W. (1914a). 'The constitution of nebulae', *Monthly Notices of the Royal Astronomical Society* **74**, 486–506.

Nicholson, John W. (1914b). 'On the nebular line λ 3729', *Monthly Notices of the Royal Astronomical Society* **74**, 623–8.

Nicholson, John W. (1914c). 'Atomic models and X-ray spectra', *Nature* **92**, 583–4.

Nicholson, John W. (1914d). 'Atomic structure and the spectrum of helium', *Philosophical Magazine* **28**, 90–103.

Nicholson, John W. (1914e). 'The spectra of helium and hydrogen', *Monthly Notices of the Royal Astronomical Society* **74**, 425–42.

Nicholson, John W. (1914f). 'The high-frequency spectra of the elements and the structure of the atom', *Philosophical Magazine* **27**, 541–64.

Nicholson, John W. (1923). 'The quantum theory and isotopes', *Philosophical Magazine* **45**, 801–17.

Niessen, Karel F. (1923). 'Zur Quantentheorie des Wasserstoffmolekülions', *Annalen der Physik* **70**, 129–34.

Nishina, Yoshio (1925). 'On the L-absorption edges of the elements from Sn (50) to W (74) and their relation to the atomic constitution', *Philosophical Magazine* **49**, 521–37.

Nisio, Sigeko (1969). 'X-rays and atomic structure at the early stages of the old atomic theory', *Japanese Studies in the History of Science* no. 8, 55–76.

Nisio, Sigeko (1973). 'The formation of the Sommerfeld theory of 1916', *Japanese Studies in the History of Science* no. 12, 39–78.

Nisio, Sigeko (2000). 'Ishiwara Jun's quantum theory, 1911–1915', *Historia Scientiarum* **10**, 120–9.

Nordheim, Lothar (1923). 'Zur Quantentheorie des Wasserstoffmoleküls', *Zeitschrift für Physik* **19**, 69–93.

Norton, John D. (2000). 'How we know about electrons', pp. 67–97 in Robert Nola and Howard Sankey, eds, *After Popper, Kuhn and Feyerabend*. London: Kluwer Academic.

Nye, Mary J. (1972). *Molecular Reality: A Perspective on the Scientific Work of Jean Perrin*. London: MacDonald.

Okada, Tomoji (2002). 'Spatial quantization and the discovery of the Bohr magneton', pp. 287–94 in Helge Kragh, Geert Vanpaemel, and Pierre Marage, eds, *History of Modern Physics*. Turnhout: Brepols.

Olmstead, P. S. and Karl T. Compton (1923). 'Radiation potentials of atomic hydrogen', *Physical Review* **22**, 559–65.

Oseen, Carl W. (1915). 'Das Bohrsche Atommodell und die Maxwellschen Gleichungen', *Physikalische Zeitschrift* **16**, 395–405.

Oxley, A. E. (1920). 'On Langmuir's theory of atoms', *Nature* **105**, 105.

Oxley, A. E. (1921). 'Magnetism and atomic structure, I', *Proceedings of the Royal Society* A **98**, 264–74.

Pais, Abraham (1991). *Niels Bohr's Times, in Physics, Philosophy, and Polity*. Oxford: Clarendon Press.

Paneth, Fritz (1922). 'Das periodische System der chemischen Elemente', *Ergebnisse der Exakten Naturwissenschaften* **1**, 362–403.

Parson, Alfred L. (1915). 'A magneton theory of the structure of the atom', *Smithsonian Miscellaneous Collections*, no. 65, 1–80.

Paschen, Friedrich (1916). 'Bohr's Heliumlinien', *Annalen der Physik* **50**, 901–40.

Pauli, Wolfgang (1919). 'Merkurperihelbewegung und Strahlenablenkung in Weyls Gravitationstheorie', *Verhandlungen der Deutschen physikalischen Gesellschaft* **21**, 742–50.

Pauli, Wolfgang (1920). 'Theoretische Bemerkungen über den Diamagnetismus einatomiger Gase', *Zeitschrift für Physik* **2**, 201–5.

Pauli, Wolfgang (1921). 'Zur Theorie der Dielektrizitätskonstante zweiatomiger Dipolgase', *Zeitschrift für Physik* **6**, 319–327.

Pauli, Wolfgang (1922). 'Über das Modell des Wasserstoffmolekülions', *Annalen der Physik* **68**, 177–240.

Pauli, Wolfgang (1923). 'Über die Gesetzmässigkeiten des anomalen Zeemaneffektes', *Zeitschrift für Physik* **16**, 155–64.

Pauli, Wolfgang (1924). 'Zur Frage der theoretische Deutung der Satelitten einiger Spektrallinien und ihrer Beeinflussung durch magnetische Felder', *Naturwissenschaften* **12**, 741–3.

Pauli, Wolfgang (1925a). 'Über den Einfluss der Geschwindigkeitsabhängigkeit der Elektronenmasse auf den Zeemaneffekt', *Zeitschrift für Physik* **31**, 373–85.

Pauli, Wolfgang (1925b). 'Über den Zusammenhang des Anschlusses der Elektronengruppen im Atom mit der Komplexstruktur der Spektren', *Zeitschrift für Physik* **31**, 765–85.

Pauli, Wolfgang (1926a). 'Quantentheorie', pp. 1–278 in Hans Geiger and Karl Scheel, eds, *Handbuch der Physik* **23**, part 1. Berlin: Springer-Verlag.

Pauli, Wolfgang (1926b). 'Über das Wasserstoffspektrum vom Standpunkt der neuen Quantenmechanik', *Zeitschrift für Physik* **36**, 336–63.

Pauli, Wolfgang (1945). 'Niels Bohr on his 60th birthday', *Reviews of Modern Physics* **17**, 97–101.

Pauli, Wolfgang (1946). 'Remarks on the history of the exclusion principle', *Science* **103**, 213–15.

Pauli, Wolfgang (1958). *Theory of Relativity*. Oxford: Pergamon Press.

Pauli, Wolfgang (1994). 'Rydberg and the periodic system of the elements', pp. 73–7 in Charles P. Enz and Karl von Meyenn, eds, *Wolfgang Pauli: Writings on Physics and Philosophy*. Berlin: Springer-Verlag.

Pauling, Linus (1926). 'The quantum theory of the dielectric constant of hydrogen chloride and similar gases', *Physical Review* **27**, 568–77.

Pauling, Linus (1927). 'The influence of a magnetic field on the dielectric constant of a diatomic dipole gas', *Physical Review* **29**, 145–60.

Pearson, Karl (1885). 'On a certain atomic hypothesis', *Transactions of the Cambridge Philosophical Society* **14**, 71–120.

Pearson, Karl (1891). 'Ether squirts', *American Journal of Mathematics* **13**, 309–72.

Peddie, William (1914). 'On the structure of the atom', *Philolosophical Magazine* **27**, 257–68.

Peierls, Rudolf (1988). 'Rutherford and Bohr', *Notes and Records of the Royal Society* **42**, 229–41.

Perrin, Jean (1901). 'Les hypothèses moléculaires', *Revue Scientifique* **15**, 449–61.

Perrin, Jean (1913). *Les Atomes*. Paris: Librairie Félix Alcan.

Petruccioli, Sandro (1993). *Atoms, Metaphors and Paradoxes: Niels Bohr and the Construction of a New Physics*. Cambridge: Cambridge University Press.

Pfund, August H. (1924). 'The emission of nitrogen and hydrogen in the infrared', *Journal of the Optical Society of America* **9**, 193–6.

Pickering, Edward (1896). 'Stars having peculiar spectra. New variable stars in Crux and Cygnus', *Astrophysical Journal* **4**, 369–70.

Planck, Max (1913). *Vorlesungen über die Theorie der Wärmestrahlung*, second edition. Leipzig: Barth.

Planck, Max (1915). 'Bemerkungen über die Emission von Spektrallinien', *Sitzungsberichte der Preussischen Akademie der Wissenschaften Berlin*, 909–13.

Planck, Max (1919). 'Dissoziationswärme des Wasserstoffs nach dem Bohr-Debyeschen Modell', *Sitzungsberichte der Preussischen Akademie der Wissenschaften Berlin*, 914–31.

Planck, Max (1922). *Physikalische Rundblicke: Gesammelte Reden und Aufsätze*. Leipzig: Hirzel.

Planck, Max (1923). 'Die Bohrsche Atomtheorie', *Naturwissenschaften* **11**, 535–7.

Plaskett, Harry H. (1922). 'The Pickering series and Bohr's atom', *Journal of the Royal Astronomical Society of Canada* **16**, 137–49.

Plaskett, John S. (1915). 'The sidereal universe', *Journal of the Royal Astronomical Society of Canada* **9**, 37–56.

Poincaré, Henri (1958). *The Value of Science*. New York: Dover Publications.

Popper, Karl R. (1963). *Conjectures and Refutations: The Growth of Scientific Kowledge*. London: Routledge and Kegan Paul.

Popper, Karl R. (1982a). *The Open Universe: An Argument for Indeterminism*. London: Routledge.

Popper, Karl R. (1982b). *Quantum Theory and the Schism of Physics*. London: Unwin Hyman.

Porter, Theodore M. (2004). *Karl Pearson: The Scientific Life in a Statistical Age*. Princeton: Princeton University Press.

Quédec, Pierre (1988). 'Weiss' magneton: The sin of pride or a venial mistake?', *Historical Studies in the Physical and Biological Sciences* **18**, 349–75.

Radder, Hans (1982a). 'Between Bohr's atomic theory and Heisenberg's matrix mechanics: a study of the role of the Dutch physicist H. A. Kramers', *Janus* **69**, 223–52.

Radder, Hans (1982b). 'An immanent criticism of Lakatos' account of the "degenerating phase" of Bohr's atomic theory', *Zeitschrift für Allgemeine Wissenschaftstheorie* **13**, 99–109.

Radder, Hans (1983). 'Kramers and the Forman thesis', *History of Science* **21**, 165–79.

Radder, Hans (1991). 'Heuristics and the generalized correspondence principle', *British Journal for the Philosophy of Science* **42**, 195–226.

Raman, Chandrasekhara V. (1928). 'A change of wavelength in light scattering', *Nature* **121**, 619.

Rau, Heinrich (1914). 'Über die Lichterregung durch langsame Kathodenstrahlen', *Sitzungsberichte der Physikalisch-Medizinischen Gesellschaft Würzburg*, no. 2, 20–7.

Rayleigh, Lord (1906). 'On electrical vibrations and the constitution of the atom', *Philosophical Magazine* **11**, 117–23.

Redtenbacher, Ferdinand (1857). *Das Dynamidensystem: Grundzüge einer Mechanischen Physik*. Mannheim: Friedrich Bassermann.

Reiche, Fritz (1920). 'Zur Theorie der Rotationsspektren', *Zeitschrift für Physik* **1**, 283–93.

Reiche, Fritz (1921). *Die Quantentheorie: Ihr Ursprung und die Entwicklung*. Berlin: Julius Springer.

Reiche, Fritz and Adolf Smekal (1918). 'Zur Theorie der Röntgenspektren', *Annalen der Physik* **57**, 124–44.

Reiche, Fritz and Willy Thomas (1925). 'Über die Zahl der Dispersionselektronen, die einem stationären Zustand zugeordnet sind', *Zeitschrift für Physik* **34**, 410–525.

Reid, Constance (1976). *Courant in Göttingen and New York. The Story of an Improbable Mathematician*. New York: Springer-Verlag.

Richardson, Owen (1914). *The Electron Theory of Matter*. Cambridge: Cambridge University Press.

Richardson, Owen (1927). 'The present state of atomic physics', *Proceedings of the Physical Society* **39**, 171–88.

Richter, Steffen (1979). *Wolfgang Pauli. Die Jahre 1918–1930: Skizzen zu einer Wissenschaftlichen Biographie*. Aarau: Verlag Sauerländer.

Riecke, Eduard (1915). 'Bohrs Theorie der Serienspektren von Wasserstoff und Helium', *Physikalische Zeitschrift* **16**, 222–7.

Righi, Augusto (1909). *Modern Theory of Physical Phenomena: Radio-Activity, Ions, Electrons*. London: Macmillan.

Ritz, Walther (1908). 'On a new law of the series spectra', *Astrophysical Journal* **28**, 237–43.

Robertson, Peter (1979). *The Early Years: The Niels Bohr Institute 1921–1930*. Copenhagen: Akademisk Forlag.

Robotti, Nadia (1983). 'The spectrum of ξ Puppis and the historical evolution of empirical data', *Historical Studies in the Physical Sciences* **14**, 123–46.

Robotti, Nadia (1986). 'The hydrogen and the old quantum mechanics', *Rivista Internationale di Storia della Scienza* **3**, 45–102.

Robotti, Nadia (1990). 'Quantum numbers and electron spin: The history of a discovery', *Archives Internationales d'Histoire des Sciences* **40**, 305–31.

Robotti, Nadia (1992). 'The Zeeman effect in hydrogen and the old quantum theory', *Rivista Internationale di Storia della Scienza* **29**, 809–31.

Robotti, Nadia and Francesca Pastorino (1998). 'Zeeman's discovery and the mass of the electron', *Annals of Science* **55**, 161–83.

Romer, Alfred (1970). *Radiochemistry and the Discovery of Isotopes*. New York: Dover Publications.

Röseberg, Ulrich (1985). *Niels Bohr: Leben und Werk eines Atomphysiker*. Berlin: Akademie-Verlag.

Rosenberger, Ferdinand (1965). *Die Geschichte der Physik*, 3 vols. Hildesheim: Georg Olms Verlagsbuchhandlung.

Rosenfeld, Léon (1931). 'Zur korrespondenzmässigen Behandlung der Linienbreite', *Zeitschrift für Physik* **71**, 273–8.

Rosenfeld, Léon, ed. (1963). *Niels Bohr: On the Constitution of Atoms and Molecules*. Copenhagen: Munksgaard.

Rosenfeld, Léon (1973). 'The wave-particle dilemma', pp. 251–63 in Jagdish Mehra, ed., *The Physicist's Conception of Nature*. Dordrecht: Reidel.

Rosenfeld, Léon and Erik Rüdinger (1967). 'The decisive years, 1911–1918', pp. 38–73 in Stefan Rozental, ed., *Niels Bohr: His Life and Work as Seen by His Friends and Colleagues*. New York: Interscience Publishers.

Rosseland, Svein (1923a). 'Zur Quantentheorie der radioaktiven Zerfallsvorgänge', *Zeitschrift für Physik* **14**, 173–81.

Rosseland, Svein (1923b). 'Origin of radioactive disintegration', *Nature* **111**, 357.

Rozental, Stefan, ed. (1967). *Niels Bohr: His Life and Work as Seen by His Friends and Colleagues*. New York: Interscience Publishers.

Rubinowicz, Adalbert (1918). 'Bohrsche Frequenzbedingung und Erhaltung des Impulsmomentes', *Physikalische Zeitschrift* **19**, 441–5, 465–74.

Rud Nielsen, Jens (1963). 'Memories of Niels Bohr', *Physics Today* **16** (October), 22–30.

Rud Nielsen, Jens, ed. (1972). *Niels Bohr. Collected Works*, vol. 1. Amsterdam: North-Holland.

Rud Nielsen, Jens, ed. (1976). *Niels Bohr. Collected Works*, vol. 3. Amsterdam: North-Holland.

Rud Nielsen, Jens, ed. (1977). *Niels Bohr. Collected Works*, vol. 4. Amsterdam: North-Holland.

Russell, Bertrand (1927). *The ABC of Atoms*. New York: E. P. Dutton & Co.

Russo, Arturo (1981). 'Fundamental research at Bell Laboratories: The discovery of electron diffraction', *Historical Studies in the Physical Sciences* **12**, 117–60.

Rutherford, Ernest (1906). *Radioactive Transformations*. London: Constable & Co.

Rutherford, Ernest (1911). 'The scattering of α– and β–particles and the structure of the atom', *Philosophical Magazine* **21**, 669–88.

Rutherford, Ernest (1912). 'The origin of β and γ rays from radioactive substances', *Philosophical Magazine* **24**, 453–62.

Rutherford, Ernest (1913a). *Radioactive Substances and their Radiations*. Cambridge: Cambridge University Press.

Rutherford, Ernest (1913b). 'The structure of the atom', *Nature* **92**, 423.

Rutherford, Ernest (1914a). 'The structure of the atom', *Philosophical Magazine* **27**, 488–98.

Rutherford, Ernest (1914b). 'The structure of the atom', *Scientia* **16**, 337–51.

Rutherford, Ernest (1922). 'Identification of a missing element', *Nature* **109**, 781.

Rutherford, Ernest (1923). 'The electrical structure of matter', Report, British Association for the Advancement of Science, 1–24.

Rutherford, Ernest (1934). 'The periodic law and its interpretation', *Journal of the Chemical Society* **133**, 635–42.

Rutherford, Ernest and John M. Nuttall (1913). 'Scattering of α particles by gases', *Philosophical Magazine* **26**, 702–12.

Rydberg, Johannes R. (1913). 'Untersuchungen über das System der Grundstoffe', *Lunds Universitets Årsskrift* **9**, no. 18.

Saha, Meghnad (1920). 'Ionization in the solar chromosphere', *Philosophical Magazine* **40**, 472–88.

Sauer, Tilman (2007). 'Einstein and the early theory of superconductivity', *Archive for History of Exact Sciences* **61**, 159–211.

Sauer, Tilman and Ulrich Majer, eds (2009). *David Hilbert's Lectures on the Foundations of Physics 1915–1927*. Dordrecht: Springer.

Scerri, Eric R. (1994). 'Prediction of the nature of hafnium from chemistry: Bohr's theory and quantum theory', *Annals of Science* **51**, 137–50.

Scerri, Eric R. (1998). 'Popper's naturalized approach to the reduction of chemistry', *International Studies in the Philosophy of Science* **12**, 33–44.

Scerri, Eric R. (2007). *The Periodic Table: Its Story and its Significance*. Oxford: Oxford University Press.

Scherrer, Paul (1962). 'Personal reminiscences', pp. 642–6 in Paul P. Ewald, ed., *Fifty Years of X-Ray Diffraction*. Utrecht: International Union of Crystallography.

Schidlof, Arthur (1911). 'Zur Aufklärung der universellen elektrodynamischen Bedeutung der Planckschen Strahlungskonstanten h', *Annalen der Physik* **35**, 90–100.

Schirrmacher, Arne (2009). 'Bohrsche Bahnen in Europa: Bilder und Modelle zur Vermittlung des modernen Atoms', pp. 73–82 in Charlotte Bigg and Jochen Hennig, eds, *Atombilder: Ikonografie des Atoms in Wissenschaft und Öffentlichkeit des 20. Jahrhunderts*. Göttingen: Wallstein Verlag.

Schönbeck, Charlotte (1982). 'Atommodelle um 1900', pp. 67–96 in Charlotte Schönbeck, ed., *Atomvorstellungen im 19. Jahrhundert*. Paderborn: Ferdinand Schöningh.

Schott, George A. (1904). 'On the kinetics of a system of particles illustrating the line and band spectrum', *Philosophical Magazine* **8**, 384–7.

Schott, George A. (1907). 'On the electron theory of matter and on radiation', *Philosophical Magazine* **13**, 189–213.

Schott, George A. (1918). 'On Bohr's hypothesis of stationary states of motion and the radiation from an accelerated electron', *Philosophical Magazine* **36**, 243–61.

Schrödinger, Erwin (1921). 'Versuch der modellmässigen Deutung des Terms der scharfen Nebenserien', *Zeitschrift für Physik* **4**, 347–57.

Schrödinger, Erwin (1924). 'Bohrs neue Strahlungshypothese und der Energiesatz', *Naturwissenschaften* **12**, 720–4.

Schwarzschild, Karl (1914). 'Bemerkung zur Aufspaltung der Spektrallinien im elektrischen Feld', *Verhandlungen der Deutschen Physikalischen Gesellschaft* **16**, 20–4.

Schwarzschild, Karl (1916). 'Zur Quantenhypothese' *Sitzungsberichte der Preussischen Akademie der Wissenschaften, Berlin*, 548–68.

Seeliger, Rudolf (1914a). [Review of Bohr 1913a, b, c], *Beiblätter der Annalen der Physik* **38**, 802–4.

Seeliger, Rudolf (1914b). 'Moderne Anschauungen über die Entstehung der Spektrallinien und der Serienspektren', *Naturwissenschaften* **2**, 285–90, 309–14.

Serwer, Daniel (1977). 'Unmechanischer Zwang: Pauli, Heisenberg, and the rejection of the mechanical atom, 1923–1925', *Historical Studies in the Physical Sciences* **8**, 189–256.

Seth, Suman (2004). 'Quantum theory and the electromagnetic world-view', *Historical Studies in the Physical and Biological Sciences* **35**, 67–95.

Seth, Suman (2007). 'Crisis and the construction of modern theoretical physics', *British Journal for the History of Science* **40**, 25–51.

Seth, Suman (2009). '*Zweideutigkeit* about "Zweideutigkeit": Sommerfeld, Pauli, and the methodological origins of quantum mechanics', *Studies in History and Philosophy of Modern Physics* **40**, 303–15.

Seth, Suman (2010). *Crafting the Quantum: Arnold Sommerfeld and the Practice of Theory, 1890–1926*. Cambridge, Mass.: MIT Press.

Shapere, Dudley (1977). 'Scientific theories and their domains', pp. 518–65 in Frederick Suppe, ed., *The Structure of Scientific Theories*. Urbana, Ill.: University of Illinois Press.

Shore, Steven N. (2003). 'Macrocosmos/microcosmos: celestial mechanics and old quantum theory', *Historia Mathematica* **30**, 494–513.

Sidgwick, Nevil V. (1923). 'Co-ordination compounds and the Bohr atom', *Journal of the Chemical Society, Transactions*, **123**, 725–30.

Siegbahn, Manne (1925). *The Spectroscopy of X-Rays*. London: Oxford University Press.

Silberstein, Ludwik (1920a). 'Contribution to quantum theory of spectrum emission', *Philosophical Magazine* **39**, 46–66.

Silberstein, Ludwik (1920b). *Report on the Quantum Theory of Spectra*. London: Adam Hilger.

Sinclair, Steve B. (1987). 'J. J. Thomson and the chemical atom: From ether vortex to atomic decay', *Ambix* **34**, 89–116.

Singh, Rajinder (2002). 'C. V. Raman and the discovery of the Raman effect', *Physics in Perspective* **4**, 399–420.

Slater, John C. (1923). 'Radiation and atoms', *Nature* **113**, 307–8.

Slater, John C. (1925a). 'A quantum theory of optical phenomena', *Physical Review* **25**, 395–428.

Slater, John C. (1925b). 'The nature of radiation', *Nature* **116**, 278.

Slater, John C. (1967). 'Quantum physics in America between the wars', *International Journal of Quantum Chemistry* **1 S**, 1–23.

Slater, John C. (1975). *Solid-State and Molecular Theory: A Scientific Biography*. New York: Wiley.

Smekal, Adolf (1921). 'Über Stark's Kritik der Bohrschen Theorie der Lichtemission', *Verhandlungen der Deutschen Physikalische Gesellschaft* **2**, 20–2.

Smekal, Adolf (1923). 'Zur Quantentheorie der Dispersion', *Naturwissenschaften* **11**, 873–5.

Snelders, Harry A. M. (1974). 'A bio-bibliography of the Dutch amateur physicist A. J. van den Broek (1870–1926)', *Janus* **61**, 59–72.

Soddy, Frederick (1913). 'Intra-atomic charge', *Nature* **92**, 399–400.

Soddy, Frederick (1914). 'Radioactivity', *Annual Progress Report to the Chemical Society* **10**, 262–88.

Soddy, Frederick (1917). 'Radioactivity', *Annual Progress Report to the Chemical Society* **13**, 245–72.

Sommerfeld, Arnold (1915a). 'Die allgemeine Dispersionsformel nach den Bohrschen Modell', pp. 549–84 in *Elster-Geitel Festschrift (Arbeiten aus den Gebieten der Physik, Mathematik, Chemie)*. Braunschweig: Vieweg & Sohn.

Sommerfeld, Arnold (1915b). 'Zur Theorie der Balmerschen Serie', *Sitzungsberichte der Bayerischen Akademie der Wissenschaften München*, 425–58.

Sommerfeld, Arnold (1916a). 'Zur Theorie des Zeeman-Effekts des Wasserstofflinien, mit einem Anhang über den Stark-Effekt', *Physikalische Zeitschrift* **17**, 491–507.

Sommerfeld, Arnold (1916b). 'Zur Quantentheorie der Spektrallinien, III: Theorie der Röntgenspektren', *Annalen der Physik* **51**, 125–67.

Sommerfeld, Arnold (1916c). 'Zur Quantentheorie der Spektrallinien', *Annalen der Physik* **51**, 1–94.

Sommerfeld, Arnold (1918). 'Atombau und Röntgenspektren', *Physikalische Zeitschrift* **19**, 297–307.

Sommerfeld, Arnold (1919). *Atombau und Spektrallinien*, first edn. Braunschweig: Vieweg & Sohn.

Sommerfeld, Arnold (1920a). 'Schwebende Fragen der Atomphysik', *Physikalische Zeitschrift* **21**, 619–21.

Sommerfeld, Arnold (1920b). 'Ein Zahlenmysterium in der Theorie des Zeemaneffektes', *Naturwissenschaften* **8**, 61–4.

Sommerfeld, Arnold (1921). 'Zur Kritik der Bohrschen Theorie der Licht-Emission', *Jahrbuch der Radioaktivität und Elektronik* **17**, 417–29.

Sommerfeld, Arnold (1922a). *Atombau und Spektrallinien*, third edn. Braunschweig: Vieweg & Sohn.

Sommerfeld, Arnold (1922b). 'Quantentheoretische Umdeutung der Voigtschen Theorie des anomalen Zeemaneffektes vom D-Linientypus', *Zeitschrift für Physik* **8**, 257–72.

Sommerfeld, Arnold (1923). 'The model of the neutral helium atom', *Journal of the Optical Society of America and Review of Scientific Instruments* **7**, 509–15.

Sommerfeld, Arnold (1924a). 'Grundlagen der Quantentheorie und des Bohrschen Atommodelles', *Naturwissenschaften* **12**, 1047–9.

Sommerfeld, Arnold (1924b). *Atombau und Spektrallinien*, fourth edn. Braunschweig: Vieweg & Sohn.

Sommerfeld, Arnold (1925). 'Zur Theorie des periodischen Systems', *Physikalische Zeitschrift* **26**, 70–4.

Sommerfeld, Arnold (1929). *Atombau und Spektrallinien: Wellenmechanischer Ergänzungsband*. Braunschweig: Vieweg & Sohn.

Sommerfeld, Arnold and Albrecht Unsöld (1926). 'Über das Spektrum des Wasserstoffs', *Zeitschrift für Physik* **36**, 259–76.

Sommerfeld, Arnold and Fritz Bopp (1951). 'Fifty years of quantum theory', *Science* **113**, 85–92.

Sommerfeld, Arnold and Werner Heisenberg (1922). 'Eine Bemerkung über relativistiche Röntgendubletts und Linienschärfe', *Zeitschrift für Physik* **10**, 393–8.

Soné, Také (1920). 'On the magnetic susceptibilities of hydrogen and some other gases', *Philosophical Magazine* **39**, 305–50.

Soon Im, Gyeong (1995). 'The formation and development of the Ramsauer effect', *Historical Studies in the Physical and Biological Sciences* **25**, 269–300.

Soon Park, Buhm (2001). 'A principle written in diagrams: The *Aufbau* principle for molecules and its visual representations, 1927–1932', pp. 179–98 in Ursula Klein, ed., *Tools and Modes of Representation in the Laboratory Sciences*. Dordrecht: Kluwer Academic.

Sopka, Katherine R. (1988). *Quantum Physics in America: The Years Through 1935*. New York: American Institute of Physics.

Spencer, J. F. (1913). [Abstracts of Bohr 1913b and 1913c], *Journal of the Chemical Society, Abstracts* **104**, ii 689 and ii 943–4.

Stachel, John (1993). 'The other Einstein: Einstein contra field theory', *Science in Context* **6**, 275–90.

Stachel, John (2002). *Einstein from 'B' to 'Z'*. Boston: Birkhäuser.

Stark, Johannes (1910). *Prinzipien der Atomdynamik, Teil I: Die Elektrischen Quanten*. Leipzig: Hirzel.

Stark, Johannes (1911). *Prinzipien der Atomdynamik, Teil II: Die Elementare Strahlung*. Leipzig: Hirzel.

Stark, Johannes (1913a). 'Über das dreiatomige Wasserstoffmolekül nach Hypothese und Erfahrung', *Zeitschrift für Elektrochemie* **19**, 862–3.

Stark, Johannes (1913b). 'Observation of the separation of spectral lines by an electric field', *Nature* **92**, 401.

Stark, Johannes (1914a). 'Bemerkung zum Bogen- und Funkenspektrum des Heliums', *Verhandlungen der Deutschen Physikalischen Gesellschaft* **16**, 468–74.

Stark, Johannes (1914b). *Elektrische Spektralanalyse Chemischer Atome*. Leipzig: Hirzel.

Stark, Johannes (1915). *Prinzipien der Atomdynamik. Teil III: Die Elektrizität im Chemischen Atom*. Leipzig: Hirzel.

Stark, Johannes (1916). 'Tatsachen und Folgerungen über Zahl und Koppelung von Elektronen im Wasserstoffatom', *Annalen der Physik* **50**, 53–88.

Stark, Johannes (1917). 'Erfahrung und Bohrsche Theorie der Wasserstoffspektren', *Annalen der Physik* **54**, 111–16.

Stark, Johannes (1920). 'Zur Kritik der Bohrschen Theorie der Lichtemission', *Jahrbuch der Radioaktivität und Elektronik* **17**, 161–73.

Stark, Johannes (1922). *Der Gegenwärtige Krisis in der Deutschen Physik*. Leipzig: Barth.

Stephenson, Reginald J. (1967). 'The scientific career of Charles Glover Barkla', *American Journal of Physics* **35**, 140–52.

Stern, Otto (1921). 'Ein Weg zur experimentellen Prüfung der Richtungsquantelung im Magnetfeld', *Zeitschrift für Physik* **7**, 249–53.

Stewart, G. Walther (1914). 'The content and structure of the atom', *Science* **40**, 661–3.

Stolzenburg, Klaus, ed. (1984). *Niels Bohr. Collected Works*, vol. 5. Amsterdam: North-Holland.

Stone, A. Douglas (2005). 'Einstein's unknown insight and the problem of quantizing chaos', *Physics Today* **58** (August), 37–43.

Stoner, Edmund C. (1924). 'The distribution of electrons among atomic levels', *Philosophical Magazine* **48**, 719–26.

Stranges, Anthony N. (1982). *Electrons and Valence: Development of the Theory, 1900–1925*. College Station: Texas A&M University Press.

Strutt, Robert J. (1968). *Life of John William Strutt, Third Baron Rayleigh*. Madison: University of Wisconsion Press.

Stuewer, Roger H. (1975). *The Compton Effect: Turning Point in Physics*. New York: Science History Publications.

Stuewer, Roger H. (1983). 'The nuclear electron hypothesis', pp. 19–68 in William R. Shea, ed., *Otto Hahn and the Rise of Nuclear Physics*. Dordrecht: Reidel.

Sutherland, William (1899). 'Cathode, Lenard and Röntgen rays', *Philosophical Magazine* **47**, 269–84.

Sutherland, William (1901). 'The cause of the structure of spectra', *Philosophical Magazine* **2**, 245–74.

Svidzinsky, Anatoly A., Marlan O. Scully, and Dudley R. Herschbach (2005). 'Bohr's 1913 molecular model revisited', *Proceedings of the National Academy of Science* **102**, 11985–8.

Szavassi, Arthur (1918). 'Über das Bohrsche Atommodell', *Physikalische Zeitschrift* **19**, 505–8.

Tanona, Scott (2004). 'Uncertainty in Bohr's response to the Heisenberg microscope', *Studies in History and Philosophy of Modern Physics* **35**, 483–507.

Ter Haar, Dirk (1967). *The Old Quantum Theory*. Oxford: Pergamon Press.

Ter Haar, Dirk (1998). *Master of Modern Physics: The Scientific Contributions of H. A. Kramers*. Princeton: Princeton University Press.

Thomsen, H. P. Julius (1895). 'Systematische Gruppierung der chemischen Elemente', *Zeitschrift für Anorganische Chemie* **9**, 190–3.

Thomson, Joseph J. (1883). *A Treatise on the Motion of Vortex Rings*. London: Macmillan.

Thomson, Joseph J. (1897). 'Cathode rays', *Philosophical Magazine* **44**, 293–316.

Thomson, Joseph J. (1899). 'On the masses of the ions at low pressures', *Philosophical Magazine* **48**, 547–67.

Thomson, Joseph J. (1904a). 'On the structure of the atom', *Philosophical Magazine* **7**, 237–65.

Thomson, Joseph J. (1904b). *Electricity and Matter*. London: Constable & Co.

Thomson, Joseph J. (1904c). 'On the structure of the atom: an investigation of the stability and periods of oscillation of a number of corpuscles arranged at equal intervals around the circumference of a circle', *Philosophical Magazine* **7**, 237–65.

Thomson, Joseph J. (1905). 'The structure of the atom', *Proceedings of the Royal Institution* **18**, 1–15.

Thomson, Joseph J. (1906). 'On the number of corpuscles in an atom', *Philosophical Magazine* **11**, 769–81.

Thomson, Joseph J. (1907). *The Corpuscular Theory of Matter*. London: Constable & Co.

Thomson, Joseph J. (1910). 'The scattering of rapidly moving electrified particles', *Proceedings of the Cambridge Philosophical Society* **15**, 465–71.

Thomson, Joseph J. (1912). 'Further experiments on positive rays', *Philosophical Magazine* **24**, 209–53.

Thomson, Joseph J. (1913a). 'On the structure of the atom', *Philosophical Magazine* **26**, 792–9.

Thomson, Joseph J. (1913b). 'Rays of positive electricity', *Proceedings of the Royal Society* A **89**, 1–20.

Thomson, Joseph J. (1913c). *Rays of Positive Electricity and their Application to Chemical Analyses*. New York: Longmans, Green and Co.

Thomson, Joseph J. (1914). 'The forces between atoms and chemical affinity', *Philosophical Magazine* **27**, 757–89.

Thomson, Joseph J. (1919). 'On the origin of spectra and Planck's law', *Philosophical Magazine* **37**, 419–46.

Thomson, Joseph J. (1921). 'La structure de l'atome', pp. 1–74 in Robert B. Goldschmidt, Maurice de Broglie, and Frederick A. Lindemann, eds, *La Structure de la Matière. Rapports et Discussions du Conseil de Physique Tenu à Bruxelles du 27 au 31 octobre 1913 sous les Auspices de l'Institut International de Physique Solvay*. Paris: Solvay Institute.

Thomson, Joseph J. (1936). *Recollections and Reflections*. London: G. Bell and Sons.

Thomson, William (1904). 'Plan of a combination of atoms having the properties of polonium or radium', *Philosophical Magazine* **8**, 528–34.

Thorsen, Jens, ed. (1987). *Niels Bohr. Collected Works*, vol. 8. Amsterdam: North-Holland.

Thyssen, Pieter and Koen Binnemans (2011). 'Accomodation of the rare earths in the periodic table: A historical analysis', pp. 1–94 in Karl A. Gschneider, ed., *Handbook on the Physics and Chemistry of Rare Earths*, vol. 41. Burlington: Academic Press.

Tolman, Richard C. (1922). 'Review of the present status of the two forms of quantum theory', *Journal of the Optical Society of America and Review of Scientific Instruments* **6**, 211–28.

Tolman, Richard C. (1925). 'Note on the problem of the quantitative formulation of Bohr's correspondence principle', *Philosophical Magazine* **49**, 130–6.

Tolmasquim, A. T. and I. C. Moreira (2002). 'Einstein in Brazil: The communication to the Brazilian Academy of Sciences on the constitution of light', pp. 229–42 in Helge Kragh, Geert Vanpaemel, and Pierre Marage, eds, *History of Modern Physics.* Turnhout: Brepols.

Trenn, Thaddeus J., ed. (1975). *Radioactivity and Atomic Theory. Annual Progress Reports on Radioactivity 1904–1920 to the Chemical Society by Frederick Soddy F.R.S.* London: Taylor & Francis.

Tunzelmann, George W. De (1913). [Abstract of Bohr 1913b], *Science Abstracts, Series A: Physics* **16**, 628–31.

Tunzelmann, George W. De (1914). [Abstract of Bohr 1913c], *Science Abstracts, Series A: Physics* **17**, 6–8.

Urbain, Georges (1922). 'Les numéros atomiques du néo-ytterbium, du lutécium et du celtium', *Comptes Rendus* **174**, 1349–51.

Urbain, Georges and Alexandre Dauvillier (1923). 'On the element of atomic number 72', *Nature* **111**, 218.

Van Brakel, J. (1985). 'The possible influence of the discovery of radio-active decay on the concept of physical probability', *Archive for History of Exact Sciences* **31**, 369–85.

Van Delft, Dirk (2007). *Freezing Physics: Heike Kamerlingh Onnes and the Quest for Cold.* Amsterdam: Koninklijke Nederlandse Akademie van Wetenschappen.

Van den Broek, Antonius J. (1913). 'Intra-atomic charge', *Nature* **92**, 372–3.

Van den Broek, Antonius J. (1914). 'The structure of atoms and molecules', *Nature* **93**, 241–2.

Van der Bijl, Hendrik J. (1917). 'Theoretical considerations concerning ionization and single-lined spectra', *Physical Review* **10**, 546–56.

Van der Waerden, Bartel L. (1960). 'Exclusion principle and spin', pp. 199–244 in Markus Fierz and Victor F. Weisskopf, eds, *Theoretical Physics in the Twentieth Century.* New York: Interscience Publishers.

Van der Waerden, Bartel L., ed. (1967). *Sources of Quantum Mechanics.* New York: Dover Publications.

Van Leeuwen, Hendrika J. (1916). 'Die Instabilität des Bohr-Debyeschen Wassterstoff-Moleküls und die Dispersion von Wasserstoff', *Physikalische Zeitschrift* **17**, 196–8.

Van Leeuwen, Hendrika J. (1921). 'Problèmes de la théorie électronique du magnétisme', *Journal de Physique et le Radium* **2**, 361–77.

Van Spronsen, J. W. (1969). 'Gustavus Detlef Hinrichs discovered, one century ago, the periodic system of chemical elements', *Janus* **56**, 47–62.

Van Urk, Theodor (1923). 'Die Normalbahnen des Serienelektrons der Alkalien', *Zeitschrift für Physik* **13**, 268–74.

Van Vleck, John H. (1922). 'The normal helium atom and its relation to the quantum theory', *Philosophical Magazine* **44**, 842–69.

Van Vleck, John H. (1923). 'Note on the quantum theory of the helium arc spectrum', *Physical Review* **21**, 372–3.

Van Vleck, John H. (1924). 'The absorption of radiation by multiply periodic orbits, and its relation to the correspondence principle and the Rayleigh-Jeans law. Part I', *Physical Review* **24**, 330–46.

Van Vleck, John H. (1926). *Quantum Principles and Line Spectra.* Washington D.C.: National Research Council.

Van Vleck, John H. (1928). 'The correspondence principle in the statistical interpretation of quantum mechanics', *Proceedings of the National Academy of Sciences* **14**, 176–88.

Van Vleck, John H. (1929). 'The new quantum mechanics'. *Chemical Reviews* **5**, 467–507.

Van Vleck, John H. (1932). *The Theory of Electric and Magnetic Susceptibilities.* Oxford: Oxford University Press.

Vegard, Lars (1917). 'Der Atombau auf Grundlage der Röntgenspektren', *Berichte der Deutschen Physikalischen Gesellschaft* **19**, 344–53.

Vegard, Lars (1918). 'The X-ray spectra and the constitution of the atom', *Philosophical Magazine* **35**, 293–326.

Vegard, Lars (1919). 'On the X-ray spectra and the constitution of the atom, II', *Philosophical Magazine* **37**, 237–80.

Vickers, Peter (2008). 'Bohr's theory of the atom: Content, closure and consistency', http://philsci-archive.pitt.edu/4005.

Vickers, Peter (2011). 'Historical magic in old quantum theory? *European Journal for Philosophy of Science* **2**. DOI 10.1007/s13194-010-0010-6.

Walker, James (1922). 'The role of the physicist in the development of chemical theory', *Journal of the Chemical Society* **121**, 735–45.

Waller, Ivar (1927). 'Scattering of radiation from atoms', *Philosophical Magazine* **4**, 1228–37.

Walter, L. H. (1914). [Abstract of Bohr 1913d], *Science Abstracts, Series A: Physics* **17**, 277.

Walter, Maila L. (1990). *Science and Cultural Crisis: An Intellectual Biography of Percy Williams Bridgman (1882–1961)*. Stanford: Stanford University Press.

Warburg, Emil (1913). 'Bemerkungen zu der Aufspaltung der Spektrallinien im elektrischen Feld', *Verhandlungen der Deutschen Physikalischen Gesellschaft* **15**, 1259–66.

Warburg, Emil, ed. (1915). *Physik*, Part 3.3.1 of *Die Kultur der Gegenwart*. Leipzig: B. G. Teubner.

Warwick, Andrew (1993). 'Frequency, theorem and formula: Remembering Joseph Larmor in electromagnetic theory', *Notes and Records of the Royal Society* **47**, 49–60.

Weaire, Denis and S. O'Connor (1987). 'Unfulfilled renown: Thomas Preston (1860–1900) and the anomalous Zeeman effect', *Annals of Science* **44**, 617–44.

Weber, Wilhelm (1871). 'Über elektrodynamische Massenbestimmungen, insbesondere über das Princip der Erhaltung der Energie', *Abhandlungen der Sächsischen Gesellschaft der Wissenschaften (Leipzig)* **10**, 1–62.

Webster, David L. (1918). 'The scattering of α-particles as evidence of the Parson magneton hypothesis', *Journal of the American Chemical Society* **40**, 375–9.

Webster, David L. and Leigh Page (1921). 'A general survey of the present status of the atomic structure problem', *Bulletin of the National Research Council* **2**, 335–98.

Weinert, Friedel (1994). 'Wrong theory—right experiment: The significance of the Stern-Gerlach experiments', *Studies in History and Philosophy of Modern Physics* **26**, 75–86.

Weisskopf, Victor F. (1970). 'Three steps in the structure of matter', *Physics Today* **23** (August), 17–24.

Wendt, Gerald and Robert Landauer (1920). 'Triatomic hydrogen', *Journal of the American Chemical Society* **42**, 930–46.

Wentzel, Gregor (1922). 'Fortschritte der Atom- und Spektraltheorie', *Ergebnisse der Exakten Naturwissenschaften* **1**, 298–314

Wereide, Thorstein (1917). 'Maxwells Gleichungen und die Atomstrahlung', *Annalen der Physik* **52**, 276–82.

Werner, Sven (1925). 'On the spark spectrum of lithium', *Nature* **116**, 574.

Wheaton, Bruce R. (1978). 'Philipp Lenard and the photoelectric effect, 1899–1911', *Historical Studies in the Physical Sciences* **9**, 299–322.

Whitaker, M. A. B. (1985). 'Planck's first and second theories and the correspondence principle', *European Journal of Physics* **6**, 266–70.

Whittaker, Edmund T. (1922). 'On the quantum mechanism in the atom', *Proceedings of the Royal Society, Edinburgh* **42**, 129–46.

Wien, Wilhelm (1915). 'Theorie der Wärmestrahlung', pp. 209–22 in Emil Warburg, ed., *Physik*, Part 3.3.1 of *Die Kultur der Gegenwart*. Leipzig: B.G. Teubner.

Wilson, David (1983). *Rutherford: Simple Genius*. Cambridge, Mass.: MIT Press.

Wilson, Ernest D. (1916). *The Structure of the Atom*. Chicago: University of Chicago Libraries.

Wilson, Harold A. (1911). 'The number of electrons in the atom', *Philosophical Magazine* **21**, 718–22.

Wilson, William (1915). 'The quantum-theory of radiation and line spectra', *Philosophical Magazine* **29**, 795–802.

Wilson, William (1956). 'John William Nicholson, 1881–1955', *Biographical Memoirs of Fellows of the Royal Society* **2**, 209–14.

Wise, M. Norton (1981). 'German concepts of force, energy, and the electromagnetic ether: 1845–1880', pp. 269–307 in Geoffrey N. Cantor and Michael J. S. Hodge, eds, *Conceptions of Ether: Studies in the History of Ether Theories 1740–1900*. Cambridge: Cambridge University Press.

Wise, M. Norton (1990). 'How do sums count? On the cultural origin of statistical causality', pp. 395–426 in Lorenz Krüger, Lorraine J. Daston, and Michael Heidelberger, eds, *The Probabilistic Revolution, Volume I: Ideas in History*. Cambridge, Mass.: MIT Press.

Wolfke, Mieczyslaw (1916). 'Strahlungseigenschaften des Debyeschen Modells eines Wasserstoff-Moleküls', *Physikalische Zeitschrift* **17**, 71–2.

Wood, Robert W. (1911). *Physical Optics*. New York: Macmillan.

Wright, William H. (1915). 'On the occurrence of the line 4686 A and the related series of lines in the spectra of the planetary nebulae', *Proceedings of the National Academy of Sciences* **1**, 266–9.

Yagi, Eri (1964). 'On Nagaoka's Saturnian atomic model', *Japanese Studies in the History of Science* **3**, 29–47.

Zahn, Charles T. (1924). 'The electric moment of gaseous molecules of halogen hydrides', *Physical Review* **24**, 400–17.

Zapffe, Carl A. (1969). 'Gustavus Hinrichs, precursor of Mendeleev', *Isis* **60**, 461–76.

Zoubek, G. and B. Lauth (1992). 'Zur Rekonstruktion des Bohrschen Forschungsprogramms', *Erkenntnis* **37**, 223–47, 249–73.

INDEX

The manufacturer's authorised representative in the EU for product safety is Oxford University Press España S.A. of El Parque Empresarial San Fernando de Henares, Avenida de Castilla, 2 - 28830 Madrid (www.oup.es/en or product.safety@oup.com). OUP España S.A. also acts as importer into Spain of products made by the manufacturer.
Printed and bound by CPI Group (UK) Ltd, Croydon, CR0 4YY
12/09/2026
02215930-0006